Table of Contents

1

STRUCTURE AND BONDING IN ORGANIC COMPOUNDS

KEYS TO THE CHAPTER

Atomic Structure and Properties

Two periodic trends are important to understanding the physical and chemical properties of organic compounds. They are electronegativity and atomic radius.

The electronegativity scale is an index of the attraction of an atom for an electron. It increases from left to right in a period and from bottom to top in a group of the periodic table. The order of electronegativities for the three most common elements in organic molecules, excluding hydrogen, is C < N < O. Their electronegativity values differ by 0.5 between neighboring elements in this part of the second period. There is a more pronounced difference between second and third period elements. Thus, fluorine and chlorine differ by 1.0, as do oxygen and sulfur. The order of the electronegativity values of the halogens is I < Br < Cl.

Ionic and Covalent Bonds

There are two main classes of bonds. Ionic bonds predominate in inorganic compounds, but covalent bonds are much more important in organic chemistry. When positive and negative ions combine to form an ionic compound, the charges of the cations and anions must be balanced to give a neutral compound. For ionic compounds, the cation is named first and then the anion. Thus, ammonium sulfide contains $(NH_4)_2$ and S^{2-}. Two ammonium ions are required to balance the charge of one sulfide ion, so the formula of ammonium sulfide is $(NH_4)_2S$. Parentheses enclose a polyatomic ion when a formula unit contains two or more of that ion, and the subscript is placed outside the parentheses.

A covalent bond forms when two nuclei are simultaneously attracted to the same pair of electrons. Carbon usually forms covalent bonds to other elements. The stability of Lewis structures is attributed to the octet rule that states that second row elements tend to form associations of atoms with eight electrons (both shared and unshared) in the valence shell of all atoms of the molecule.

One or more pairs of electrons can be shared between carbon atoms. Single, double, and triple bonds are linked one, two, and three pairs of electrons, respectively. In applying the octet rule, the bonding electrons are counted twice. That is, each atom "owns" the bonding electrons, so they count toward the total of eight for each atom.

With the exception of bonds to carbon and to hydrogen, carbon forms polar covalent bonds to other elements. The degree of polarity depends on the difference in the electronegativity values of the bonded atoms. The direction of the bond moment is indicated by an arrow with a cross at the end opposite the arrow head. The symbols δ^+ and δ^- indicate the partially positive and partially negative atoms of the bonded atoms.

Strategy for Writing Lewis Structures

When we write a Lewis structure, we first need to know how many electrons are in a molecule based and where they are located.

Consider vinyl chloride, C_2H_3Cl, which is used to produce polymers for commercial products such as PVC pipes. It contains a total of 18 electrons. Hydrogen forms only one bond in all compounds. Chlorine also forms one bond to carbon. The basic skeleton of the molecule is shown below.

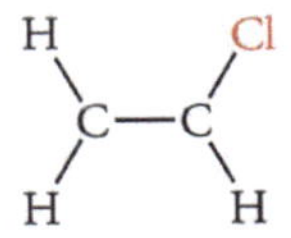

The molecular skeleton accounts for eight electrons; two per single bond. Each carbon atom still needs two more electrons to complete its octet, and the chlorine atom needs six. The six electrons on chlorine form three lone pairs. Each carbon contributes one electron to the single bond. Each carbon has four electrons, and each donates one more to form a double bond.

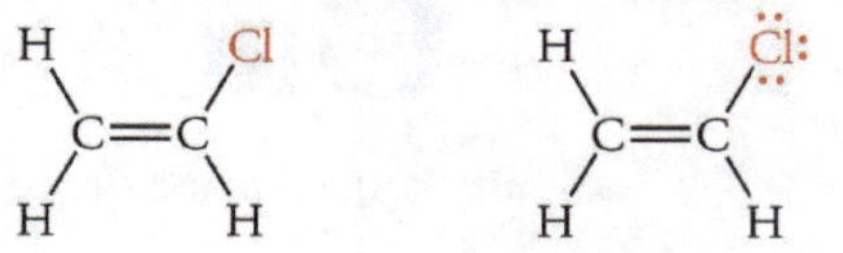

Formal Charge

We determine formal charges in several steps.

1. Count the total number of valence electrons for each atom in the molecule.
2. Each atom "owns" its nonbonded electron pairs.
3. Electrons in bonds are shared equally between the bonded atoms; in a single bond each atom gets one electron, in a double bond it gets two, and so forth.
4. If an atom has more electrons in the bonded structure than it would have if neutral, it has a formal negative charge; if it has fewer electrons than it would have as a neutral atom, it has a formal positive charge.

A few simple rules make it easy to determine the formal charge in most cases by inspection. For example, if nitrogen has three bonds—regardless of the combination of single, double, or triple bonds—and a pair of electrons, then it has no formal charge. If there are four bonds to nitrogen—regardless of the combination of single, double, or triple bonds—the nitrogen atom has a formal +1 charge. Similarly, if oxygen has two bonds—regardless of the combination of single or double bonds—and two pairs of electrons, then it has no formal charge. If there are three bonds to oxygen—regardless of the combination of single or double bonds—the oxygen atom has a formal +1 charge. The structure shown below contains an oxygen atom with a +1 formal charge; the entire species has a net +1 charge.

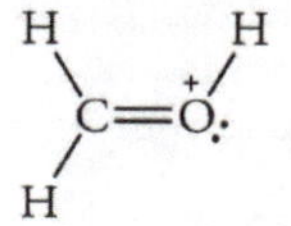

Resonance Theory

For most compounds, one Lewis structure describes the distribution of electrons and the types of bonds in a molecule. However, for some species a single Lewis structure does not provide an adequate description of bonding. Resonance structures provide a bookkeeping device to describe the delocalization of electrons, giving structures that cannot be adequately described by a single Lewis structure. Such bonding is described using two or more resonance contributors that differ only in the location of the electrons. The positions of the nuclei are unchanged. The actual structure of a molecule that is pictured by resonance structures has characteristics of all the resonance contributors.

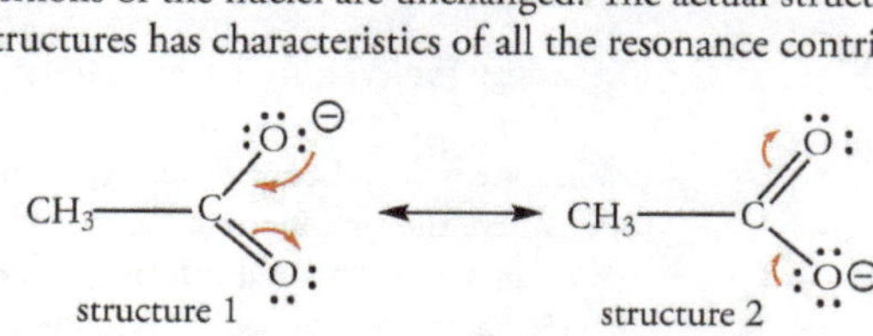

Curved arrows are used to show the movement of electrons to transform one resonance contributor into another. The electrons move from the position indicated by the tail of the arrow toward the position shown by the head.

The degree to which various resonance forms contribute to the actual structure in terms of the properties of the bonds and the location of charge is not the same for all resonance forms. The overriding first rule is that the Lewis octet must be considered as a first priority. After that, the location of charge on atoms of appropriate electronegativity can be considered.

Valence-Shell Electron-Pair Repulsion Theory

Like charges repel each other, so the electron pairs surrounding a central atom in a molecule should repel each other and move as far apart as possible. We use valence-shell electron-pair repulsion (VSEPR) theory to predict the shapes of molecules. VSEPR theory allows us to predict whether the geometry around any given atom is tetrahedral, trigonal planar, or linear.

Using VSEPR theory requires that regions of electron density be considered regardless of how many electrons are contained in the region. Thus, a single-bonded pair or two pairs of electrons in a double bond are considered as "equal." The following rules cover most cases.

1. Two regions containing electrons around a central atom are 180° apart, producing a linear arrangement.
2. Three regions containing electrons around a central atom are 120° apart, producing a trigonal planar arrangement.
3. Four regions containing electrons around a central atom are 109.5° apart, producing a tetrahedral arrangement.

The electron pairs around a central atom may be bonding electrons or nonbonding electrons, and both kinds of valence-shell electron pairs must be considered in determining the shape of a molecule. When all of the electron pairs are arranged to minimize repulsion, we look at the molecule to see how the atoms are arranged in relation to each other. The geometric arrangement of the atoms determines the bond angles.

Consider the structure of an isocyanate group in methylisocyanate.

methyl isocyanate

The nitrogen atom has three regions containing electrons around it. They are a single bond, a double bond, and a nonbonded pair of electrons. So, these features will have a trigonal planar arrangement, and the R—N=C bond angle is 120°. The isocyanate carbon atom has two groups of electrons around it—two double bonds—so they will have a linear arrangement. The N=C=O bond angle is 180°.

Dipole Moments

The polarity of a molecule is given by its dipole moment. The dipole moment depends upon both the polarity of individual bonds and the arrangement of those bonds in the molecule. In some molecules, the dipole moments are pointed in opposite directions so that they cancel one another. As a result, there is no net resultant dipole moment. In other molecules, the dipole moments may reinforce each other or partially cancel, causing a net dipole moment.

Atomic and Molecular Orbitals

Atomic orbitals are mathematical equations that describe the discrete, quantized energy levels of atoms. They are described as 1s, 2s, 2p, and so forth. Each atomic orbital can contain a maximum of two electrons with opposite spins. The square of the equation for an atomic orbital gives the probability of finding an electron within a given region of space.

The concepts developed for atomic orbitals can be extended to molecular orbitals that extend across a molecule. Molecular orbitals are linear combinations of atomic orbitals, which represent the distribution of electrons over two or more atoms. The important concepts are summarized below.

1. The number of molecular orbitals must equal the number of atomic orbitals used to generate them.
2. Molecular orbitals, as Well as atomic orbitals, are represented by wave functions whose value may be positive or negative and is a function of geometry.
3. There are two types of bonding molecular orbitals to hydrogen and to second row elements, called sigma (σ) and pi (π). Hydrogen forms only one σ bond.
4. Molecular orbitals can be bonding or antibonding.

The Hydrogen Molecule

The 1s orbitals of two hydrogen atoms can combine in two ways to give molecular orbitals. One of these is a bonding σ orbital; the other is an antibonding, σ* orbital. Bonding molecular orbitals have lower energy (are more stable) than the original atomic orbitals. Antibonding molecular orbitals have higher energy (are less stable) than the original atomic orbitals. The bonding σ orbital holds two electrons, and the antibonding σ* orbital is empty.

Bonding in Carbon Compounds

The strongest bonds between carbon atoms and other atoms are σ bonds that result from overlap of atomic orbitals along the internuclear axis. Side-by-side overlap of p orbitals leads to a less stable π bond.

Atomic orbitals are combined (mixed) to give hybridized atomic orbitals. These orbitals account for the geometry and properties of molecules, and they follow the rules for VSEPR theory.

sp³ Hybridization of Carbon in Methane

Bonding in methane can be regarded as the formation of covalent bonds between an sp³-hybridized carbon atom and 1s orbital of hydrogen atoms. An sp³-hybrid orbital is constructed from mixing the 2s orbital of an excited state carbon atom, which contains one electron, with three 2p orbitals, each of which also contains one electron. The resulting sp³-hybrid orbitals point at the corners of a tetrahedron. Each of them forms a σ bond with the 1s orbital of a hydrogen atom.

The term % s character is used to describe the contribution of the atomic orbitals to a hybridized orbital. Thus an sp³-hybrid orbital has 25% s character.

sp³ Hybridization of Carbon in Ethane

Ethane and other organic compounds containing four single bonds to carbon atoms consist of sigma bonds to sp³-hybridized carbon atoms arranged at tetrahedral angles to one another. In ethane, two sp³ hybrid orbitals overlap to give a σ bond. The other three sp³ hybrid orbitals on each carbon make σ bonds to hydrogen atoms.

Groups of atoms can rotate about a sigma bond without breaking the bond. The resulting conformations are different temporary arrangements of atoms that still maintain their bonding arrangement.

sp² Hybridization of Carbon in Ethene

The sp² hybrid orbitals of carbon occur in compounds such as ethene that contain a double bond. The overlap of these orbitals with one another or with other orbitals such as an s orbital of hydrogen gives a sigma (σ) bond. The three sp² hybrid orbitals are coplanar and lie 120° to one another. They have 33% s character because they are formed from one 2s orbital and two 2p orbitals. An sp² hybridized carbon also has a 2p orbital that can form a π bond with a neighboring carbon atom in ethene or to a carbon atom in methanal. The σ bond in ethene and other alkenes is stronger than the π bond because there is less orbital overlap in the π bond.

sp Hybridization of Carbon in Ethyne

The sp hybrid orbitals of carbon occur in compounds such as ethyne that contain a triple bond. The overlap of these orbitals with one another or with other orbitals such as an s orbital of hydrogen gives a sigma bond. The sp hybrid orbitals are at 180° to one another. They have 50% s character because they are formed from one 2s orbital and one 2p orbital. Each time there are two sp hybrid orbitals about a carbon atom, there are also two remaining p orbitals that form two π bonds with a neighboring atom, as in the case of another carbon atom in ethyne or a nitrogen atom in cyano compounds.

Effect of Hybridization on Bond Length and Bond Strength

With increasing % s character, the electrons within a hybrid orbital are held closer to the nucleus of the atom. As a consequence, the bond lengths decrease as the % s character increases. And, the strength of the bond increases as % s character increases.
1. C—H bond strengths: ethane (sp³) < ethene (sp²) < ethyne (sp).
2. C—H bonds lengths: ethyne < ethene < ethane.

Hybridization of Nitrogen

Hybridization is not a phenomenon restricted to carbon. It applies to other atoms as well. The only difference is in the number of electrons that are distributed in the orbitals. Nitrogen, a Group VA element, has five valence electrons.

An sp^3-hybridized nitrogen has three half-filled orbitals that can form σ bonds and one filled sp^3 orbital that is a nonbonding electron pair. The orbital containing the nonbonding electron pair and the three half-filled orbitals the bonding are directed to the corners of a tetrahedron. However, the geometry of such molecules is pyramidal, like ammonia, because the position of the atoms, not the electron pairs, defines the molecular geometry.

An sp^2-hybridized nitrogen atom can form three σ bonds and one π bond. The geometry of sp^2 hybridized nitrogen is trigonal planar, and the bond angles around the nitrogen are 120°.

An sp-hybridized nitrogen atom can form two σ bonds with sp orbitals and two π bonds with its half-filled 2p orbitals.

Hybridization of Oxygen

The difference between the hybridization of oxygen compared to nitrogen and carbon is in the number of electrons that are distributed in the orbitals. Oxygen, a Group VIA element, has six valence electrons.

An sp^3-hybridized oxygen atom has two electrons in each of two sp^3 orbitals and one electron in each of the remaining two sp^3 orbitals. The bonded and nonbonded electron pairs are directed to the corners of a tetrahedron. However, the shape of molecules like water is angular.

An sp^2-hybridized oxygen atom has two electrons in two filled sp^2 orbitals and one half-filled sp^2-orbital. The sixth electron is in a 2p orbital, which can form a π bond. Note that the bond angle for σ bonds to sp^2-hybridized orbitals is 120°.

End of Chapter Exercises

Atomic Properties

1.1 How many valence shell electrons are in each of the following elements?
   (a) N   (b) F   (c) C   (d) O
   (e) Cl   (f) Br   (g) S   (h) P

   Answers: (a) 5   (b) 7   (c) 4   (d) 6   (e) 7   (f) 7   (g) 6   (h) 5

1.2 Which of the following atoms has the higher electronegativity? Which has the larger atomic radius?
   (a) Cl or Br   (b) O or S   (c) C or N   (d) N or O   (e) C or O

   Answers: electronegativity: (a) Cl > Br   (b) O > S   (c) N > C   (d) O > N   (e) O > C
   Answers: atomic radius: (a) Br > Cl   (b) S > O   (c) C > N   (d) N > O   (e) C > O

Ions and Ionic Compounds

1.3 Write a Lewis structure for each of the following ions.
   (a) OH^- (b) CN^- (c) H_3O^+ (d) NO_3^-

Answers:

(a) $^-:\ddot{O}H$ (b) $^-:C\equiv N$ (c) $H-\overset{+}{\underset{H}{\ddot{O}}}-H$ (d) $H-\overset{+}{\underset{H}{N}}-H$ (e) $^-:\ddot{O}-\overset{+}{N}=\ddot{O}$
                                                                                   $\quad\quad\quad\quad:\ddot{O}:^-$

1.4 Write a Lewis structure for each of the following ions.
   (a) NO_2^- (b) SO_3^- (c) NH_2^- (d) CO_3^-

Answers:

(a) $^-:\ddot{O}-\ddot{N}=\ddot{O}$ (b) $\ddot{O}=S=\ddot{O}$ (c) $\ddot{O}=S=\ddot{O}$ (d) $H-\ddot{N}-H$ (e) $^-:\ddot{O}-C=\ddot{O}$
                                         $\quad\quad:\underset{}{O}:_-$ $\quad\quad:\underset{}{O}:_-$ $\quad\quad\quad\quad:\ddot{O}:^-$

With $:\ddot{O}:^-$ above part (c).

Lewis Structures of Covalent Compounds

1.5 Write a Lewis structure for each of the following compounds.
   (a) NH_2OH (b) CH_3CH_3 (c) CH_3OH (d) CH_3NH_2 (e) CH_3Cl (f) CH_3SH

Answers:

(a) $H-\overset{H}{\underset{H}{N}}-\ddot{O}:$ (b) $H-\overset{H}{\underset{H}{C}}-\overset{H}{\underset{H}{C}}-H$ (c) $H-\overset{H}{\underset{H}{C}}-\ddot{O}:$

(d) $H-\overset{H}{\underset{H}{C}}-\overset{H}{\underset{}{\ddot{N}}}-H$ (e) $H-\overset{H}{\underset{H}{C}}-\ddot{C}l:$ (f) $H-\overset{H}{\underset{H}{C}}-\ddot{S}:$

1.6 Write a Lewis structure for each of the following compounds.

(a) HCN   (b) HNNH   (c) CH$_2$NH   (d) CH$_3$NO   (e) CH$_2$NOH   (f) CH$_2$NNH$_2$

Answers:

(a) H—C≡N: (b) H—N̈=N̈—H   (c) H—C̈=N̈—H (with H above C)

(d) H—C—N̈=Ö: (CH$_3$ with H above and below, C bonded to N)   (e) H—C=N—Ö—H (with H above C)   (f) H—C=N—N—H (with H above C and H above right N)

1.7 Add any required unshared pairs of electrons that are missing from the following formulas.

(a) CH$_3$—C(=O)—OH   (b) CH$_3$—C(=O)—OCH$_3$ (c) H—C(=O)—NHCH$_3$

(d) CH$_3$—S—CH=CH$_2$ (e) CH$_3$—C(=N—H)—CH$_3$ (f) N≡C—CH$_2$—C≡N

Answers:

(a) CH$_3$—C(=Ö:)—ÖH   (b) CH$_3$—C(=Ö:)—ÖCH$_3$ (c) H—C(=:Ö)—N̈HCH$_3$

(d) CH$_3$—S̈—CH=CH$_2$ (e) CH$_3$—C(=:N—H)—CH$_3$ (f) :N≡C—CH$_2$—C≡N:

1.8 Add any required unshared pairs of electrons that are missing from the following formulas.

(a) CH$_3$—C(=O)—Cl   (b) CH$_3$—O—CH=CH$_2$ (c) CH$_3$—C(=O)—SH

(d) CH$_3$—CH(—O—CH$_3$)—O—CH$_3$ (e) NH$_2$—C(=O)—O—CH$_3$ (f) CH$_3$—O—CH$_2$—O—CH$_3$

Answers:

(a) CH$_3$—C(=:O:)—C̈l: (b) CH$_3$—Ö—CH=CH$_2$ (c) CH$_3$—C(=:O:)—S̈H

(d) CH$_3$—CH(—Ö—CH$_3$)—Ö—CH$_3$ (e) NH$_2$—C(=:O:)—Ö—CH$_3$ (f) CH$_3$—Ö—CH$_2$—Ö—CH$_3$

1.9 Using the number of valence electrons in the constituent atoms and the given arrangement of atoms in the compound, write the Lewis structure for each of the following molecules.

(a) (CH$_2$)C—N—C(H)—H with H's   (b) Cl—C(=O)—Cl

(c) H—N(H)—C(=O)—N(H)—H   (d) H—C(H)(H)—S(=O)—O—H

7

Answers:

(a) $CH_2{=}\overset{\cdot\cdot}{N}{-}CH_3$ (b) $:\overset{\cdot\cdot}{\underset{\cdot\cdot}{Cl}}{-}\overset{\overset{\displaystyle :O:}{\|}}{C}{-}\overset{\cdot\cdot}{\underset{\cdot\cdot}{Cl}}:$ (c) $\overset{\cdot\cdot}{N}H_2{-}\overset{\overset{\displaystyle :O:}{\|}}{C}{-}\overset{\cdot\cdot}{N}H_2$ (d) $CH_3{-}\overset{\overset{\displaystyle :S:}{\|}}{C}{-}\overset{\cdot\cdot}{\underset{\cdot\cdot}{O}}{-}H$

1.10   Using the number of valence electrons in the constituent atoms and the given arrangement of atoms in the compound, write the Lewis structure for each of the following molecules.

(a)

$$H{-}\overset{\overset{\displaystyle H}{|}}{\underset{\underset{\displaystyle H}{|}}{C}}{-}S{-}S{-}\overset{\overset{\displaystyle H}{|}}{\underset{\underset{\displaystyle H}{|}}{C}}{-}H$$

(b)

$$H{-}\overset{\overset{\displaystyle H}{|}}{\underset{\underset{\displaystyle H}{|}}{C}}{-}\overset{\overset{\displaystyle O}{\|}}{\underset{\underset{\displaystyle H}{|}}{C}}{-}S{-}H$$

(c)

$$H{-}\overset{\overset{\displaystyle H}{|}}{\underset{\underset{\displaystyle H}{|}}{C}}{-}O{-}\overset{\overset{\displaystyle H}{|}}{\underset{\underset{\displaystyle H}{|}}{C}}{-}Cl$$

(d)

$$H{-}\overset{\overset{\displaystyle H}{|}}{\underset{\underset{\displaystyle H}{|}}{C}}{-}O{-}\overset{\overset{\displaystyle O}{\|}}{C}{-}\overset{\overset{\displaystyle H}{|}}{N}{-}H$$

Answers:

(a) $CH_3{-}\overset{\cdot\cdot}{\underset{\cdot\cdot}{S}}{-}\overset{\cdot\cdot}{\underset{\cdot\cdot}{S}}{-}CH_3$ (b) $CH_3{-}\overset{\overset{\displaystyle :\overset{\cdot\cdot}{O}}{\|}}{C}{-}\overset{\cdot\cdot}{\underset{\cdot\cdot}{S}}{-}H$

(c) $CH_3{-}\overset{\cdot\cdot}{\underset{\cdot\cdot}{O}}{-}\overset{\overset{\displaystyle H}{|}}{\underset{\underset{\displaystyle H}{|}}{C}}{-}\overset{\cdot\cdot}{\underset{\cdot\cdot}{Cl}}:$ (d) $CH_3{-}\overset{\cdot\cdot}{\underset{\cdot\cdot}{O}}{-}\overset{\overset{\displaystyle :O:}{\|}}{C}{-}\overset{\cdot\cdot}{\underset{\underset{\displaystyle H}{|}}{N}}{-}H$

1.11   Two compounds used as dry cleaning agents have the molecular formulas C_2Cl_4 and C_2HCl_3. Write the Lewis structures for each compound.

Answers:

(a)

$$\overset{:\overset{\cdot\cdot}{\underset{\cdot\cdot}{Cl}}}{\underset{:\overset{\cdot\cdot}{Cl}:}{}}\diagdown C{=}C\diagup\overset{:\overset{\cdot\cdot}{\underset{\cdot\cdot}{Cl}}:}{\underset{:\overset{\cdot\cdot}{Cl}:}{}}$$

(b)

$$\overset{:\overset{\cdot\cdot}{\underset{\cdot\cdot}{Cl}}}{\underset{:\overset{\cdot\cdot}{Cl}:}{}}\diagdown C{=}C\diagup\overset{:\overset{\cdot\cdot}{\underset{\cdot\cdot}{Cl}}:}{\underset{H}{}}$$

1.12   Acrylonitrile, a compound used to produce fibers for rugs, is represented by the formula CH_2CHCN. Write the Lewis structure for the compound.

Answer:

$$\overset{H}{\underset{H}{}}\diagup C{=}C\diagdown\overset{H}{\underset{C{\equiv}N:}{}}$$

Formal Charge

1.13   Assign the formal charges for the atoms other than carbon and hydrogen in each of the following species.

(a) H—Ö—C≡N: (b) H—Ö—N≡C:

(c) 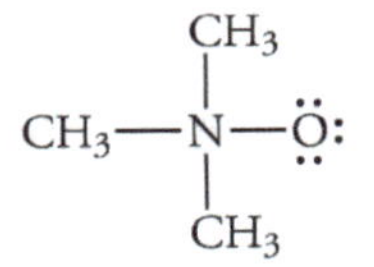 CH_3—N—Ö:
with CH₃ above and CH₃ below N

(d) CH_3—N̈=N=N̈:

Answers:

(a) H—Ö—C≡N: (b) H—Ö—N⁺≡C⁻:

(c) 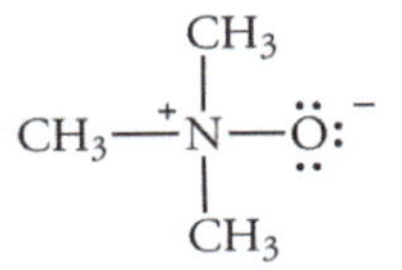 CH_3—N⁺—Ö:⁻
with CH₃ above and CH₃ below N

(d) CH_3—N̈=N⁺=N̈:⁻

(a) none of the atoms has a formal charge
(b) nitrogen is +1; carbon is −1
(c) nitrogen is +1; oxygen is −1
(d) nitrogen atoms from left to right have 0, +1, and −1 formal charges

1.14   Assign the formal charges for the atoms other than carbon and hydrogen in each of the following species.

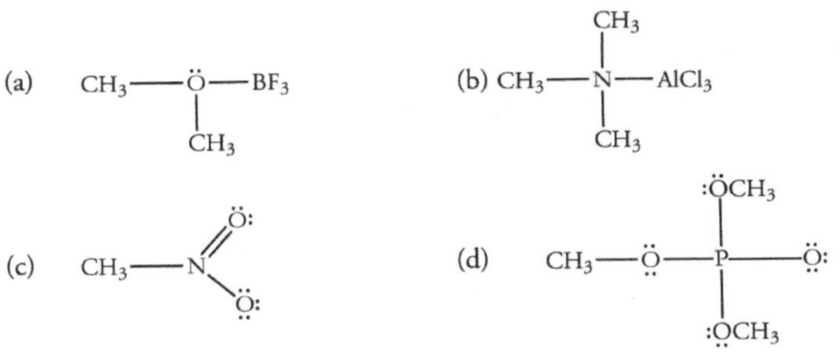

Answers:
(a) oxygen is +1; boron is −1
(b) nitrogen is +1; aluminum is −1
(c) nitrogen is +1; singly bonded oxygen atom is −1
(d) phosphorus is +1; oxygen atom on the right is −1

(a) CH_3—Ö—BF_3
with CH₃ below O

(b) CH_3—N—$AlCl_3$
with CH₃ above and CH₃ below N

(c) CH_3—N
with Ö: (double bonded) and Ö: (single bonded)

(d) CH_3—Ö—P—Ö:
with :ÖCH₃ above and :OCH₃ below P

1.15   All of the following species are isoelectronic, that is, they have the same number of electrons bonding the same number of atoms. Determine which atoms have a formal charge. Calculate the net charge for each species.

Answers:
(a) carbon is −1; oxygen is +1; total charge is 0
(b) nitrogen is zero; oxygen is +1; total charge is +1
(c) carbon is −1; nitrogen is 0; total charge is −1
(d) both carbon atoms are −1; total charge is −2

(a) :C≡O: (b) :N≡O: (c) :C≡N: (d) :C≡C:

1.16   All of the following species are isoelectronic, that is, they have the same number of electrons bonding the same number of atoms. Determine which atoms have a formal charge. Calculate the net charge for each species.

(a) :N̈=N=N̈:   (b) :Ö=N=Ö:

Answers:
(a) central nitrogen atom is +1; the other nitrogen atoms are each −1; the total charge is −1
(b) nitrogen atom is +1; both oxygen atoms are 0; the total charge is +1

1.17   The following species are isoelectronic. Determine which atoms have a formal charge. Calculate the net charge for each species.

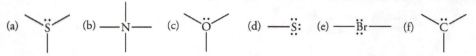

Answers:
(a) S is +1   (b) N is +1
(c) O is +1   (d) S is −1
(e) Br is +1   (f) C is −1

1.18   The following species are isoelectronic. Determine which atoms have a formal charge. Calculate the net charge for each species.

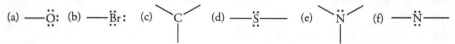

Answers:
(a) O is −1   (b) Br is 0
(c) C is +1   (d) S is 0
(e) N is 0   (f) N is −1

1.19   Acetylcholine, a compound involved in the transfer of nerve impulses, has the following structure. What is the formal charge on the nitrogen atom? What is the net charge of acetylcholine?

Answer:
The nitrogen atom has a charge of +1; the total charge is +1.

$$\underset{\underset{CH_3}{|}}{\overset{\overset{CH_3}{|}}{CH_3-N}}-CH_2CH_2-\overset{..}{\underset{..}{O}}-\overset{\overset{\overset{..}{O}:}{\|}}{C}-CH_3$$

acetylcholine

1.20   Sarin, a nerve gas, has the following structure. What is the formal charge of the phosphorus atom?

Answer:
The phosphorus atom has a charge of +1.

$$CH_3-\underset{\underset{CH_3}{|}}{\overset{\overset{CH_3}{|}}{C}}-\overset{..}{\underset{..}{O}}-\underset{\underset{:F:}{|}}{\overset{\overset{:\overset{..}{O}:}{|}}{P}}-CH_3$$

Sarin

Resonance

1.21   The small amounts of cyanide ion contained in the seeds of some fruits are eliminated from the body as SCN⁻. Draw two possible resonance forms for the ion. Which atom has the formal negative charge in each form?

$$^-:\overset{..}{\underset{..}{S}}-C\equiv N: \quad \longleftrightarrow \quad :\overset{..}{\underset{..}{S}}=C=\overset{..}{N}:^-$$

Answer:
Sulfur has a −1 charge on the left; nitrogen has a −1 charge on the right.

1.22   Are the following pairs contributing resonance forms of a single species? Formal charges are not shown and have to be added.

(a) $:\overset{.}{N}=N=\overset{.}{N}:$ and $:\overset{.}{N}-N\equiv N:$ (b) $H-C\equiv N-\overset{..}{\underset{..}{O}}:$ and $H-\overset{.}{C}=N=\overset{..}{O}:$

Answers:

(a) $^-:\overset{.}{N}=\overset{+}{N}=\overset{.}{N}:^- \quad \longleftrightarrow \quad :N\equiv\overset{+}{N}-\overset{..}{\underset{..}{N}}:^{2-}$

(b) $H-\overset{+}{C}\equiv N-\overset{..}{\underset{..}{O}}:^- \quad \longleftrightarrow \quad H-\overset{.}{\underset{.}{C}}=\overset{+}{N}=\overset{..}{O}:$

1.23   Write the resonance structure that results when electrons are moved in the direction indicated by the curved arrows for the following amide. Calculate any formal charges that result.

Answer:

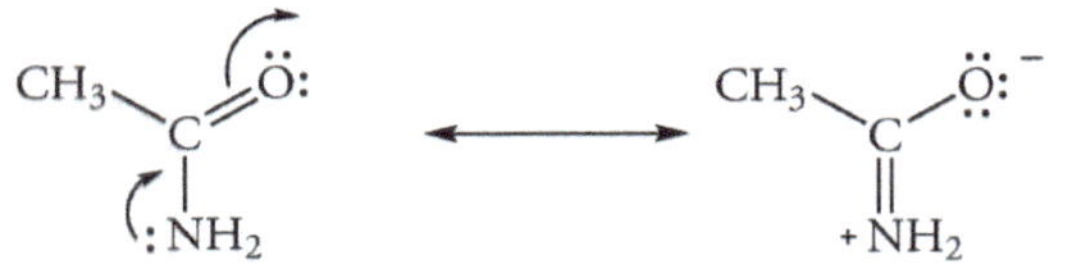

CH_3–C=O: ⟷ CH_3–C–O:⁻
     $:NH_2$ $^+NH_2$

1.24   Write the resonance structure that results when electrons are moved in the direction indicated by the curved arrows for acetate. Calculate any formal charges that result.

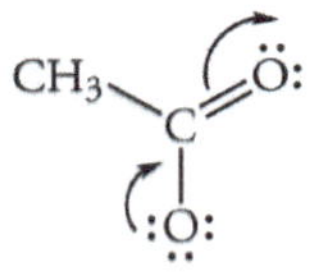

CH_3–C=O:
     $:O:$

Answer:

CH_3–C=O: ⟷ CH_3–C–O:⁻
     $:O:^-$ $:O:$

1.25   Write the resonance structure that results when electrons are moved in the direction indicated by the curved arrow for the following electron-deficient ion. To what extent do each of the two resonance forms contribute to the structure of the ion?

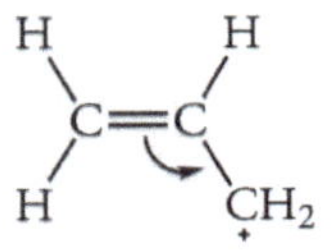

Answer:
The alternate resonance form is structurally equivalent to the given resonance form and both contribute equally.

1.26   Write the resonance structure that results when electrons are moved in the direction indicated by the curved arrows for the diazomethane. Do each of the two resonance forms contribute equally to the structure of the ion?

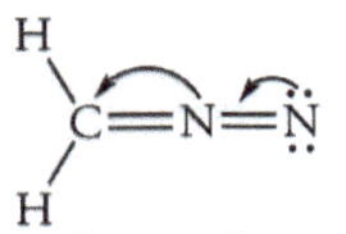

diazomethane

Answer:

Answer:
The alternate resonance form has a negative charge on the carbon atom rather than the nitrogen atom. Because nitrogen is more electronegative than carbon, the original resonance form contributes to a larger extent.

Molecular Shapes

1.27   Based on VSEPR theory, what is the expected value of the indicated bond angle in each of the following compounds?

(a) C—C—N in CH_3—C≡N   (b) C—O—C in CH_3—O—CH_3

(c) C—N—C in CH_3—NH—CH_3 (d) C—C—C in CH_3—C≡C—H

1.28   Based on VSEPR theory, what is the expected value of the indicated bond angle in each of the following ions?

(a) C—O—H in CH_3—OH_2^+ (b) C—N—H in CH_3—NH_3^+

(c) O—C—O in $CH_3CO_2^-$ (d) C—O—C in $(CH_3)_2OH^+$

1.29   Based on VSEPR theory, what is the expected value of the indicated bond angle in each of the following compounds?

(a) C—O—C in CH_3—C(=O)—OCH_3 (b) O—C—N in H—C(=O)—NH_2

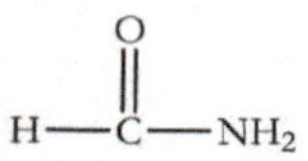

(c) O—C—O in CH_3—C(=O)—OCH_3

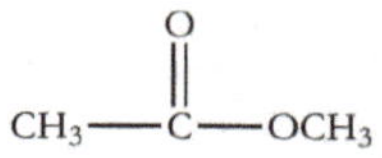

1.30   Based on VSEPR theory, what is the expected value of the indicated bond angle in each of the following compounds?

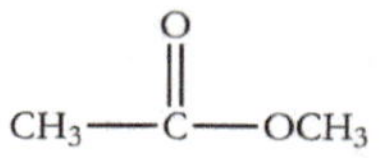

(a) C—O—C in CH_3—C(=O)—OCH_3 (b) O—C—N in H—C(=O)—NH_2

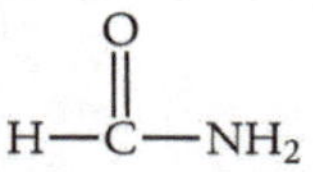

(c) O—C—O in CH_3—C(=O)—OCH_3

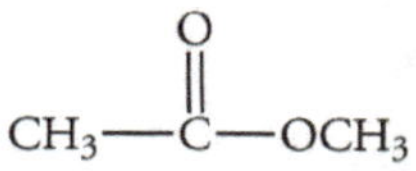

Dipole Moments

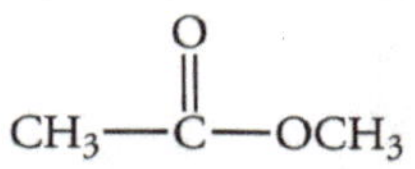

1.31   Fluorine is more electronegative than chlorine, but the dipole moment for a C—F bond (1.4 D) is less than the dipole moment for a C—Cl bond (1.5 D). Explain why this is so.

Answer: The carbon—fluorine bond is much shorter than the carbon—chlorine bond.

1.32   Arrange the following bond moments in order of decreasing polarity: H—N, H—O, H—S. Explain the trend that you predict.

Answer:

H — S > H — N > H — O; the difference in electronegativity of the atoms in the O—H bond is larger than that of the atoms in the N—H bond. There is a substantially smaller difference in electronegativity of the atoms in the S—H bond and the bond has a small polarity.

1.33   The dipole moments of both CO, and CS, are zero. However, SCO has a dipole moment. Explain why. Draw the structure of SCO and then an arrow indicating the direction of the dipole moment.

Answer:

The C=O and C=S bond moments are not equal, so they don't cancel each other. The net dipole moment is toward the more electronegative oxygen atom.

1.34   Which compound has the larger dipole moment, acetone or phosgene? Explain why.

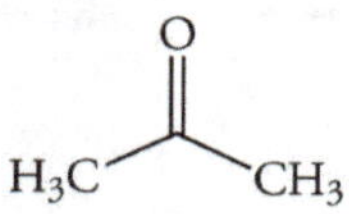

acetone 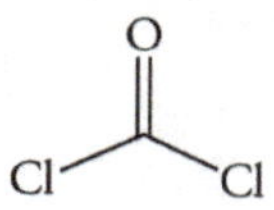   phosgene

Answer:
Acetone has the larger dipole moment because the bond moments of the two C—Cl bonds oppose the C=O dipole moment in phosgene.

1.35   Which compound has the larger dipole moment, *cis*- or *trans*-1,2-dibromoethene? Explain why.

cis-1,2-dibromoethene   *trans*-1,2-dibromoethene

Answer:
The net resultant of the dipole moments of the carbon—bromine bonds in the *cis* isomer is toward the side of the molecule containing the two carbon—bromine bonds. The bond moments of the carbon—bromine bonds in the *trans* isomer are opposed and therefore cancel one another.

1.36   The dipole moment of chlorobenzene (C_6H_5Cl) is 1.56 D and that of nitrobenzene ($C_6H_5NO_2$) is 3.97 D. The dipole moment of *para*-chloronitrobenzene is 2.57 D. What does this value indicate about the direction of the moments of the two groups with respect to the benzene ring?

    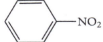   

chlorobenzene   nitrobenzene   *p*-chloronitrobenzene

Answer:
The two dipole moments must oppose one another to give a resultant that is less than the large dipole bond moment value. The dipole moment of the carbon—chlorine bond is toward the chlorine atom. Thus, the dipole moment of the carbon—nitrogen bond must be toward nitrogen.

Hybridization

1.37   What is the hybridization of each carbon atom in each of the following compounds?

Answers:
(a) from left to right: sp^3, sp^2
(b) from left to right: sp^3, sp^2, sp^2
(c) from left to right: sp^3, sp^2
(d) from left to right: sp^3, sp^2, sp^3

(a) CH_3—C(=O)—H   (b) CH_3O—CH=CH_2 (c) CH_3—C(=O)—OH

(d) CH_3—C(=O)—OCH_3

1.38   What is the hybridization of each carbon atom in each of the following compounds?

Answers:
(a) from left to right: sp^2, sp^3
(b) from left to right: sp^3, sp^2, sp^2
(c) from left to right: sp^3, sp^2, sp^3
(d) from left to right: sp, sp^3, sp

(a) H—C(=O)—NH—CH_3 (b) CH_3NH—CH=CH_2

(c) CH_3—C(=N—H)—CH_3 (d) N≡C—CH_2—C≡N

1.39   What is the hybridization of the oxygen atom in each compound in Exercise 1.37?

Answers:
(a) sp^2
(b) sp^3
(c) double-bonded oxygen is sp^2; single-bonded is sp^3
(d) double-bonded oxygen is sp^2; single-bonded is sp^3

1.40   What is the hybridization of the oxygen atom in each compound in Exercise 1.38?

Answers:
(a) sp^3
(b) sp^3
(c) sp^2
(d) sp

1.41   Carbocations and carbanions are unstable organic species with a positive and a negative charge, respectively, on the carbon atom. What is the hybridization of the carbon atom in each ion? What are the H—C—H bond angles?

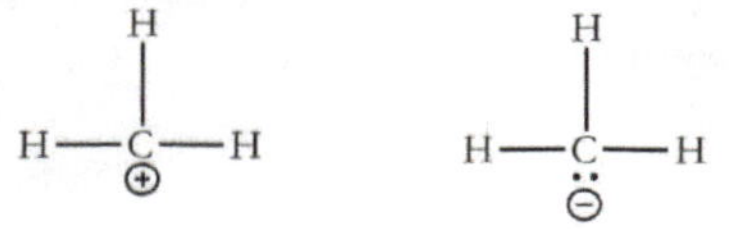

Answer:

The carbocation is sp^2 hybridized, and the bond angles are 120°. The carbanion is sp^3 hybridized, and the bond angles are 109°.

1.42   Assuming that all of the valence electrons are paired and located in hybrid orbitals, what is the H—C—H bond angle in the reactive species CH_2?

Answer:

The three pairs of electrons, one nonbonded and two bonded, are in a common plane at 120° to one another.

1.43   Write the Lewis structure of CO_2. What is the hybridization of the carbon atom? What is the hybridization of the oxygen atoms?

$$:\ddot{O}\!\!=\!\!C\!\!=\!\!\ddot{O}:$$

Answer:

The carbon atom is sp hybridized. The oxygen atoms are sp^2 hybridized.

1.44   Write the Lewis structure of NO_2^+, the nitronium ion. What is the hybridization of the nitrogen atom? What is the hybridization of the oxygen atoms?

$$:\ddot{O}\!\!=\!\!\overset{+}{N}\!\!=\!\!\ddot{O}:$$

Answer: The nitrogen atom is sp hybridized. The oxygen atoms are sp^2 hybridized.

1.45   Phosgene ($COCl_2$) is a poisonous gas. Write its Lewis structure and determine the hybridization of the carbon atom.

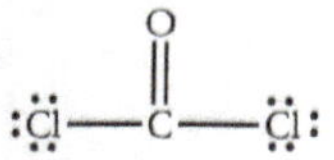

Answer:

The carbon atom is sp^2 hybridized.

1.46   Carbamic acid is an unstable substance that decomposes to form carbon dioxide and ammonia. Based on the following Lewis structure, what are the hybridizations of the carbon atom and the two oxygen atoms?

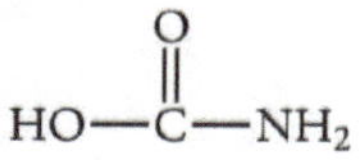

Answer:

The carbon atom is sp^2 hybridized. The double-bonded oxygen atom is sp^2; single-bonded oxygen atom is sp^3 hybridized.

Bond Lengths

1.47   The oxygen—hydrogen bond length in both hydrogen peroxide (HO—OH) and hydroxylamine (NH_2—OH) are the same; 96 pm. Explain why.

Answer:

Bond lengths between common sets of atoms tend to be the same and do not depend markedly on the other atoms of the structure.

1.48   The C=N bond length of methyleneimine (CH_2=NH) is 127 pm. Compare this value to the C=C bond length of ethene (133 pm) and suggest a reason for the difference.

Answer:

The C=N bond is shorter than the C=C bond because the atomic radius of nitrogen is smaller than the atomic radius of carbon.

1.49   The nitrogen—oxygen bond lengths of hydroxylamine (NH_2—OH)) and the nitronium ion (NO_2)$^+$ are 145 and 115 pm, respectively. Write their Lewis structures and explain why the bond lengths differ.

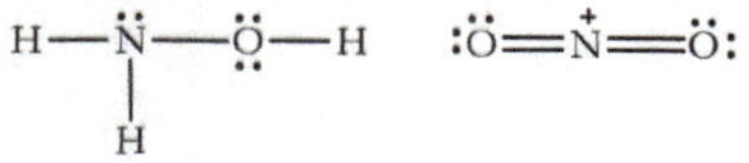

1.50   The C—F bond length of CF_4 is 138 pm. The estimated bond length of CF_3^+ is 127 pm. Suggest a reason for the difference between these two values.

Answer:
Each of the three contributing resonance forms in CF_3^+ has a C=F bond which contributes to the overall shortening of the carbon–fluorine bonds in the resonance hybrid structure.

1.51   The carbon–carbon single bond lengths of propane and propene are 154 and 151 pm, respectively. Why do these values differ?

Answer:
The bonds are sp^3–sp^3 and sp^3–sp^2 hybridized, respectively. An sp^2-hybridized atom holds the bonding pair of electrons closer to the nucleus, and this leads to a shortening of the bond.

1.52   The carbon—oxygen bond length of dimethyl ether is 142 pm. Predict the lengths of each of the two carbon—oxygen bonds in methyl vinyl ether.

Answer:
The bond to the CH_3 group should also be 142 pm. The bond of oxygen to the CH group should be shorter than 142 pm because an sp^2-hybridized atom holds the bonding pair of electrons closer to the nucleus, and this leads to a shortening of the bond.

Bond Angles

1.53   What is the C—N—H bond angle in each of the following species?

Answer: (a) 109° (b) 120°

1.54   What is the C—O—H bond angle of protonated methanal?

Answer: 120°

1.55   Diimide (HNNH) is a reactive reducing agent. Draw its Lewis structure. Compare its Lewis structure with that of ethene. Compare the hybridization of the two compounds. What is the H—N—N bond angle in diimide?

Answer:
The hybridization of both the carbon atoms in ethene and the nitrogen atoms in diimide is sp^2. The H—N—N bond angle is 120°.

1.56   What is the H—C—H bond angle in allene (CH_2=C=CH_2)? What is the C—C—C bond angle? What is the hybridization of each atom?

Answer:

The H—C—H bond angle is 120°. The C—C—C bond angle is 180°. The hybridization of both terminal atoms is sp^2; the hybridization of the central carbon atom is sp.

1.57   What is the Cl—C—Cl bond angle of the CCl_3^- ion, an intermediate formed by treating CCl_3H with base?

Answer: 109°

1.58   What is the O—N—O bond angle of the nitronium ion (NO_2)$^+$, a reactive intermediate in reactions with benzene compounds?

Answer: 180°

---

2 PART I: FUNCTIONAL GROUPS AND THEIR PROPERTIES

KEYS TO PART I OF THE CHAPTER

2.1 Functional Groups

Functional groups are structural features of organic compounds other than carbon–carbon single bonds and carbon–hydrogen single bonds. Multiple bonds between carbon atoms and bonds from carbon to atoms such as oxygen, nitrogen, sulfur, and the halogens are components of functional groups. When learning the features of the functional groups, we must pay attention both to their composition and to their structure and bonding.

As we proceed with our study of organic chemistry, we will find that functional groups behave chemically in ways that we can predict based on the number and type of bonds to carbon in each functional group. The chemistry of organic molecules depends on the functional groups that they contain. The only functional groups that do not contain atoms other than carbon and hydrogen contain carbon–carbon multiple bonds, as in ethene (ethylene) and ethyne (acetylene). Benzene, which also contains multiple carbon–carbon bonds, belongs to a separate class of compounds called aromatic hydrocarbons.

2.2 Functional Groups Containing Oxygen

Several types of functional groups contain oxygen. Compounds with a carbon–oxygen and an oxygen–hydrogen bond are **alcohols**. Compounds with two carbon–oxygen bonds are **ethers**.
The oxygen atom forms double bonds to carbon in a **carbonyl group** in several functional groups. If the remaining two single bonds are to other carbon atoms, the compound is a **ketone**. If there is one single bond to a carbon atom and one to a hydrogen atom, the compound is an **aldehyde**.
Compounds with a single bond from an oxygen atom to a carbonyl group are found in **carboxylic acids** and **esters**. In carboxylic acids, the second bond to that oxygen atom is to a hydrogen atom; in esters it is to another carbon atom. Note that a carboxylic acid is not an aldehyde, a ketone, or an alcohol. Both the carbonyl group and the hydroxyl group *together* are considered as a single functional group when they share a common carbon atom.

2.3 Functional Groups Containing Nitrogen

Nitrogen can form functional groups that contain single bonds in amines, double bonds in imines, and triple bonds in nitriles. A nitrogen atom bonded to a carbonyl group is an amide. The amide nitrogen atom may be bonded to any combination of hydrogen atoms or carbon atoms.

2.4 Functional Groups Containing Sulfur or Halogens

Sulfur occurs in functional groups that parallel those of alcohols and ethers. These sulfur-containing compounds are **thiols** and **thioethers**. Halogens can be bonded to sp^3-hybridized carbon atoms or to the sp^2-hybridized carbon atom of a carbonyl group.

2.5 Structural Formulas

Molecular formulas identify the total number of atoms of each element in a molecule. They tell us nothing about the structure of the molecule. **Structural formulas** show how the atoms in the molecule are arranged and which atoms are bonded to each other. A complete structural formula shows every bond. A condensed formula abbreviates the structure by omitting some or all of the bonds and indicating the number of atoms bonded to each carbon atom with subscripts.

Several conventions are used to represent structures in varying degrees of detail and in shorthand form. In general, make sure that each atom has the appropriate number of bonds. **Condensed structural formulas** leave out some bonds, and the bonded atoms are written close to each other. In general, atoms bonded to a carbon atom are usually written right after the carbon atom.

2.6 Bond-Line Structures

This section introduces a "shorthand" skill that helps us show the details of chemical structure. Remember that there is a carbon atom at every intersection of two or more lines and at the end of every line. Also remember that there are four bonds to every carbon atom. The bonds from one carbon atom to other carbon atoms and to atoms of other elements are easy to identify; the bonds to hydrogen atoms are not visible in the bond-line structure, and we must carefully account for them. Bond-line structures are a better and faster way to record structural formulas than writing both atoms and bonds.

Remember that the chemistry occurs at the functional groups. Consider, for example, the structure of diphepanol, which is used as a cough suppressant. Can you identify the functional groups? Can you write its molecular formula?

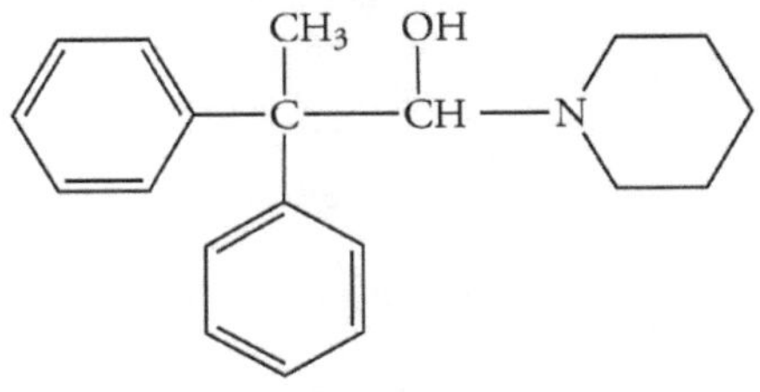

diphepanol

The oxygen atom in diphepanol is part of a hydroxyl group, so the functional group is an alcohol. The nitrogen atom is bonded only to carbon atoms, so it is an amine. The molecular formula is $C_{20}H_{25}NO$.

Here's another example. What are the oxygen-containing functional groups in the herbicide with the commercial name 2,4-D? Its structure is shown below.

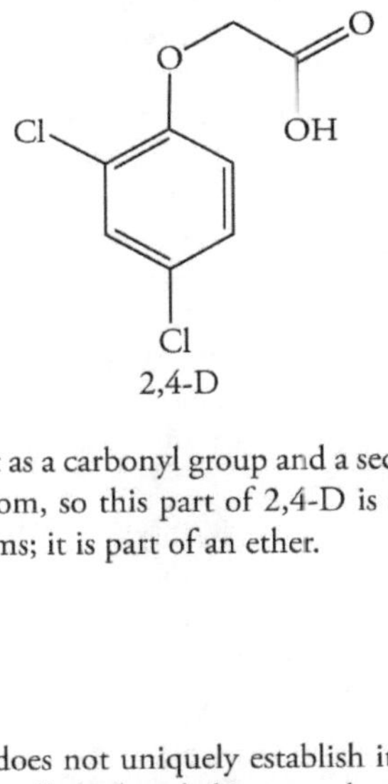

2,4-D

One of the oxygen atoms is present as a carbonyl group and a second as a hydroxyl group. They both are bonded to the same carbon atom, so this part of 2,4-D is a carboxylic acid. The third oxygen atom is bonded to two carbon atoms; it is part of an ether.

2.7 Isomers

The composition of a compound does not uniquely establish its structure. For all but the simplest molecules, a group of atoms can usually be bonded in several ways to give different structures called **constitutional isomers**. Distinguishing between structures that are isomers and those that are merely different representations of the same molecule requires practice.

There are many ways to write the structural formula of an organic compound. Two structural formulas with the same molecular formula may look so different that they appear at first glance to represent isomers. To determine if two structures represent isomers, carefully check the bonding sequence in each formula. If the sequence of bonded atoms is the same, the structural formulas represent two views of the same compound. If the sequence of bonded atoms is different, the two structural formulas represent isomers.

PART II: IDENTIFICATION OF FUNCTIONAL GROUPS BY INFRARED SPECTROSCOPY

2.8 Spectroscopy

The energy of light is directly proportional to its frequency; $E = hv$. Wavelength and frequency are inversely proportional and are related by $\lambda = c/v$, where c is the speed of light. As the wavelength of the electromagnetic radiation increases, the corresponding frequency decreases. Spectroscopy is used to probe the physical changes in a molecule as the result of absorption of energy. In infrared spectroscopy, the energy absorbed can change the extent to which a bond stretches or bends.

2.9 Infrared Spectroscopy

Infrared spectroscopy is extremely valuable because it allows us to confirm the presence (and sometimes more importantly the absence) of functional groups. The infrared spectrum is displayed so that absorptions of energy are related to wavelength or wavenumber. The energy of the absorption is indicated by an inverted "peak" pointed down from a baseline.

Infrared absorptions correspond to the stretching of a bond or the bending of a bond angle. The strength of the bond is given by a force constant. Multiple bonds have higher force constants, and their absorbances occur at higher energy. The energy required to stretch a bond is also related to the atomic mass of the bonded atoms. Bonds to hydrogen such as C—H, O—H, and N—H require higher energy than bonds such as C—C, O—C, and N—C.

The amount of energy required to stretch a specific bond in an organic molecule depends on the nature of the bonded atoms and the type of bond between them. The full interpretation of the IR spectrum of a molecule is difficult, but certain functional groups have characteristic absorptions which can be used to propose a structure for an unknown compound.

The spectrum of an unknown compound can be established by comparison to the spectrum of a known compound. If the spectrum of the unknown compound has *all* of the same absorption peaks as a compound of known structure, then the two samples are identical. If the "unknown" has one or more peaks that differ from the spectrum of a known, then the two compounds are not identical, or some impurity in the unknown sample is causing the extra absorptions. If the unknown lacks even one absorption peak that is present in the known structure, then the "unknown" has a different structure than the known one.

2.10 Identifying Hydrocarbons

The energy for the absorbance for a C—H bond depends on the % s character of the bond. With increased % s character, the electrons are more tightly held by an atom, so a bond to that atom requires higher energy to stretch. This difference is used to detect alkene and alkynes providing they have C—H bonds as well as C=C or C≡C bonds.

2.11 Identifying Oxygen-Containing Compounds

The presence of a carbon–oxygen double bond is easily detected by its characteristic strong absorption near the middle of an IR spectrum at a wavenumber of about 1700 cm^{-1}. The exact location is controlled by the extent to which a dipolar resonance form contributes to the structure. If the dipolar resonance form is stabilized by atoms bonded to the carbonyl group, then the C—O bond has more single bond character, and the energy required to stretch the bond is smaller. Alcohols and ethers both contain C—O bonds that are difficult to confirm unambiguously in IR spectra. However, the presence of an O—H bond in an alcohol is easily detected by a strong absorption on the left of the spectrum in the energy range 3400–3600 cm^{-1}.

2.12 Identifying Nitrogen-Containing Compounds

The presence of an N—H bond in an amine is easily detected. Primary amines have two N—H absorbances that occur over a range from 3250 to 3550 cm^{-1}. Secondary amines have a single N—H absorbance that occurs over a range from 3250 to 3550 cm^{-1}. C—N bond stretching occurs in the 1000–1250 cm^{-1} region. C—N peaks are weak. In contrast, the C≡N absorbance, which occurs at around 2250 cm^{-1} is very strong.

2.13 Bending Deformations

The fingerprint region of the presence of the IR spectrum contains many kinds of bending modes. Some of these are readily identified. They provide clues about the substitution pattern on benzene rings.

Functional Groups

2.1   Identify the functional groups contained in each of the following structures.
        (a) caprolactam, a compound used to produce a type of nylon

Answer: amide

caprolactam

        (b) civetone, a compound in the scent gland of the civet cat

Answer: ketone and double bond

civetone

        (c) DEET, the active ingredient in some insect repellents

Answer: amide and benzene ring

DEET

2.2   Identify the oxygen-containing functional groups in each of the following compounds.
        (a) isopimpinellin, a carcinogen found in diseased celery

Answer: three ethers, ester, and benzene ring

isopimpinellin

        (b) aflatoxin B_1, a carcinogen found in moldy foods

Answer: three ethers, ketone, ester, two double
bonds, and benzene ring

aflatoxin B_1

22

(c) penicillin G, an antibiotic first isolated from a mold.

Answer: two amides, thioether, carboxylic acid, and benzene ring

penicillin G

Molecular Formulas

2.3   Write the molecular formula for each of the following.
(a) $CH_3-CH_2-CH_2-CH_2-CH_3$ (b) $CH_3-CH_2-CH_2-CH_3$ (c) $CH_2=CH-CH_2-CH_3$
(d) $CH_3-CH_2-C\equiv C-H$ (e) $CH_3-CH_2-CH_2-CH=CH_2$ (f) $CH_3-CH_2-C\equiv C-CH_3$

Answers: (a) C_5H_{12} (b) C_4H_{10} (c) C_4H_8
(d) C_5H_8 (e) C_5H_{10} (f) C_5H_8

2.4   Write the molecular formula for each of the following.
(a) $CH_3CH_2CH_2CH_2CH_2CH_2CH_2CH_2CH_3$ (b) $CH_3CH_2CH_2CH_2CH_2CH_2CH_2CH_3$ (c) $CH_3CH_2C\equiv CH$
(d) $CH_3CH_2C\equiv CCH_3$ (e) $CH_3CH_2CH_2CH=CHCH_3$ (f) $CH_2=CHCH_2CH_3$

Answers: (a) C_9H_{20} (b) C_8H_{18} (c) C_4H_6 (d) C_5H_8 (e) C_6H_{12} (f) C_5H_8

2.5   Write the molecular formula for each of the following.
(a) $CH_3-CH_2-CHCl_2$ (b) $CH_3-CCl_2-CH_3$ (c) $Br-CH_2-CH_2-Br$
(d) $CH_3-CHBr-CHBr_2$ (e) $CH_3-CF_2-CH_2F$ (f) $F-CH_2-CHF-CH_2-F$

Answers: (a) $C_3H_6Cl_2$ (b) $C_3H_6Cl_2$ (c) $C_2H_4Br_2$ (d) $C_3H_5Br_3$ (e) $C_3H_5F_3$ (f) $C_3H_5F_3$

2.6   Write the molecular formula for each of the following.
(a) $CH_3-CH_2-CH_2-OH$ (b) $CH_3-CH_2-O-CH_2-CH_3$ (c) CH_3-CH_2-SH
(d) $CH_3-CH_2-S-CH_3$ (e) $CH_3-CH_2-CH_2-NH_2$ (f) $CH_3-CH_2-NH-CH_3$

Answers: (a) C_3H_8O (b) $C_4H_{10}O$ (c) C_2H_6S (d) C_3H_8S (e) C_3H_9N (f) C_3H_9N

Structural Formulas

2.7   For each of the following, write a condensed structural formula in which only the bonds to hydrogen are not shown.

Answers: (a) $Br-CH_2-CH_2-Br$
(b) $CH_3-CH_2-CH_2-CH_2-CH_3$
(c) $CH_3-CH_2-CH_2-SH$
(d) $CH_3-CH_2-CH_2-CH_2-NH_2$

2.8   For each of the following, write a condensed structural formula in which only the bonds to hydrogen are not shown.

Answers:
(a) CH_3—CH_2—CH_2—NH—CH_3

(b) CH_3—CH_2—CH_2—O—CH_3

(c) CH_3—CH_2—CH_2—CH_2—CCl_3

(d) CH_3—CH_2—NH—CH_2—CH_3

2.9   Write a condensed structural formula in which no bonds are shown for each of the structures in problem 2.7.

Answers: (a) $BrCH_2CH_2Br$
             (b) $CH_3CH_2CH_2CH_2CH_3$
             (c) $CH_3CH_2CH_2SH$
             (d) $CH_3CH_2CH_2CH_2NH_2$

2.10   Write a condensed structural formula in which no bonds are shown for each of the structures in problem 2.8.

Answers: (a) $CH_3CH_2CH_2NHCH_3$
         (b) $CH_3CH_2CH_2OCH_3$
         (c) $CH_3CH_2CH_2CH_2CCl_3$
         (d) $CH_3CH_2NHCH_2CH_3$

2.11   Write a complete structural formula, showing all bonds, for each of the following condensed formulas.
        (a) $CH_3CH_2CH_2CH_3$ (b) $CH_3CH_2CH_2Cl$ (c) $CH_3CHClCH_2CH_3$
        (d) $CH_3CH_2CHBrCH_3$ (e) $CH_3CH_2CHBr_2$ (f) $CH_3CBr_2CH_2CH_2CH_3$

Answers:

24

2.12   Write a complete structural formula, showing all bonds, for each of the following condensed formulas.
    (a) $CH_3CH_2CH_3$ (b) $CH_3CH_2CHCl_2$ (c) $CH_3CH_2CH_2CH_2SH$

    (d) $CH_3CH_2C\equiv CCH_3$ (e) $CH_3CH_2OCH_2CH_3$ (f) $CH_3CH_2CH_2C\equiv CH$

Answers:

(a)
```
      H   H   H
      |   |   |
  H — C — C — C — H
      |   |   |
      H   H   H
```

(b)
```
      H   H   Cl
      |   |   |
  H — C — C — C — Cl
      |   |   |
      H   H   H
```

(c)
```
      H   H   H   H
      |   |   |   |
  H — C — C — C — C — S — H
      |   |   |   |
      H   H   H   H
```

(d)
```
      H   H               H
      |   |               |
  H — C — C — C≡C — C — H
      |   |               |
      H   H               H
```

(e)
```
      H   H        H   H   H
      |   |        |   |   |
  H — C — C — O — C — C — C — H
      |   |        |   |   |
      H   H        H   H   H
```

(f)
```
      H   H   H
      |   |   |
  H — C — C — C — C≡C — H
      |   |   |
      H   H   H
```

Bond-Line Structures

2.13   What is the molecular formula for each of the following bond-line representations?

Answers:
(a) $C_9H_{16}O$ (b) $C_7H_{16}O$
(c) $C_8H_{17}Br$ (d) $C_{11}H_{18}$

(a)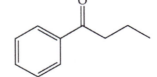

(b)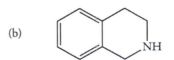

(c)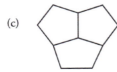

(d)

2.14   What is the molecular formula for each of the following bond-line structures?

Answers:
(a) $C_{10}H_{12}O$ (b) $C_9H_{11}N$
(c) $C_{10}H_{16}$ (d) $C_4H_6O_2$

(a)

(b)

(c)

(d)

2.15   What is the molecular formula for each of the following bond-line structures?
        (a) a scent marker of the red fox

Answer:
(a) $C_6H_{12}S$

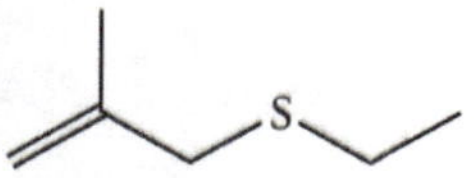

        (b) a compound responsible for the odor of the iris

Answer:
(b) $C_{12}H_{20}O$

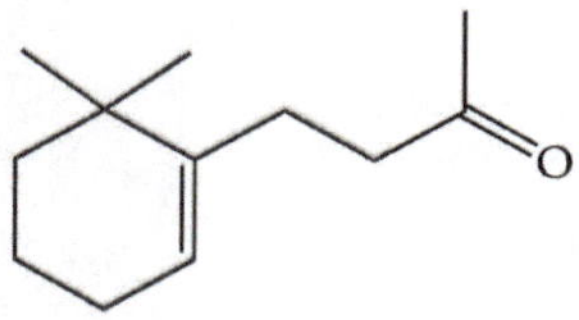

        (c) a defense pheromone of some ants

Answer:
(c) $C_{15}H_{22}O$

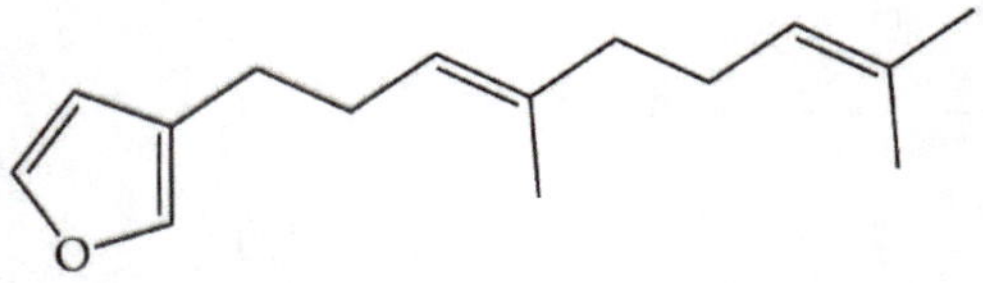

2.16   What is the molecular formula for each of the following bond-line structures?
        (a) a compound found in clover and grasses

Answer:
(a) $C_9H_6O_2$

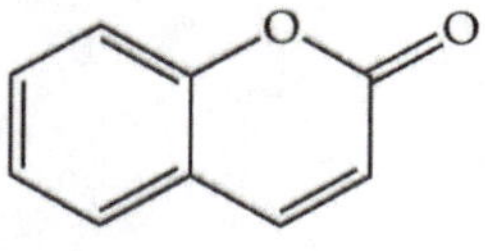

        (b) an oil found in citrus fruits

Answer:
(b) $C_{10}H_{18}O$

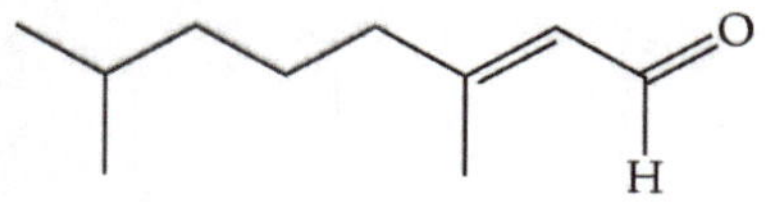

        (c) a male sex hormone

Answer:
(c) $C_{19}H_{28}O_2$

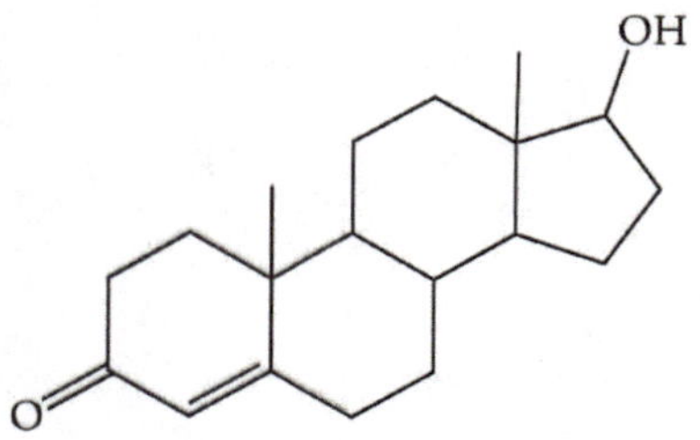

Isomerism

2.17   Indicate whether the following pairs of structures are isomers or different representations of the same compound.

Answers:
(a) different representations for the same structure
(b) different representations for the same structure
(c) isomers

(a) 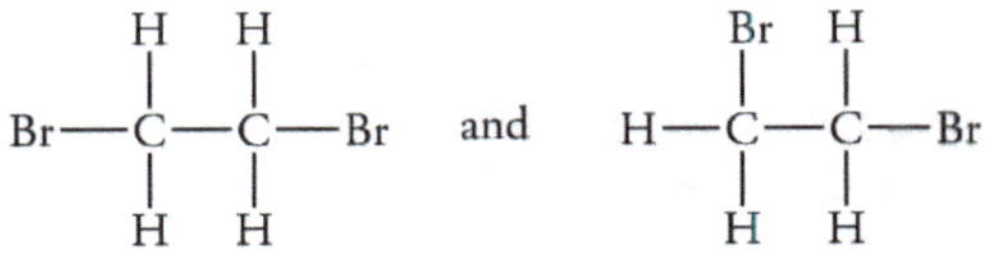   Br—C—C—Br   and   H—C—C—Br
        (with H, H above; H, H below for first; Br, H above and H, H below for second)

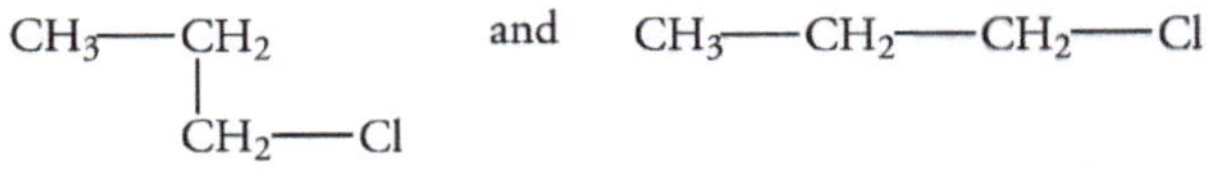

(b) CH₃—CH₂   and   CH₃—CH₂—CH₂—Cl
         |
         CH₂—Cl

(c) CH₃—CH—Cl   and   CH₃—CH₂—CH₂—Cl
         |
         CH₃

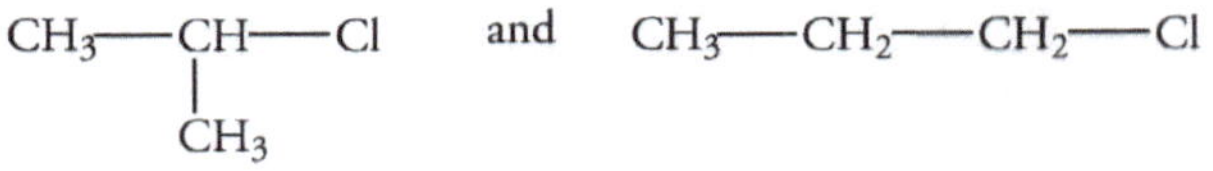

2.18   Indicate whether the following pairs of structures are isomers or different representations of the same compound.

Answers:
(a) isomers
(b) isomers
(c) isomers

(a) 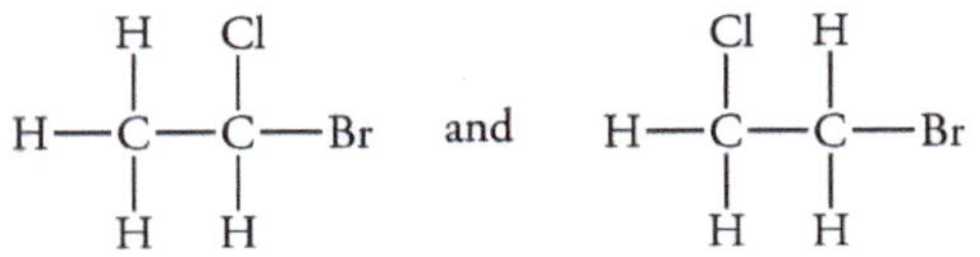   H—C—C—Br   and   H—C—C—Br
      (H, Cl above; H, H below for first; Cl, H above and H, H below for second)

(b) CH₃—CH₂   and   CH₃—CH—CH₃
         | |
         CH₂—Cl   Cl

(c) CH₃—CH—CH₂—Cl   and   CH₃—CH₂—CH₂—CH₂—Cl
         |
         CH₃

27

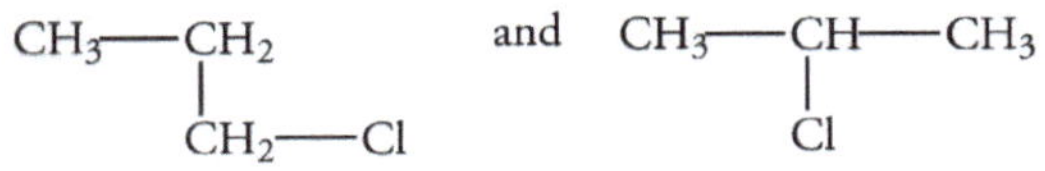

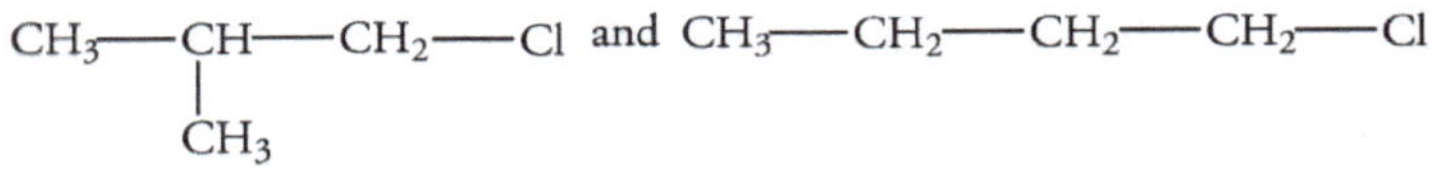

2.19 There are two isomers for each of the following molecular formulas. Draw their structural formulas.

(a) $C_2H_2Br_2$ (b) C_2H_6O (c) C_2H_4BrCl

(d) C_2H_7Cl (e) C_2H_7N (f) $C_2H_3Br_3$

Answers:

(a)
```
     H   H                    H   Br
     |   |                    |   |
Br—C—C—Br     and     H—C—C—Br
     |   |                    |   |
     H   H                    H   H
```

(b)
```
     H   H                    H       H
     |   |                    |       |
H—C—C—O     and     H—C—O—C—H
     |   |   |                |       |
     H   H   H                H       H
```

(c)
```
     H   Br                   H   H
     |   |                    |   |
H—C—C—Cl     and     Cl—C—C—Br
     |   |                    |   |
     H   H                    H   H
```

(d)
```
     H   Cl  H                H   H   H
     |   |   |                |   |   |
H—C—C—C—H     and     H—C—C—C—Cl
     |   |   |                |   |   |
     H   H   H                H   H   H
```

(e)
```
     H   H   H                H   H   H
     |   |   |                |   |   |
H—C—N—C—H     and     H—C—C—N—H
     |       |                |   |
     H       H                H   H
```

(f)
```
     H   Br                   H   Br
     |   |                    |   |
H—C—C—Br     and     Br—C—C—Br
     |   |                    |   |
     H   Br                   H   H
```

2.20 There are three isomers for each of the following molecular formulas. Draw their structural formulas.

(a) $C_2H_3Br_2Cl$ (b) C_3H_8O (c) C_3H_8S

Answers:

(a)
```
     H   Cl              H   H              Br  H
     |   |               |   |              |   |
H—C—C—Br     Br—C—C—Br     Br—C—C—Cl
     |   |               |   |              |   |
     H   Br              H   Cl             H   H
```

(b)
```
     H   H       H              H   OH  H              H   H   H
     |   |       |              |   |   |              |   |   |
H—C—C—O—C—H     H—C—C—C—H     H—C—C—C—OH
     |   |       |              |   |   |              |   |   |
     H   H       H              H   H   H              H   H   H
```

(c)
```
     H   H       H              H   SH  H              H   H   H
     |   |       |              |   |   |              |   |   |
H—C—C—S—C—H     H—C—C—C—H     H—C—C—C—SH
     |   |       |              |   |   |              |   |   |
     H   H       H              H   H   H              H   H   H
```

Infrared Spectroscopy

2.21 How can infrared spectroscopy be used to distinguish between propanone and 2-propen-1-ol?

Answer: The carbonyl group of propanone (acetone) has a strong absorption at 1749 cm^{-1}. 2-Propene-1-ol (allyl alcohol) has an absorption for the carbon–carbon double bond at 1645 cm^{-1} and an absorption for the oxygen-hydrogen bond at 3400 cm^{-1}.

2.22 How can infrared spectroscopy be used to distinguish between 1-pentyne and 2-pentyne?

Answer: 1-Pentyne is a terminal alkyne, so its sp-hybridized C—H bond has an absorption in the 3450 cm^{-1}, and another strong C≡C absorption at 2120 cm^{-1}. 2-Pentyne, which is an internal alkyne, does not have a C—H absorption at 3450 cm^{-1}. Also, the C≡C absorption is so weak that it is barely visible.

2.23 The carbonyl stretching vibration of ketones is at a longer wavelength than the carbonyl stretching vibration of aldehydes. Suggest a reason for this observation.

Answer: The longer wavelength absorption (smaller wavenumber) corresponds to a lower energy vibration. The dipolar resonance form of a ketone is more stable than that of an aldehyde because the extra alkyl group donates electron density. The increased contribution of the resonance form with a carbon–oxygen single bond means that the ketone carbonyl bond absorption requires less energy.

2.24 The carbonyl stretching vibrations of esters and amides occur at 1735 and 1670 cm^{-1}, respectively. Suggest a reason for this difference.

Answer: Both oxygen and nitrogen are inductively electron withdrawing, and they destabilize the dipolar resonance form of the carbonyl group. Since oxygen is more electronegative than nitrogen, this effect is larger for oxygen, so the dipolar resonance form of an ester is less stable that of an amide. The relative ability of the two atoms to donate electrons by resonance is also important. Because nitrogen donates electrons by resonance more effectively than oxygen, there is an increased contribution of a dipolar resonance form for the amide.

2.25 An infrared spectrum of a compound with molecular formula $C_4H_8O_2$ has an intense, broad band between 3500 and 3000 cm^{-1} and an intense peak at 1710 cm^{-1}. Which of the following compounds best fits these data?

I: $CH_3CH_2CO_2CH_3$ II: $CH_3CO_2CH_2CH_3$ III: $CH_3CH_2CH_2CO_2H$

Answer: The absorptions correspond to an O—H and a carbonyl group, respectively. Only the carboxylic acid group of III has both structural features. The other two compounds are esters that would have an absorption corresponding to a carbonyl group but, because esters do not have an O—H group, would have no absorption in the 3500–3000 cm^{-1} region.

2.26   Explain why the carbonyl stretching vibrations of the following two esters differ.

$$CH_2{=}CH{-}CH_2{-}\overset{\overset{\displaystyle O}{\|}}{C}{-}O{-}CH_3 \qquad CH_3{-}CH{=}CH{-}\overset{\overset{\displaystyle O}{\|}}{C}{-}O{-}CH_3$$

1735 cm^{-1} 1720 cm^{-1}

Answer: The carbonyl group of the second compound is conjugated with a double bond. As a result, there is some contribution of a resonance form in which the carbon–oxygen bond has single bond character. The increased contribution of the resonance form with a carbon–oxygen single bond means that the carbonyl bond absorption requires less energy.

2.27   Explain how the two isomeric nitration products of isopropylbenzene can be distinguished using infrared spectroscopy.

Answer: The ortho nitro isomer has four adjacent C—H bonds, and the out-of plane bending of these bonds occurs at 748 cm^{-1}. The para nitro isomer has two sets of two adjacent C—H bonds, and the out-of plane bending occurs at 866 cm^{-1}.

3 INTRODUCTION TO ORGANIC REACTION MECHANISMS

3.1 Acid–Base Reactions

The properties of acids are characterized by K_a and pK_a values. Stronger acids have large K_a values and small pK_a values. For example, alcohols and carboxylic acids have pK_a values in the 16 and 5 range, respectively. The properties of bases are characterized by K_b and pK_b values. Stronger bases have large K_b values and small pK_b values.

Acid–base reactions proceed to favor the weaker of the two possible acids (or the weaker of the two bases). The equilibrium constant for the overall reaction is given by a quotient of two equilibrium constants. Thus, we need to determine the number of powers of 10 by which the equilibrium constants differ. If that difference is 5 powers of 10, for example, then the equilibrium constant is either 10^{-5} or 10^5 depending on your analysis of whether the reaction is favorable or unfavorable.

A curved arrow convention considers the movement of electrons from the tail of the arrow to a point indicated by the arrowhead. Many organic reactions can be described as "have pair-will share." In organic reactions, one species with a nonbonded (or bonded) pair of electrons "donates" an electron pair to an electron-deficient species by forming a covalent bond between the two species.

3.2 Chemical Equilibrium and Equilibrium Constants

Much of the discussion of organic chemical reactions centers on the "driving force" that refers to the magnitude of the equilibrium constant and the change in free energy for the reaction. In the case of reactions with small equilibrium constants, the reaction conditions are usually adjusted to shift the position of equilibrium by taking advantage of Le Chatelier's principle. For example, if an equilibrium constant is small, the equilibrium position can be shifted to the right by removing the products.

3.3 pH and pK Values

The properties of acids are characterized by K_a and pK_a values. Stronger acids have large K_a values and small pK_a values. The properties of bases are characterized by K_b and pK_b values. Stronger bases have large K_b values and small pK_b values.

The equilibrium constant for an acid–base reaction lies on the side of the weaker of the two possible acids (or the weaker of the two bases). The equilibrium constant for the overall reaction is given by the ratio of the two equilibrium constants.

$$HA + B^- \rightleftharpoons A^- + HB$$

$$K_{eq} = \frac{K_{HA}}{K_{HB}}$$

3.4 Effect of Structure on Acidity

Four factors—periodic trends, resonance effects, inductive effects, and hybridization effects—influence acidity. The strength of an acid, HA, depends in part upon the strength of the H—A bond. The bond strength decreases as we move down a column of the periodic table. Because bond strength is inversely related to the acidity, the acidity of the halogen acids increases in the order HF < HCl < HBr < HI. For the same reasons, H_2O is a weaker acid than H_2S.

Acidity increases from left to right in a given row of the periodic table. The order of increasing acidity is $CH_4 < NH_3 < H_2O < HF$. This trend reflects the stabilization of the negative charge, which varies directly with the electronegativity of the atom of the conjugate base. That is, the order of increasing strength of conjugate bases is $F^- < OH^- < NH_2^- < CH_3^-$. Stabilizing the negative charge in the conjugate base increases K_a. One way the conjugate base is stabilized is by delocalization of the negative charge over two or more atoms. This effect is called *resonance stabilization*. When the conjugate base of an acid is resonance stabilized, acid strength increases substantially. For example, both methanol and ethanoic acid ionize to form conjugate bases with a negative charge on oxygen. However, ethanoic acid is **ten billion** (10^{10}) times more acidic than methanol.

Any atom or group of atoms in an organic molecule that withdraws electron density from the bond between hydrogen and another atom—such as carbon, oxygen, or nitrogen—increases its acidity by an inductive effect.

The acidity of hydrocarbons is related to the hybridization of the carbon atom of the C—H bond. The K_a of a carbon acid increases in the order $sp^3 < sp^2 < sp$. The order of acidities parallels the contribution of the lower energy of the 2s orbital to the hybrid orbitals in the σ bond.

3.5 Standard Free Energy Changes in Chemical Reactions

The standard Gibbs free energy change ($\Delta G°$) is the energy change that occurs in going from the reactants to the products.

$$\Delta G°_{rxn} = \Delta G°_f \text{ (products)} - \Delta G°_f \text{ (reactants)}$$

When the products are more stable than the reactants, $\Delta G°_{rxn}$ is negative, and the reaction is *exergonic*. If the reactants less stable than the products, $\Delta G°_{rxn}$ is positive, and the reaction is *endergonic*.

The following equation describes the relation between the standard free energy change, $\Delta G°_{rxn}$ and the equilibrium constant.

$$\Delta G°_{rxn} = -2.303RT \log K_{eq}$$
$$R = 8.314 \text{ kj kelvin}^{-1} \text{ mole}^{-1} (1.987 \text{ cal kelvin}^{-1} \text{ mole}^{-1})$$
$$T = \text{absolute temperature (kelvin)}$$

3.6 Standard Enthalpy Changes in Chemical Reactions

The heat released or absorbed in a reaction at constant pressure is the **enthalpy change**, $\Delta H°_{rxn}$. If heat flows out of the reaction into the surroundings, the reaction is *exothermic*. For an exothermic reaction, $\Delta H°_{rxn} < 0$. If heat flows into the reaction from the surroundings, the reaction is *endothermic*. For an endothermic reaction, $\Delta H°_{rxn} > 0$.

When a bond forms, energy is released; the process is exothermic. Conversely, breaking a bond requires energy; the process is endothermic. Therefore, the energy change for a chemical reaction reflects the differences in the energies of the bonds that are broken and formed. If the products of a reaction contain less stored energy than the reactants, the net difference is released as heat, $\Delta H°_{rxn}$. The magnitude of the standard enthalpy change for a reaction depends only on the difference in enthalpy between the products and reactants.

3.7 Bond Dissociation Energies

The bond dissociation energy is the energy required—an endothermic process—to break a bond and form two atomic or molecular fragments, each with one electron of the original shared pair. Thus, a very stable bond has a large bond dissociation energy—more energy must be added to cleave the bond. A high bond dissociation energy means that the bond (and molecule) is of low energy and stable. Bond energies depend on the number of bonds between atoms. Even though π bonds are weaker than σ bonds, a double bond, which consists of a σ and π is bond, is stronger than a single bond because there are two bonds.

3.8 Introduction to Reaction Mechanisms

A mechanism is the series of steps that occur as a reactant is converted to a product. The complexity of mechanisms covers a wide range from one-step, concerted mechanisms to complex multistep mechanisms in which a series of intermediates form on the pathway from reactants to products.

Classifying the type of bond cleavage in a particular mechanism requires us to look carefully at the reactant and product, to identify which bond breaks, and how it breaks. Bond cleavage is homolytic if the two resulting fragments each retain one electron from the bond; fragments containing an unpaired electron are called radicals and are highly reactive. For example, *tert*-butylbromide can break into a homolytic process to give a *tert*-butyl **radical** and a bromine atom, each of which has a single, unpaired electron.

Bond cleavage is heterolytic if both electrons in the bond stay with one fragment; the fragment retaining the electron pair acquires a −1 charge, and the other fragment a +1 charge. When a carbon atom is bonded to an electronegative element such as Br, heterolytic cleavage gives the electron pair to the bromine and leaves the carbon atom to which it was bound with a positive charge. The carbon fragment is a **carbocation**, as the following equation shows.

When a carbon atom is bonded to an electropositive element such as H, heterolytic cleavage releases the electropositive species as a cation and leaves the electron with the carbon fragment, which becomes a **carbanion.**

3.9 Structures and Stabilities of Reactive Carbon Intermediates

The stability of a carbon intermediate depends on the number of electrons about the carbon atom and the identity of the attached groups. Both carbocations and radicals, which are electron deficient, are stabilized by larger numbers of alkyl groups. Carbanions already have a sufficient number of electrons, and the supply of additional electron density by attached carbon groups is counterproductive. Thus, the order of stability of carbanions is opposite to that of carbocations. Carbocation and radical stability decrease in the order tertiary > secondary > primary >> methyl.

3.10 Reaction Rate Theory

Reactions occur via one or more transition states in which the bonding patterns correspond to neither the reactants nor the products. The transition state occurs at a maximum point on the minimum energy pathway. This point is at the top of a two-dimensional reaction coordinate diagram. Reactions with a high activation energy occur at a slower rate than those with a lower activation energy. Increasing the temperature increases reaction rates because a larger fraction of molecules possess an energy equal to or greater than the activation energy and can achieve the transition state structure as the temperature increases.

Multistep reactions have more than one transition state, and the lower energy species that forms between transition states is an intermediate. Catalysts provide for a mechanism that occurs via a transition state with a lower activation energy.

The Hammond postulate states that strongly exothermic reactions occur via transition states that more closely resemble the reactant structure. Endothermic reactions occur via transition states that more closely resemble the product. Transition state structures cannot be determined experimentally. The structure of reactants and products is known, and it is often possible to elucidate the structure of intermediates. The structure of the transition state is estimated using the Hammond postulate.

3.11 Stability and Reactivity

The term "stability" of a compound is related to standard free energy change for making a compound from its elements, that is $\Delta G^\circ_{formation}$. If we compare two closely related structural isomers, the one with the more negative $\Delta G^\circ_{formation}$ is more stable. The term stability is used to describe reactants, products, and even intermediates.

The term *reactivity* refers to the *rate* at which a compound reacts. Therefore, reactivity refers to the activation energy required for that substance to form a particular transition state. We must refer to a specific reaction to discuss reactivity. Two compounds can have opposite reactivities depending upon the specific kind of reaction they are undergoing.

 **End of Chapter Exercises**

Acids and Bases

3.1   Write the structure of the conjugate acid of each of the following species.

(a) H—O—O—H   (b) NH_2—NH_2 (c) CH_3—S—CH_3

(d) CH_3—O—CH_3 (e) CH_3—NH_2 (f) CH_3—OH

Answers:

(a) H—Ö—Ö⁺—H   (b) H—N—N⁺—H with H substituents

(c) H—C—S⁺—C—H with H substituents

(d) H—C—Ö⁺—C—H with H substituents

(e) H—C—N⁺—H with H substituents

(f) H—C—Ö⁺—H with H substituents

3.2   Write the structure of the conjugate base of each of the following species.

(a) CH_3—SH   (b) CH_3—NH_2 (c) CH_3—O—SO_3H

(d) CH_2=CH_2 (e) HC≡CH   (f) CH_3—CN

Answers:

(a) H—C—S:⁻ with H substituents

(b) H—C—N:⁻ with H substituents

(c) H—C—Ö—S—Ö:⁻ with H and O substituents

(d) C=C^- ethylene anion with H substituents

(e) H—C≡C:⁻

(f) H—C̈—C≡N: with H substituent

3.3   Write the structure of the conjugate acid of each of the following species.

(a) CH_2=O   (b) CH_3—NH—CH_3 (c) CH_2=NH   (d) CH_3—C(=O)—OH

Answers:

(a) C=Ö⁺ with H substituents

(b) H—C—N⁺—C—H with H substituents

(c) C=N⁺ with H substituents

(d) H—C—C with Ö⁺—H and :O—H substituents

3.4 Write the structure of the conjugate acid of each of the following species.

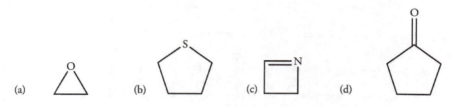

(a)   (b)   (c)   (d)

Answers:

(a)   (b)   (c)   (d)

3.5 Identify the Lewis acid and Lewis base in each of the following reactions.

(a) CH_3—CH_2—Cl + $AlCl_3$ ⟶ CH_3—CH_2+ + $AlCl_4^-$

(b) CH_3—CH_2—SH + CH_3O^- ⟶ CH_3—CH_2—S^- + CH_3OH

(c) CH_3—CH_2—OH + NH_2^- ⟶ CH_3—CH_2—OH^- + NH_3

(d) $(CH_3)_2N^-$ + CH_3OH ⟶ $(CH_3)_2NH$ + CH_3O^-

Answers:

(a) CH_3—CH_2—Cl is the Lewis base; $AlCl_3$ is the Lewis acid.
(b) CH_3—CH_2—SH is the Lewis acid; CH_3—O^- is the Lewis base
(c) CH_3—CH_2—OH is the Lewis acid; NH_2^- is the Lewis base
(d) $(CH_3)_2N^-$ is the Lewis base; CH_3—OH is the Lewis acid

3.6 Identify the Lewis acid and Lewis base in each of the following reactions.

(a) $(CH_3)_2O$ + HI ⟶ $(CH_3)_2OH+$ + I^-

(b) CH_3—CH_2+ + H_2O ⟶ CH_3—CH_2—OH_2+

(c) CH_3—CH=CH_2 + HBr ⟶ $(CH_3)_2CH+$ + Br^-

(d) CH_3—C≡CH + CH_3NH^- ⟶ CH_3—C≡C^- + CH_3NH_2

Answers:

(a) $(CH_3)_2O$ is the Lewis base; HI is the Lewis acid.
(b) CH_3—CH_2^+ is the Lewis acid; H_2O is the Lewis base.
(c) CH_3—CH=CH_2 is the Lewis base; HBr is the Lewis acid.
(d) CH_3—C≡CH is the Lewis acid; CH_3N^- is the Lewis base.

Equilibrium Constant Expressions

3.7 Write the equilibrium constant expression for the reaction of ethanal and methanol to give an acetal.

$$CH_3CHO + 2\ CH_3OH \rightleftharpoons CH_3CH(OCH_3)_2 + H_2O$$

Answer: $$K_{eq} = \frac{[CH_3CH(OCH_3)_2][H_2O]}{[CH_3CHO][CH_3OH]^2}$$

3.8   Write the equilibrium constant expression for the reaction of acetylene (C_2H_2) to give cyclooctatetraene (C_8H_8).

Answer:

$$K_{eq} = \frac{[C_8H_8]}{[C_2H_2]^4}$$

3.9   How do the equilibrium constant expressions differ for the hydrolysis reaction of ethyl ethanoate (written right to left) and the esterification reaction of ethanol and ethanoic acid (written left to right)? What is the equilibrium constant for the hydrolysis reaction?

Answer:

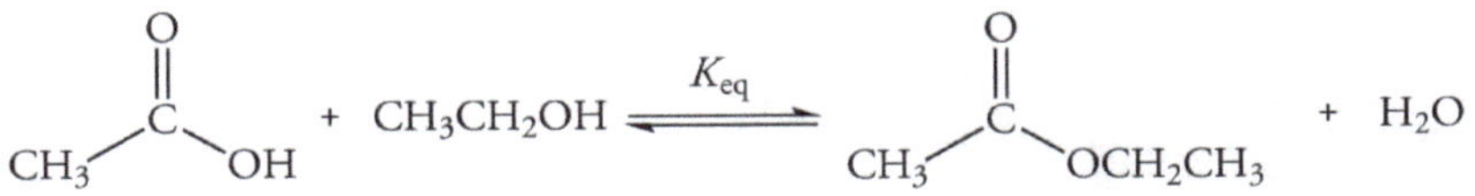

ethanoic acid   ethanol   ethyl ethanoate

$$K_{equilibrium} = \frac{[CH_3CO_2CH_2CH_3]\,[H_2O]}{[CH_3CO_2H]\,[CH_3CH_2OH]} = 4.0$$

The hydrolysis reaction written above is the reverse of the esterification reaction. Thus, the equilibrium constant expression for hydrolysis is the reciprocal of the equilibrium constant expression for esterification. The value of the equilibrium constant for hydrolysis is 0.25.

3.10   At equilibrium, the yield of the condensation product of acetone is about 5%. Calculate the equilibrium constant for the reaction.

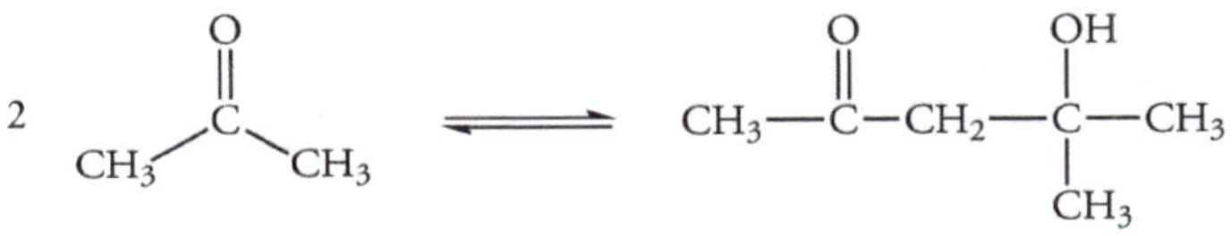

Answer:

For an initial concentration of acetone equal to x mole liter^{-1}, the theoretical concentration of product for a complete reaction would be $0.5\,x$ mole liter^{-1}. For a 5% yield, the actual concentration is $0.025\,x$ mole liter^{-1}. The equilibrium concentration of reactant is $0.95\,x$ mole liter^{-1} because two moles of reactant are required to give one mole of product. The equilibrium constant is approximately $0.028\,x^{-1}$ mole^{-1} liter.

$$K_{eq} = \frac{[0.025x]}{[0.95x]^2}$$

pH and pK Values

3.11   Without reference to tables of pK_a values, predict the position of the following equilibrium.

Answer: $CH_3—CH_2—SH + CH_3—O^- \rightleftharpoons CH_3—CH_2—S^- + CH_3—OH$

The acid dissociation constants of organic compounds containing atoms within a common group of the periodic table bonded to hydrogen increase down the column. Thus, thiols are more acidic than alcohols. The equilibrium position lies on the side of the equation containing the weaker acid. Therefore, the position of the above equilibrium is to the right, where $CH_3—OH$ is located in the above equilibrium, and $K_{eq} > 1$.

3.12   Without reference to tables of pK_a values, predict the position of the following equilibrium.

$$CH_3—CH_2—CO_2H + CH_3—O^- \xrightarrow{K_{eq}} CH_3—CH_2—CO_2^- + CH_3—OH$$

Answer:
The equilibrium position lies on the side of the equation containing the weaker acid. The acids are methanol located on the right, and acetic acid, which is located on the left of the above equation. Acetic acid is the stronger acid because its conjugate base is resonance stabilized. Thus, the position of the equilibrium is on the right, where the weaker acid, methanol, is found.

3.13   The approximate pK_a values of CH_4 and CH_3OH are 49 and 16, respectively. Which is the stronger acid? Will the equilibrium position of the following reaction lie to the left or to the right?

$$CH_4 \quad + \quad CH_3{-}O^- \quad \xrightleftharpoons{\quad K_{eq} \quad} \quad CH_3^- \quad + \quad CH_3{-}OH$$

Answer:

The equilibrium position lies on the side of the equation containing the weaker acid. The acids are methanol located on the right, and acetic acid, which is located on the left of the above equation. Acetic acid is the stronger acid because its conjugate base is resonance stabilized. Thus, the position of the equilibrium is on the right, where the weaker acid, methanol, is found.

3.14   The approximate pK_a values of NH_3 and CH_3OH are 36 and 16, respectively. Which is the stronger acid? Will the equilibrium position of the following reaction lie to the left or to the right?

$$CH_3{-}OH \ + \ NH_2{}^- \quad \xrightleftharpoons{\quad K_{eq} \quad} \quad NH_3 \ + \ CH_3{-}O^-$$

Answer:

Methanol is the stronger acid by a factor of 10^{33} in K_a. The equilibrium position lies on the left side of the equation, which contains the weaker acid, methane. The equilibrium constant is 10^{-33}.

Structure and Acid Strength

3.15 Write the structures of the two conjugate acids of hydroxylamine ($NH_2{-}OH$). Which is the more acidic?

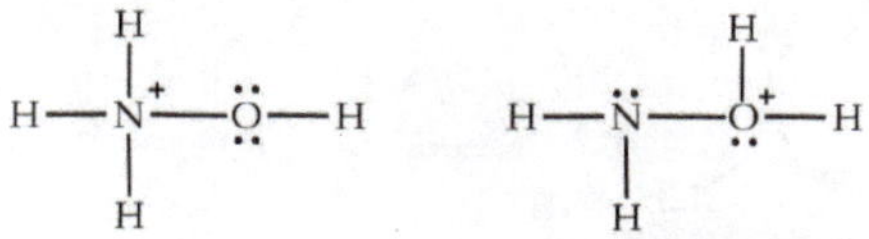

stronger acid

Answer:

The acidity of hydrogen atoms bonded to atoms contained in similarly structured compounds increases from left to right within a period of the periodic table. For example, H_2O is a stronger acid than NH_3. The conjugate acid with a proton located on the oxygen atom of hydroxylamine must be a stronger acid than the conjugate acid with a proton located on the nitrogen atom.

3.16 Write the structures of the two conjugate bases of hydroxylamine ($NH_2{-}OH$). Which is the more basic?

stronger base

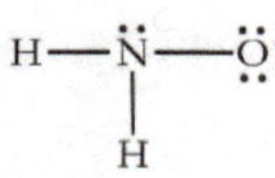

Answer:

The basicity of atoms contained in similarly structured compounds decreases from left to right within a period of the periodic table. For example, NH_2^- is a stronger base than OH^-. The conjugate base with the charge located on the nitrogen atom of hydroxylamine must be a stronger base than the conjugate base with a charge on the oxygen atom.

3.17 Which is the stronger acid, chloroethanoic acid ($ClCH_2CO_2H$) or bromoethanoic acid ($BrCH_2CO_2H$)? Explain your answer.

Answer:

The inductive electron withdrawal by chlorine is larger than that of bromine because chlorine is more electronegative than bromine. Therefore, electron density is pulled away from the O—H group, and the acidity of chloroacetic acid is greater than the acidity of bromoacetic acid.

3.18 Which acid has the larger pK_a, chloroethanoic acid ($ClCH_2CO_2H$) or dichloroethanoic acid (Cl_2CHCO_2H)? Explain your answer.

Answer:
Dichloroacetic acid is a stronger acid than chloroacetic acid because the two chlorine atoms inductively withdraw more electron density from the O—H group than a single chlorine atom. The pK_a of dichloroacetic acid is therefore smaller than the pK_a of chloroacetic acid.

3.19 Based on the pK_a values of substituted butanoic acids (Section 3.4), predict the pK_a of 4-chlorobutanoic acid.

Answer:
The pK_a of the substituted chlorobutanoic acids increases with increasing distance separating the chlorine atom and the acidic site. Thus, the pK_a of the 4-chloro compound is greater than 4.02, the pK_a of the 3-chloro compound. It is also less than the pK_a of butanoic acid, which is 4.82.

3.20 Explain the trends in the pK_a values of the following ammonium ions.

$$CH_3CH_2CH_2CH_2\overset{+}{N}H_3 \qquad\qquad CH_3-O-CH_2CH_2CH_2CH_2\overset{+}{N}H_3$$

$$pK_a = 10.6 \qquad\qquad\qquad\qquad pK_a = 9.9$$

$$N\equiv C-CH_2CH_2CH_2CH_2\overset{+}{N}H_3$$

$$pK_a = 7.8$$

Answer:
The order of decreasing pK_a values indicates that the groups bonded to the nitrogen atom of the ammonium ions increase in ability to inductively withdraw electron density. Although oxygen is more electronegative than nitrogen, the nitrile has a triple bond and is a much more polar group.

3.21 Explain why the hydrogen of the CH_3 of propene is more acidic than hydrogen of the CH_3 of propane.

Answer:
The conjugate base of propane has its negative charge localized on a single carbon atom. The conjugate base of propene has its negative charge delocalized over two carbon atoms, as shown by two contributing resonance structures.

3.22 Ethanonitrile (CH_3CN) is a stronger acid than ethane. Explain why.

Answer:
The conjugate base of ethane has its negative charge localized on a single carbon atom. The conjugate base of ethanonitrile has its negative charge delocalized with some of the charge located on the more electronegative nitrogen atom as shown in one of the two contributing resonance structures.

3.23 The pK_a of acetic acid (CH_3CO_2H) is 4.8. Explain why the carboxylic acid group of amoxicillin (pK_a = 2.4), a synthetic penicillin, is more acidic than acetic acid, whereas the carboxylic acid group of indomethacin (pK_a = 4.5), an anti-inflammatory analgesic used to treat rheumatoid arthritis, is of comparable acidity.

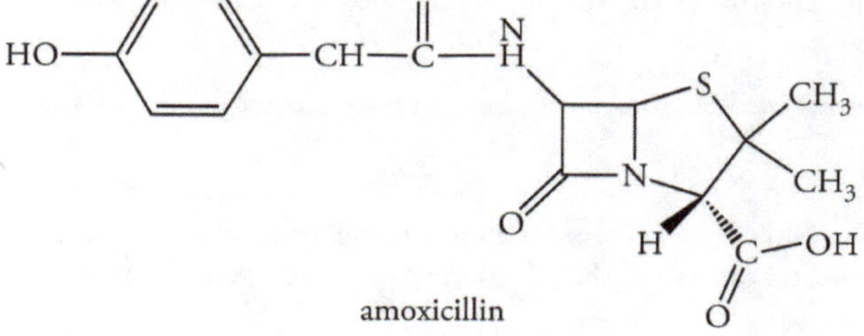

amoxicillin

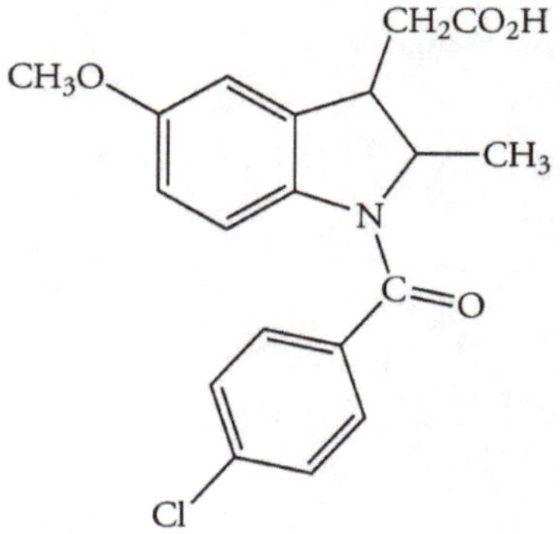

Indomethaxan

Answer:

In amoxicillin, the acidic —CO_2H group is bonded to a carbon atom that is also bonded to a nitrogen atom that inductively withdraws electron density and increases the acidity of the O—H group. In indomethacin, the —CO_2H group is bonded to a carbon atom that is not directly bonded to any electronegative groups. The nitrogen atom in indomethacin is one atom farther removed than that in amoxicillin.

3.24 The pK_a of the OH group of phenobarbital is 7.5, whereas the pK_a of CH_3OH is 16. Explain why phenobarbital is significantly more acidic.

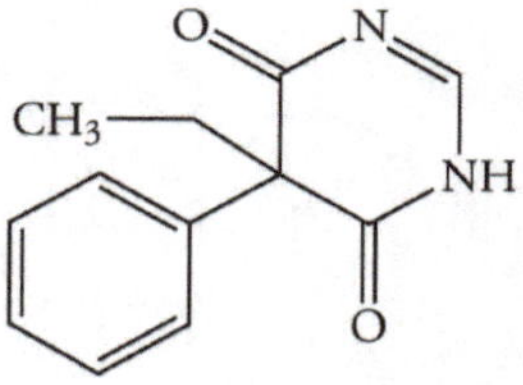

phenobarbital

Answer:

The greatly increased acidity of the O—H group in phenobarbital reflects the resonance stabilization of the conjugate base, in which the charge is delocalized over two oxygen atoms.

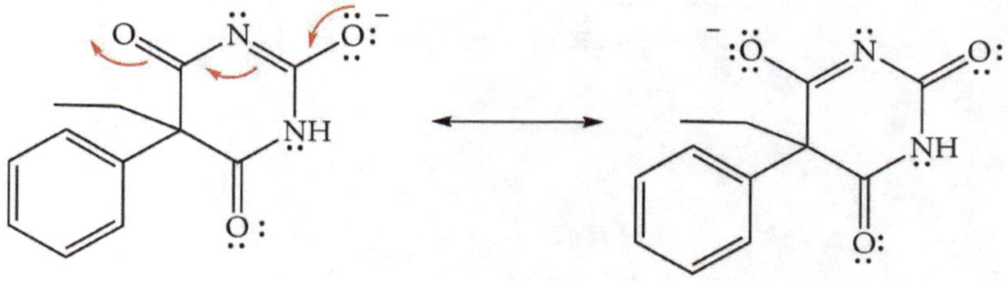

3.25 The N—H bond of ammonia is not very acidic (pK_a = 33). However, the pK_a for the N—H bond of sulfanilamide, a sulfa drug, is 10.4. Suggest a reason for the higher acidity of sulfanilamide.

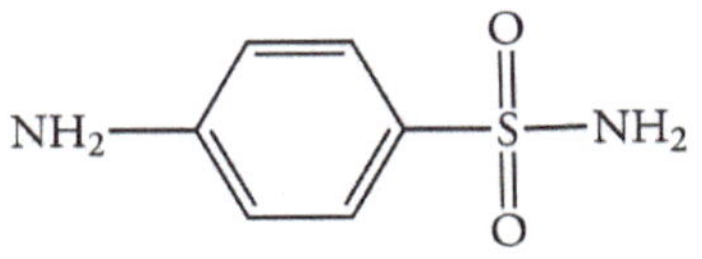

sulfanilamide

Answer:

The nitrogen atom is bonded to a sulfur atom that has two oxygen atoms, which are electronegative, and hence withdraw electron density from the N—H bond.

3.26 The pK_a of sulfadiazine, a sulfa drug, is 6.5. Why is this compound more acidic than sulfanilamide?

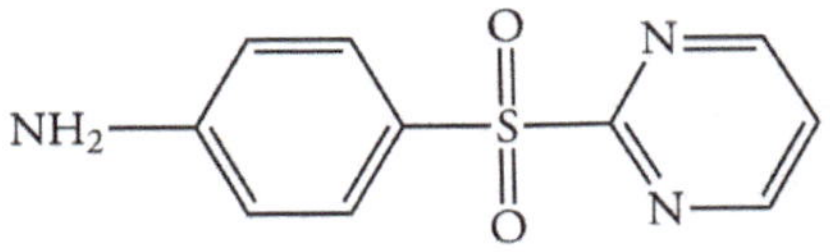

sulfadiazine

Answer:
The nitrogen atom is bonded to a carbon atom of a ring that has two nitrogen atoms, which are electronegative, and hence withdraw electron density from the N—H bond.

Equilibrium Constant and Free Energy

3.27 A reaction has K_{eq} = 1 × 10^{-5}. Are the products more or less stable than the reactants? Is the reaction exergonic or endergonic?

Answer: For an equilibrium constant less than 1, the products are less stable than the reactants. Such a reaction is endergonic.

3.28 Which reaction would be exergonic, one with K_{eq} = 100 or one with K_{eq} = 0.01?

Answer: The reaction with K = 100 proceeds further to completion and is more exergonic than a reaction with K = 0.01.

3.29 Can a reaction have K_{eq} = 1? What relationship would exist between the free energies of the reactants and products?

Answer: Yes, a reaction can have K_{eq} =1 and have products and reactants of equal stability—that is, $\Delta G°_{rxn}$ = 0.

3.30 Which reaction has an equilibrium constant greater than 1, one with $\Delta G°_{rxn}$ = + 15 kJ mole^{-1} or one with $\Delta G°_{rxn}$ = −15 kJ mole^{-1}?

Answer: Spontaneous reactions are exergonic and have $\Delta G°_{rxn}$ < 0. The reaction with $\Delta G°_{rxn}$ = −15 kJ mole^{-1} is exergonic.

3.31 The $\Delta G°_{rxn}$ for the following reaction is +2 kJ mole^{-1}. What is K_{eq} at 25 °C?

Answer: Using the relationship $\Delta G°_{rxn}$ = −2.303 RT log K, the equilibrium constant is 0.4. Remember that 25 °C is 298 K and that $\Delta G°_{rxn}$ must be expressed in kJ mole^{-1}.

$$CH_3SH \ + \ HBr \ \rightleftharpoons \ CH_3Br \ + \ H_2S$$

3.32 The equilibrium constant for the isomerization of butane to 2-methylpropane is 4.9. What is $\Delta G°_{rxn}$?

Answer: Using the relationship $\Delta G°_{rxn}$ = −2.303 RT log K, the $\Delta G°_{rxn}$ = −3.9 J mole^{-1}.

Bond Cleavage and Reaction Intermediates

3.33 Write the structure of the radical formed by abstraction of a hydrogen atom by a chlorine atom for each of the following compounds.

(a) CH_3CH_3 (b) CH_3Cl (c) CH_2Cl_2

Answers:

(a)
```
     H   H
     |   |
 H — C — C •
     |   |
     H   H
```

(b)
```
     H
     |
 H — C •
     |
     Cl
```

(c)
```
     Cl
     |
 H — C •
     |
     Cl
```

3.34 Write the structures of all possible radicals formed by abstraction of a hydrogen atom by a chlorine atom for each of the following compounds.

Answers:

(a)
```
     H   H   H
     |   |   |
 H — C — C — C •          H — C — C — C — H
     |   |   |                |   •   |
     H   H   H                H   H   H
```
   n-propyl radical   *iso*-propyl radical

(b)
```
     H   H   H   H
     |   |   |   |
 H — C — C — C — C •      H — C — C — C — C — H
     |   |   |   |            |   |   •   |
     H   H   H   H            H   H   H   H
```
   n-butyl radical   *sec*-butyl radical

(c)
```
     H        H                H   H   H
     |        |                |   |   |
 H — C — C — C — H         H — C — C — C •
     |   •    |                |   |   |
     H   CH₃  H                H   CH₃ H
```
   tert-butyl radical   *iso*-butyl radical

3.35 The oxygen–chlorine bond of methyl hypochlorite (CH_3—O—Cl) can cleave heterolytically. Based on the electronegativity values of chlorine and oxygen, predict the charges on the cleavage products.

Answer: Oxygen is more electronegative than chlorine. Thus, the electrons of the O—Cl bond will remain with oxygen and the products should be CH_3O^- and Cl^+.

3.36 2-Chloropropane reacts with the Lewis acid $AlCl_3$ to give $AlCl_4^-$ and a carbon intermediate. What is the intermediate?

Answer: $AlCl_3$ combines with Cl^- to give $AlCl_4^-$. Thus, the C—Cl bond cleaves heterolytically. The intermediate is a carbocation $(CH_3)_2CH^+$.

3.37 Hydrogen peroxide (H—O—O—H) reacts with a proton to give a conjugate acid, which undergoes heterolytic, oxygen–oxygen bond cleavage to yield water. What is the second product?

Answer: Heterolytic cleavage that places the O—O bonding electrons on the oxygen atom of water leaves a cation with the positive charge on the oxygen atom of the HO group.

3.38 Benzoyl peroxide is used in creams to control acne. It is an irritant that causes proliferation of epithelial cells. It undergoes a homolytic cleavage of the oxygen–oxygen bond. Write the structure of the product, indicating all of the electrons present on all of the oxygen atoms.

Answer: Homolytic cleavage leaves one electron of the pair of electrons in the O—O bond with each of the two equivalent radical fragments, giving the structures shown below.

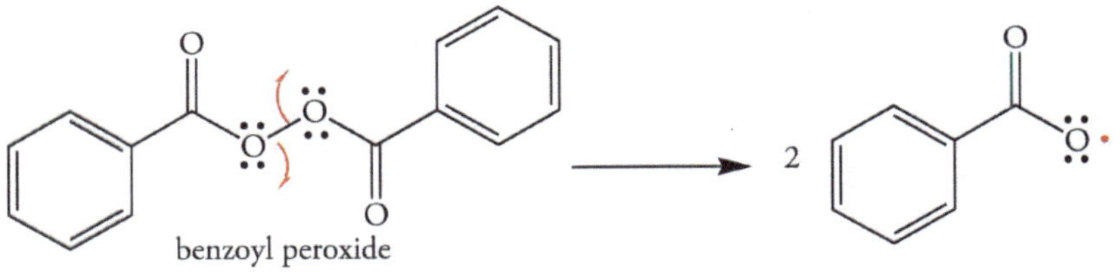

benzoyl peroxide

Stability of Reactive Intermediates

3.39 Arrange the following intermediates in order of increasing stability.

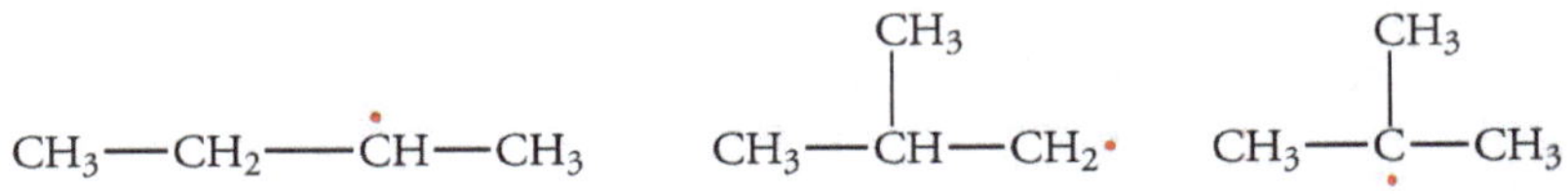

I   II   III

Answer: The order of increasing stability is II < I < Ill, which corresponds to primary < secondary < tertiary.

3.40 Arrange the following intermediates in order of increasing stability.

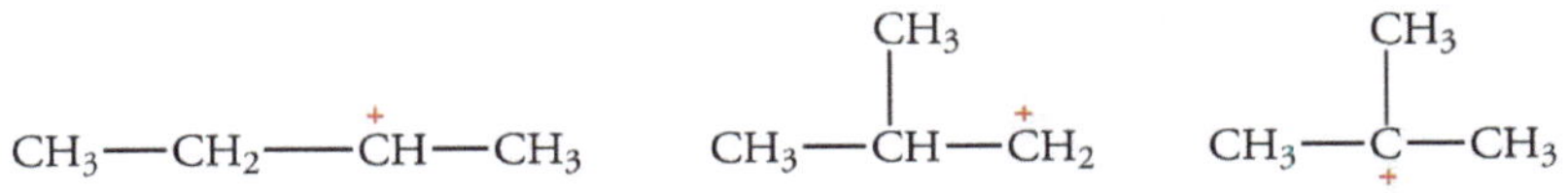

I   II   III

Answer: The order of increasing stability is I < III < II, which corresponds to primary < secondary < tertiary.

3.41 Explain why more energy is required for heterolytic bond cleavage of the carbon–bromine bond of 1-bromopropane than is needed to cleave the carbon–bromine bond of 2-bromopropane.

$CH_3CH_2CH_2Br$ $CH_3CHBrCH_3$
1-bromopropane   2-bromopropane

Answer: Heterolytic cleavage of a C—Br bond produces a carbocation and a bromide ion. The secondary carbocation derived from 2-bromopropane is more stable than the primary carbocation derived from 1-bromopropane, so formation of the primary carbocation requires more energy.

3.42 Explain why less energy is required to cleave the carbon–chlorine bond of 3-chloropropene than that needed to cleave the carbon–chlorine bond of 1-chloropropane.

$CH_2=CHCH_2Cl$
3-cloropropene

$CH_3CH_2CH_2Cl$
1-chloropropane

Answer: Heterolytic cleavage of a C—Cl bond produces a carbocation and a chloride ion. A resonance-stabilized primary carbocation is derived from 3-chloropropene and is thus more stable than the primary carbocation derived from 1-chloropropane, in which the charge is localized.

resonance-stabilized allyl carbocation

3.43 Chloroform ($CHCl_3$) reacts with a strong base in an unusual elimination reaction to give dichlorocarbene (CCl_2). Write the Lewis structure for this species. What features of the chlorine atoms might stabilize this carbene compared to CH_2?

Answer: The dichlorocarbene is electron deficient; there are only four bonding electrons and a lone pair of electrons about the carbon atom. However, either of the chlorine atoms can share one of its lone pairs of electrons in contributing resonance forms. The delocalization of electrons makes CCl_2 more stable than CH_2.

3.44 Draw the Lewis structure of OH^+. (a) How does it differ from OH^-? Is OH^+ a nucleophile or an electrophile?

Answers:
(a) OH^+ has one less unshared electron pair than hydroxide.
(b) OH^+ is an electron-deficient species, so it is an electrophile.

Activation Energy and Rates of Reaction

3.45 Given the following information about two reactions, which one will occur at the faster rate at a common temperature?

Reaction	ΔH°_{rxn}	E_a
A ⟶ X	−120 kJ mole^{-1}	+100 kJ mole^{-1}
B ⟶ Y	−100 kJ mole^{-1}	+120 kJ mole^{-1}

Answer: The reaction converting A to X has the lower activation energy (E_a), so it proceeds at the faster rate.

3.46 Given the information given in Exercise 3.44, which one is more exothermic?

Answer: The reaction converting A to X has the more negative ΔH°_{rxn}, so it is more exothermic.

3.47 Given the activation energies for the following free radical reactions, which one occurs at the faster rate?

$CH_4 + F \longrightarrow CH_3{\cdot} + HF \qquad E_a = 5$ kJ mole^{-1}

$CH_4 + Cl{\cdot} \longrightarrow CH_3{\cdot} + HCl \qquad E_a = 16$ kJ mole^{-1}

Answer: The reaction of methane with a fluorine atom has the lower activation energy (E_a), so it proceeds at the faster rate.

3.48 Consider the activation energies for the following nucleophilic substitution reactions. Which reaction occurs at the faster rate?

$$CH_3-I + Cl^- \longrightarrow CH_3-Cl + I^- \quad E_a = 104 \text{ kJ mole}^{-1}$$

$$CH_3-I + Br^- \longrightarrow CH_3-Br + I^- \quad E_a = 96 \text{ kJ mole}^{-1}$$

Answer: The reaction of bromide ion with iodomethane has the lower activation energy (E_a), so it proceeds at the faster rate.

Kinetic Order of Reaction

3.49 Sodium cyanide reacts with chloroethane by the following equation. When the concentration of cyanide ion is tripled, the reaction rate triples. When the concentration of chloroethane doubles, the reaction rate doubles. What is the overall kinetic order of the reaction? Write the rate equation for the reaction.

$$N\equiv C:^- + CH_3-CH_2-Cl \longrightarrow CH_3-CH_2-C\equiv N: + Cl^-$$

Answer: The rate of reaction is first order in both chloroethane and cyanide ion, so it is second order overall. The rate equation is:

$$\text{rate} = k\,[CH_3CH_2Cl][CN^-].$$

3.50 Reaction of *tert*-butyl alcohol with concentrated HBr gives *tert*-butyl bromide. When the concentration of the alcohol is doubled, the reaction rate doubles. When the concentration of acid is tripled, the reaction rate triples. If more bromide ion in the form of sodium bromide is added, the rate is unaffected. What is the kinetic order with respect to each reactant? What is the overall kinetic order of the reaction?

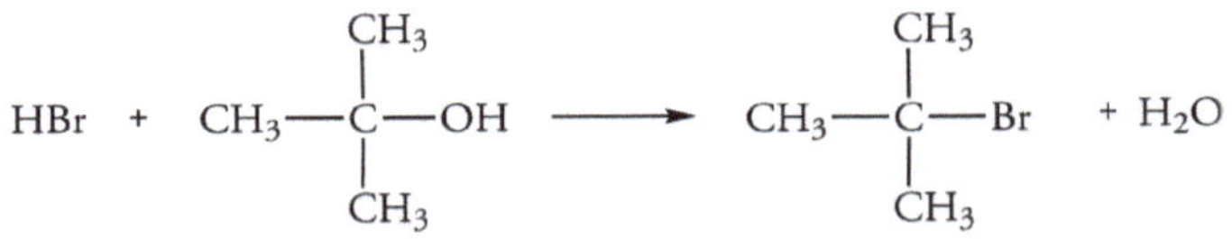

Answer: The rate of reaction is first order in both *tert*-butyl alcohol and hydrogen ion and is zero order in bromide ion, so it is second order overall. The rate law is:

$$\text{rate} = k\,[(CH_3)3OH][H^+].$$

Reaction Mechanisms

3.51 Identify the processes of bond cleavage and bond formation for each of the following reactions.

Answers:
(a) Homolytic cleavage of a C—H bond and homogenic formation of a H—Br bond.
(b) Homolytic cleavage of a Br—Br bond and homogenic formation of a C—Br bond.

3.52 Identify the processes of bond cleavage and bond formation for each of the following reactions.

(a) 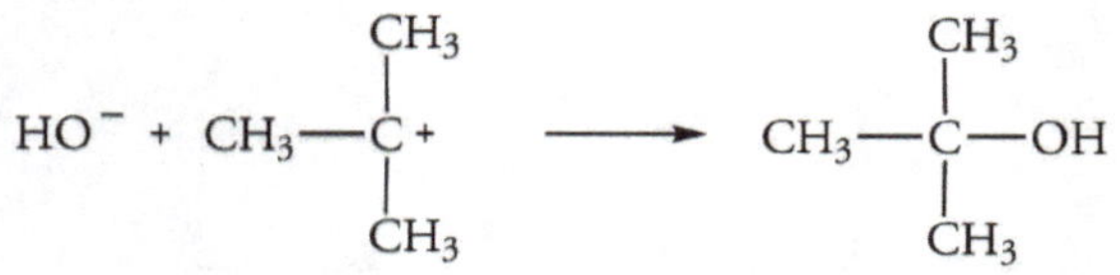 HO^- + CH_3—C(CH_3)(CH_3)$^+$ ⟶ CH_3—C(CH_3)(CH_3)—OH

(b) 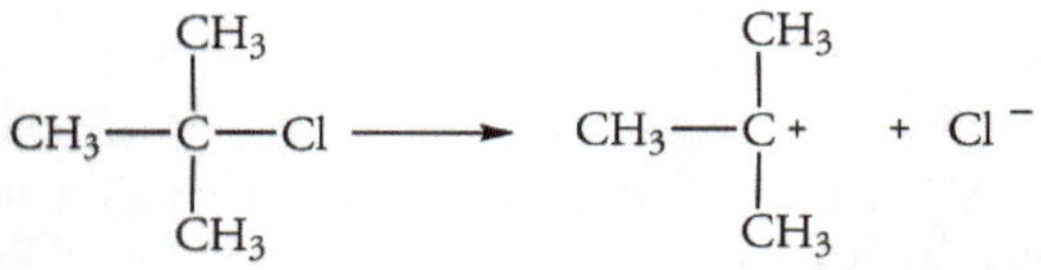 CH_3—C(CH_3)(CH_3)—Cl ⟶ CH_3—C(CH_3)(CH_3)$^+$ + Cl^-

Answers:
(a) Heterogenic formation of a C—O bond.
(b) Heterolytic cleavage of a C—Cl bond.

3.53 In the presence of a strong acid, *tert*-butyl alcohol acts as a base. The resulting conjugate acid produces water and an intermediate. Write the structure of the intermediate. What type of bond cleavage occurs?

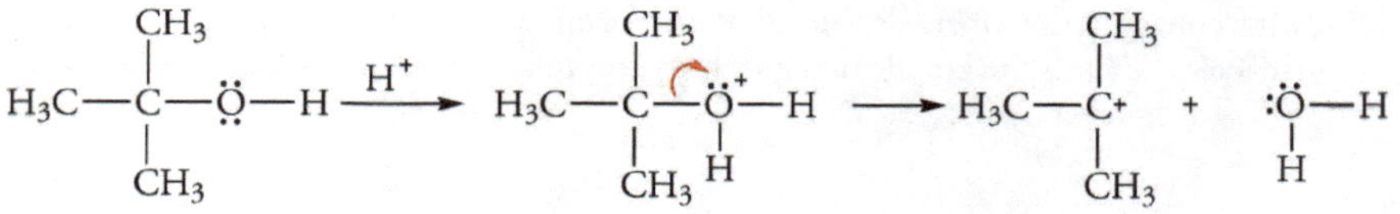

H_3C—C(CH_3)(CH_3)—Ö—H $\xrightarrow{H^+}$ H_3C—C(CH_3)(CH_3)—Ö$^+$—H(H) ⟶ H_3C—C(CH_3)(CH_3)$^+$ + :Ö—H(H)

Answer:
Heterogenic cleavage of a C—O bond.

3.54 Dimethyl ether (CH_3—O—CH_3) can be prepared by adding a strong base such as NaH to methanol (CH_3OH) and then adding iodomethane (CH_3I) to the reaction mixture. Write plausible steps for this reaction.

Answer:

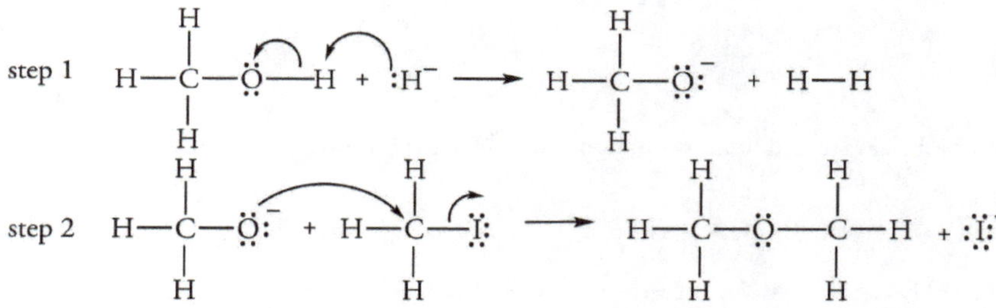

step 1 H—C(H)(H)—Ö—H + :H$^-$ ⟶ H—C(H)(H)—Ö:$^-$ + H—H

step 2 H—C(H)(H)—Ö:$^-$ + H—C(H)(H)—Ï: ⟶ H—C(H)(H)—Ö—C(H)(H)—H + :Ï:$^-$

Reaction Coordinate Diagrams

3.55 What are the differences between a reaction intermediate and a transition state.

Answer: An intermediate, although short lived, can be detected experimentally. A transition state is a transient species whose structure can only be postulated.

3.56 A reaction occurs in three steps. How many transition states are there? How many intermediates form?

Answer: There are three transitions states—one for each step. There are two intermediates. One is formed from step 1 and reacts in step 2. The second intermediate is formed from step 2 and reacts in step 3.

3.57 Draw a reaction coordinate diagram for a two-step exothermic reaction in which the second step is a rate-determining.

Answer:

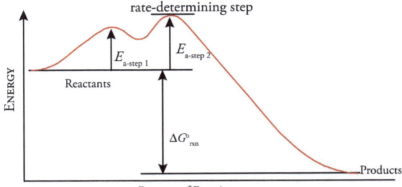

Hammond Postulate

3.58 The $\Delta H°_{rxn}$ for abstracting each of the possible hydrogen atoms of propane by a bromine atom is indicated below. Based on the data and the fact that the starting materials are the same, what might be surmised about the relative energies of activation for the two reactions? Do the transition states more closely resemble the reactants or the products?

CH_3—CH_2—CH_3 + Br CH_3—CH_2—$CH_2^{\cdot}$ + HBr $\Delta H° = +42$ kJ mole^{-1}

CH_3—CH_2—CH_3 + Br CH_3—$\overset{\cdot}{CH}$—CH_3 + HBr $\Delta H° = +29$ kJ mole^{-1}

Answer: The activation energy must be larger for the first reaction listed because it is the more endothermic reaction. The transition state for this reaction must more closely resemble the product than for the second reaction.

3.59 The $\Delta H°_{rxn}$ for abstracting each of the possible hydrogen atoms of propane by a bromine atom is indicated below. Based on the data and the fact that the starting materials are the same, what might be surmised about the relative energies of activation for the two reactions? Do the transition states more closely resemble the reactants or the products?

CH_3—CH_2—CH_3 + Cl· CH_3—CH_2—$CH_2^{\cdot}$ + HCl $\Delta H° = -20$ kJ mole^{-1}

CH_3—CH_2—CH_3 + Cl· CH_3—$\overset{\cdot}{CH}$—CH_3 + HCl $\Delta H° = -33$ kJ mole^{-1}

Answer: The activation energy must be smaller for the second reaction because it is the more exothermic reaction. The transition state for this reaction must more closely resemble the reactant than for the first reaction.

ALKANES AND CYCLOALKANES STRUCTURES AND REACTIONS

4.1 Classes of Hydrocarbons

In this section, we encountered new terms that we will continue to use throughout the text. **Hydrocarbons** contain only carbon and hydrogen. **Saturated hydrocarbons** contain only carbon–carbon single bonds; **unsaturated hydrocarbons** contain carbon–carbon multiple bonds. **Alkanes** have only carbon atoms bonded in chains of atoms. **Cycloalkanes** have only carbon atoms bonded in a ring of atoms. Compounds without rings are acyclic; compound with rings are cyclic. Other atoms may be found in some rings. Atoms other than carbon within rings are **heteroatoms**, and the compounds are **heterocyclic**.

4.2 Alkanes

Normal alkanes consist of a continuous chain of carbon atoms; **branched alkanes** have some carbon atoms bonded to more than two other carbon atoms. The general formula for an alkane is C_nH_{2n+2}, whether it is normal or branched. If the number of carbon atoms are known, the number of hydrogen atoms and the molecular formula are known. Thus, inspecting the molecular formula and comparing it to the reference molecular formula expected for an alkane provides a clue about the identity of other structural features. The C_nH_{2n+2} formula is the reference.

A carbon atom is classified as primary (1°), secondary (2°), or tertiary (3°) when it has 1, 2, or 3 alkyl groups, respectively, bonded to it. A carbon atom is quaternary (4°) when it has 4 alkyl groups bonded to it.

4.3 Nomenclature of Alkanes

The nomenclature rules in this section form the foundation on which we will base all other nomenclature. Here is a brief summary.

1. Locate the longest carbon chain, called the parent chain.
2. Identify the groups that are substituents attached to the parent chain.
3. Number the parent chain to give the branching carbon atoms and other substituents the lowest possible numbers.
4. Use a prefix to the name of the parent chain to identify the name and location of all branches and other substituents.
5. Each substituent must be assigned a number to indicate its position. Thus, if two methyl groups are bonded to C-2 in a chain of carbon atoms, the name "2-dimethyl" as part of the prefix is incorrect; two methyl groups bonded to C-2 must be designated as 2,2-dimethyl. To determine the numbering of the substituents to use in the prefix, choose the point of first difference.
6. List the names of substituents alphabetically. Note that the prefixes di, tri, etc., do not affect the alphabetic method of listing alkyl groups. For example, ethyl is listed before dimethyl because it is the "e" of ethyl that takes precedence over the "m" of methyl.
7. The most common alkyl groups are methyl, ethyl, propyl, isopropyl, butyl, *sec*-butyl, isobutyl, and *tert*-butyl.

4.4 Conformations of Alkanes

The study of the chemical and physical properties of different conformations of compounds, called conformational analysis, formed a basis for understanding the relationships between structure and properties. The energy difference between the conformation of a molecule in its most stable conformation and that required for the molecule in the transition state affects the rates of reactions.

Conformations of Ethane

The conformation of a molecule refers to different arrangements of atoms in a molecule that result from rotation about carbon–carbon sigma bonds. The conformations (conformers) of ethane have low energy forms that are "staggered" and high energy conformations that are "eclipsed." The staggered conformation is the most stable; the eclipsed conformation is the least stable. In general, the bonding electron pairs of the carbon–hydrogen bonds of neighboring carbon atoms tend to stay as far apart as possible.

Newman projection formulas give us a method for conveying three-dimensional information in two dimensions. The energy difference between one staggered conformation and another is equal to that required to get past an eclipsed conformation. The energy difference is called **torsional strain**. For each carbon–hydrogen bond, the contribution to the torsional strain is 4.2 kJ mole^{-1}.

Conformations of Propane

Rotation about a carbon–carbon bond of propane is similar to that of ethane. However, the eclipsed conformation now has a hydrogen–methyl interaction in addition to two hydrogen–hydrogen interactions, so the barrier to rotation is larger. The resulting increase in energy is attributed to van der Waals repulsion between atoms. Although small in this case, the **steric hindrance** between atoms is larger when atoms are larger or brought closer together.

Conformations of Butane

With butane, the comparison of conformations becomes more interesting. In fact, the concepts introduced here are important to the understanding of many other phenomena such as the stability of cycloalkanes and many reactions that we will encounter later. There are two nonequivalent staggered conformations—the *anti* conformation and the *gauche* conformation, which have a **dihedral angle**, or **torsion angle** of 180° and 60°, respectively. The *anti* conformation is the more stable because there is van der Waals repulsion between the two methyl groups in the gauche conformation.

anti butane

gauche butane

4.5 Cycloalkanes

Cycloalkanes, as their name tells us, contain rings. Some cyclic compounds have atoms that are shared between two or more rings. These are **spirocyclic, bridged-ring,** and **fused-ring compounds**.

The general formula for an alkane is C_nH_{2n+2}. Each ring in a compound reduces the number of hydrogen atoms by 2 relative to an alkane because a ring contains an extra carbon–carbon bond and, therefore, two fewer carbon–hydrogen bonds. Thus, the general formula for cycloalkanes with one ring is C_nH_{2n}, the formula for compounds with two rings is C_nH_{2n-2}, and so on.

Geometric isomers can result when two or more substituents are attached to the ring at different carbon atoms. If the substituents are on the same side of the ring, the compound is the *cis* isomer. When substituents are on opposite sides of the ring, the compound is the *trans* isomer. Geometric isomers are one type of **stereoisomer**.

Cycloalkanes are named by prefixing the term *cyclo-* to a name giving the number of carbon atoms in the ring. The number 1 carbon atom is selected based on the importance of a functional group or alkyl group attached to the ring. The direction of numbering is selected to give the lowest combination of numbers to the remaining substituents at the point of first difference. Geometric isomers are identified with the appropriate *cis-* or *trans-* prefix.

Small ring compounds are unstable due to **ring strain**, which is the result of the small bond angles required to maintain the structure. The most severely strained compounds are cyclopropane and cyclobutane.

4.6 Conformations of Cycloalkanes

The small ring compounds cyclopropane and cyclobutane are strained rings. Their total strain energy is a combination of bond angle strain and torsional strain. In cylcopropane, the carbon–carbon bonds are highly strained, and there is also steric strain because the carbon–hydrogen bonds of adjacent carbon atoms are eclipsed. In cyclobutane, there is considerable bond angle strain and some eclipsing strain. A little twisting decreases the torsional strain, but the bonds giving rise to the torsional strain are still close. In cyclopentane, the bond angle strain is small because the bond angles are nearly tetrahedral. However, torsional interactions still occur in the molecule. Twisting cyclopentane into an envelop conformation alleviates but does not completely eliminate torsion angle strain.

The most stable conformation of cyclohexane is a "chair" in which there are three axial bonds pointed up and three pointed down. There are also six equatorial bonds, which point out around the ring; three are pointed slightly upward and three slightly downward.

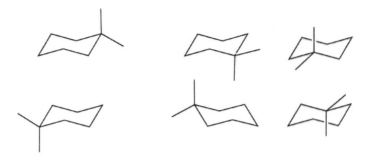

4.7 Conformation Mobility of Cyclohexane

The chair conformation of cyclohexane can change by a chair–chair interconversion, or "flip," and this process changes the orientations of all bonds. The equatorial bonds become axial and vice versa. Chair–chair interconversion passes through a "boat" conformation. Boat conformations are unstable and exist in vanishingly small amounts.

4.8 Conformations of Monosubstituted Cyclohexanes

Substituents bonded to the cyclohexane ring have a conformational preference for the equatorial position. Substituents in the axial position are sterically hindered to some degree because they are within the van der Waals radii of the axial hydrogen atoms at the C-3 and C-5 positions. This interaction is called a 1,3-diaxial interaction.

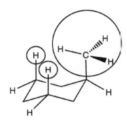

The energy differences between the axial and equatorial conformations of monosubstituted cyclohexanes are listed in Table 4.5. These values represent the magnitude of the two 1,3-diaxial interactions, and they depend on the size of the atom, the length of the bond, the polarizability of the atom, and the number of atoms bonded to the atom directly bonded to the cyclohexane ring.

4.9 Conformations of Disubstituted Cyclohexanes

In disubstituted cyclohexanes, not only the stability of the two possible conformations but also the relative stability of the geometric isomers depend on two factors. One is the inherent conformational preference of each substituent, and the other is any possible steric interaction between the two groups themselves. The *trans*-1,2-, the *cis*-1,3-, and the *trans*-1,4 dimethyl compounds are most stable in diequatorial conformations.

trans-1,2-dimethyl-
cyclohexane

cis-1,3-dimethyl-
cyclohexane

trans-1,4-dimethyl-
cyclohexane

These isomers are more stable than their respective equatorial/axial geometric isomers, because the axial methyl group has an unfavorable 1,3-diaxial interaction.

The most stable isomers for compounds with two different substituents are also the *trans*-1,2, the *cis*-1,3, and the *trans*-1,4 compounds. The difference in energy between either the isomers or the alternate conformations of each compound can be calculated by considering the conformational preferences of each group.

4.10 Polycyclic Molecules

The isomeric decalins provide the models for the rings that occur in fused ring compounds such as steroids. In *trans*-decalin, the hydrogens at the ring junction are both axial; in the *cis* isomer, one is equatorial and one is axial.

trans-decalin   *cis*-decalin

4.11 Physical Properties of Alkanes

Both alkanes and cycloalkanes have nonpolar covalent bonds. Thus, only van der Waals forces control the intermolecular interactions between neighboring molecules and those between solute and solvent molecules. Boiling points increase with increasing molecular weight and decrease with branching.

 **End of Chapter Exercises**

Molecular Formulas

4.1   Does each of the following molecular formulas for an acyclic hydrocarbon represent a saturated compound?

Answers: Acyclic saturated compounds have the general formula C_nH_{2n+2}. Only (b) and (d) meet this requirement.

(a) C_6H_{12} (b) C_5H_{12} (c) $C8H_{16}$ (d) $C10H_{22}$.

4.2   Can each of the following formulas correspond to an actual acyclic or cyclic molecule?
       (a) C_6H_{14} **Answer:** This formula is possible because it has $2n + 2$ hydrogen atoms.

       (b) $C_{10}H_{23}$ **Answer:** This formula is impossible because it has more than $2n + 2$ hydrogen atoms and also has an odd number of hydrogen atoms.

       (c) C_7H_{14} **Answer:** This formula is possible because it has $2n$ hydrogen atoms, which can result from either a cyclic structure or unsaturation.

       (d) C_5H_{14} **Answer:** This formula is impossible because it has more than $2n + 2$ hydrogen atoms.

4.3   Beeswax contains approximately 10% hentriacontane, a normal alkane with 31 carbon atoms. What is the molecular formula of hentriacontane? Write a completely condensed formula of hentriacontane.

       Answer: The number of carbon atoms represented by n in the general formula for alkanes is 31. The number of hydrogen atoms must be $2n + 2$ or 64. The completely condensed formula for hentriacontane is $CH_3(CH_2)_{29}CH_3$.

4.4   Hectane is a normal alkane with 100 carbon atoms. What is the molecular formula of hectane? Write a completely condensed formula of hectane.
Answer: For $n = 100$, the value of $2n + 2$ is 202. The molecular formula is $C_{188}H_{202}$. The completely condensed formula for this normal alkane is $CH_3(CH_2)_{98}CH_3$.

Structural Formulas

4.5   Redraw each of the following so that the longest continuous chain is written horizontally.

(a) CH_3——CH_2
                  |
              CH_2——CH_3

(b) CH_2——CH_2——CH——CH_2——CH_3
       | |
       CH_3 CH_2——CH_3

(c) CH_3——CH——CH_2——CH_3
               |
            CH_2——CH_3

(d) CH_3——CH——CH——CH_3
               | |
            CH_3 CH_2——CH_3

Answers:

(a) CH_3——CH_2——CH_2——CH_3

(b) CH_3——CH_2——CH_2——CH——CH_2——CH_3
                                |
                             CH_2——CH_3

(c) CH_3——CH_2——CH——CH_2——CH_3
                      |
                    CH_3

(d) CH_3——CH——CH——CH_2——CH_3
              | |
            CH_3 CH_3

4.6   Redraw each of the following so that the longest continuous chain is written horizontally.

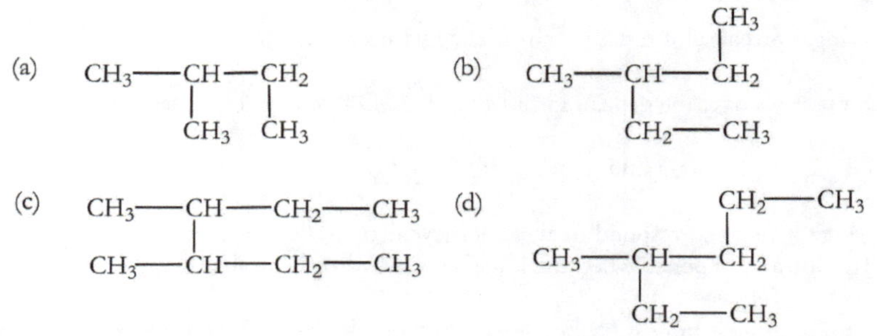

(a)   CH₃—CH—CH₂
                 | |
                CH₃   CH₃

(b)   CH₃
                  |
        CH₃—CH—CH₂
                 |
                CH₂—CH₃

(c)   CH₃—CH—CH₂—CH₃
                 |
        CH₃—CH—CH₂—CH₃

(d)   CH₂—CH₃
                  |
        CH₃—CH—CH₂
                 |
                CH₂—CH₃

Answers:

(a) CH₃—CH—CH₂—CH₃
              |
             CH₃

(b) CH₃—CH₂—CH—CH₂—CH₂—CH₃
                  |
                 CH₃

(c) CH₃—CH₂—CH—CH—CH₂—CH₃
                   | |
                  CH₃ CH₃

(d) CH₃—CH₂—CH—CH—CH₂—CH₃
                   | |
                  CH₃ CH₃

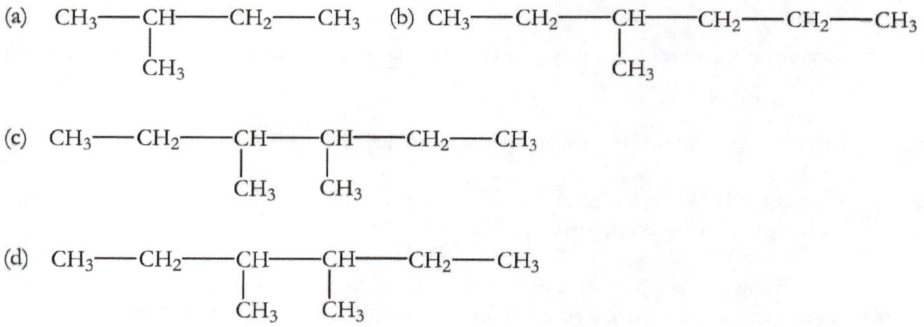

4.7   Which of the following structures represent the same compound?

CH₃—CH—CH—CH₂—CH₃
        | |
       CH₃ CH—CH₂—CH₃
        I

CH₃—CH—CH—CH₂—CH₂
        | | |
       CH₃ CH₂—CH₃   CH₃
        II

CH₃—CH—CH₂—CH₃
        |
CH₃—CH—CH₂—CH₃
        III

              CH₂—CH₂—CH₃
               |
CH₃—CH—CH—CH₃
        |
       CH₂—CH₃
        IV

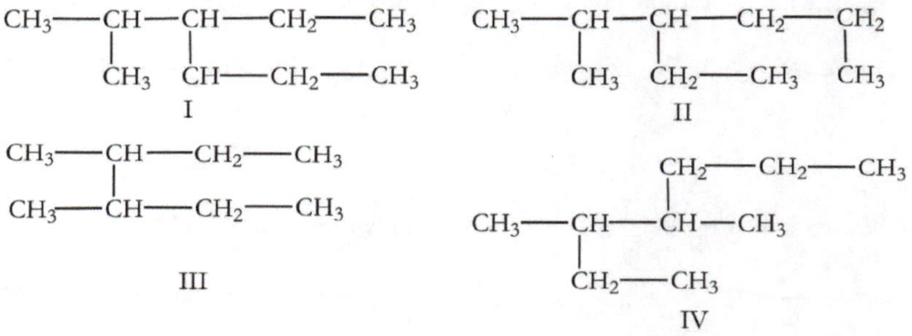

Answer: Both I and II have a chain of six carbon atoms with a methyl group at C-2 and an ethyl group at C-3. Both III and IV have a chain of seven carbon atoms with methyl groups at C-3 and C-4.

4.8   Which of the following structures represent the same compound?

CH₃—CH—CH—CH₂—CH₃
        | |
       CH₃ CH₃
        I

CH₃—CH—CH₂—CH₂
        | |
       CH₂—CH₃   CH₃
        II

CH₃—CH—CH₂—CH₃
        |
CH₃—CH—CH₃
        III

              CH₂—CH₃
               |
CH₃—CH—CH₂
        |
       CH₂—CH₃
        IV

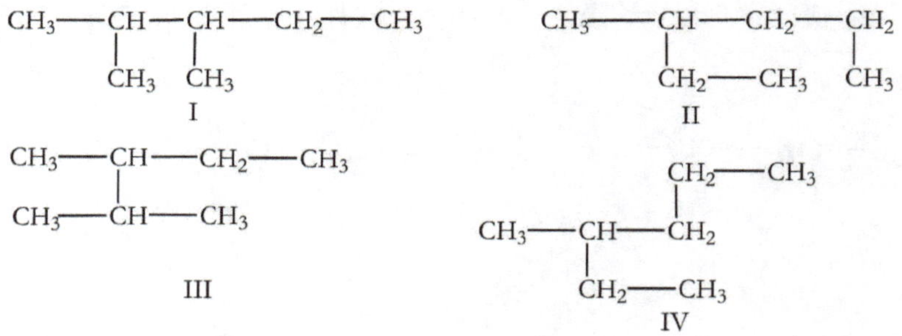

Answer: Both I and III have a chain of five carbon atoms with methyl groups at C-2 and C-3. Both III and IV have a chain of six carbon atoms with a methyl group at C-3.

Alkyl Groups

4.9 What is the common name for each of the following alkyl groups?

(a) CH_3—— (b) CH_3——CH_2
    $|$
    CH_2——

(c) CH_3——CH——CH_2——CH_3
        $|$

(d) CH_3——CH——CH_2——
        $|$
        CH_3

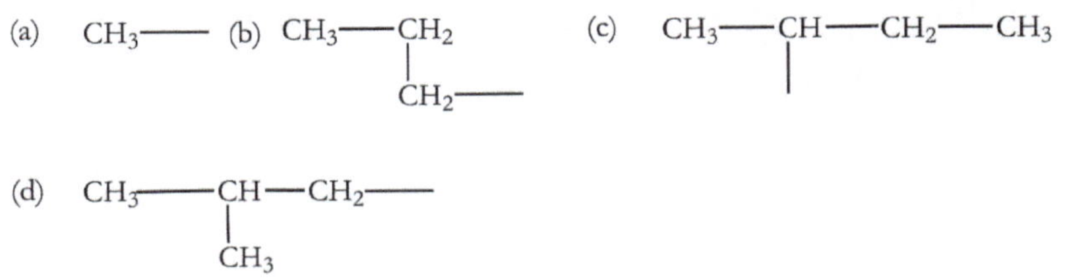

Answers: (a) methyl   (b) propyl   (c) *sec*-butyl   (d) isobutyl

4.10 What is the common name for each of the following alkyl groups?

(a) CH_3——CH_2——   (b) CH_3——CH——
                            $|$
                            CH_3

(c) CH_3——CH_2——CH_2——CH_2——

(d) CH_3
        $|$
    CH_3——C——
        $|$
        CH_3

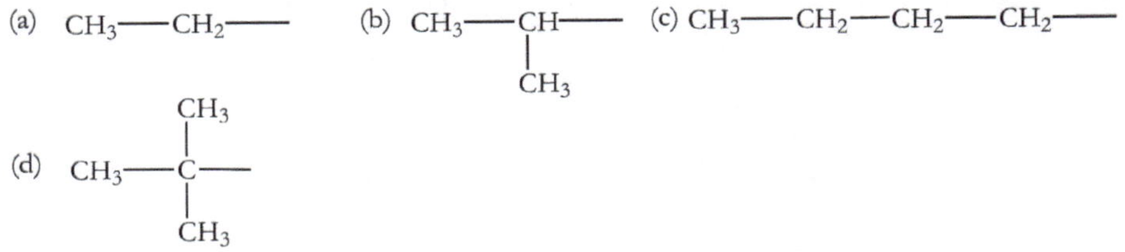

Answers: (a) ethyl   (b) isopropyl   (c) butyl   (d) *tert*-butyl

4.11 What is the common name for each of the following alkyl groups?

(a) CH_3——CH_2
            $|$
            CH_2——CH_2——

(b) CH_3——CH——CH_3
        $|$
        CH_2——

(c) CH_3——CH——CH_2——CH_3
        $|$
        CH_2——

(d) CH_3——CH——CH_2——CH_2——
        $|$
        CH_3

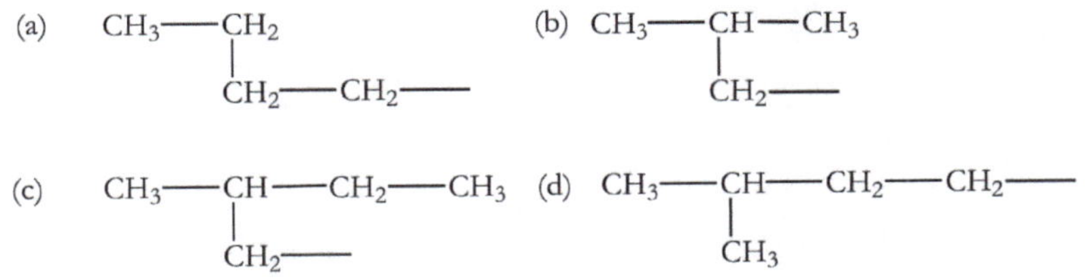

Answers: (a) butyl   (b) 2-methylpropyl   (c) 2-methylbutyl   (d) 3-methylbutyl

4.12 What is the IUPAC name for each of the following alkyl groups?

(a) CH_3——CH——CH_2——
        $|$
        CH_3

(b) CH_3——CH——CH_2——
        $|$
        CH_2——CH_3

(c) CH_3——CH——CH_2——CH_2——
        $|$
        CH_2——CH_3

(d) $|$
    CH_3——C——CH_2——CH_3
        $|$
        CH_3

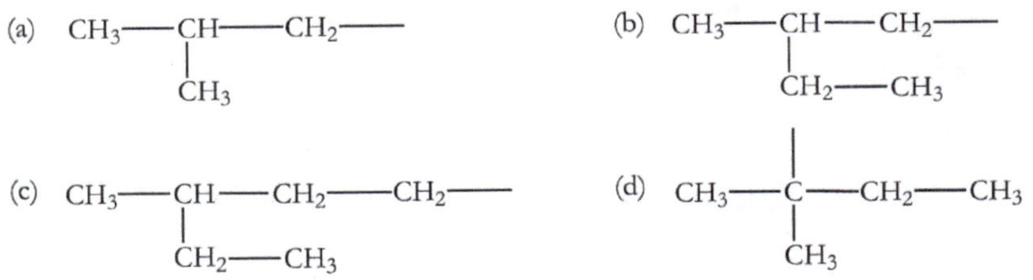

Answers: (a) 2-methylpropyl   (b) 2-methylbutyl   (c) 3-methylpentyl   (d) 1,1-dimethylpropyl

4.13   The spermicide octoxynol-9 is used in diverse contraceptive products. Name the alkyl group to the left of the benzene ring.

octoxynol-9

Answer: 1,1,3,3-tetramethylbutyl

4.14   The name vitamin E actually refers to a series of closely related compounds called tocopherols. Name the complex alkyl group present in α-tocopherol.

Answer: 4,8,12-trimethyltridecyl

Nomenclature of Alkanes

4.15   Give the IUPAC name for each of the following compounds.

Answers: (a) 2-methylbutane (b) 3-methylhexane (c) 2-methylpentane (d) 3-methylpentane (e) 2-methylpentane (f) 5-methylnonane

4.16   Give the IUPAC name for each of the following compounds.

Answers: (a) 2-methylhexane (b) 2,5-dimethylhexane (c) 3,4-dimethylheptane (d) 3,5-dimethylheptane (e) 3-methylhexane
    (f) 3-ethylpentane

4.17   Give the IUPAC name for the following compound.

$$CH_3-CH_2-CH_2-CH_2-CH-CH_2-CH_2-CH_2-CH_2-CH_3$$
$$CH_3-CH_2-CH-CH_2-CH_3$$

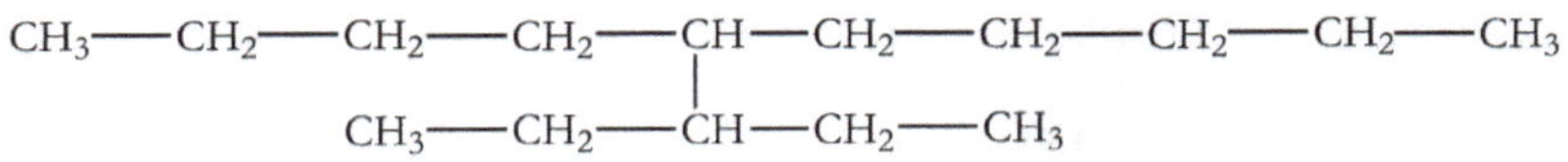

Answer: 5-(1-ethylpropyl)decane

4.18   Give the IUPAC name for the following compound.

$$CH_3-CH_2-CH_2-CH_2-CH-CH_2-CH_2-CH_2-CH_3$$
$$CH_3-C-CH_3$$
$$CH_2-CH_3$$

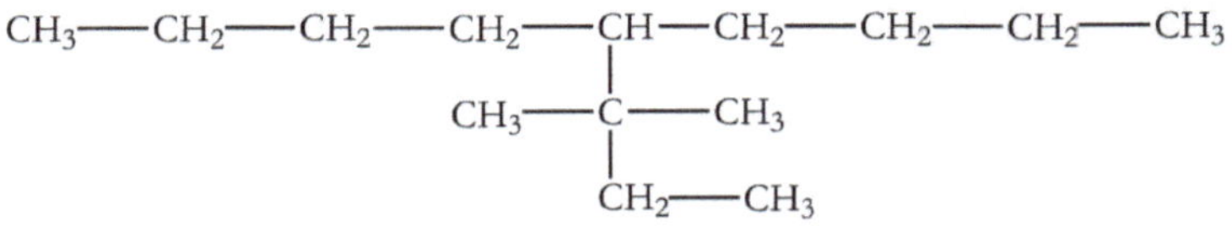

Answer: 5-(1,1-dimethylpropyl)nonane

4.19   Write the structural formula for each of the following compounds.
        (a) 3-methylpentane   (b) 3,4-dimethylhexane   (c) 2,2,3-trimethylpentane
        (d) 4-ethylheptane   (e) 2,3,4,5-tetramethylhexane

Answers: (a) 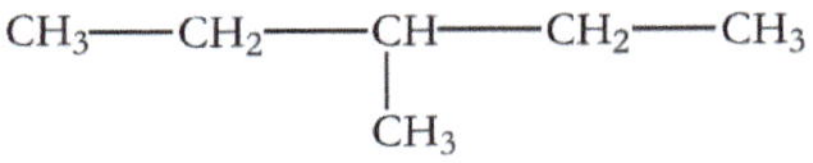 $CH_3-CH_2-CH-CH_2-CH_3$
                                   $|$
                                  CH_3

            (b) 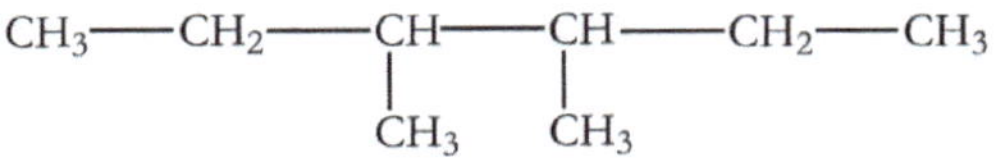 $CH_3-CH_2-CH-CH-CH_2-CH_3$
                                $|$ $|$
                               CH_3 CH_3

                      CH_3
                       $|$
            (c) 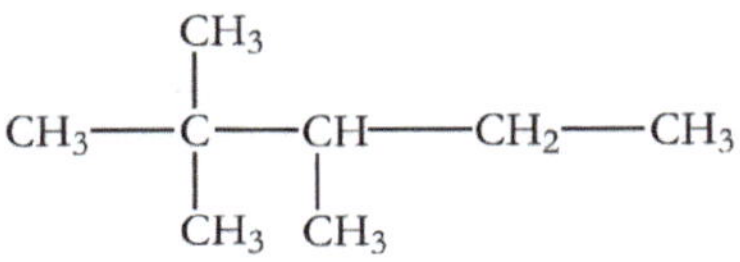 $CH_3-C-CH-CH_2-CH_3$
                       $|$ $|$
                      CH_3 CH_3

            (d) 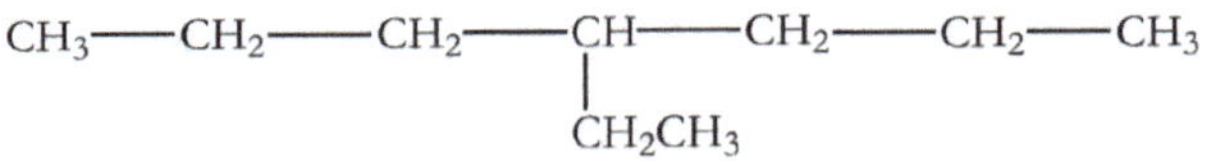 $CH_3-CH_2-CH_2-CH-CH_2-CH_2-CH_3$
                                   $|$
                                CH_2CH_3

            (e) 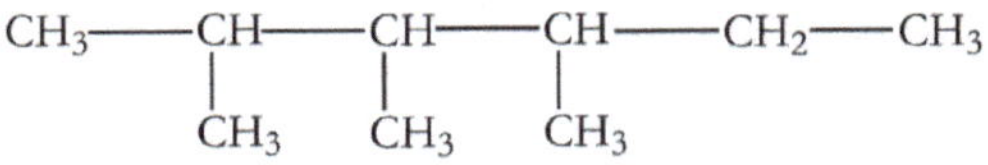 $CH_3-CH-CH-CH-CH_2-CH_3$
                        $|$ $|$ $|$
                       CH_3 CH_3 CH_3

4.20   Write the structural formula for each of the following compounds.
      (a) 2-methylpentane   (b) 3-ethylhexane (c) 2,2,4-trimethylhexane
      (d) 2,4-dimethylheptane   (e) 2,2,3,3-tetramethylpentane

Answers: (a)

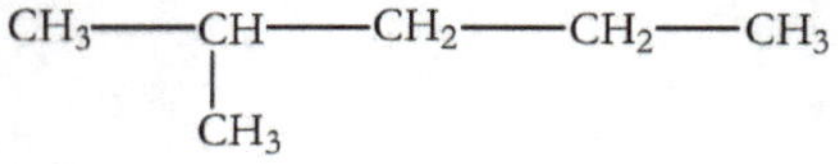

$$CH_3-\underset{\underset{CH_3}{|}}{CH}-CH_2-CH_2-CH_3$$

(b)

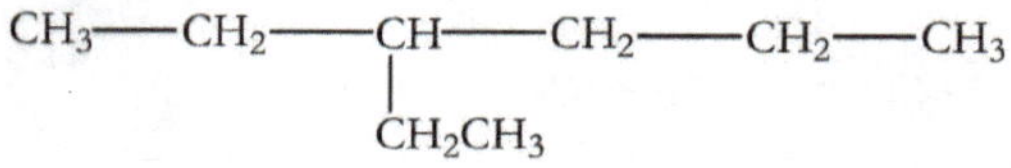

$$CH_3-CH_2-\underset{\underset{CH_2CH_3}{|}}{CH}-CH_2-CH_2-CH_3$$

(c)

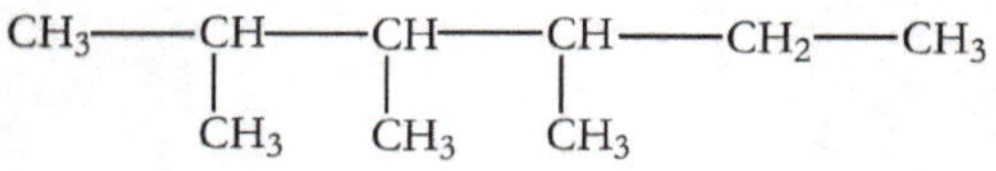

$$CH_3-\underset{\underset{CH_3}{|}}{CH}-\underset{\underset{CH_3}{|}}{CH}-\underset{\underset{CH_3}{|}}{CH}-CH_2-CH_3$$

(d)

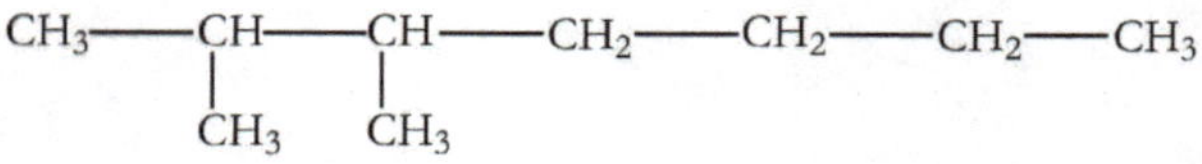

$$CH_3-\underset{\underset{CH_3}{|}}{CH}-\underset{\underset{CH_3}{|}}{CH}-CH_2-CH_2-CH_2-CH_3$$

(e)

$$CH_3-\overset{\overset{CH_3}{|}}{\underset{\underset{CH_3}{|}}{C}}-\overset{\overset{CH_3}{|}}{\underset{\underset{CH_3}{|}}{C}}-CH_2-CH_3$$

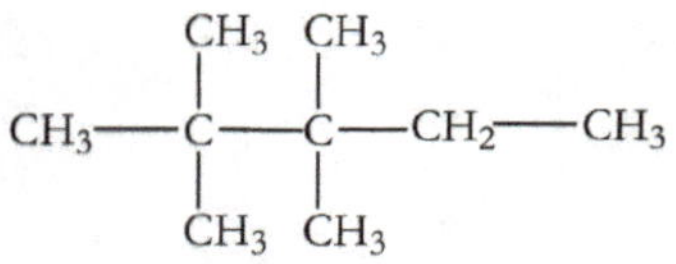

4.21   Write the structural formula for each of the following compounds.
      (a) 4-(1-methylethyl)heptane   (b) 5-(1,1-dimethylethyl)nonane   (c) 5-(1-methylpropyl)decane

Answers: (a)

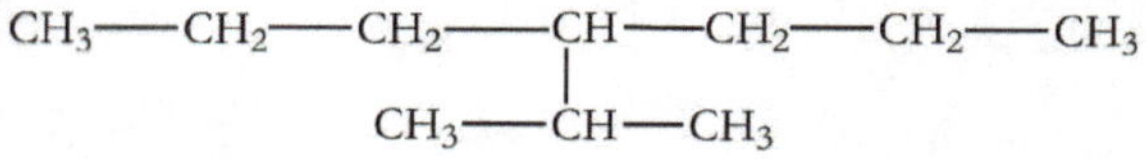

$$CH_3-CH_2-CH_2-\underset{\underset{CH_3-CH-CH_3}{|}}{CH}-CH_2-CH_2-CH_3$$

(b)

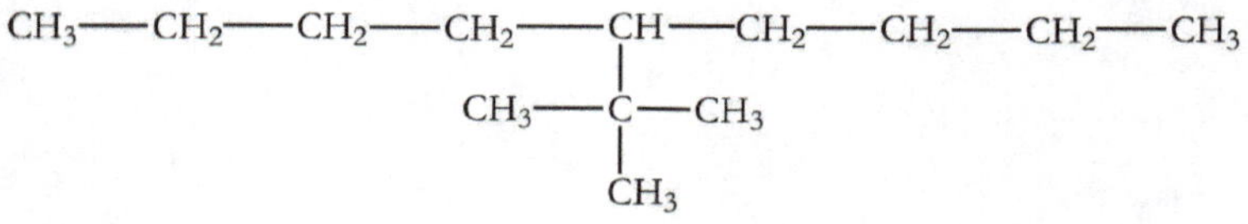

$$CH_3-CH_2-CH_2-CH_2-\underset{\underset{\underset{CH_3}{|}}{\underset{CH_3-C-CH_3}{|}}}{CH}-CH_2-CH_2-CH_2-CH_3$$

(c)

$$CH_3-CH_2-CH_2-CH_2-\underset{\underset{CH_3-CH-CH_2-CH_3}{|}}{CH}-CH_2-CH_2-CH_2-CH_2-CH_3$$

58

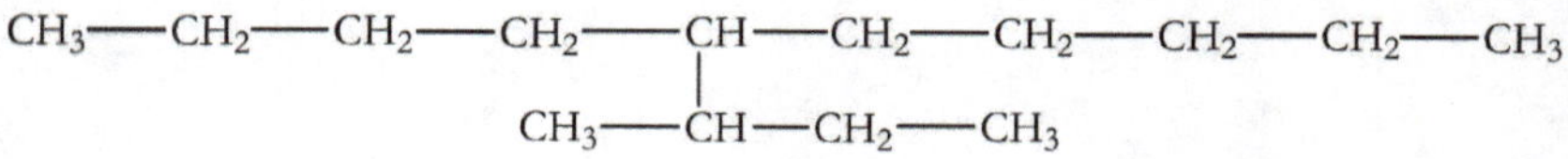

4.22   Write the structural formula for each of the following compounds.
        (a) 5-(2-methylpropyl)nonane   (b) 4-butyl nonane   (c) 5-(2,2-dimethylpropyl)decane

Answers: (a) CH_3—CH_2—CH_2—CH_2—CH—CH_2—CH_2—CH_2—CH_3
                                                                    |
                                                            CH_2—CH—CH_3
                                                                        |
                                                                       CH_3

          (b) CH_3—CH_2—CH_2—CH_2—CH—CH_2—CH_2—CH_2—CH_3
                                                                    |
                                                            CH_2—CH_2—CH_2—CH_3

          (c) CH_3—CH_2—CH_2—CH_2—CH—CH_2—CH_2—CH_2—CH_2—CH_3
                                                                |
                                                       CH_3—C—CH_2—CH_3
                                                                |
                                                               CH_3

Isomers

4.23   There are nine isomeric C_7H_{16} compounds. Name the isomers that have a single methyl group as a branch.

Answers: 2-methylhexane and 3-methylhexane

4.24   There are nine isomeric C_7H_{16} compounds. Name the isomers that have two methyl groups as branches and are named as dimethyl-substituted pentanes.

Answers: 2,2-dimethylpentane, 3,3-dimethylpentane, 2,3-dimethylpentane, and 2,4-dimethylpentane

Classification of Carbon Atoms

4.25   Classify each carbon atom in the following compounds as primary, secondary, or tertiary.

        (a) CH_3—CH_2—CH_2—CH_2—CH_3 (b) CH_3—CH_2—CH—CH_2—CH_3
                                                                            |
                                                                           CH_3

                        CH_3
                         |
        (c) CH_3—C—CH_2—CH_3 (d) CH_3—CH—CH—CH_3
                         | | |
                        CH_3 CH_3 CH_3

Answers: 1°   2°   2°   2°   1°   1°   2°   3°   2°   1°
          (a) CH_3—CH_2—CH_2—CH_2—CH_3 (b) CH_3—CH_2—CH—CH_2—CH_3
                                                                            |
                                                                           CH_3
                                                                            1°

                1°   3°   3°   1°
          (c) CH_3—CH—CH—CH_3
                         | |
                        CH_3 CH_3
                         1°   1°

4.26   Classify each carbon atom in the following compounds as primary, secondary, tertiary, or quaternary.

(a) 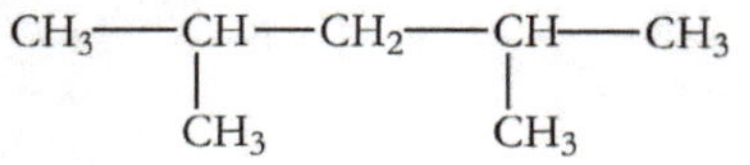 CH_3—CH—CH_2—CH—CH_3 (b) CH_3—CH_2—CH_2—CH_3
         | |
        CH_3 CH_3

(c) 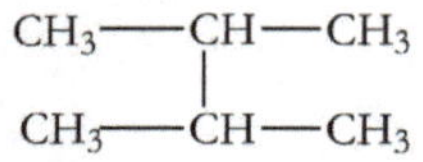 CH_3—CH—CH_3 (d) 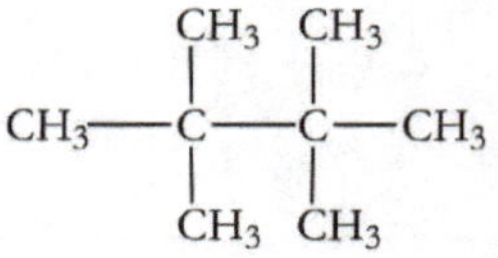 CH_3—C—C—CH_3 (with CH_3 CH_3 above and CH_3 CH_3 below)
           |
   CH_3—CH—CH_3

Answers:

(a) 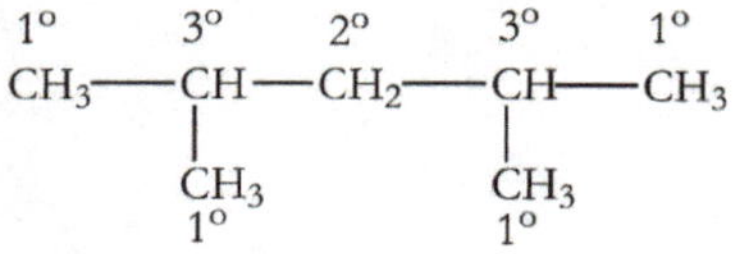 $1°$ $3°$ $2°$ $3°$ $1°$
CH_3—CH—CH_2—CH—CH_3
     | |
   CH_3 CH_3
    $1°$ $1°$

(b) 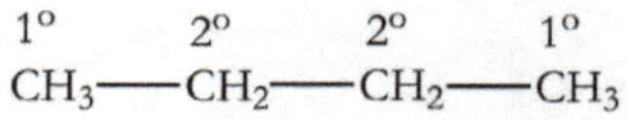 $1°$ $2°$ $2°$ $1°$
CH_3—CH_2—CH_2—CH_3

(c) 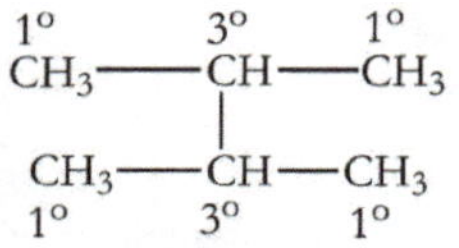 $1°$ $3°$ $1°$
CH_3—CH—CH_3
      |
CH_3—CH—CH_3
$1°$ $3°$ $1°$

(d) 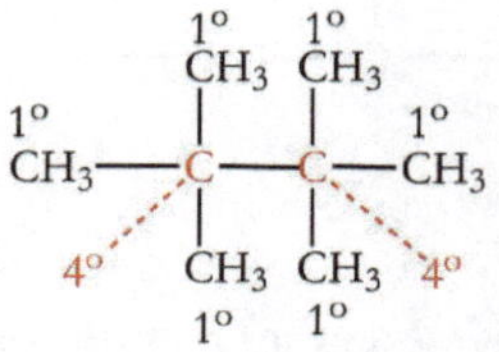 $1°$ $1°$
  CH_3 CH_3
$1°$ | | $1°$
CH_3—C—C—CH_3
$4°$ | | $4°$
  CH_3 CH_3
  $1°$ $1°$

4.27   Draw the structure of a compound with molecular formula C_5H_{12} that has one quaternary and four primary carbon atoms.

Answer:
    CH_3
     |
CH_3—C—CH_3
     |
    CH_3

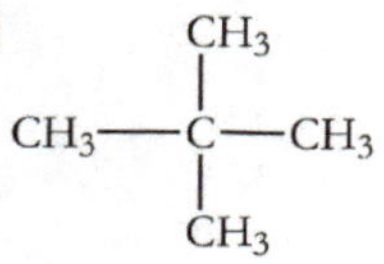

4.28   Draw the structure of a compound with molecular formula C_6H_{14} that has two tertiary and four primary carbon atoms.

Answer: $1°$ $3°$ $1°$
CH_3——CH——CH_3
       |
CH_3——CH——CH_3
$1°$ $3°$ $1°$

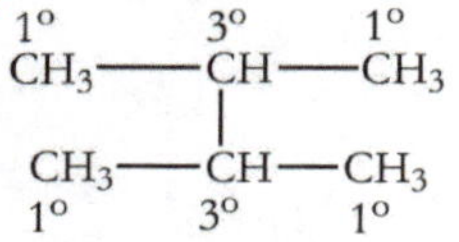

2,3-dimethylbutane

4.29   Determine the number of primary, secondary, tertiary, and quaternary carbon atoms in each of the following compounds.

Answers:
(a) four primary and one quaternary
(b) three primary, one secondary, and one tertiary
(c) three primary, two secondary, and one tertiary
(d) four primary and two tertiary

(a) 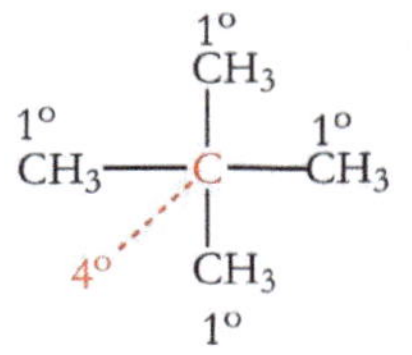
$$1^\circ$$
$$CH_3$$
$$1^\circ \quad | \quad 1^\circ$$
$$CH_3 \!-\! C \!-\! CH_3$$
$$4^\circ \quad | $$
$$CH_3$$
$$1^\circ$$

(b) 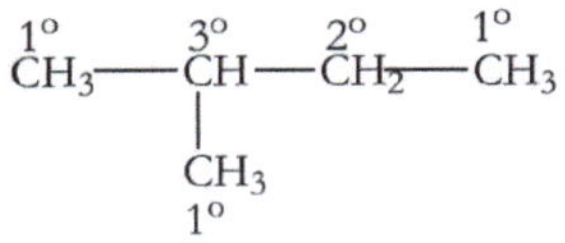
$$1^\circ \qquad 3^\circ \qquad 2^\circ \qquad 1^\circ$$
$$CH_3 \!-\! CH \!-\! CH_2 \!-\! CH_3$$
$$|$$
$$CH_3$$
$$1^\circ$$

(c)
$$1^\circ \qquad 3^\circ \qquad 2^\circ \qquad 2^\circ \qquad 1^\circ$$
$$CH_3 \!-\! CH \!-\! CH_2 \!-\! CH_2 \!-\! CH_3$$
$$|$$
$$CH_3$$
$$1^\circ$$

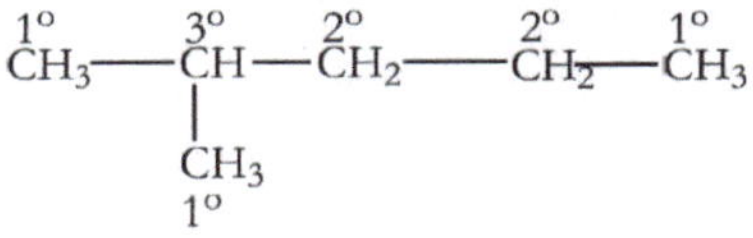

(d)
$$1^\circ \qquad 3^\circ \qquad 1^\circ$$
$$CH_3 \!-\! CH \!-\! CH_3$$
$$|$$
$$CH_3 \!-\! CH \!-\! CH_3$$
$$1^\circ \quad 3^\circ \quad 1^\circ$$

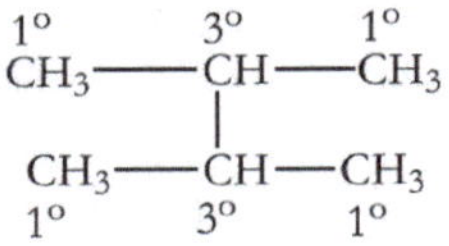

4.30   Determine the number of primary, secondary, tertiary, and quaternary carbon atoms in each of the following compounds.

Answers:
(a) six primary, one secondary, and two quaternary
(b) four primary, one secondary, and two tertiary
(c) three primary, two secondary, and one tertiary
(d) five primary and three tertiary

(a)
$$CH_3 \qquad\qquad CH_3$$
$$| \qquad\qquad\qquad |$$
$$CH_3 \!-\! C \!-\! CH_2 \!-\! C \!-\! CH_3$$
$$| \qquad\qquad\qquad |$$
$$CH_3 \qquad\qquad CH_3$$

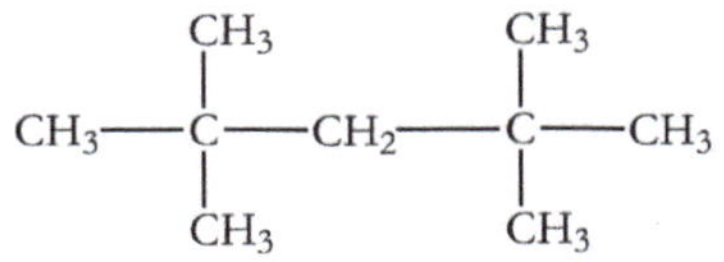

(b) 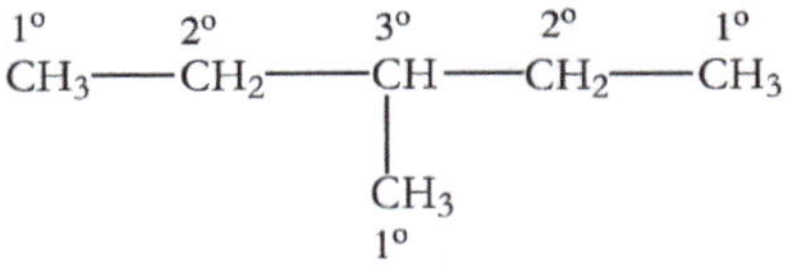
$$1^\circ \qquad 2^\circ \qquad 3^\circ \qquad 2^\circ \qquad 1^\circ$$
$$CH_3 \!-\! CH_2 \!-\! CH \!-\! CH_2 \!-\! CH_3$$
$$|$$
$$CH_3$$
$$1^\circ$$

(c)
$$CH_3 \!-\! CH_2 \!-\! CH \!-\! CH_2 \!-\! CH_3$$
$$|$$
$$CH_3$$

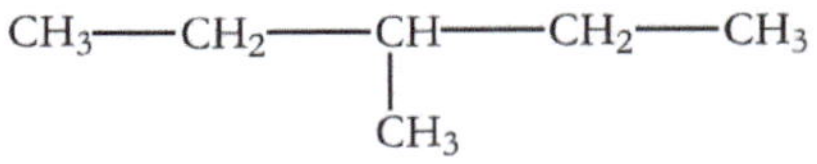

(d)
$$CH_3 \!-\! CH \!-\! CH \!-\! CH \!-\! CH_2 \!-\! CH_3$$
$$| \qquad | \qquad |$$
$$CH_3 \quad CH_3 \quad CH_3$$

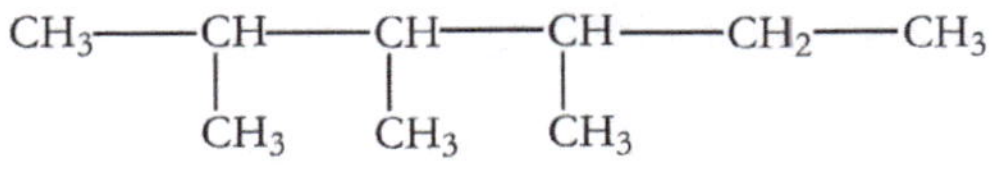

Cycloalkanes

4.31   Write condensed planar formulas for each of the following compounds.

(a) chlorocyclopropane   (b) 1,1-dimethylcyclobutane   (c) cyclooctane

Answers:

(a)   (b)    (c)

4.32   Write condensed planar formulas for each of the following compounds.

(a) bromocyclobutane   (b) 1,1-dichlorocyclopropane   (c) cyclopentane

Answers:

(a)   (b)   (c)

4.33   Name each of the following compounds.

(a) 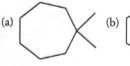   (b)    (c) 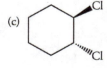   (d)

Answers: (a) 1,1-dimethylcycloheptane   (b) cyclodecane   (c) *trans*-1,2-dichlorocyclohexane   (d) 1,1-dichlorocyclohexane

4.34   Name each of the following compounds.

Answers: (a) butylcyclooctane
(b) isopropylcyclohexane
(c) *cis*-1,3-dimethylcyclohexane
(d) 1,1,3-trichlorocyclohexane   (a) 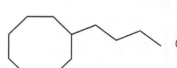   (b)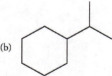

(c) 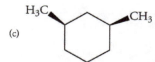   (d)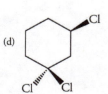

4.35   A saturated refrigerant has the molecular formula C_4F_8. Draw structural formulas for two possible isomers of this compound.

Answers:

(a)   (b)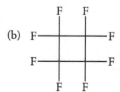

4.36   How many isomeric saturated hydrocarbons have the molecular formula C_5H_{10}?

Answers: Five compounds have this formula: (i) cyclopentane, (ii) methylcyclobutane, (iii) 1,1-dimethylcyclopropane, (iv) *cis*-1,2-dimethylcyclopropane, and (v) *trans*-1,2-dimethylcyclopropane

Bicyclic compounds
4.37   What is the molecular formula of each of the following compounds?

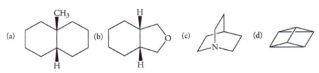

Answers: (a) $C_{10}H_{16}$ (b) $C_{10}H_{12}O$ (c) $C_7H_{13}N$ (d) C_6H_6

4.38   What is the molecular formula of each of the following compounds?

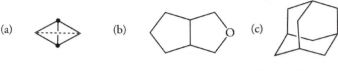

Answers: (a) $C_{10}H_{18}O$ (b) $C_{10}H_{15}$ (c) $C_{10}H_{14}$ (d) C_8H_{14}

Polycyclic compounds
4.39   How many rings are present in each of the following polycyclic compounds?

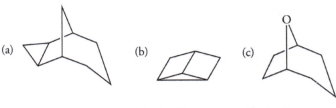

Answers: (a) tricyclic   (b) bicyclic   (c) tricyclic

4.40   How many rings are present in each of the following polycyclic compounds?

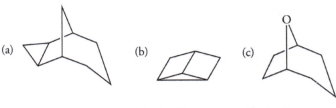

Answers: (a) pentacyclic   (b) tricyclic   (c) bicyclic

Properties of Hydrocarbons

4.41   Which of the isomeric C_8H_{18} compounds has the highest boiling point? Which has the lowest boiling point?

Answer: Octane has the highest boiling point. All of its isomers have branches and should have lower boiling points. 2,2,3,3-Tetra-methylbutane has the largest number of branches and should have the lowest boiling point.

4.42   The boiling point of methylcyclopentane is lower than the boiling point of cyclohexane. Suggest a reason why.

Answer: Methylcyclopentane compared to cyclohexane has a more compact structure and the difference results from the same phenomena observed for branched alkanes.

Newman Projection Formulas of alkanes

4.43   Draw the Newman projection of the staggered conformation of 2,2-dimethylpropane around the C-1 to C-2 bond.

Answer:

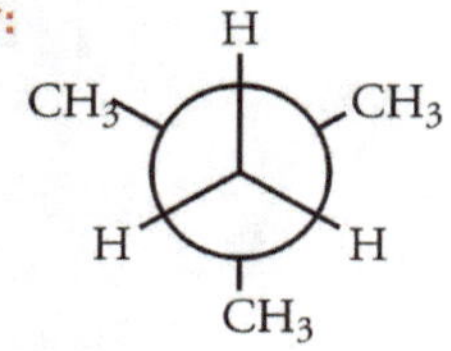

4.44   Draw the Newman projections of the two possible staggered conformations of 2,3-dimethylbutane around the C-2 to C-3 bond.

Answer:

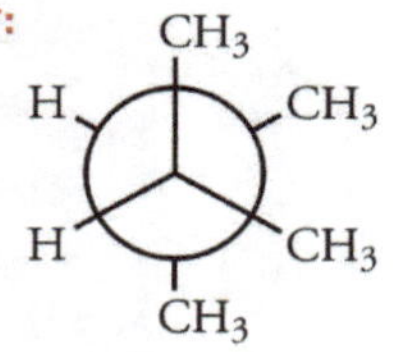

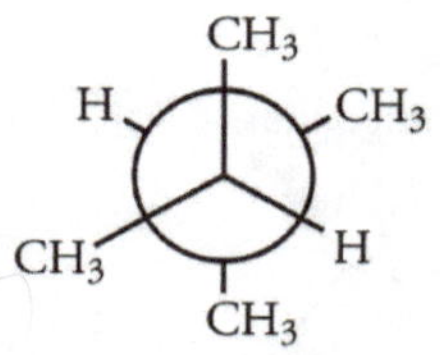

4.45   Draw the Newman projections of the two possible staggered conformations of 2-methylbutane around the C-2 to C-3 bond. Which is the more stable?

Answer:

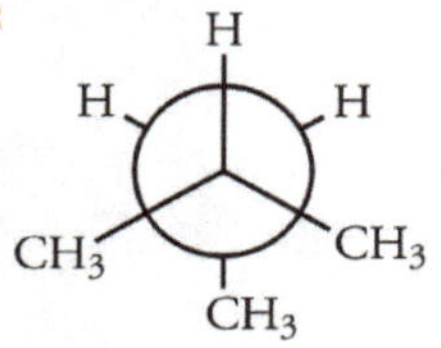

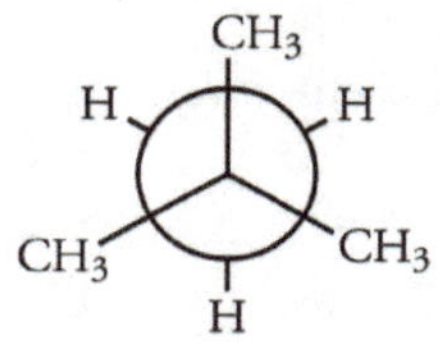

more stable

4.46   Draw the Newman projections of the two possible staggered conformations of 2,2-dimethylpentane around the C-3 to C-4 bond. Which is the more stable?

Answer:

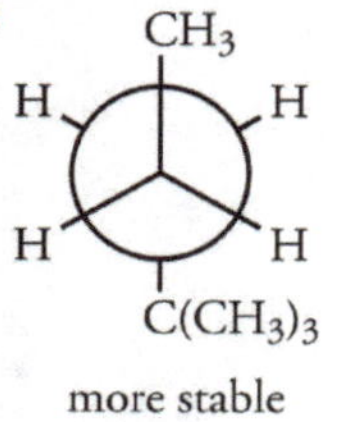

more stable

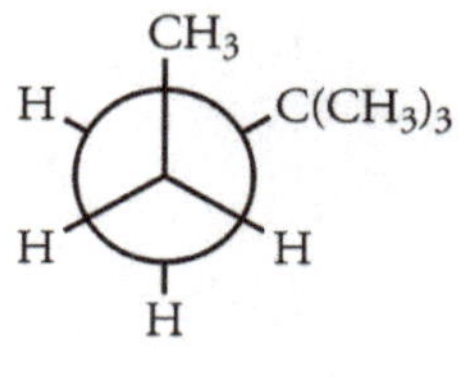

Stabilities of Acyclic Conformations

4.47   Do you expect the barrier to rotation around the central bond for CH_3—CH_2—SiH_2—CH_3 to be smaller or larger than the barrier to rotation for butane? Why?

Answer: The silicon–carbon bond length is longer than a carbon–carbon bond. Thus, the distance between the two methyl groups is larger in the silicon compound than in butane. The barrier to rotation should be lower in the silicon compound because all of the sets of eclipsing bonded pairs of electrons are separated by a larger distance.

4.48   Draw a potential energy diagram for rotation around the C-2 to C-3 bond of 2,2-dimethylbutane.

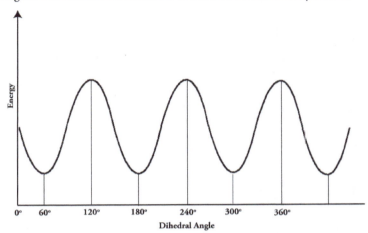

4.49   Draw a potential energy diagram for rotation around the C-2 to C-3 bond of 2-methylbutane.

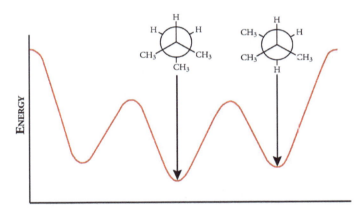

4.50   2-Chloroethanol (ClCH$_2$CH$_2$OH) is most stable in the gauche conformation. Suggest a reason for this fact.

Answer: There is an attractive interaction between chlorine and the hydroxyl group. The interaction is a dipole–dipole attraction that resembles a hydrogen bond.

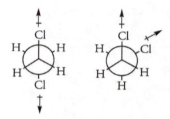

4.51   1-Chloropropane is most stable in the gauche conformation. What does this fact indicate about the interaction of chlorine and a methyl group in this compound?

Answer: There is an attractive van der Waals interaction between chlorine and the methyl group that results from the polarizability of the chlorine atom.

4.52   Draw the two staggered conformations of 1,2-dichloroethane. Which of the conformations has a dipole moment? The dipole moment of 1,2-dichloroethane is 1.1 D. Does this fact provide any information about the composition of the mixture of conformations?

Answer: The *anti* conformation does not have a dipole moment because the bond moments of the C—Cl bonds cancel. The gauche conformation has a dipole moment. The observed dipole moment indicates that some of the compound must exist in the gauche conformation.

4.53   Ethylene glycol (HOCH$_2$CH$_2$OH) forms intramolecular hydrogen bonds. Does this fact provide any information about the composition of the mixture of conformations?

Answer: The *anti* conformation cannot form intramolecular hydrogen bonds because the hydroxyl groups are widely separated. The hydroxyl groups are sufficiently close in the gauche conformation to form an intramolecular hydrogen bond. The structure has five atoms in a hydrogen-bonded "ring" counting the atoms starting from the hydroxyl hydrogen atom of one hydroxyl group to the oxygen atom of the other hydroxyl group.

Conformations of Cyclohexanes

4.54   Draw the most stable conformation of the equatorial form of methylcyclohexane showing the relationship of the methyl hydrogen atoms to the hydrogen atom at C-1.

Answer:

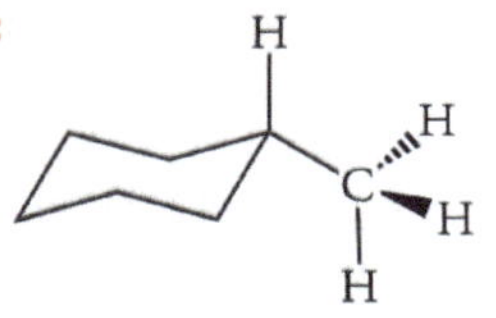

4.55   Draw the conformation of the axial form of methylcyclohexane showing the relationship of the methyl hydrogen atoms to C-3 and C-5.

Answer:

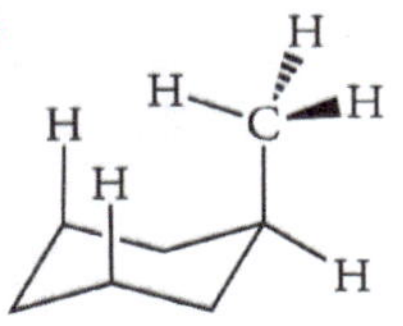

4.56   Draw the most stable chair conformation of each of the following compounds.
(a) *trans*-l-fluoro-3-methylcyclohexane   (b) *trans*-l-*tert*-butyl-3-methylcyclohexane   (c) *trans*-1,2-dimethylcyclohexane

Answers:

(a) 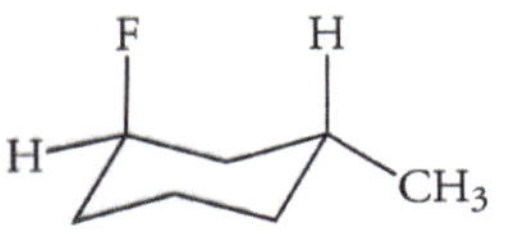   (b)

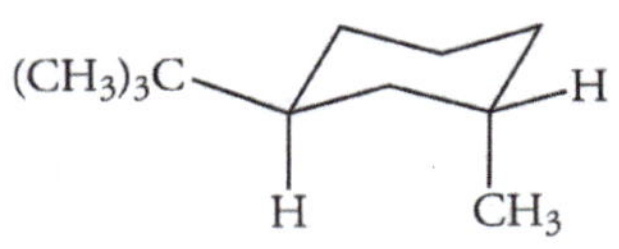

(c)

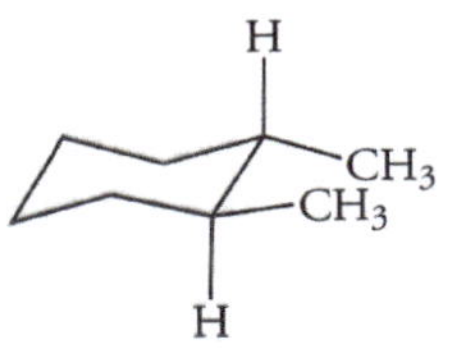

4.57   Draw the most stable chair conformation of each of the following compounds.
(a) *cis*-1,1,4-trimethylcyclohexane   (b) *trans*-1,1,3-trimethylcyclohexane (c) *cis*-1-fluoro-4-ethylcyclohexane

Answers:

(a) 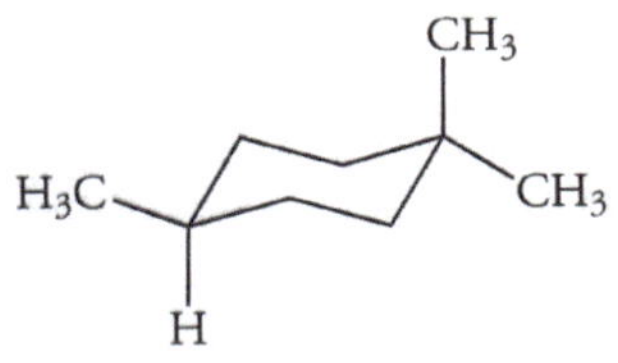   (b)

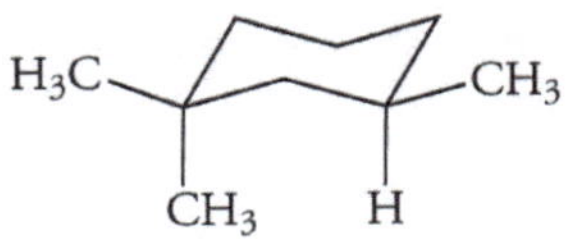

(c)

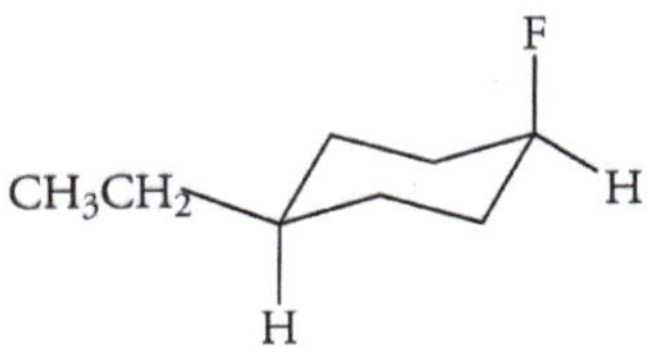

4.58   Why is the steric strain caused by the *tert*-butyl group so different from those of methyl, ethyl, and isopropyl groups?

Answer: In any conformation of the *tert*-butyl compound, there is a methyl group located over the cyclohexane ring and there is severe steric repulsion between it and the axial hydrogen atoms at C-3 and C-5.

4.59   Within experimental error, the steric strain caused by a bromine atom is the same as that of a chlorine atom. Taking into account the "size" of the atoms and the length of the carbon–halogen bond, explain these data.

Answer: The C—Br bond is longer than the C—Cl bond, so the distance separating the bromine atom from the C-3 and C-5 hydrogen atoms is larger. In addition, the bromine atom is more polarizable than the chlorine atom, and its electrons may be more easily distorted away from the steric congestion in the axial conformation.

4.60   *cis*-1,3-Cyclohexanediol is most stable in a diaxial conformation. Suggest a reason for this "unusual" stability.

Answer: The hydroxyl groups are close enough in the diaxial conformation to form an intramolecular hydrogen bond. The hydrogen-bonded structure has 6 atoms in a hydrogen-bonded "ring," counting the atoms starting from the hydroxyl hydrogen atom of one hydroxyl group to the oxygen atom of the other hydroxyl group.

4.61   *trans*-1,3-Di-*tert*-butylcyclohexane exists in a twist boat conformation, rather than a chair conformation. Why?

Answer: An axial *tert*-butyl group exists in the *trans* isomer, which has a steric repulsion of 22 kJ mole^{-1} (Table 4.6). This repulsion is eliminated in the twist boat conformation, even though it is 22 kJ mole^{-1} less stable than the chair conformation because both *tert*-butyl groups can occupy equatorial-like positions in this conformation.

4.62   The diaxial conformation of *cis*-1,3-dimethylcyclohexane is 23 kJ mole^{-1} less stable than the diequatorial conformation. Why is this value larger than twice the steric strain of a methyl group?

Answer: The steric strain results from repulsion between an axial group and the axial hydrogen atoms at C-3 and C-5. Repulsion between an axial group and larger atoms at the C-3 and C-5 positions must be substantially larger. In the diaxial conformation of *cis*-1,3-dimethylcyclohexane, that repulsion is between two methyl groups.

4.63   The diaxial conformation of *cis*-l-chloro-3-methylcyclohexane is 16 kJ mole^{-1} less stable than the diequatorial conformation. Why is this value larger than the sum of the steric strains of a chlorine atom and a methyl group?

Answer: The steric strains all due to repulsion between an axial group and the axial hydrogen atoms at C-3 and C-5. Repulsion between an axial methyl group and a larger atom at C-1, such as chlorine, is substantially larger.

Bicyclic compounds

4.64   An isomerization equilibrium between *cis*-decalin and *trans*-decalin can be established by heating the mixture to about 300 °C in the presence of a palladium catalyst. The *trans* isomer predominates. Why is the *trans* isomer more stable than the *cis* isomer?

Answer: The ring junction in *cis*-decalin is equatorial axial, so there are always unfavorable 1,3-diaxial interactions; but, in *trans*-decalin, the ring junction is diequatorial, and there are fewer unfavorable steric interactions.

Steroids

4.65   Examine the structure of an A/B (*trans*) steroid skeleton and determine whether each of the following is in an equatorial or axial location.

(a) a 2α hydroxyl group   (b) a 3α chlorine atom   (c) a 6α amino (—NH₂) group
(d) an 11β bromine atom   (e) a 12β cyano group

Answers: (a), (c), and (e) have their groups in equatorial positions.

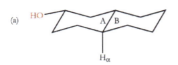

2α hydroxyl group

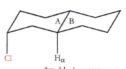

3α chlorine atom

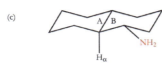

6α amino group

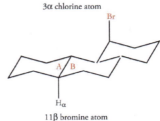

11β bromine atom

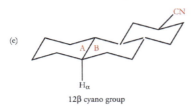

12β cyano group

4.66   Examine the structure of an A/B (*cis*) steroid skeleton and determine whether each of the following is in an equatorial or axial location.

(a) a 1β hydroxyl group   (b) a 3α chlorine atom   (c) a 6α amino (—NH₂) group
(d) an 11α bromine atom   (e) a 12α cyano group

Answers: (b) and (d) have their groups in equatorial positions.

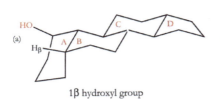

1β hydroxyl group

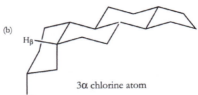

3α chlorine atom

 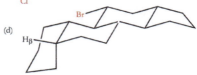

6β amino group

11α bromine atom

12α cyano group

ALKENES
STRUCTURES AND PROPERTIES

KEYS TO THE CHAPTER

5.1 Alkenes

A double bond in an alkene consists of one sigma bond and one pi bond. A triple bond in an alkyne consists of one sigma bond and two pi bonds.

On the average, the electrons in π bonds are more reactive than those in σ bonds—the subject of Chapter 6. Virtually all of the reactions of alkenes and alkynes (Chapter 7) involve the reactivity of π bonds. Two or more double bonds separated by one carbon–carbon single bond constitute conjugated double bonds. A compound may contain one set or a whole series of conjugated double bonds. Double bonds separated by more than one carbon–carbon single bond are not conjugated. The chemistry of conjugated double bonds is a more specialized subject than the reactions of alkenes, and it is discussed in Chapter 11.

5.2 Structure and Bonding of Alkenes

Functional groups are structural features of organic compounds other than carbon–carbon single bonds. Alkenes consist of a sigma bonded framework, which for the most part we can ignore, and two electrons in a π bond that results from the side-by-side overlap of two 2p orbitals. As a consequence of the π bond, the two carbon atoms and the four atoms bonded to these atoms are coplanar.

The concept of the % s character of the hybrid orbitals of an alkene is the first of many times throughout the study of organic chemistry that we will use this terminology. The % s character affects the bond energy of a σ bond. In order to apply the concept, we have to be careful to understand what is required in a chemical reaction—for example, does the process occur by homolytic or heterolytic cleavage of the bond? In the case of a bond dissociation, homolytic cleavage occurs, and each atom retains one electron in the bond. If the electrons are held more tightly by an atom as a consequence of its larger % s character, then it will take more energy to break the bond. That is why the bond dissociation energy increases with increased % s character.

Bond distances also depend on % s character. If electrons are held more tightly by the carbon atom as a result of its hybridization, then the bond length must be shorter.

Time and time again, we will find that there are advantages not only to classifying compounds by their functional groups but also to fine tuning the classification by subclasses. Fortunately, the terminology of **degree of substitution** is straightforward. Simply count the number of alkyl groups attached to the double-bonded carbon atoms.

5.3 Unsaturation Number

Before we can determine what functional groups might be part of a structure, we have to limit the possibilities based on the molecular formula. When we look at molecular formulas, we have to recognize likely possibilities for functional groups.

The presence of double or triple bonds in a structure is indicated by the **degree of unsaturation**. Each multiple bond diminishes the maximum number of hydrogen atoms by two. Each pair of "missing" hydrogen atoms can correspond to a pi bond. However, it may also signal the presence of a ring. Based on this one criterion, we can't say which structural feature is present. But we have limited the possibilities. The effect of other atoms on the unsaturation number is summed up as follows:

1.   Oxygen and sulfur have no effect.
2.   Count halogen atoms as hydrogen atoms.
3.   Add a hydrogen.

5.4 Geometric Isomerism

As a consequence of the restricted rotation about a carbon–carbon double bond, alkenes can exist as geometric isomers if two different groups are bonded to each carbon atom of the double bond. If either carbon atom has two identical groups, then only one compound is possible.

Cycloalkenes usually have a *cis* configuration. Bridging a *trans* configuration in a cycloalkene requires at least an eight-membered ring due to geometric constraints.

5.5 The *E,Z* Designation of Geometric Isomers

To decide whether a geometric isomer is *E* or *Z* we apply a set of **sequence rules**. The most important criterion for deciding on the priority of the groups is atomic number. Furthermore, just one atom of high atomic number has a higher priority than any number of atoms with lower atomic numbers. For example, —CH_2Br has a higher priority than —CCl_3 because bromine has a higher atomic number than chlorine. Another criterion is the "point of first difference", which means keep going until something important is found in a chain of atoms that makes it more important than a second group of atoms. Thus, the group —$CH_2CH_2CHFCH_2CH_3$ has a higher priority than —$CH_2CH_2CH_2CHBrCH_3$ because the fluorine atom is found sooner on the chain (at C-3) than the bromine atom (which is at C-4).

5.6 Nomenclature of Alkenes

We name alkenes by a set of rules that parallel those used for alkanes, with the added complication of the double bond and possibility of geometric isomers. Some of the rules are the same, and others are closely related to those used for alkanes. The double bond takes priority in numbering the longest chain, which must contain the double bond. Only the first number of the two carbon atoms in the double bond is used to name a compound. In cycloalkenes, one of the two carbon atoms of the double bond is given the number 1 and the numbering of the carbon atoms then proceeds through the second carbon atom of the double bond. The name does not include a number locating the double bond, but numbers are required for the location of substituents such as alkyl groups.

5.7 Physical Properties of Alkenes

As we have observed for alkanes, trends in physical properties are related to the size and structure of the compounds studied. Boiling points for a homologous series of alkenes increase with molecular weight, reflecting the increase in strength of London forces. The identity and geometry of the groups attached to the double-bonded carbon atoms determine whether an alkene is polar or nonpolar. If one isomer is polar and the other is nonpolar, the polar compound has a higher boiling point.

5.8 Oxidation of Alkenes

The heats of combustion of alkenes allow us to compare the relative stabilities of isomeric compounds. For isomers, the same number of moles of carbon dioxide and water is formed. Thus, a comparison of the heats of combustion indicates the difference in the enthalpy content of the isomers. Three generalizations can be made based on the data depicted in Figure 5.4 in the text. These are:

1.   Branched isomers are more stable than unbranched ones, so they have smaller heats of combustion.
2.   More highly substituted alkenes are more stable, so they have smaller heats of combustion.
3.   Alkenes with the *E* configuration are more stable than alkenes with the *Z* configuration, so they have smaller heats of combustion.

The increased stability of alkenes with increased substitution results from the release of electron density from the sp^3-hybridized alkyl groups toward the sp^2-hybridized atoms of the carbon–carbon double bond. The electron donating capacity of alkyl groups toward sp^2-hybridized centers is a common feature that explains many chemical reactions that we will encounter in later chapters.

5.9 Reduction of Alkenes

The reaction of an alkene with hydrogen to give an alkane, called hydrogenation, is a reduction reaction. The order of reactivity of alkenes decreases with increased substitution of the double bond. Thus, the reaction shows some regioselectivity, which means that one double bond in a compound with two or more double bonds can often be reduced in preference over another double bond. This regioselectivity, which is the tendency of a reaction to generate one isomer preferentially over another, is another concept that we will encounter many times as we proceed. Transition metal catalysts such as platinum, palladium, Adams catalyst (PtO_2), and a special form of nickel called Raney nickel can be used. Although the hydrogenation reaction is usually carried out under heterogeneous conditions, one catalyst known as the Wilkinson catalyst is used for hydrogenation of alkenes under homogeneous conditions. Neither heterogeneous nor homogeneous catalysts reduce functional groups such as the carbonyl group of ketones, carboxylic acids, or esters under the relatively mild conditions required to hydrogenate a carbon–carbon double bond.

5.10 Mechanism of Catalytic Hydrogenation

Heterogeneous catalytic hydrogenation occurs on the surface of the metal and transfers the two hydrogen atoms to the carbon atoms by a *syn* addition process. For many alkenes, there is no difference in the two faces of the double bond, so there is no stereoselectivity in the reaction. However, the environment near the double bond can sometimes make one face less accessible to the transfer of hydrogen. When hydrogen adds to one face, the atoms bonded to the carbon atoms of the double bond are "pushed" to the opposite side. Because hydrogen adds from the sterically less hindered side, the groups are forced into a more sterically hindered environment.

5.11 Heats of Hydrogenation

The measurement of the heats of hydrogenation ($\Delta H° < 0$) of isomeric alkenes is used to determine their relative stabilities. The values of $\Delta H°_{hydrogenation}$ are all negative, and the reference point is the saturated hydrocarbon. The size of the term indicates how stable the alkene is relative to its isomers. Heats of hydrogenation can be used to compare the relative stabilities of isomeric alkenes only if the same alkane results from both compounds. For example, isomers with different degrees of branching cannot be directly compared because they don't produce the same saturated hydrocarbon when hydrogenated. We can still reach some basic conclusions about alkene stability if the isomers are reasonably similar.

1. Branched isomers are more stable than unbranched isomers.
2. Alkene stability increases with increasing substitution.
3. An alkene with the *E* configuration is more stable than its *Z* isomer.

SUMMARY OF REACTIONS

1. Heterogeneous Catalytic Hydrogenation

(a)

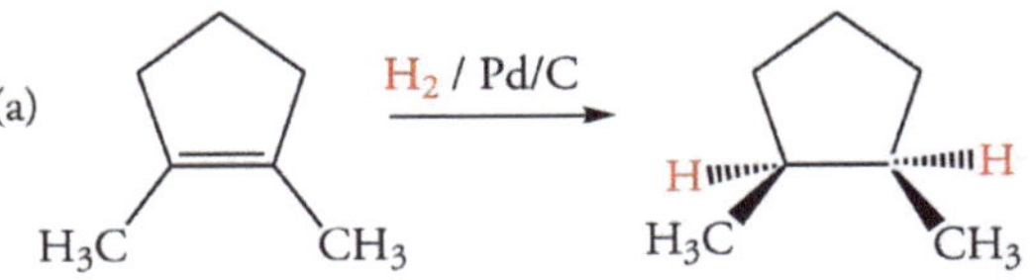

H_2 / Pd/C

(b)

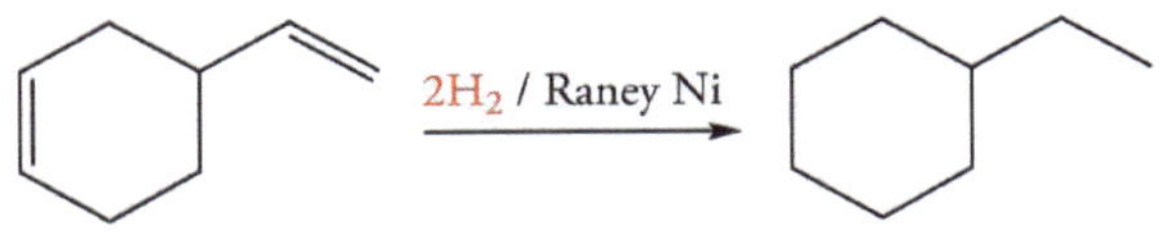

$2H_2$ / Raney Ni

(c)

H_2 / Pd/C
1 atm

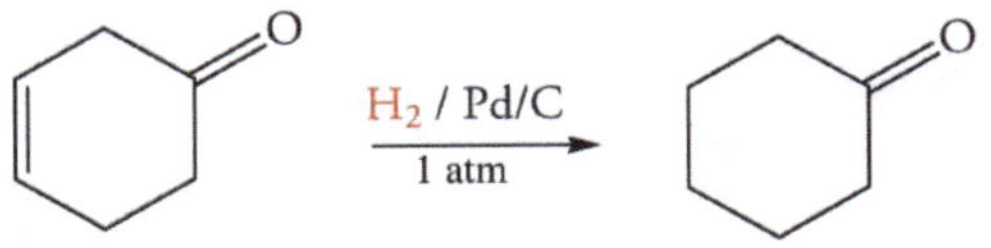

(d)

H_2 / PtO$_2$

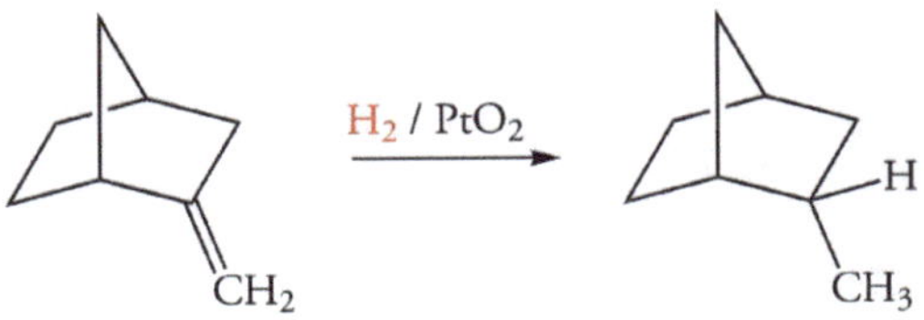

2. Homogeneous Catalytic Hydrogenation

H_2
$[(C_6H_5)_3P]_3RhClH$

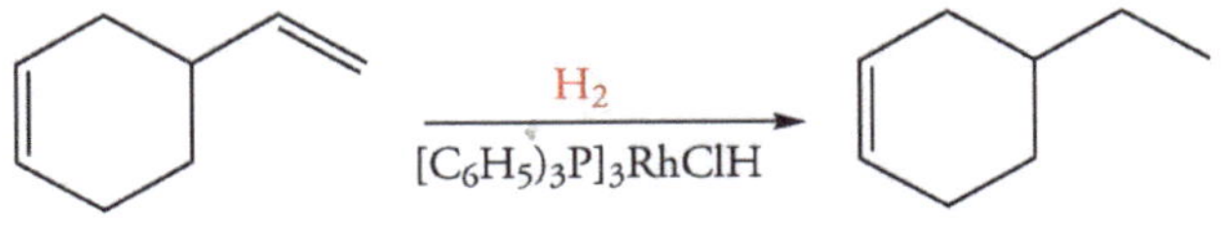

 **End of Chapter Exercises**

Molecular Formulas

5.1   What is the molecular formula for a compound with each of the following structural features?
          (a)   six carbon atoms and one double bond
          (b)   five carbon atoms and two double bonds
          (c)   seven carbon atoms, a ring, and one double bond

Answers: (a) C_6H_{12} (b) C_5H_8 (c) C_7H_{12}

5.2   What is the molecular formula for a compound with each of the following structural features?
          (a)   four carbon atoms and two double bonds
          (b)   ten carbon atoms and two rings
          (c) ten carbon atoms, two rings, and five double bonds

Answers: (a) C_4H_6 (b) $C_{10}H_{18}$ (c) $C_{10}H_8$

5.3   Write the molecular formula for each of the following compounds.

Answers: (a) C_8H_{16} (b) C_6H_{12}
          (c) $C_{10}H_{16}$ (d) C_8H_{14}

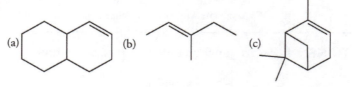

(a)   (b)   (c)   (d)

5.4   Write the molecular formula for each of the following compounds.

Answers: (a) C_8H_{12} (b) $C_{10}H_{14}$
          (c) C_7H_{10} (d) C_7H_{14}

(a)   (b)   (c)   (d)

Classification of Alkenes

5.5   Classify each double bond in the alkenes in Exercise 5.3 by its substitution pattern.
Answers: (a) trisubstituted (b) trisubstituted (c) disubstituted   (d) trisubstituted

5.6   Classify each double bond in the alkenes in Exercise 5.4 by its substitution pattern.
Answers: (a) di- and trisubstituted   (b) di- and trisubstituted   (c) disubstituted   (d) trisubstituted

5.7   Indicate the degree of substitution of the double bond in each of the following compounds.

Answer: (a) trisubstituted   (a)   Cholesterol, a steroid

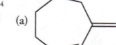
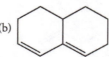

(c)   Saffrole, a carcinogen found in sassafras root

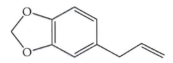

Answer: (c) monosubstituted

(d)   Tamoxifen, a drug used in the treatment of breast cancer

$OCH_2CH_2N(CH_3)_2$

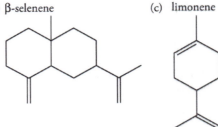

Answer: (d) tetrasubstituted

5.8   Indicate the degree of substitution of all double bonds in each of the following compounds, polyenes found in natural oils.

Answers: (a) di-, tri, and trisubstituted
         (b) both disubstituted
         (c) di- and trisubstituted

(a) zingiberene   (b) β-selenene   (c) limonene

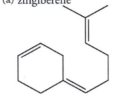

Unsaturation Number

5.9   Calculate the unsaturation number for each of the following compounds.
       (a) camphor, $C_{10}H_{16}O$ (b) nicotine, $C_{10}H_{14}N_2$
       (c) vitamin B6, $C_8H_9NO_2$ (d) hexachlorophene, $C_{13}H_6O_2Cl_6$

Answers: (a) 3   (b) 5   (c) 5   (d) 8

5.10   Calculate the unsaturation number for each of the following compounds.
        (a) β-carotene, $C_{40}H_{56}$ (b) amphetamine, $C_9H_{13}N$ (c) DDT, $C_{11}H_9Cl_5$ (d) aspirin, $C_9H_8O_4$

Answers: (a) 13   (b) 5   (c) 5   (d) 6

5.11   Calculate the unsaturation number for each of the following compounds.
        (a) vitamin A, $C_{20}H_{30}O$ (b) sucrose, $C_{12}H_{22}O_{11}$ (c) vitamin B2, $C_{17}H_{20}N_4O_6$ (d) saccharin, $C_7H_5NO_3S$

Answers: (a) 2   (b) 2   (c) 10   (d) 6

5.12   Calculate the unsaturation number for each of the following compounds.
        (a) L-dopa, $C_9H_{11}NO_4$ (b) prontosil, $C_{12}H_{13}N_5O_2S$
        (c) testosterone, $C_{19}H_{28}O_2$ (d) phenobarbital, $C_{12}H_{12}N_2O_3$

Answers: (a) 5   (b) 9   (c) 6   (d) 8

Geometric Isomers

5.13 Which of the following molecules can exist as *cis* and *trans* isomers?
(a) $CH_3CH{=}CHBr$ (b) $CH_2{=}CHCH_2Br$ (c) $CH_3CH{=}CHCH_2Cl$ (d) $(CH_3)_2C{=}CHCH_3$

Answer: only (a) and (c)

5.14 Which of the following molecules can exist as *cis* and *trans* isomers?
(a) $CH_3CH{=}CBr_2$ (b) $CH_2{=}CHCHBr_2$ (c) $CH_3CH{=}CHCHCl_2$ (d) $CH_3CH_2CH{=}C(CH_3)_2$

Answer: only (c)

5.15 Which of the following molecules can exist as *cis* and *trans* isomers?
(a) 1-hexene (b) 3-heptene (c) 4-methyl-2-pentene (d) 2-methyl-2-butene

Answer: only (b) and (c)

5.16 Which of the following molecules can exist as *cis* and *trans* isomers?
(a) 3-methyl-l-hexene (b) 3-ethyl-3-heptene (c) 2-methyl-2-pentene (d) 3-methyl-2-pentene

Answer: only (d)

E,Z System of Nomenclature

5.17 Select the group with the highest priority in each of the following sets.
(a) $—CH(CH_3)_2$ $—CHClCH_3$, $—CH_2CH_2Br$

Answer: $—CH(CH_3)_2$

(b) $—CH_2CH{=}CH_2$, $—CH_2CH(CH_3)_2$, $—CH_2C{\equiv}CH$

Answer: $—CH_2C{\equiv}CH$

(c) $—OCH_3$, $—N(CH_3)_2$, $—C(CH_3)_3$

Answer: $—OCH_3$

5.18 Select the group with the highest priority in each of the following sets.

Answers:

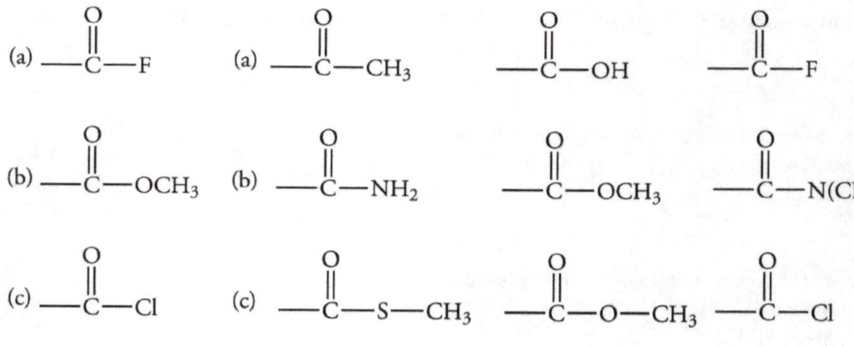

5.19   Assign the *E* or *Z* configuration to each of the following antihistamines.

Answer: (a) *Z* (a) pyrrobutamine

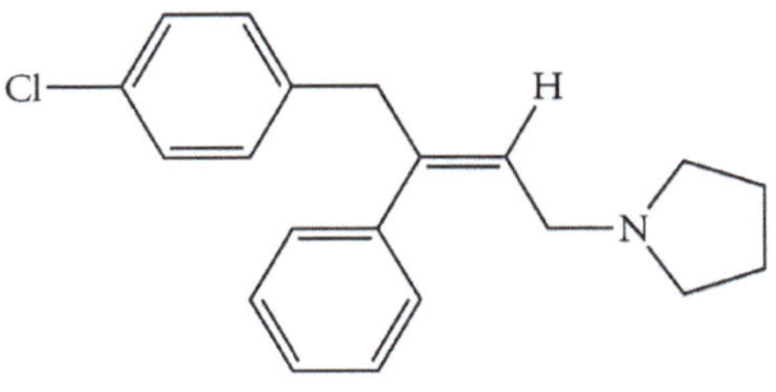

Answer: (b) *Z* (b) triprolidine

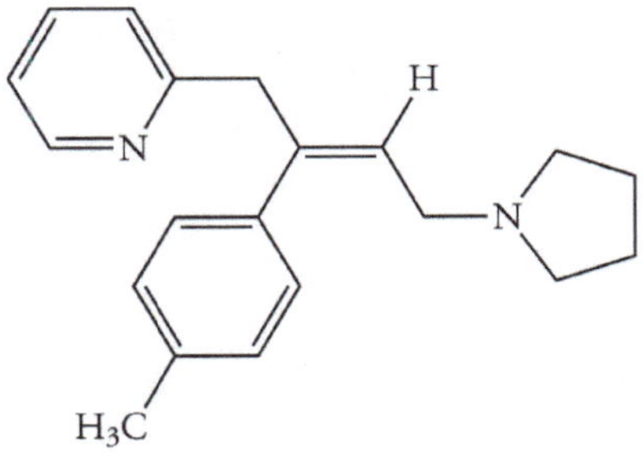

Answer: (c) *Z* (c) chloroprothixene

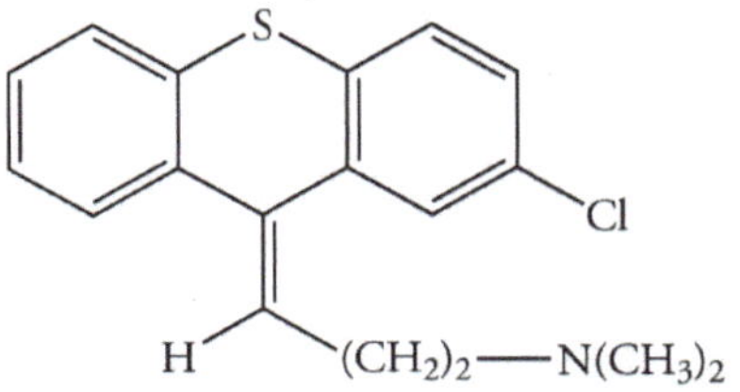

5.20   Assign the *E* or *Z* configuration to each of the following hormone antagonists used to control cancer.

(a)   Chlomiphene

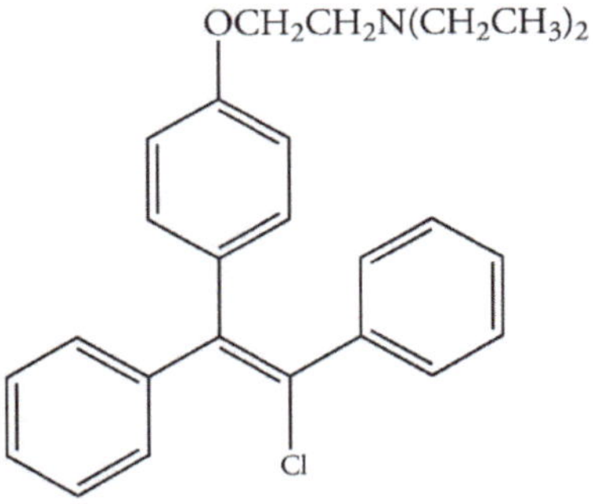

Answer: (a) *E*

(b)   Tamoxifen

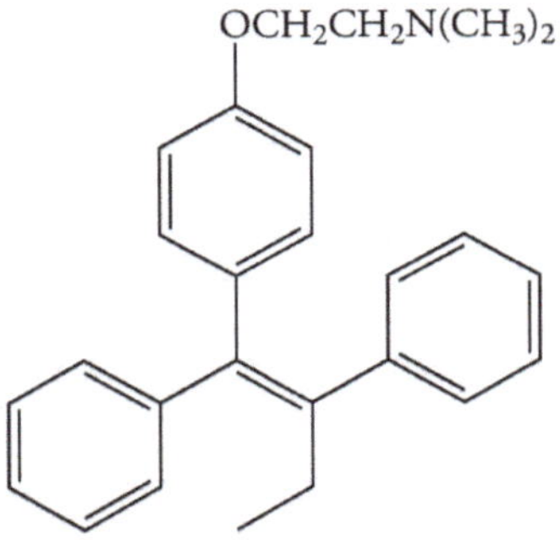

Answer: (b) *Z*

(c)   Nitromifene

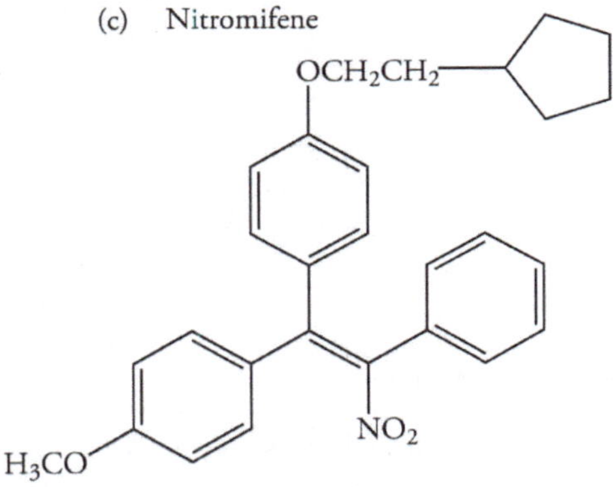

Answer: (c) *E*

5.21   Draw the structural formula for each of the following pheromones with the indicated configuration.
(a) sex pheromone of Mediterranean fruit fly, *E* isomer

$$CH_3CH_2CH{=}CH(CH_2)_4CH_2OH$$

Answer: (a) *E*

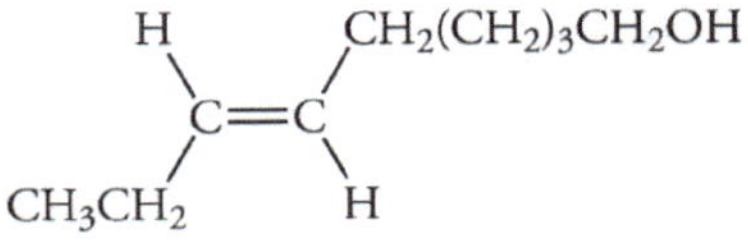

(b) sex pheromone of honey bee, *E* isomer

$$CH_3CO(CH_2)_4CH_2CH{=}CHCO_2H$$

Answer: *E*

(b)

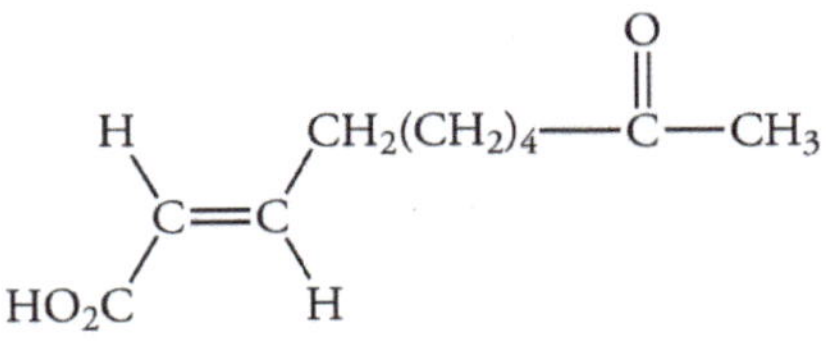

(c) defense pheromone of termite, *E* isomer

$$CH_3(CH_2)_{12}CH=CHNO_2$$

Answer: (c) *E*

5.22   Assign the configuration at all double bonds where geometrical isomerism is possible in each of the following sex pheromones.

Answer: (a) left to right: *Z*, *E*   (a) European vine moth

Answer: (b) left to right: *E*, *Z*   (b) pink bollworm moth

Answer: (c) *Z*   (c) Japanese beetle

Nomenclature of Alkenes

5.23   Name each of the following compounds.

Answers: (a) 2-methyl-1-propene   (b) 2,3-dimethyl-2-butene
        (c) 2-methyl-2-butene   (d) (*E*)-2,3-dichloro-2-pentene

5.24   Name each of the following compounds.

Answers: (a) (*E*)-1,2-dibromo-1-propene   (b) 1-bromo-3-methyl-2-butene
        (c) 3-chloro-2-methyl-2-pentene   (d) (*Z*)-3,4-difluoro-3-hexene

5.25   Name each of the following compounds.

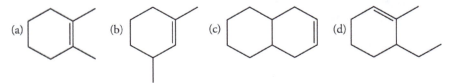

(a)   (b)   (c)   (d)

Answers: (a) 1,2-dimethylcyclohexene   (b) 1,3-dimethylcylclohexene
            (c) cyclodecene   (d) 6-ethyl-1-methylcyclohexene

5.26   Name each of the following compounds.

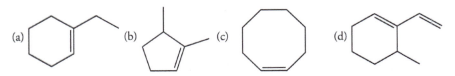

(a)   (b)   (c)   (d)

Answers: (a) 1-ethylcyclohexene   (b) 1,5-dimethylcyclopentene
            (c) cyclooctene   (d) 1,6-diethylcyclohexene

5.27   Draw a structural formula for each of the following compounds.
        (a) 2-methyl-2-pentene   (b) 1-hexene
        (c) (Z)-2-methyl-3-hexene (d) (E)-5-methyl-2-hexene

Answers:

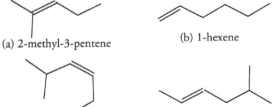

(a) 2-methyl-3-pentene   (b) 1-hexene

(c) Z-2-methyl-3-hexene   (d) E-5-methyl-2-hexene

5.28   Draw a structural formula for each of the following compounds.
        (a) (E)-1-chloropropene   (b) (Z)-2,3-dichloro-2-butene
        (c) 3-chloropropene   (d) 4-chloro-2,4-dimethyl-2-hexene

Answers:

$$\underset{CH_3}{\overset{H}{}}C=C\underset{H}{\overset{Cl}{}}$$

$$\underset{CH_3}{\overset{Cl}{}}C=C\underset{CH_3}{\overset{Cl}{}}$$

(a) E-1-chloropropene   (b) Z-2,3-dichloro-2-butene

$$\underset{H}{\overset{H}{}}C=C\underset{H}{\overset{CH_2Cl}{}}$$

(c) 3-chloropropene   (d) 4-chloro-2,4-dimethyl-2-hexene

5.29   Draw a structural formula for each of the following compounds.
        (a) cyclohexene   (b) 1-methylcyclopentene
        (c) 1,2-dibromocyclohexene   (d) 4,4-dimethylcyclohexene

Answers:

(a) cyclohexene

(b) 1-methylcyclopentene

(c) 1,2-dibromocyclohexene

(d) 4,4-dimethylcyclohexene

5.30   Draw a structural formula for each of the following compounds.
        (a) cyclopentene   (b) 3-methylcyclohexene
        (c) 1,3-dibromocyclopentene   (d) 3,3-dichlorocyclopentene

Answers:

(a) cyclopentene

(b) 3-methylcyclohexene

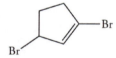

(c) 1,3-dibromocyclopentene

(d) 3,3-dichlorocyclopentene

Physical Properties

5.31   The dipole moment of hexane is 0.09 D, but the dipole moment of 1-hexene is 0.4 D. Explain the reason for the difference.
Answer: The carbon–carbon double bond of 1-hexene is not symmetrically substituted, and the single alkyl group is electron donating to the sp^2-hybridized carbon atom.

5.32   Which isomer of 2-butene has the larger dipole moment?
Answer: The bond moments of the methyl groups bonded to the sp^2-hybridized carbon atoms of *cis*-2-butene reinforce one another and there is a net dipole moment. In the *trans* isomer, the two bond moments are opposed; therefore, they cancel, and there is no dipole moment.

5.33   The dipole moment of 2-methylpropene is 0.5 D, but the dipole moment of 1-butene is 0.3 D. Explain why these values differ.
Answer: The two bond moments of the methyl groups bonded to the sp^2-hybridized carbon atom are additive in 2-methylpropene and are larger than the bond moment of the single ethyl group of 1-butene.

5.34   The dipole moment of chloroethene is 1.4 D. Predict the dipole moment of *cis*-1,2-dichloroethene.
Answer: The dipole moment should be larger than 1.4 D but less than 2.8 D because there will be some partial cancellation of the component of the dipole moment along the carbon–carbon bond axis. There will be reinforcement of the component of the dipole moment perpendicular to the bond axis.

5.35   *cis*-1-Bromopropene has a higher boiling point than *cis*-1-chloropropene but has the smaller dipole moment. Explain why.

Answer: The bromine atom is more polarizable and the resulting London forces are stronger than those resulting from the chlorine atom.

5.36   The boiling points of 1-hexene and 2,3-dimethyl-2-butene are 63.5 and 73 °C, respectively. Suggest a reason for this difference.

Answer: 2,3-Dimethyl-2-butene has a small, compact shape which allows close approach of molecules and results in increased London forces. The conformational flexibility of 1-hexene results in a larger effective volume and, therefore, a higher boiling point.

Heats of Combustion of Alkenes

5.37   The difference between the heats of combustion of *cis-* and *trans*-2-butenes is about 4.2 kJ mole^{-1}, but the difference between those of *cis-* and *trans*-4,4-dimethyl-2-pentenes is about 16 kJ mole^{-1}. Explain why these two values differ significantly.

Answer: There is substantial van der Waals repulsion of a methyl group and a *tert*-butyl group in *cis*-4,4-dimethyl-2-pentene which makes this isomer much less stable than the *trans* isomer. The van der Waals repulsion between the two methyl groups in *cis*-2-butene is significantly smaller.

5.38   The difference between the heats of combustion of *cis-* and *trans*-2,2,5,5-tetramethyl-3-hexenes is about 40 kJ mole^{-1}. Explain this very large difference.

Answer: There is a very large van der Waals repulsion of two *tert*-butyl groups in the *cis* isomer.

5.39   Which of the following two isomers should have the larger heat of combustion? Explain why.

Answer: Compound II has the larger angle strain resulting from the carbon–carbon double bond in the four-membered ring. Compound II is less stable and releases more energy in a combustion reaction.

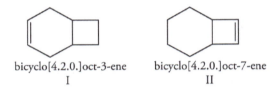

bicyclo[4.2.0]oct-3-ene   bicyclo[4.2.0]oct-7-ene
I   II

5.40   Although 1-methylcyclopropene is a trisubstituted alkene and methenecyclopropane is a disubstituted alkene, the heat of combustion of 1-methylcyclopropene is larger by about 42 kJ mole^{-1}. Explain why.

Answer: Both sp^2-hybridized carbon atoms of 1-methylcyclopropene are part of the three-membered ring, giving rise to larger angle strain than the single sp^2-hybridized carbon atom of methenecyclopropane.

methylenecyclopropane   1-methylcyclopropene

5.41   Arrange the following compounds in order of increasing heats of combustion: 3-methyl-1-butene, 2-methyl-1-butene, 2-methyl-2-butene.

Answer: 2-methyl-2-butene < 2-methyl-1-butene < 3-methyl-1-butene, which is the order of decreasing degree of alkyl substitution of the double bond.

5.42   Arrange the following compounds in order of increasing heats of combustion.

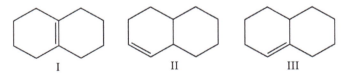

I   II   III

Answer: I < Ill < II, which is the order of decreasing degree of alkyl substitution of the double bond.

Hydrogenation of Alkenes

5.43   How many moles of hydrogen gas will react at atmospheric pressure with each of the following compounds?
(a) 1,4-cyclooctadiene   (b) 4-vinylcyclohexene
(c) 2,4-dimethyl-1,4-pentadiene   (d) 2-methyl-1,3-cyclohexadiene

Answers: (a) 2   (b) 2   (c) 2   (d) 2

5.44   How many moles of hydrogen gas will react at atmospheric pressure with each of the following compounds?

Answer: (a) 6   (a) Vitamin A, contained in freshwater fish

Answer: (b) 3   (b) zingiberene, found in oil of ginger

(c) ergosterol, a form of vitamin D

Answer: (c) 3

5.45   Oil of marjoram contains α-terpinene, whose molecular formula is $C_{10}H_{16}$. Hydrogenation using the Adams catalyst yields $C_{10}H_{20}$. How many double bonds and how many rings do the α-terpinene contain?

Answer: There are two double bonds because two moles of H_2 are added to the molecular formula. There is one ring because the unsaturation number of the product is 1.

5.46   The wax found on apples contains α-farnesene, whose molecular formula is $C_{15}H_{26}$. Hydrogenation using palladium on charcoal yields $C_{15}H_{32}$. How many double bonds and how many rings do the α-farnesene contain?

Answer: There are three double bonds because three moles of H_2 are added to the molecular formula. There are no rings because the unsaturation number of the product is 0.

Hydrogenation of Alkenes

5.47   Consider each compound of the following pairs of isomeric hydrocarbons and determine whether or not there should be a substantial difference in their heats of hydrogenation. Explain why. Indicate the compound with the higher heat of hydrogenation where possible.

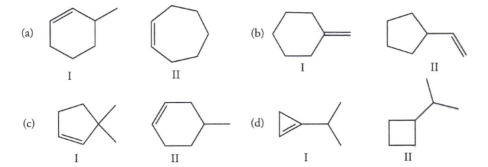

Answers: (a) There is no difference in the heats of hydrogenation because the degree of substitution of the double bonds is identical.
(b) The heat of hydrogenation of methenecyclohexane should be smaller because it is disubstituted whereas the isomeric vinyl cyclopentane is monosubstituted.
(c) There is no difference in the heats of hydrogenation because the degree of substitution of the double bonds is identical.
(d) The cyclopropene compound should have the larger heat of hydrogenation because the strain of the double bond in the smaller ring makes the compound less stable.

5.48   There are three isomeric methylcyclopentenes. Which compound has the smallest heat of hydrogenation?

Answer: 1-Methylcyclopentene has the smallest heat of hydrogenation because it has a trisubstituted double bond. The 3-methyl and 4-methyl compounds have disubstituted double bonds.

5.49   There are three isomeric methylcyclopentenes. Which compound has the smallest heat of hydrogenation?

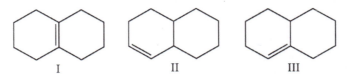

Answer: I < Ill < II, which is the order of decreasing degree of substitution of the double bond.

5.50   The standard heat of hydrogenation of of bicyclo[4.2.0]oct-7-ene is larger than that of the isomeric bicyclo[4.2.0]oct-3-ene. Based on this information, which compound is more stable? What feature of the structures of the two compounds is responsible for this difference in stability?

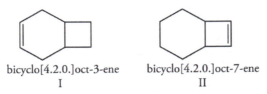

bicyclo[4.2.0.]oct-3-ene   bicyclo[4.2.0.]oct-7-ene
          I   II

Answer: The compound with a double bond in a six-membered ring is more stable. The compound with a double bond in the four-membered ring is more strained and of higher energy than the isomer, so hydrogenation of the cyclobutene will release a larger amount of energy.

5.51   Although ethylidenecyclohexane and 1-ethylcyclohexene are both trisubstituted alkenes, the latter compound predominates in an equilibrium reaction. Based on this information, which compound has the larger heat of hydrogenation?

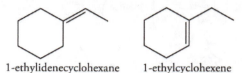

1-ethylidenecyclohexane   1-ethylcyclohexene

Answer: Both compounds yield the same saturated compound when hydrogenated. Since 1-ethylcyclohexene predominates in an equilibrium mixture, it is more stable. Thus, ethylidenecyclohexane has a larger heat of hydrogenation.

5.52   The data given in Exercise 5.51 support the observation that isomers with double bonds within rings (endocyclic) are more stable than isomers with double bonds between a carbon in the ring and a carbon outside the ring (exocyclic). Explain why the heats of hydrogenation of 1-methylcyclohexane and methenecyclohexane, which are −107 and −116 kJ mole⁻¹, respectively, cannot be used to support this generalization.

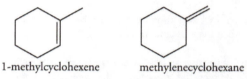

1-methylcyclohexene   methylenecyclohexane

Answer: The degree of substitution is not the same, so they cannot be directly compared.

5.53   The heat of hydrogenation of the bicyclic hydrocarbon shown below is approximately 270 kJ mole⁻¹. Why is this value so much larger than those listed in Table 5.3 for alkenes?

Answer: The double bond is present in "two" strained cyclobutene rings, which increases the energy of the compound and makes it un stable.

5.54   Explain why the $\Delta H°_{reaction}$ for the following two isomerization reactions are negative. Why is the $\Delta H°_{reaction}$ for the second reaction more negative than for the first reaction?

$\Delta H°_{rxn}$ = −3.7 kJ mole⁻¹

$\Delta H°_{rxn}$ = −16.2 kJ mole⁻¹

Answer: There is an increase in the degree of substitution from reactant to product which makes both reactions favorable. However, the four-membered ring compound has more strain in the product as the result of having two sp²-hybridized carbon atoms in the ring. Thus, the reaction is less favorable.

5.55   The difference between the heats of hydrogenation of (E)- and (Z)-4,4-dimethyl-2-pentenes is approximately 14.6 kJ mole⁻¹. Compare this difference with the difference between the heats of hydrogenation of (E)- and (Z)-2-butenes. Why do the two values differ?

Answer: The van der Waals repulsion between the methyl and *tert*-butyl groups in (Z)-4,4-dimethyl-2-pentene is much greater than between two methyl groups is (Z)-2-butene, so the difference between the (E) and (Z) isomers is greater also.

5.56   The heats of hydrogenation of the geometric isomers of 2,2,5,5-tetramethyl-3-hexene differ by 39 kJ mole⁻¹. Explain why this difference is so large compared to other geometric isomers.

Answer: There is a very large van der Waals repulsion of two *tert*-butyl groups in the *cis* isomer which makes the compound far less stable.

Stereochemistry and Stereoselectivity of Hydrogenation

5.57 Write the structure of the product obtained by hydrogenating the following diester using PtO_2 and hydrogen gas at atmospheric pressure.

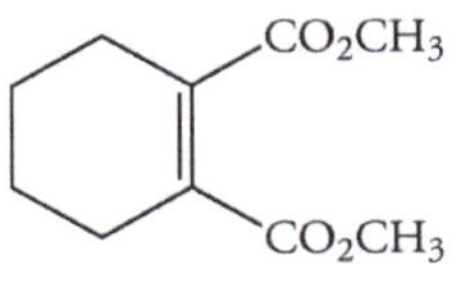

Answer:

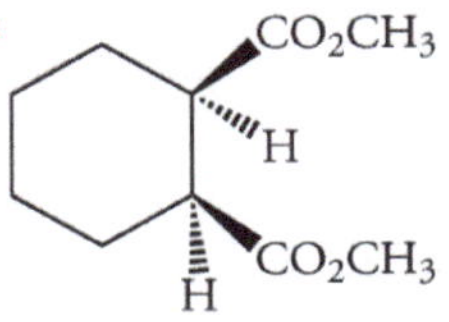

5.58 Write the product obtained by the catalytic hydrogenation of the sex pheromone of the European vine moth at atmospheric pressure using PtO_2 and hydrogen gas.

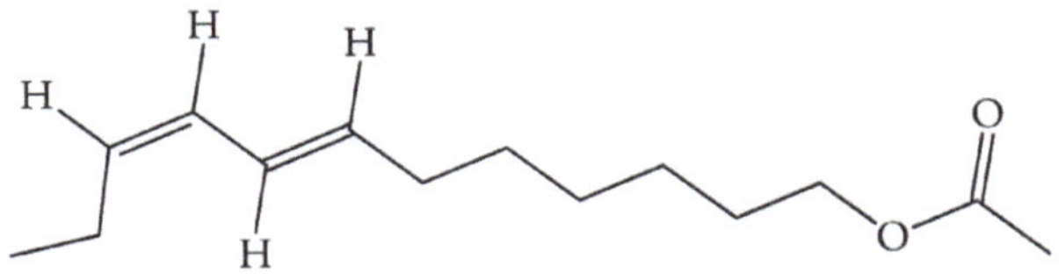

Answer: $CH_3(CH_2)_{10}CH_2—O—C(=O)—CH_3$

5.59 Deuterium gas can be used to deuterate compounds using the Adams catalyst. The reaction proceeds by the same mechanism as for hydrogenation. Write the product of the reaction of 1-ethyl-2-methylcyclohexene with D_2.

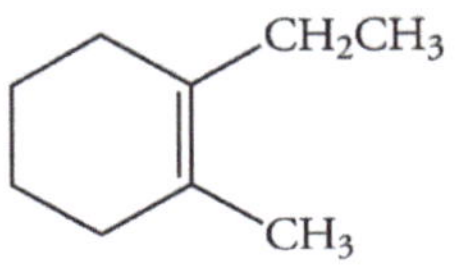

Answer:

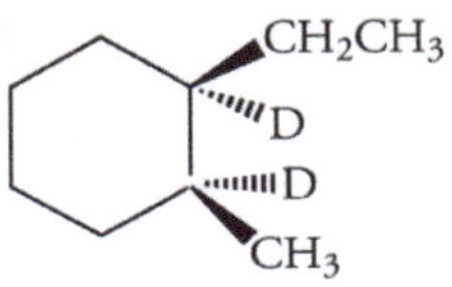

5.60 Which of the two isomeric caranes is the major product of the hydrogenation of 3-carene using the Adams catalyst?

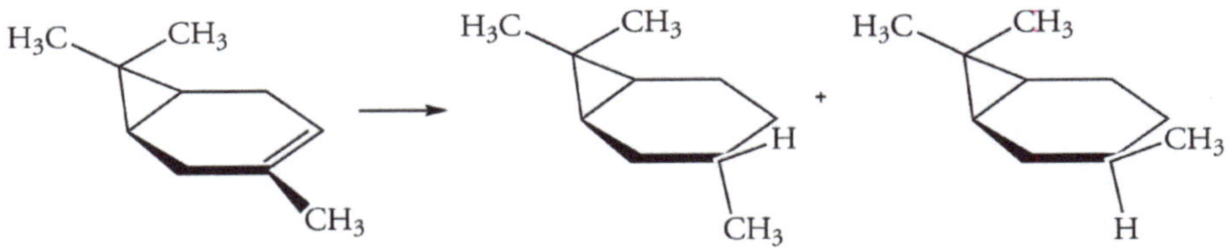

3-carene

Answer: The second product, with the methyl group and the cyclopropane ring on the same side of the six-membered ring, is the major product because hydrogen is added from the sterically less hindered side, which is the face opposite the cyclopropane ring.

5.61 Which of the double bonds of limonene is hydrogenated at the faster rate? Comment on the likelihood that selective hydrogenation may occur.

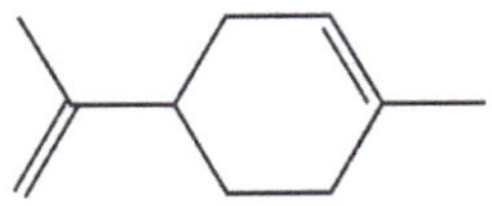

limonene

Answer: The disubstituted double bond will be hydrogenated somewhat faster than the trisubstituted double bond, but selective hydrogenation is not likely.

5.62   Explain why the hydrogenation of compound I occurs at a faster rate than the hydrogenation of compound II.

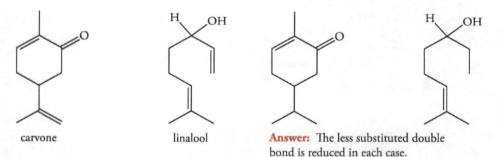

Answer: Approach of hydrogen from the "top" of compound I occurs much more easily than for compound II, where the *tert*-butyl group sterically hinders this face of the double bond.

5.63   Evaluate the degree of substitution of the double bonds of bisabolene and determine whether stereoselective reduction of a double bond is possible.

bisabolene

Answer: There are two trisubstituted double bonds and a tetrasubstituted double bond. All should react slowly in a hydrogenation reaction, and there should be little selectivity.

5.64   Write the structure of the product obtained by catalytic reduction of each of the following compounds using the Wilkinson catalyst and one molar equivalent of hydrogen gas.

carvone   linalool   **Answer:** The less substituted double
                                                bond is reduced in each case.

6 ALKENES ADDITION REACTIONS

6.1 Characteristics of Addition Reactions

The reactivity of alkenes depends on the properties of the π bond, whose electrons that are larger distances from carbon atoms than the electron in σ bonds. The π electrons are more open to attack by reagents than the σ bonds, which are located in a more protected environment between atoms. The chemistry of the π bond is most easily understood using Lewis acid–base conventions. The bonding electron pair in the pi bond can act as a Lewis base and is susceptible to attack by Lewis acids. Lewis acids are electrophiles. Electrophiles react with double bonds in addition reactions. The electrophile reacts with π electrons to form a bond to a carbon atom, and as a result, the π bond is broken.

Addition reactions of alkenes result in the incorporation of two atoms or groups of atoms on adjacent carbon atoms which share the double bond. Reagents such as HBr and H_2O add a hydrogen atom and a second species, which in these cases are bromine and the hydroxyl group, respectively. Reagents such as Br_2 can provide two equivalent groups, in which case one bromine atom adds to each of the two carbon atoms in the original π bond.

The addition reaction of a reagent X—Y to an alkene results in the destruction of one σ bond and one π bond, between two carbon atoms and between X and Y. However, two σ bonds form: one carbon atom bonds to X and the other carbon atom bonds to Y. Carbon–carbon π bonds are weaker than σ bonds. Thus, the addition reaction is exothermic because the net result is the replacement of a weak π bond by a stronger σ bond. All the common reagents we discussed in this chapter give addition products in exothermic reactions.

Different reactants add to the π bond in two different ways known as ***anti* addition** and ***syn* addition**. Thus, the geometry of the added groups in the final products gives information about the reaction pathway. This type of study is most easily done with cycloalkenes because the products are geometric isomers.

Although several types of reagents are discussed in this chapter, the most common theme is that of attack by electrophiles. As a consequence of the attack of an electrophile on the π electrons to form a bond to the carbon atom, the other carbon atom becomes electron deficient and has a positive charge. The subsequent chemistry of this carbocation is one of the common threads of addition reactions initiated by electrophiles.

6.2 Addition of Hydrogen Halides

Hydrogen halides react with alkenes to give alkyl halides. This process is called electrophilic addition because the addition of an electrophilic proton initiates the reaction. Addition of hydrogen halides is **regiospecific**. That is, not only does the hydrogen halide specifically react with the π bond, but also it gives only one of the two possible products. This result is summarized by **Markovnikov's rule**. *HX compounds add to double bonds so that the hydrogen atom bonds to the carbon atom of the double bond containing the largest number of directly bonded hydrogen atoms.* Therefore, we can easily predict the product of a reaction of hydrogen halides to a double bond.

6.3 Mechanistic Basis of Markovnikov's Rule

The proposed mechanism of electrophilic addition explains Markovnikov's rule. A proton (or other electrophlie species) adds to a π bond to give a carbocation. Thus, the direction of the addition—that is, its regiospecificity—depends on the stability of the two possible carbocations that could form. The hydrogen atom adds to the carbon atom of the double bond containing the largest number of directly bonded hydrogen atoms. This is the least substituted carbon atom. As a consequence, the carbocation formed must have the positive charge on the more substituted carbon atom. The reaction *rates* for the two possible ways that a hydrogen ion can add are controlled by the energy of the two possible transition states, and *not* by the stability of the intermediate. The first step in the addition reaction is "uphill," it is endothermic. The structure of the possible intermediate carbocations resembles the two possible transition states. Thus, the stability of a carbocations reflects the stability of the transition state leading to it.

6. 4 Carbocation Rearrangements

Organic chemistry would be a simple subject if all of the atoms of a reactant stayed in place in the conversion to product in reactions such as an electrophilic addition reaction. However, reactive intermediates such as carbocations can undergo rearrangement reactions. Either a hydride ion or an alkyl group with a negative charge can move from a center adjacent to the carbocation center to form a new bond and generate a positive charge at that adjacent center. The driving force for such reactions is the formation of a more stable carbocation. Hydride ion shifts occur when a secondary or primary carbocation is generated adjacent to a tertiary center. Shifts of alkyl groups most commonly occur when a carbocation is generated at a carbon atom adjacent to a quaternary center. Any of the alkyl groups bonded to that quaternary carbon can migrate. Thus, a mixture of products can result. Even ring bonds of cycloalkanes can migrate, resulting in rings of different size than the reactant.

6.5 Hydration of Alkenes

The mechanism of the addition of water to a double bond is similar to that of the addition of hydrogen halides. A hydrogen ion first attacks the π bond, followed by capture of the carbocation by the nucleophilic oxygen atom of water. Note that it is water that reacts, *not* a hydroxide ion. Remember that the reaction occurs under acidic conditions, in which case the most available nucleophilic material is water. Subsequent loss of a proton from the oxonium ion gives a product which appears to have been the result of addition of a hydrogen ion and a hydroxide ion.

The reaction is readily reversible; the reverse reaction is dehydration. The dehydration reaction is much like viewing a film in reverse. The same things happen but in reverse sequence. That is, the mechanism for the forward is the same as the mechanism of the reverse reaction. This the **principle of microscopic reversibility**.

6.6 Addition of Halogens

The mechanism for addition of halogens such as chlorine or bromine to a double bond differs from the mechanism for the addition of hydrogen halides. The reaction rate is sensitive to the same features of the alkenes. As in the addition of hydrogen halides or water, compounds that give more stable carbocations are also more reactive in addition of bromine or chlorine. Thus, the carbon atom of the double bond bears some positive charge in the transition state. However, rearrangement products are not observed, and the stereochemistry of the product of the reaction is the result of a net *anti* addition.

Halogenation occurs by way of a cyclic bromonium or chloronium ion in which the positive charge is largely on the halogen atom but is distributed to some extent to the two carbon atoms. Of the two carbon atoms, the more substituted atom has the greater positive charge. The charge is insufficient to induce rearrangement reactions but sufficient to distinguish between the two sites for subsequent attack by a nucleophile. This difference in charge distribution is shown the formation of halohydrins, where water attacks the more substituted carbon atom because it has the higher positive charge.

6.7 Addition of Carbenes

Carbenes, which contain divalent carbon atoms, are electron-deficient species and behave as electrophiles toward the π bond of alkenes. However, they are electrically neutral, and there is no associated nucleophile. The reaction is a simultaneous formation of two σ bonds to give a cyclopropane ring. The Simmons–Smith reagent is used as a convenient way to form cyclopropane rings. The stereochemistry of the groups of the alkene are unaltered in the product cyclopropane.

6.8 Epoxidation of Alkenes

The addition of an oxygen atom to a double bond results in a cyclic ether known as an **epoxide.** A number of peroxide reagents are available for the formation of epoxides. In every case, the reaction proceeds by a cyclic-concerted process in which the stereochemistry of the groups bonded to the original carbon atoms of the double bond are unchanged in the product.

6.9 Dihydroxylation of Alkenes

Dihydroxylation reactions convert alkenes to vicinal diols (glycols). The reaction of alkenes with osmium tetroxide occurs by concerted cyclic mechanism that simultaneously places one oxygen atom on each carbon atom of the original double bond. Subsequent reaction of the cyclic intermediate generates the two hydroxyl groups. The stereochemistry of the addition reaction is *syn*.

6.10 Ozonolysis of Alkenes

Alkenes react with ozone to give intermediates that subsequently react under work-up conditions to give two products that result from cleaving both bonds of the original carbon–carbon double bond. Those two carbon atoms are identified in the product by the location of oxygen atoms. Under reductive work-up conditions, either aldehydes or ketones can result. Under oxidative work-up conditions, either carboxylic acids or ketones can form.

SUMMARY OF REACTIONS

1. Electrophilic Addition of Hydrogen Halides

2. Addition of Water (Hydration)

3. Addition of Halogens

4. Addition of Carbenes

5. Epoxidation of Alkenes

6. Dihydroxylation of Alkenes

$$ \text{(alkene)} + KMnO_4 \xrightarrow{\text{pH 7}} \text{(diol)} + MnO_2 $$

7. Ozonolysis of Alkenes

$$ \xrightarrow[\text{Zn / H}^+]{O_3} \quad + \quad CH_2{=}O $$

$$ \xrightarrow[\text{H}_2\text{O}_2]{O_3} \quad + \; HCO_2H $$

8. The Grubbs Reaction

$$ \xrightarrow{\text{Grubbs Catalyst}} \quad + \quad CH_2{=}CH_2 $$

(E)-2-pentene

End of Chapter Exercises

Syn and Anti Addition

6.1   The indirect hydration of an alkene using a procedure called hydroboration–oxidation transforms 1-methylcyclohexene into *trans*-2-methylcyclohexanol. Describe the stereochemistry of the net addition reaction.

Answer: The net result of the reaction is *syn* addition; it is *anti*-Markovnikov.

Hydroboration–oxidation

trans-2-methyl-cyclohexanol

6.2   Reaction of 1,2-dimethylcyclopentene with potassium permanganate yields the following compound. What is the stereochemistry of the net addition reaction?

Answer: The hydroxyl groups are *cis* because it is *syn* addition.

KMnO₄

Electrophiles and Markovnikov Addition

6.3   Predict the structure of the addition product of IN₃ and 1-pentene. The mechanism occurs by electrophilic attack followed by capture of the carbocation by a nucleophile.

Reaction:

I—N₃

Explanation: The heterolytic cleavage of the I—N bond of the IN₃ compound gives I⁺ and N₃⁻. Addition of the electrophile I⁺ at C-1 of 1-pentene yields a secondary carbocation at C-2 which is captured by the nucleophile N₃⁻. The structure of the product is shown above.

6.4   Based on the information given in the following equation, outline the mechanism of the reaction of the reagent INCO.

Mechanism:

INCO

NCO⁻

Explanation: Heterolytic cleavage of the I—N bond of the INCO compound gives I⁺ and NCO⁻. Addition of the electrophile I⁺ at the double bond from face of the ring opposite the axial methyl group gives a cyclic iodonium ion. To achieve net *trans* addition, the nucleophilic NCO⁻ ion must attack at the indicated atom to open the ring of the cyclic iodonium ion via a *trans* diaxial arrangement. Attack at the other carbon atom would give the diequatorial isomer.

6.5   Write the product of the reaction of HBr with each of the following compounds.

(a) 2-methyl-l-butene   (b) 2-methyl-2-butene   (c) (Z)-2-hexene   (d) (E)-3-methyl-2-pentene

Answers:

(a)

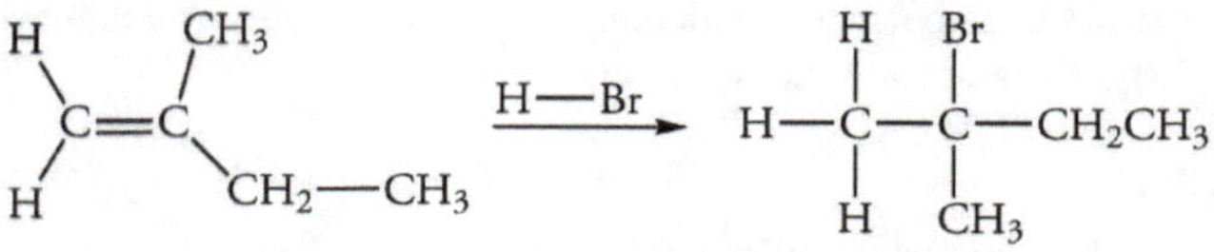

(b)

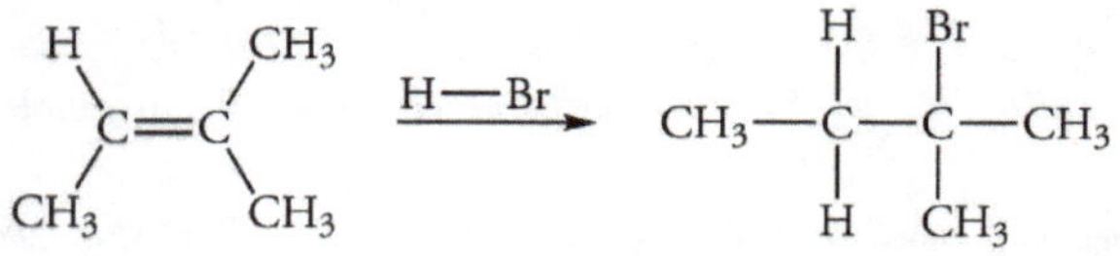

(c)

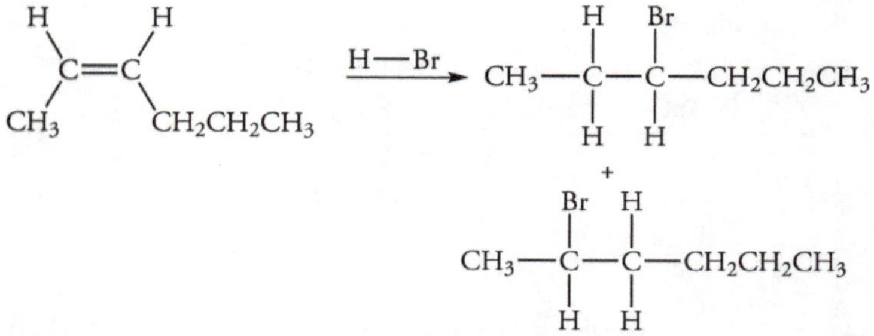

(d)

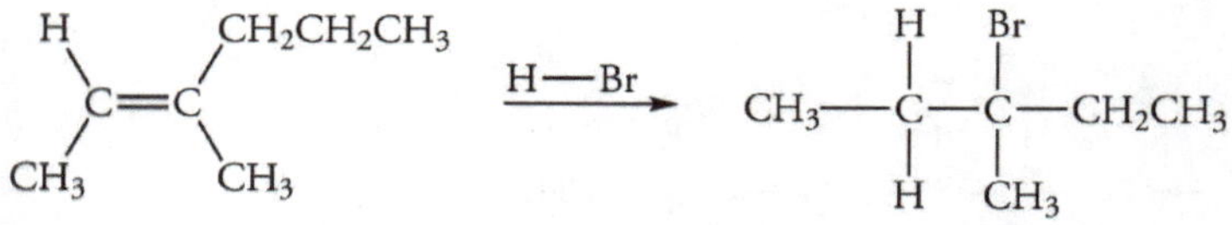

6.6   Write the product of the reaction of HBr with each of the following compounds.

(a)   (b) 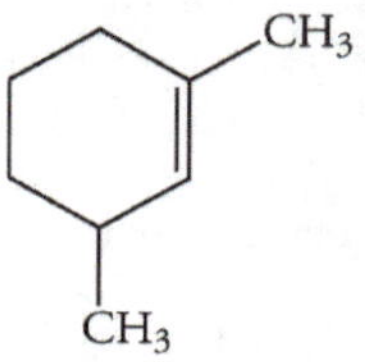   (c) 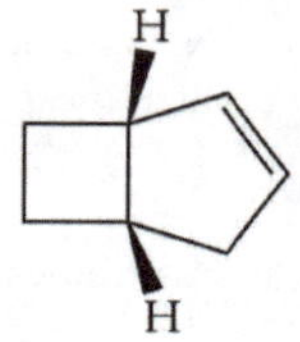   (d)

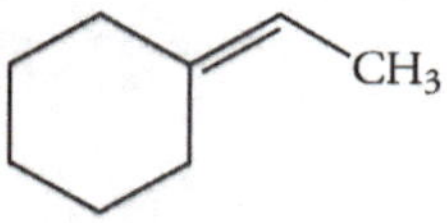

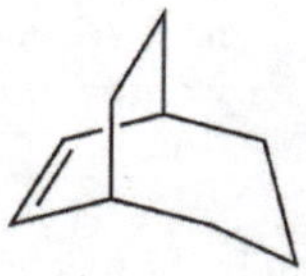

Answers:

(a)   (b) 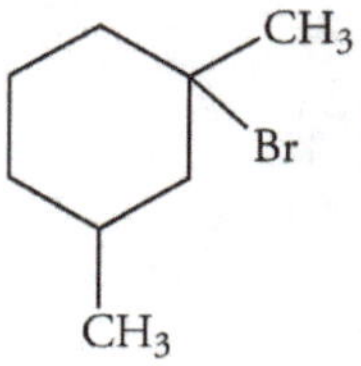   (c) 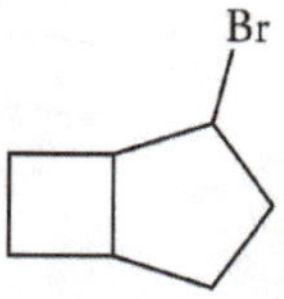   (d)

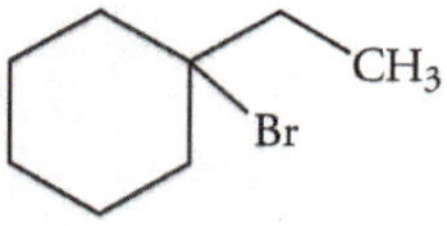

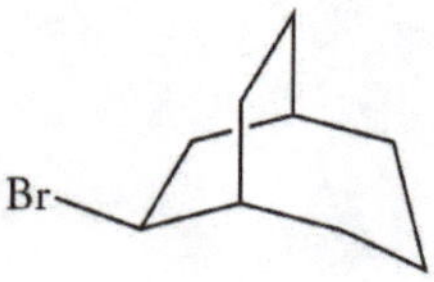

6.7   Reaction of 1,6-dimethylcyclohexene with HBr by an electrophilic addition mechanism yields two products. What are the two compounds?

Answer: The reaction proceeds through a planar carbocation intermediate that can be attacked by bromide from either side to give two geometric isomers.

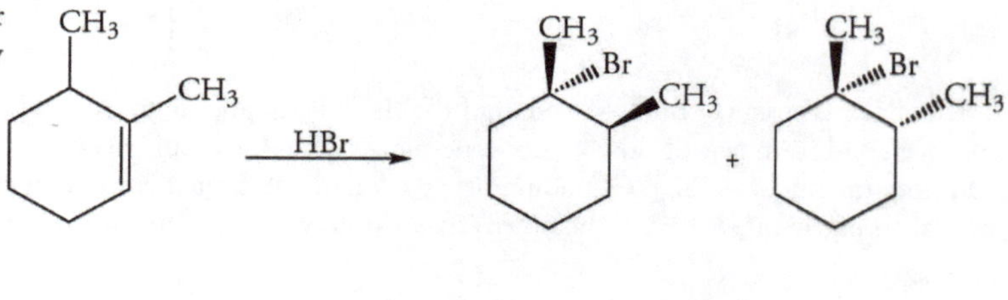

6.8 Reaction of 1,2-dimethylcyclohexene with HCl by an electrophilic addition mechanism yields two products. What are the two compounds?

Reaction:

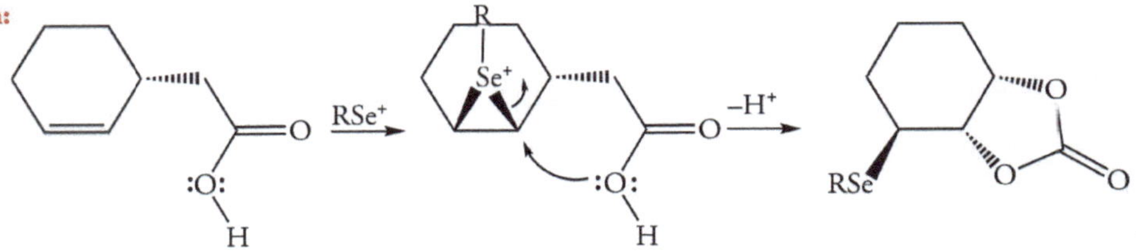

Explanation: The alkyl selenium ion, RSe^+, is an electrophile that attacks the double bond to give a cyclic selenium ion which is then opened by nucleophilic attack of the oxygen atom of the carboxyl group. Note that the oxygen and selenium atoms are *trans* in the product.

6.9 The electrophilic addition of HCl to 3,3,3-trifluoropropene gives 1-chloro-3,3,3-trifluoropropane, an anti-Markovnikov addition product. Consider the structure of the intermediate carbocations possible for the two modes of addition and suggest a reason for the observed regioselectivity.

Reaction:

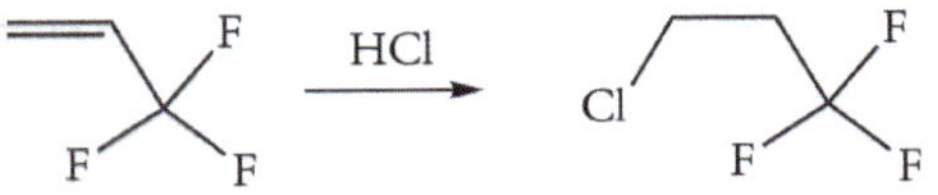

Explanation: Two carbocations are possible. Since the trifluoromethyl group is strongly electron withdrawing, the carbocation with a positive charge at C-1 is more stable than the carbocation with a positive charge at C-2.

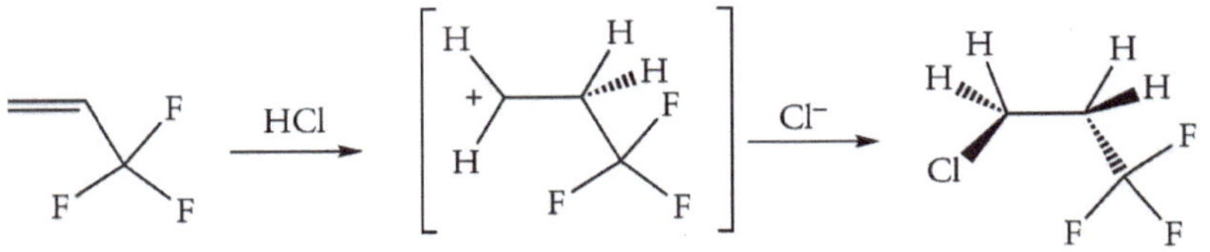

6.10 The electrophilic addition of HCl to chloroethene yields 1,1-dichloroethane. Based on resonance structures, account for the observed regioselectivity.

Reaction:

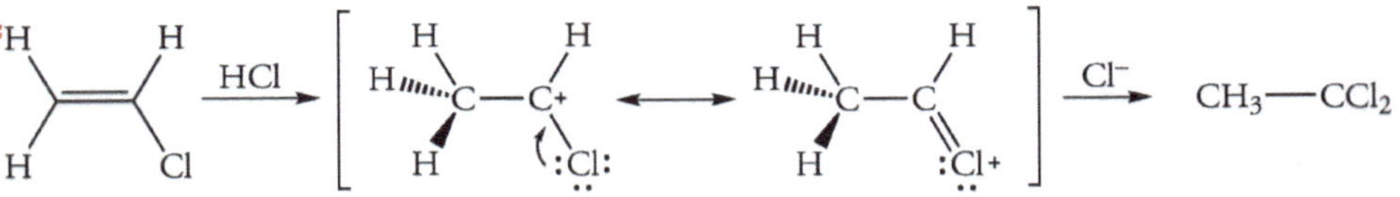

Explanation: Although chlorine is electron withdrawing, it can stabilize a carbocation by a resonance effect, which accounts for the observed regiospecificity

6.11 Reaction of 3,3-dimethy1-1-butene with HI gives a mixture of unrearranged product and rearranged product in the ratio 90:10. Account for the difference in this ratio compared to that for addition of HCl (Section 7.4).

Reaction:

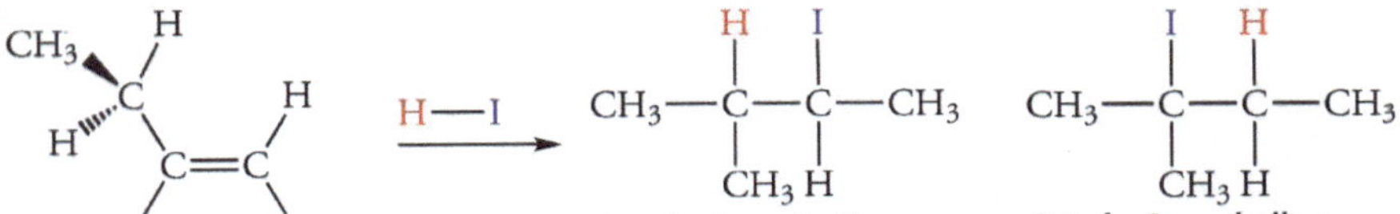

2-iodo-3-methylbutane
90%
(expected product)

2-iodo-2-methylbutane
10%
(rearranged product)

Answer: The amount of rearranged product is smaller with HI. Thus the iodide ion must capture the carbocation prior to rearrangement more efficiently than does chloride ion.

6.12 Reaction of 3,3-dimethyl-l-butene with HBr gives a mixture of two addition products in the ratio 70:30. Based on Exercise 6.11, predict the structures of the two products.

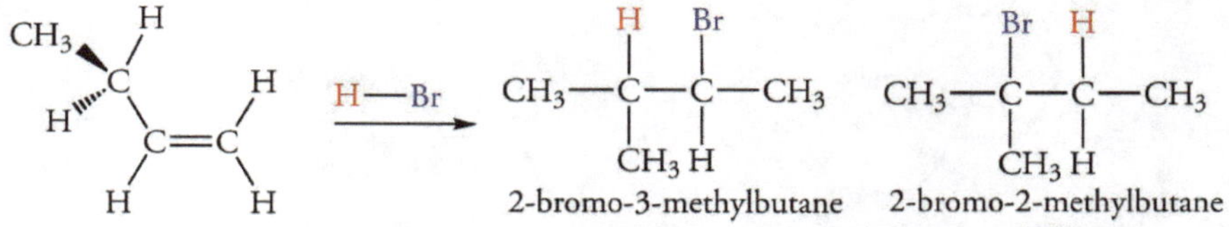

2-bromo-3-methylbutane
70%
(expected product)

2-bromo-2-methylbutane
30%
(rearranged product)

Explanation: The major product is 2-bromo-3,3-dimethylbutane, which is expected from a normal addition reaction. The bromide ion can capture the carbocation prior to its rearrangement more efficiently than does the chloride ion, but less efficiently than the iodide ion. The minor product is 2-bromo-2,3-dimethylbutane, which results from a 1,2 hydride shift of a methyl group.

Hydration of Alkenes

6.13 Write the product of hydration of each of the following compounds assuming that no rearrangement occurs.

(a) 2-methyl-l-butene   (b) 2-methyl-2-butene   (c) (*Z*)-2-hexene   (d) (*E*)-3-methyl-2-pentene

Answers:

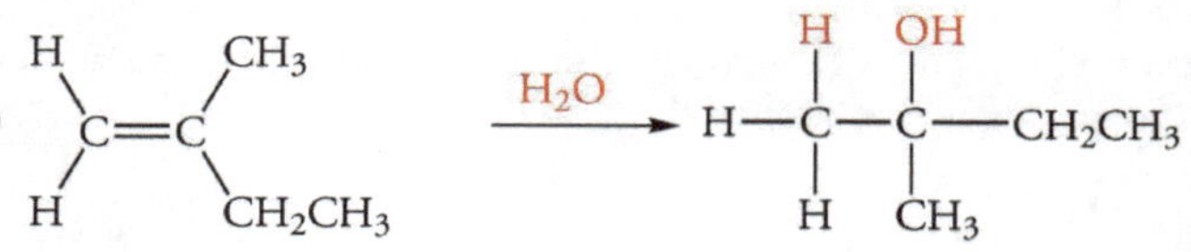

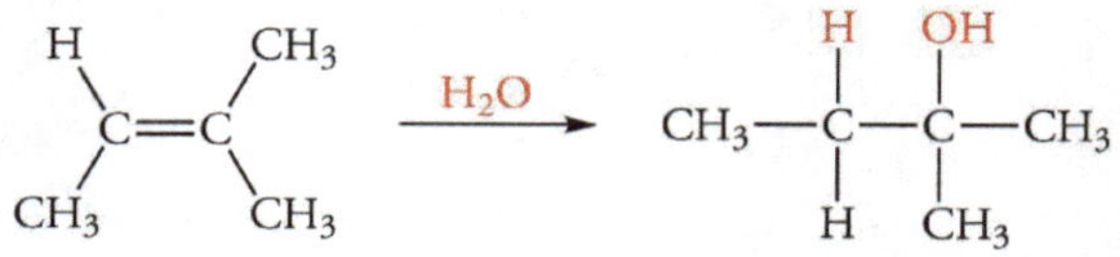

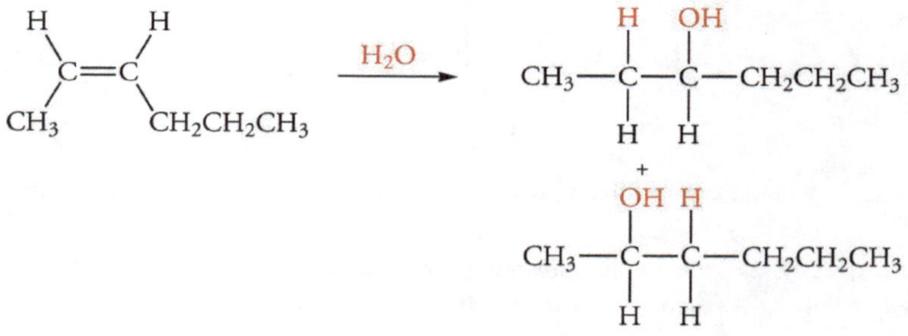

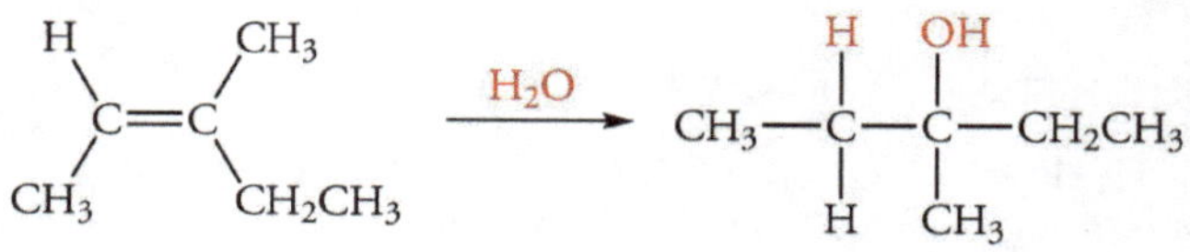

6.14   Write the product of hydration of each of the following compounds assuming that no rearrangement occurs.

(a) 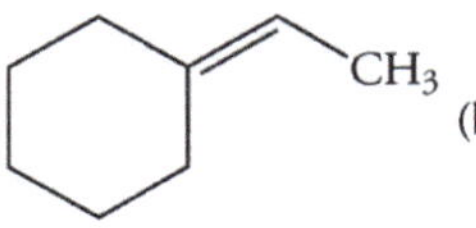   (b) 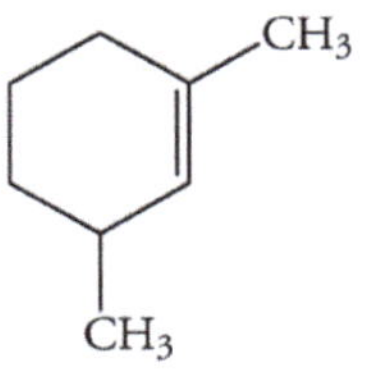   (c) 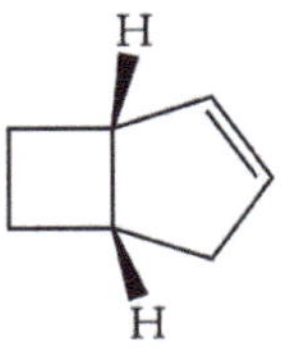   (d)

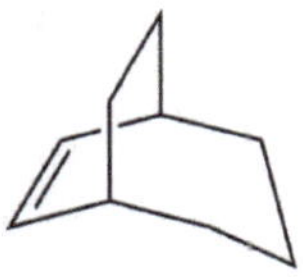

Answers:

(a)

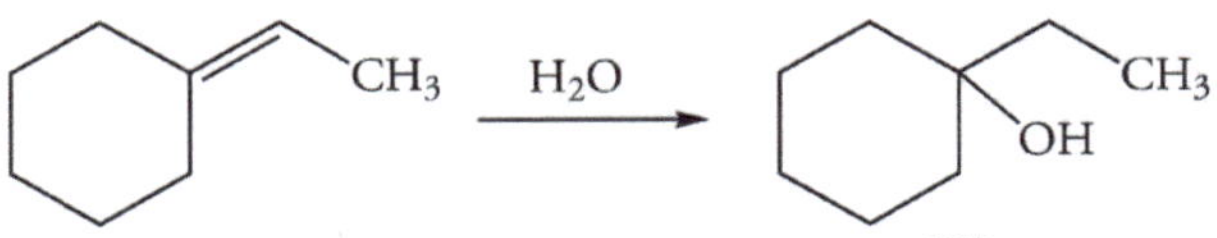

(b)

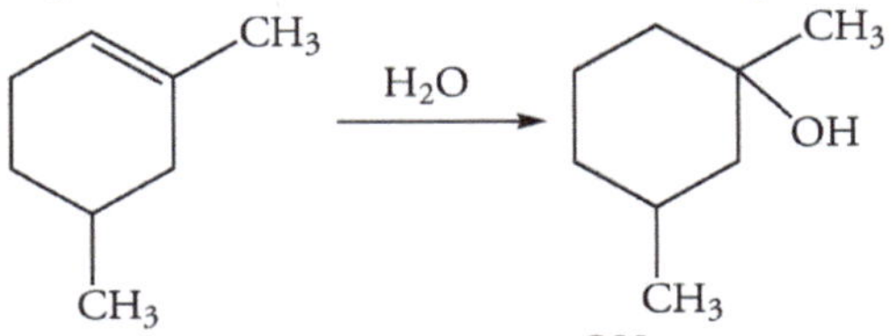

(c)

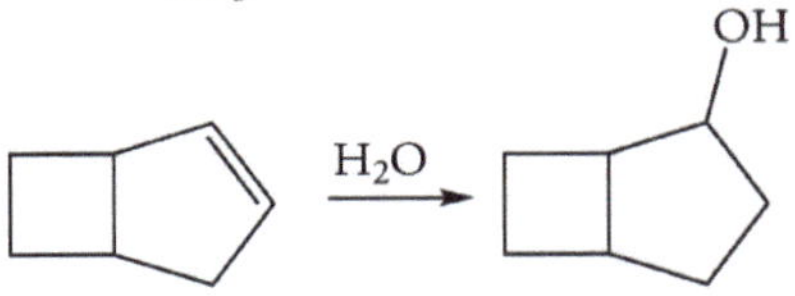

(d)

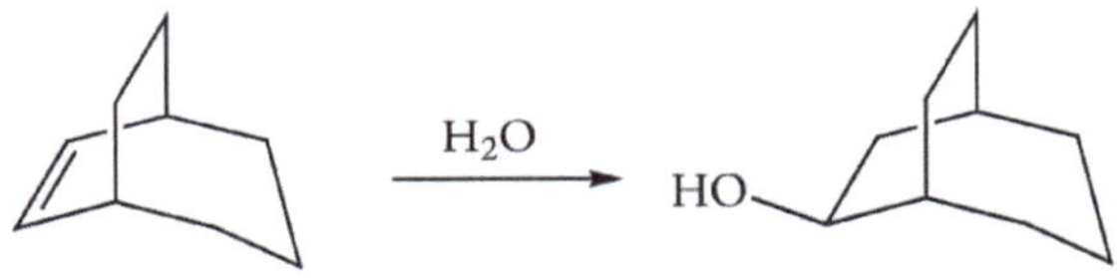

6.15   Hydration of either 2-methyl-l-butene or 2-methyl-2-butene yields the same alcohol. What is its structure? Explain why the same compound forms from both alkenes.

Answer: The same tertiary carbocation is formed from either of the two alkenes. The product is a tertiary alcohol.

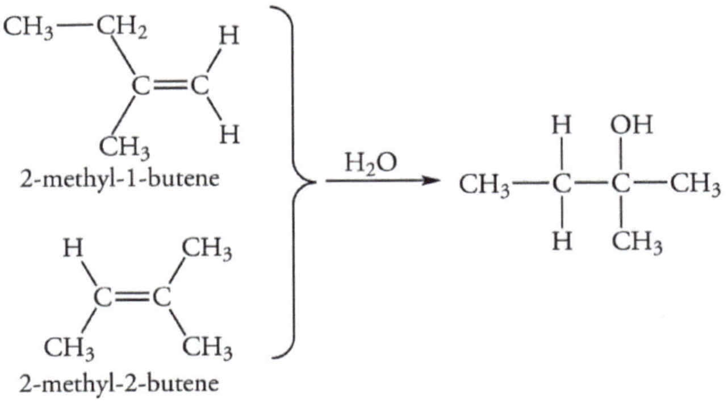

6.16   Hydration of 2,3-dimethyl-2-butene is a slower reaction than the hydration of 2,3-dimethyl-l-butene under the same reaction conditions. Suggest a possible explanation.

Answer: The structure of the tertiary carbocation is the same for both alkenes. However, the rate of the reaction depends on the difference in energy between the reactant and the transition state. The 2,3-dimethyl-2-butene is the more stable isomer because it has the more highly substituted double bond. Thus, the energy required to achieve the transition state is larger for this compound.

Addition of Bromine to Alkenes

6.17   Write the product of the reaction of Br_2 with each of the following compounds.
(a) 2-methyl-1-butene   (b) 2-methyl-2-butene   (c) (Z)-2-hexene   (d) (E)-3-methyl-2-pentene

Answers:

(a)

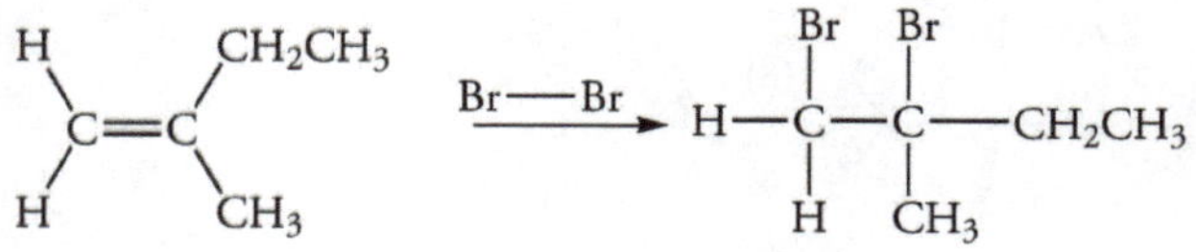

$$H,\ CH_2CH_3 \quad C=C \quad H,\ CH_3 \xrightarrow{\ Br-Br\ } H-\overset{Br}{\underset{H}{C}}-\overset{Br}{\underset{CH_3}{C}}-CH_2CH_3$$

(b)

$$H,\ CH_3 \quad C=C \quad CH_3,\ CH_3 \xrightarrow{\ Br-Br\ } CH_3-\overset{Br}{\underset{H}{C}}-\overset{Br}{\underset{CH_3}{C}}-CH_3$$

(c)

$$H,\ H \quad C=C \quad CH_3,\ CH_2CH_2CH_3 \xrightarrow{\ Br-Br\ } CH_3-\overset{Br}{\underset{H}{C}}-\overset{Br}{\underset{H}{C}}-CH_2CH_2CH_3$$

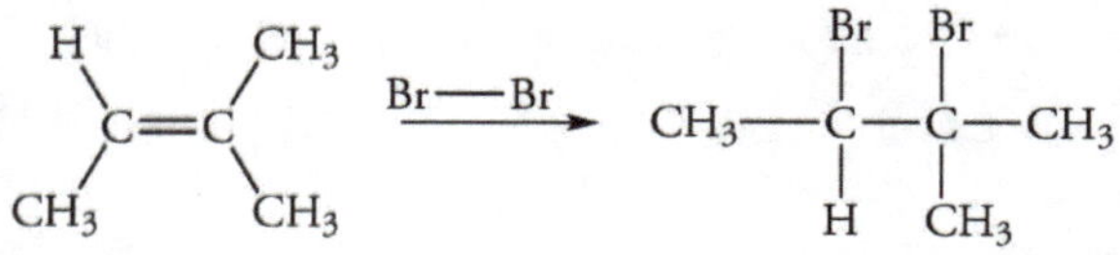

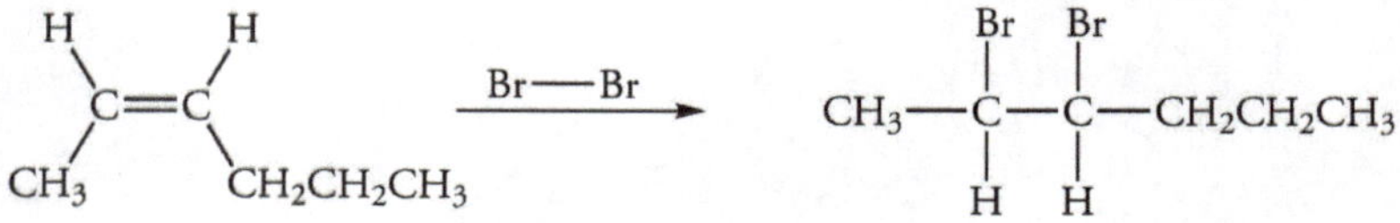

(d)

$$H,\ CH_3 \quad C=C \quad CH_3,\ CH_2CH_3 \xrightarrow{\ Br-Br\ } CH_3-\overset{Br}{\underset{H}{C}}-\overset{Br}{\underset{CH_3}{C}}-CH_2CH_3$$

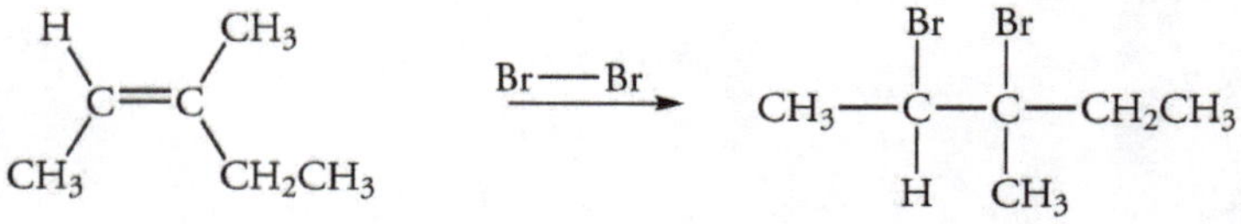

6.18   Write the product of the reaction of Br_2 with each of the following compounds.

Answers:

(a)

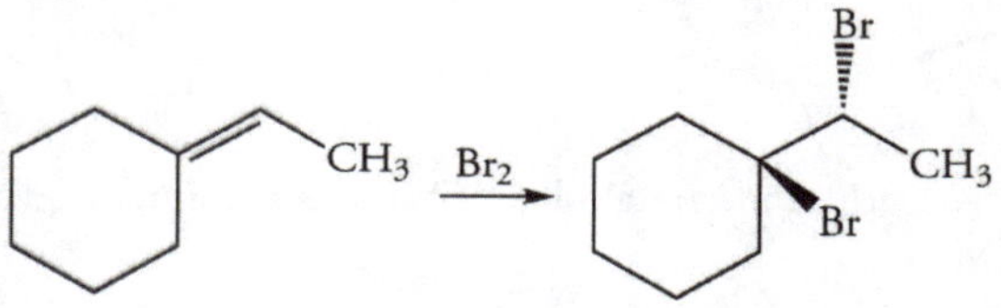

(b)

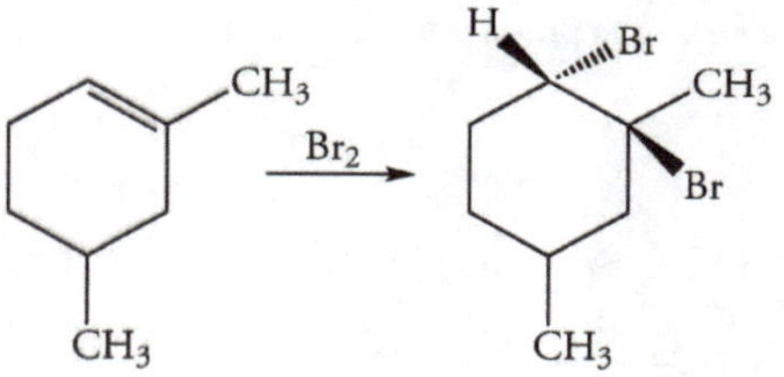

(c)

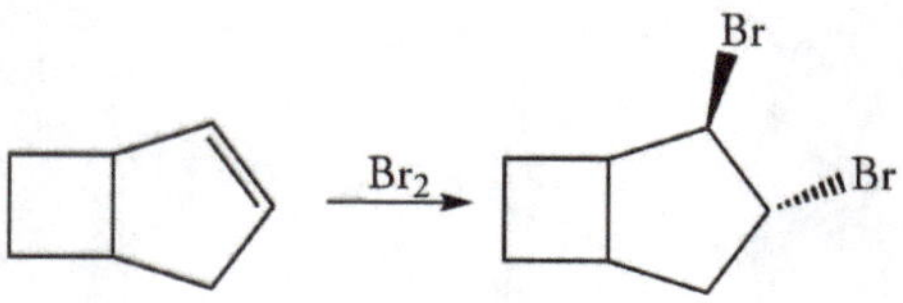

(d)

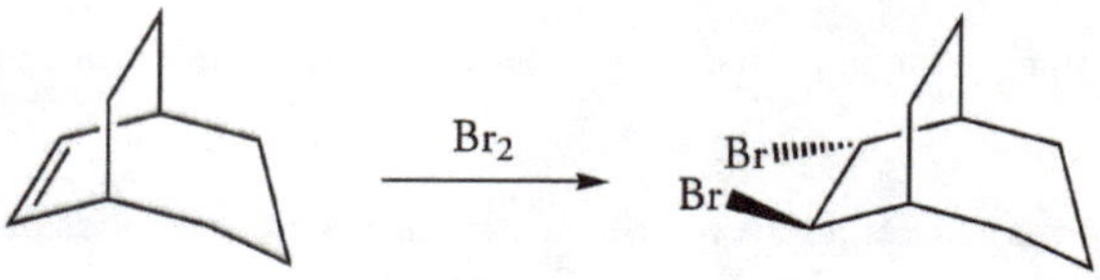

6.19   Reaction of 3-methylcyclohexene with bromine in CCl$_4$ gives a mixture of two products. Explain why two products result.

Answer: The bromide ion may attack either of two carbon atoms of the bromonium ion.

6.20   Reaction of 3-bromocyclohexene with HBr gives an "unusual" product, *trans*-1,2-dibromocyclohexane. Explain its origin using an appropriate mechanism and intermediate.

Answer: A proton adds at C-1, giving a secondary carbocation at the C-2 atom, which is then temporarily captured by the bromine atom at C-3. The intermediate is thus a bromonium ion that subsequently reacts with bromide ion.

6.21   The reaction of cyclohexene with bromine in water as the solvent yields the alcohol *trans*-2-bromocyclohexanol. Explain why.

Answer: The reaction of the bromonium ion with water is similar to the reaction of the chloronium ion with water to give a chlorohydrin. The nucleophilic water attacks on the "face" of the original alkene that is opposite the bromine atom of the bromonium ion.

6.22   Reaction of cyclohexene with an aqueous bromine solution saturated with sodium chloride gives a mixture of *trans*-2-bromocyclohexanol and a compound with the molecular formula C$_6$H$_{10}$BrCl. What is the structure of the latter compound?

Answer: The reaction of the bromonium ion with chloride ion is similar to that of its reaction with bromide ion in the reaction of bromine. The nucleophilic chloride ion attacks on the "face" of the original alkene that is opposite the bromine atom of the bromonium ion to give *trans*-1-bromo-2-chlorocyclohexane.

6.23 Bromination of 3,3-dimethyl-1-butene in methanol (CH_3OH) gives a mixture of the expected dibromo compound and a bromoether. Explain the origin of the two products.

Answer: The bromonium ion, which has a bromine atom bridging C-1 and C-2, may react with methanol as well as with Br_2, thus producing a mixture of the dibromo compound and a bromoether.

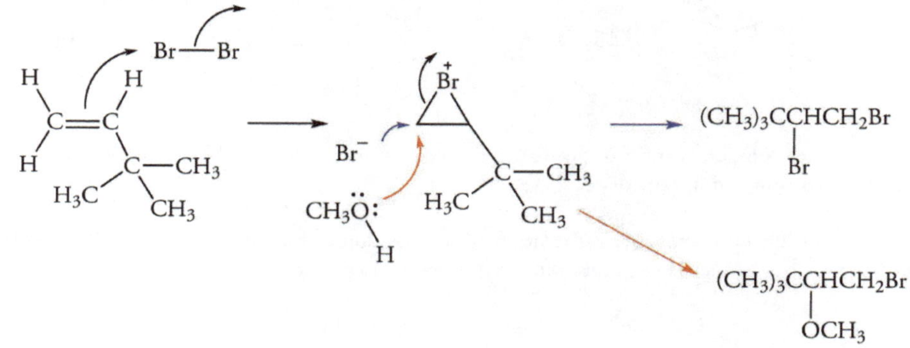

6.24 Based on the information given in Exercise 6.21, predict the structure of the chloroalcohol formed in the reaction of methylenecyclohexane with an aqueous chlorine solution.

Answer: Attack of chlorine forms a chloronium ion which has primary carbocation character at the original methylene carbon atom and tertiary carbocation character at the ring atom. Methanol is expected to attack the tertiary center to give the product shown below.

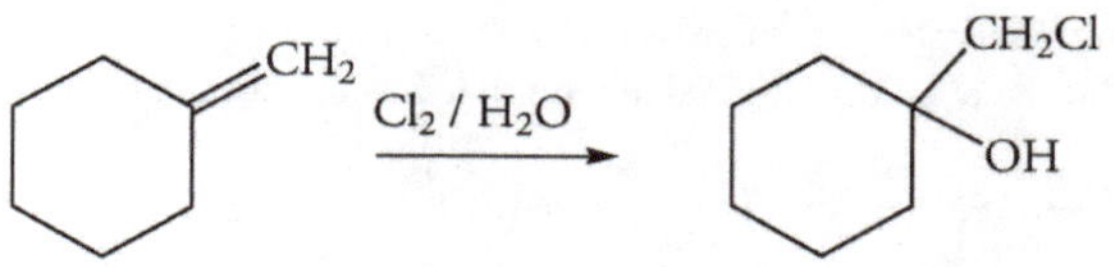

6.25 Based on the information given in Exercise 6.21, predict the structure of the chloroalcohol formed in the reaction of methylenecyclohexane with an aqueous chlorine solution.

Answer: Attack of chlorine forms a chloronium ion which has primary carbocation character at the original methylene carbon atom and tertiary carbocation character at the ring atom. Methanol is expected to attack the tertiary center to give the product shown below.

6.26 Reaction of 4-penten-l-ol with aqueous bromine gives the indicated cyclic bromoether. Write a mechanism for its formation.

Answer: The bromonium ion undergoes intramolecular attack by the oxygen atom of the hydroxyl group followed by loss of a proton.

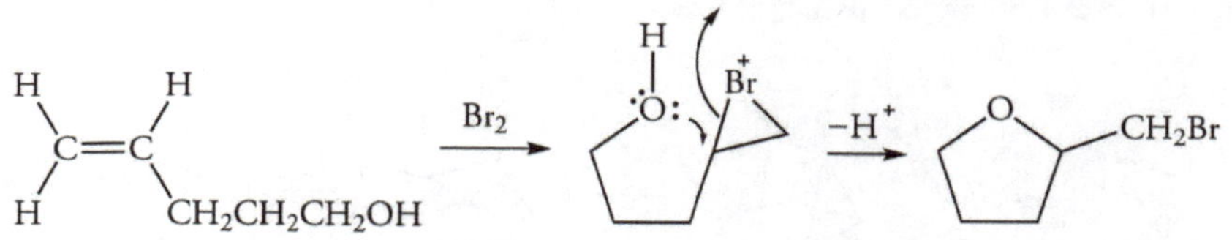

Addition of Carbenes to Alkenes

6.27 Chlorocarbene (CHCl) can be produced from dichloromethane using butyl lithium ($CH_3CH_2CH_2CH_2^-$ Li^+), but cannot be produced using potassium *tert*-butoxide. Suggest a reason why not.

Answer: The *tert*-butoxide ion is not as strong a base as butyl lithium and cannot remove the proton from dichloromethane, which is a weaker acid than trichloromethane.

6.28   Addition of dichlorocarbene to *cis*-2-butene gives a mixture of two isomeric compounds. Explain why.

Answer: The chlorocarbene can add to place the chlorine group either *cis* or *trans* to the two methyl groups.

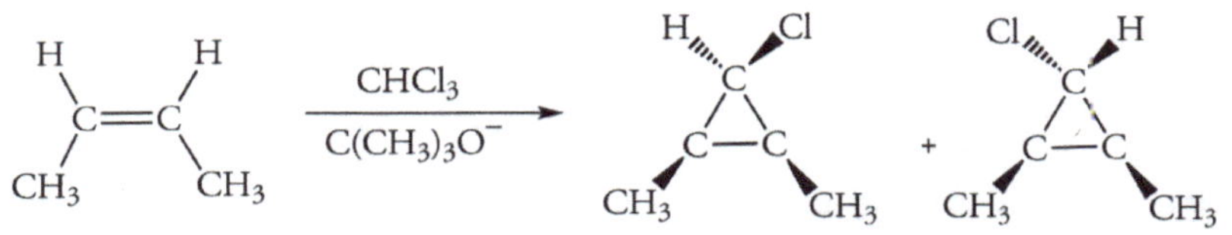

6.29   Write the products of the following reaction.

Answer:

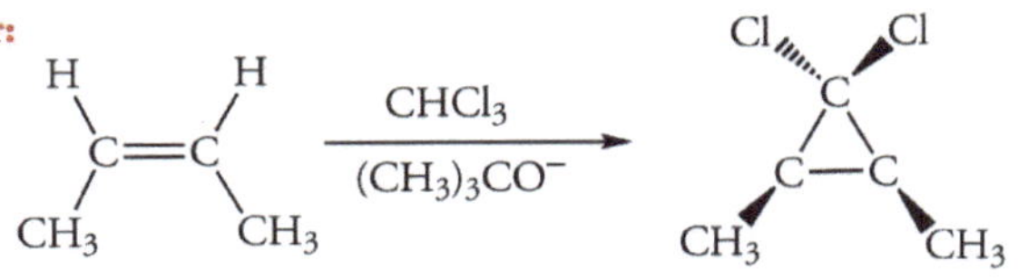

6.30   Write the products of the following reaction.

Answer:

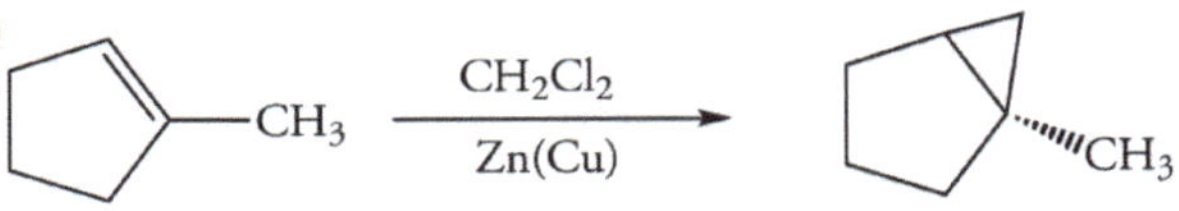

6.31   Based on the stated electrophilicity of dichlorocarbene, predict the relative reactivities of 1-butene and *trans*-2-butene with dichlorocarbene.

Answer: The methyl groups of the disubstituted *trans*-2-butene should supply electrons to the double bond and increase the availability of electrons to the electrophilic dichlorocarbene. The monosubstituted double bond of 1-butene will be less reactive.

6.32 Predict the relative electrophilicities of dichlorocarbene and chlorocarbene.

Answer: Based on inductive electron withdrawal of the electronegative chlorine atoms, dichlorocarbene should be more electrophilic.

6.33 Reaction of 1,1-dichloroethane with butyllithium does not give a carbene. Why?

Answer: The base can remove a proton from the methyl group and cause an elimination reaction to give chloroethene.

6.34 Dichlorocarbene can be formed by heating sodium trichloroacetate. Propose a mechanism for this reaction.

Answer:

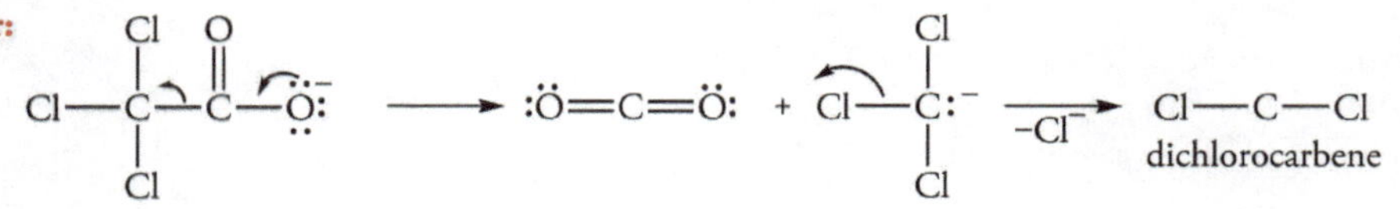

trichloroacetate

Epoxidation of Alkenes

6.35 Write the structure of the epoxide obtained from the reaction of *trans*-9-octadecen-l-ol with MCPBA.

Answer: The *trans* stereochemistry of the groups about the double bond is retained in the epoxide product.

6.36 The following epoxide is an intermediate in the synthesis of disparlure, the sex attractant of the gypsy moth. Write the structure of the unsaturated alcohol used to produce the epoxide.

Answer: The *cis* stereochemistry of the epoxide must exist about the double bond of the unsaturated alcohol.

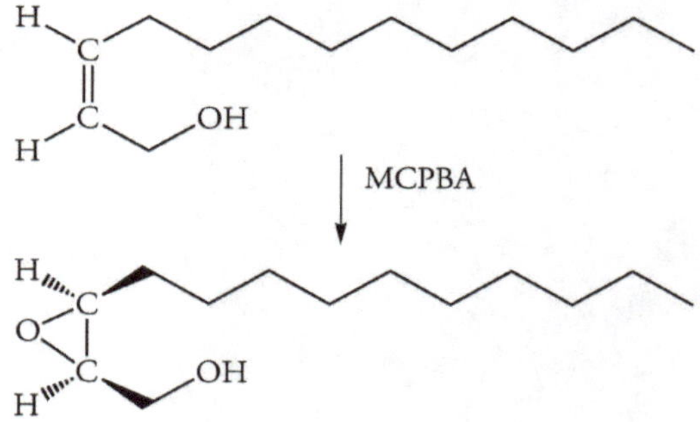

6.37 Predict which of the two isomeric epoxides will be produced from the following bicyclic unsaturated compound.

Answer: The first compound forms because the oxygen of the epoxide is delivered to the double bond from the face that is not sterically hindered by the axial methyl group.

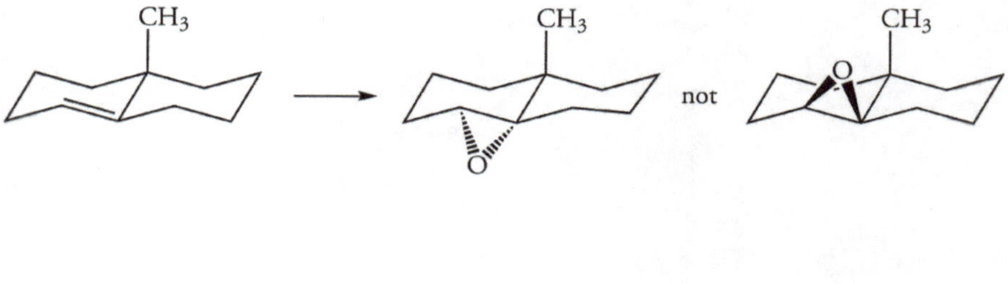

6.38 Oxidation of bicyclo[2.2.1]hept-2-ene gives the indicated epoxide. Write the structure of an alternative epoxide product and explain why this compound is not produced.

Answer: The first compound, which is *exo*, forms because the oxygen of the epoxide is delivered to the double bond from the face that is less sterically hindered.

6.39 Write the structure of the epoxide expected from the reaction of the following diene with one molar equivalent of MCPBA.

Answer: The more highly substituted double bond should react with MCPBA. The oxygen atom should be placed on the side of the ring opposite the vinyl group.

6.40 Arrange the following compounds in order of increasing rate of reaction with MCPBA.
I: 5-methyl-1-hexene   II: 3-methyl-2-hexene   III: 4-methyl-2-hexene   IV: 2,3-dimethyl-2-pentene

Answer: The order of increasing rate of reactivity is the same as the order of increasing degree of substitution, which is I < Ill < II < IV.

Dihydroxylation of Alkenes

6.41 Describe the visual appearance of the reaction that occurs when *cis*-2-pentene reacts with potassium permanganate. How could this reagent be used to distinguish between the isomers *cis*-2-pentene and cyclopentane?

Answer: The purple color of the permanganate ion disappears, and a brown precipitate of manganese dioxide results. These results are not obtained with cyclopentane.

6.42 Write the product of the reaction of vinylcyclohexane with potassium permanganate.

Answer: A diol forms when hydroxyl groups are added to the vinyl group.

6.43   The *exo* face of bicyclo[2.2.1]hept-2-ene (norbornene) is less sterically hindered than the *endo* face. Based on this information, write the product of reaction of norbornene with $KMnO_4$.

Answer:

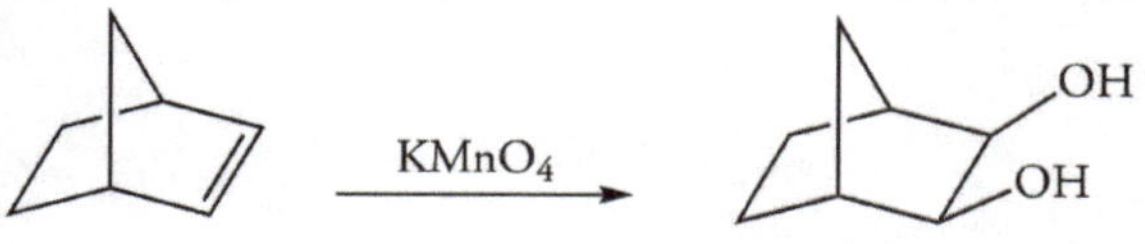

norbornene $\xrightarrow{KMnO_4}$ product (exo diol, two OH groups)

exo

6.44   Write the structure of the diol that forms when OsO_4 reacts with the following alkenes.

Answers:

(a)   cyclopentene with CH₃ $\xrightarrow[\text{2. Zn/H}^+]{\text{1. O}_3}$ cyclopentane diol with CH₃, OH, OH

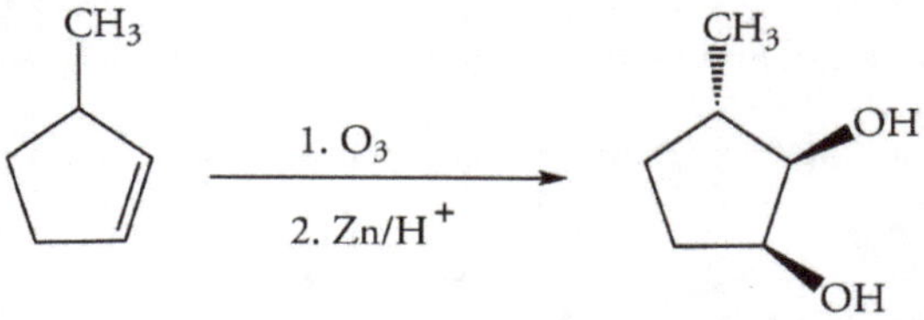

(b)   cyclohexene (CH₃, CH(CH₃)₂ substituents) $\xrightarrow[\text{2. Zn/H}^+]{\text{1. O}_3}$ diol product (HO, HO, CH₃, H, CH(CH₃)₂)

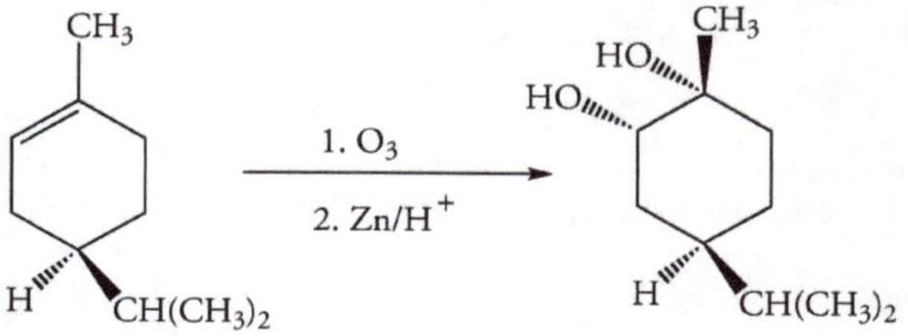

(c)   cyclohexene (CH₃, CH₃ substituents) $\xrightarrow[\text{2. Zn/H}^+]{\text{1. O}_3}$ diol product (CH₃, CH₃, OH, OH)

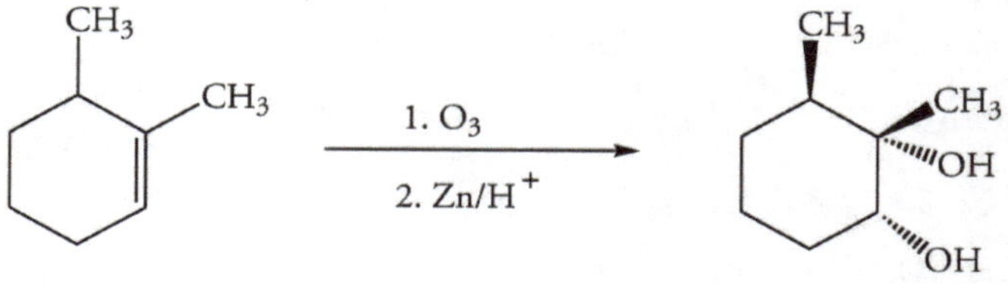

Ozonolysis of Alkenes

6.45   Write the product(s) of ozonolysis of each of the following compounds under reductive workup conditions.

Answers:

(a) CH_3, CH_2CH_3 / H, CH_2CH_3 alkene $\xrightarrow[\text{2. Zn /H}^+]{\text{1. O}_3}$ $CH_3(H)C=O + O=C(CH_2CH_3)(CH_2CH_3)$

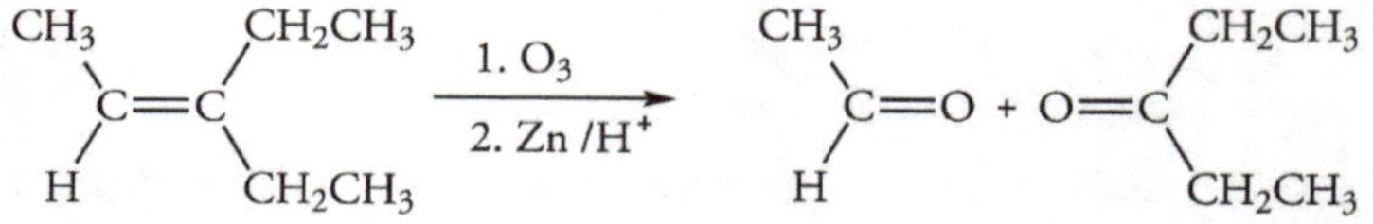

(b) CH_3, CH_3 / CH_3, CH_3 alkene $\xrightarrow[\text{2. Zn /H}^+]{\text{1. O}_3}$ $CH_3(CH_3)C=O + O=C(CH_3)(CH_3)$

(c) H, CH_3 / CH_3, CH_3 alkene $\xrightarrow[\text{2. Zn /H}^+]{\text{1. O}_3}$ $H(CH_3)C=O + O=C(CH_3)(CH_3)$

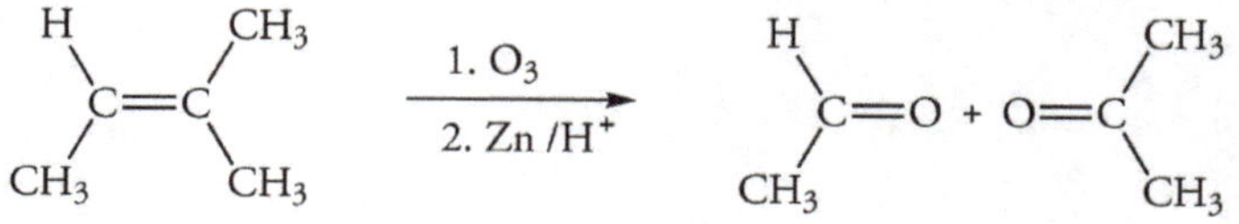

(d) H, H / CH_3, CH_2CH_3 alkene $\xrightarrow[\text{2. Zn /H}^+]{\text{1. O}_3}$ $H(CH_3)C=O + O=C(H)(CH_2CH_3)$

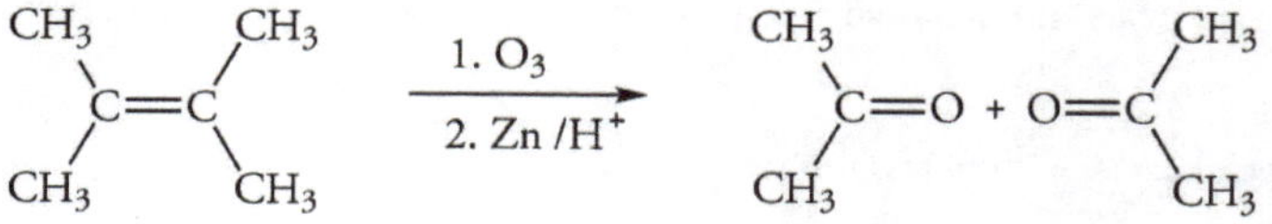

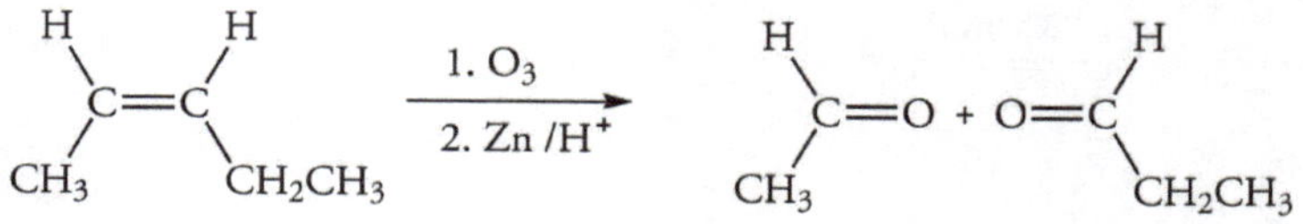

6.46   Write the product(s) of ozonolysis of each of the following compounds under oxidative workup conditions.

Answers:

(a)

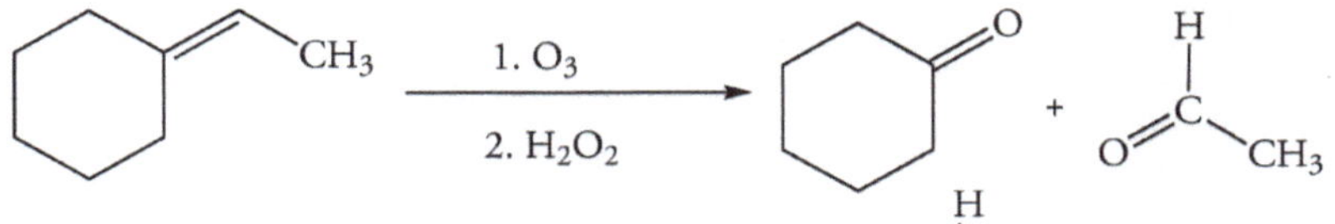

(b)

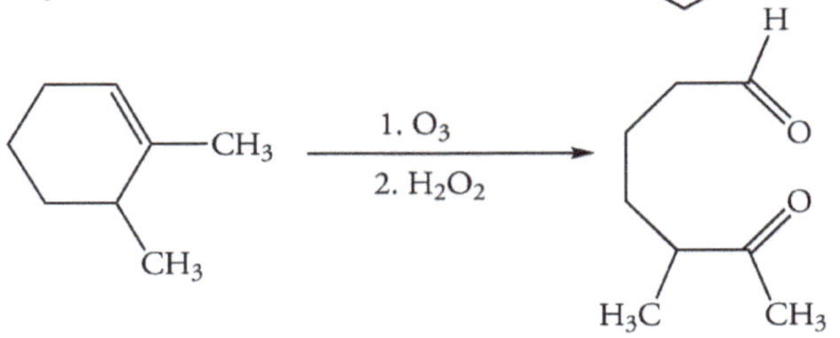

(c)

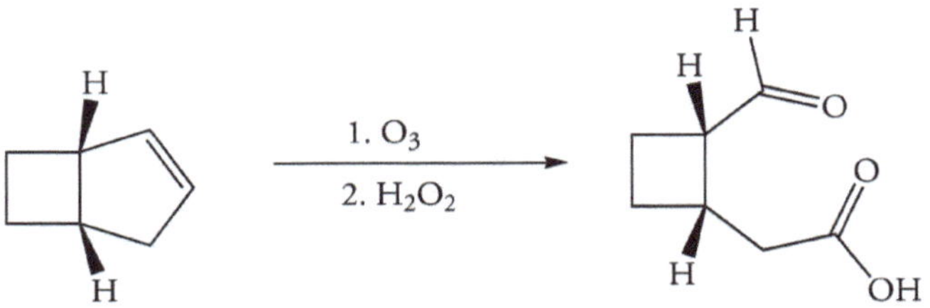

(d)

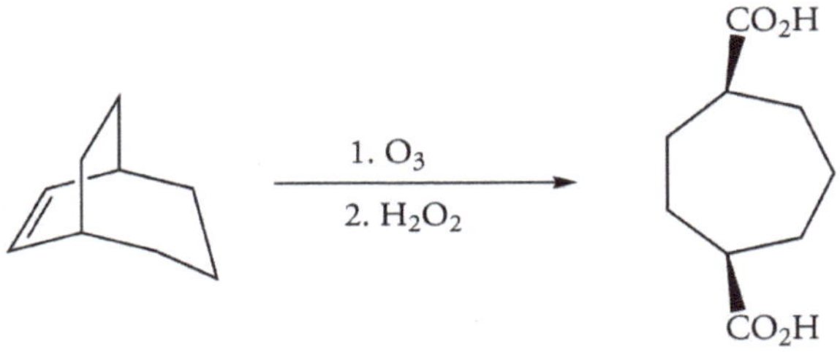

6.47   How can you distinguish between 1,3-cyclohexadiene and 1,4-cyclohexadiene based on their ozonolysis products?

Answer: 1,4-Cyclohexadiene gives only one ozonolysis product, but 1,3-cyclohexadiene gives two.

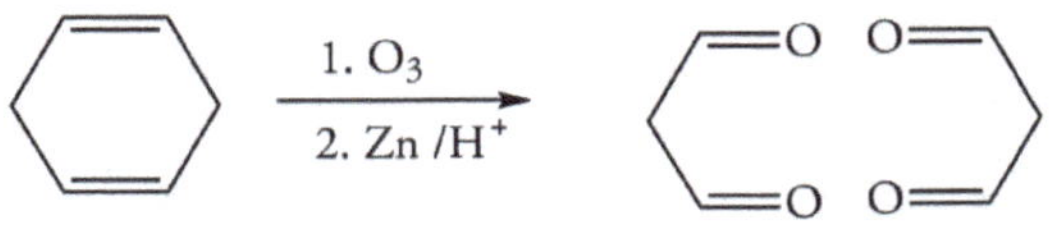

1,4-cyclohexadiene

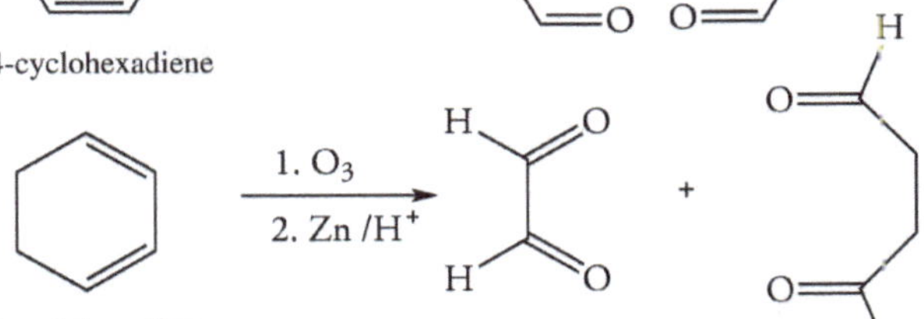

1,3-cyclohexadiene

6.48   How can you distinguish between 1,3-cyclohexadiene and 1,4-cyclohexadiene based on their ozonolysis products?

Answer: 1-Methylcyclohexene gives a dicarbonyl compound that is an aldehyde and a ketone. Both 3-methylcyclohexene and 4-methylcyclohexene give dialdehydes.

6.49   Write the products of ozonolysis using reductive workup conditions for each of the three isomeric methylcyclohexenes and classify the carbonyl group present in each product.

Answer: 1-Methylcyclohexene gives a dicarbonyl compound that is an aldehyde and a ketone. Both 3-methylcyclohexene and 4-methylcyclohexene give dialdehydes.

6.50   A hydrocarbon of molecular formula C_9H_{14} is found in sandalwood oil. Ozonolysis of the hydrocarbon followed by oxidative workup gives the following diketone. Draw the structure of the hydrocarbon.

6.51   A hydrocarbon component of a pheromone of a species of moth reacts with ozone followed by reductive workup to give the following compounds. Draw a structure of the hydrocarbon. How many geometric isomers are possible with this structure?

Answer: There are two double bonds in the pheromone and E-Z isomers are possible about each bond, giving rise to four isomers. The $6(E)$, $9(E)$ isomer is shown.

6.52   Two isomeric unsaturated carboxylic acids, oleic acid and elaidic acid, melt at 13 and 45 °C, respectively. Ozonolysis of either compound under oxidative conditions yields the following two compounds. What are possible structures of the two compounds? Why do they give the same products?

Answer: The two compounds are geometric isomers. When the double bond is cleaved, that structural distinction no longer exists and the resulting fragments are identical. Oleic acid has configuration (Z), and elaidic acid has configuration (E).

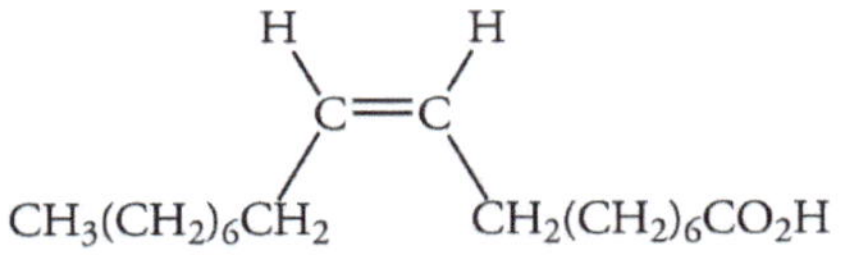

$$CH_3(CH_2)_7\overset{\displaystyle O}{\overset{\|}{C}}-OH \qquad HO-\overset{\displaystyle O}{\overset{\|}{C}}(CH_2)_7\overset{\displaystyle O}{\overset{\|}{C}}-OH$$

oleic acid (Z) isomer

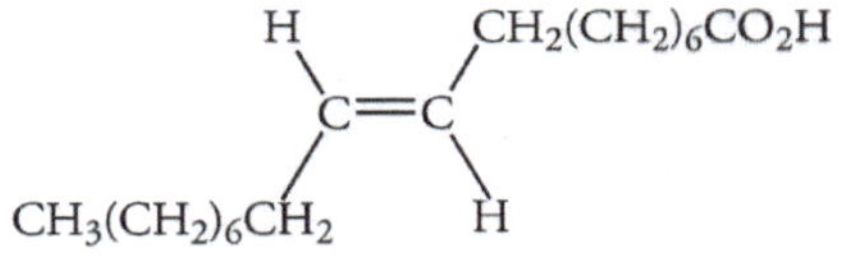

elaidic acid (E) isomer

6.53   An unsaturated fatty acid isolated from brain tissue has the molecular formula $C_{24}H_{40}O_2$. Hydrogenation yields an unbranched carboxylic acid with the molecular formula $C_{24}H_{46}O_2$. Ozonolysis of the fatty acid under reductive conditions yields two equivalents of 1,3-propanedial and one equivalent each of hexanal and an aldehydic acid with the formula $C_{12}H_{22}O_3$. Write the structure of the most stable isomer that is most consistent with these data. How many other isomeric compounds are also consistent with the data?

Answer: There are three double bonds in the compound, and E-Z isomers are possible about each double bond, so eight geometric isomers are possible. The 12(E), 15(E), and 18(E) isomer is shown.

$$CH_3(CH_2)_4\overset{\displaystyle O}{\overset{\|}{C}}H \qquad H\overset{\displaystyle O}{\overset{\|}{C}}CH_2\overset{\displaystyle O}{\overset{\|}{C}}H$$

hexanal   propanedial

$CH_3(CH_2)_3CH_2$ ⟶ $CH_2(CH_2)_9CO_2H$

107

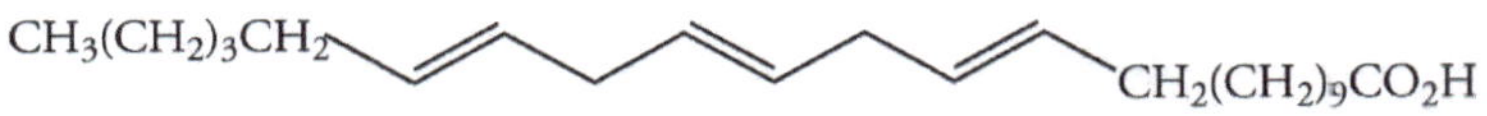

7 ALKYNES

KEYS TO THE CHAPTER

In many respects, the chemistry of alkynes closely resembles that of alkenes. Both classes of compounds have π bonds that dominate their chemical reactivity, which is largely addition of electrophiles. Moreover, at least some of the synthetic methods used to produce alkynes are the same as those used for the synthesis of alkenes, namely, elimination reactions. Terminal alkynes have one important new feature. The C—H bond of terminal alkynes is sufficiently acidic for the proton to be removed by strong bases. As a consequence, the conjugate base formed (a carbanion) is a nucleophile.

7.1 Occurrence and Uses of Alkynes

Alkynes are less common than alkenes in naturally occurring materials. The few examples cited that are of interest have multiple conjugated triple bonds. Carbon–carbon triple bonds are contained in a few drugs, including oral contraceptives.

7.2 Structure and Properties of Alkynes

A triple bond in an alkyne consists of one sigma bond and two pi bonds. As a result of the geometry of the sp hybrid orbitals, the two carbon atoms of the triple bond and the two atoms directly attached are collinear. There are two classes of alkynes—**monosubstituted** (terminal) and **disubstituted** (internal).

The greater % s character of the sp-hybridized carbon atom of alkynes strongly affects the properties of the bond of that carbon atom. The electrons in the bond are held more tightly by the carbon atom, and as a consequence, the homolytic cleavage of the C—H bond requires a greater amount of energy. The length of the C—H bond as well as bonds to other atoms is shorter than for sp^2 and sp^3 bonds of the same type.

The bond energy of the carbon–carbon triple bond reflects the less effective bonding of π electrons. The bond energy of the carbon–carbon triple bond is substantially less than three times the bond energy of a carbon–carbon single bond.

The heats of formation of alkynes containing 10 or fewer carbon atoms are positive because they contain a triple bond that is less stable than carbon–hydrogen and carbon–carbon single bonds. The heats of formation of disubstituted alkynes are less positive than the heats of formation of isomeric monosubstituted alkynes.

Alkynes are relatively nonpolar molecules, and their boiling points are controlled by London forces. Terminal alkynes have small dipole moments that are slightly larger than the dipole moments of terminal alkenes. Internal alkynes have no dipole moment. The chemical properties of alkynes are similar to the properties of alkenes. The only difference is that there are twice as many π bonds to react.

7.3 Nomenclature

Alkynes are named by selecting the longest continuous carbon chain that contains the triple bond. The chain is numbered to assign the lowest number to the first carbon atom of the triple bond. Alkyl groups and halogens are disregarded in selecting the direction of numbering unless the same number for the triple bond is obtained from either end of the chain. For compounds containing both double and triple bonds, the chain is numbered from the end nearer the first multiple bond. However, in equivalently placed multiple bonds, the double bond takes precedence over triple bonds in the direction of numbering. Compounds with both double and triple bonds are called enynes, not ynenes.

7.4 Acidity of Terminal Alkynes

Although weakly acidic, terminal alkynes can be converted to their conjugate bases called alkynide ions. The pK_a is approximately 25. Thus, a base whose conjugate acid has a pK_a greater than 25 must be used to abstract the hydrogen ion. Hydroxide ion is not sufficiently basic, but the amide ion is. Because the pK_a value of ammonia is approximately 36, the equilibrium constant for the reaction of an alkyne with amide ion is 10^7. This reaction is used to produce alkynide ions for use as nucleophiles in displacement of a halide ion from a haloalkane to synthesize alkynes.

7.5 Hydrogenation of Alkynes

When the hydrogenation of alkynes is catalyzed with finely divided platinum, palladium, or nickel, hydrogenation of alkynes is complete and produces alkanes. Hydrogenation requires one mole of hydrogen gas for each π bond in a compound, so triple bonds require two moles of hydrogen gas.

It is possible to stop the hydrogenation of alkynes after adding one molar equivalent of hydrogen, giving alkenes as the product. This is accomplished by using a specially prepared catalyst. Hydrogenation of alkynes with Lindlar catalyst produces *cis*-alkenes by *syn* addition, whereas hydrogenation using lithium in liquid ammonia produces *trans*-alkenes by *anti* addition.

7.6 Electrophilic Addition Reactions

An unsymmetrical reagent such as HBr adds to a triple bond in a characteristic way given by Markovnikov's rule. The hydrogen atom adds to the less substituted carbon atom of the triple bond. The bromine atom adds to the more substituted carbon atom of the triple bond. The addition product has the two added atoms *trans* in the resulting alkene, although the stereoselectivity may be low.

Hydrogen bromide adds more slowly to triple bonds than to double bonds. However, after one mole of hydrogen bromide has added, the resulting double bond is less reactive as a result of the electron withdrawing bromine atom. As a consequence, it is possible to obtain the product formed from the addition of one mole of HBr. When the second mole of HBr adds, the product has two hydrogen atoms added to the carbon atom that was less substituted originally. Two bromine atoms are located on the other carbon atom.

Two moles of bromine will add to compounds with triple bonds. The initial addition product has two *trans* bromine atoms. Continued addition yields a tetrabromoalkane.

Hydration of alkynes results in the Markovnikov addition of one mole of water to give an enol that rearranges to give a ketone.

7.7 Synthesis of Alkynes

Alkynes can be prepared from vicinal or geminal dihalides by a double dehydrohalogenation using a strong base such as $NaNH_2$ in liquid ammonia. Vicinal dihalides are obtained from the addition of a halogen to an alkene. The number of moles of base required for the reaction depends on the type of alkyne produced. A terminal alkyne produced in the synthesis is deprotonated by the reacting base, and thus a total of three moles of amide ion is required. Upon work up with water, the terminal alkyne forms.

1. Hydrogenation of Alkynes
A. Hydrogenation with Palladium on Carbon Catalysts

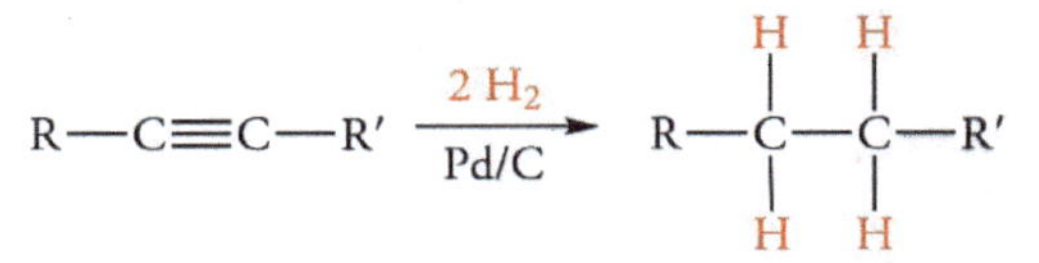

$$R—C≡C—R' \xrightarrow[\text{Pd/C}]{2\ H_2} R—\overset{\overset{H}{|}}{\underset{\underset{H}{|}}{C}}—\overset{\overset{H}{|}}{\underset{\underset{H}{|}}{C}}—R'$$

alkyne   alkane

B. *Syn* hydrogenation with Lindlar Catalyst

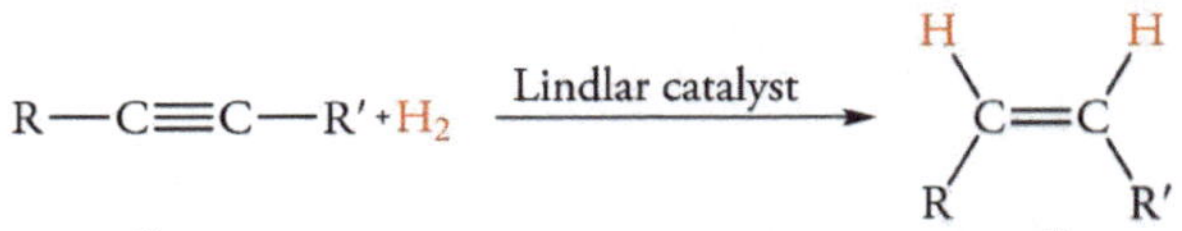

$$R—C≡C—R' + H_2 \xrightarrow{\text{Lindlar catalyst}}$$

cis-alkene

C. *Anti* Hydrogenation with Sodium in Liquid Ammonia

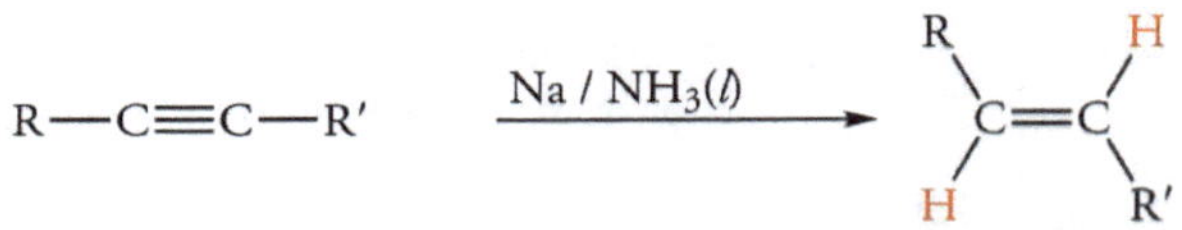

$$R—C≡C—R' \xrightarrow{Na\ /\ NH_3(l)}$$

alkyne   trans-alkene

2. Electrophilic Addition Reactions of Alkynes
A. Addition of Hydrogen Halides

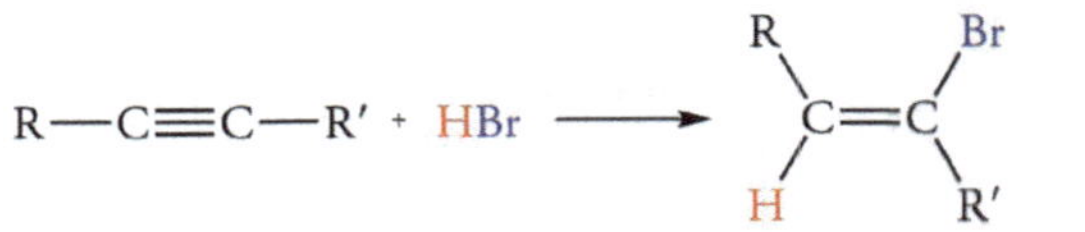

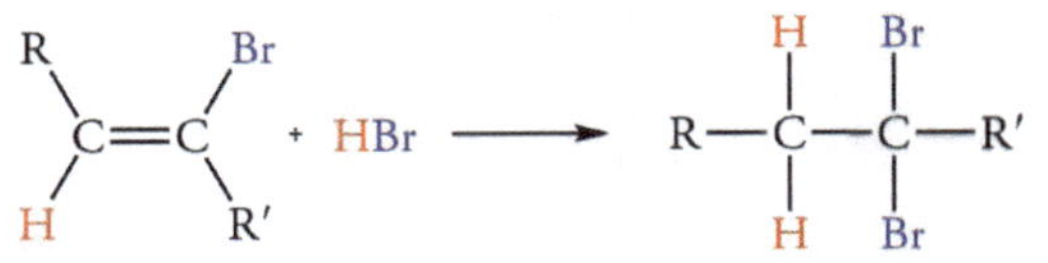

$$R—C≡C—R' + HBr \longrightarrow$$

$$\text{(alkene with R, Br, H, R')} + HBr \longrightarrow R—\overset{\overset{H}{|}}{\underset{\underset{H}{|}}{C}}—\overset{\overset{Br}{|}}{\underset{\underset{Br}{|}}{C}}—R'$$

B. Addition of Halogens

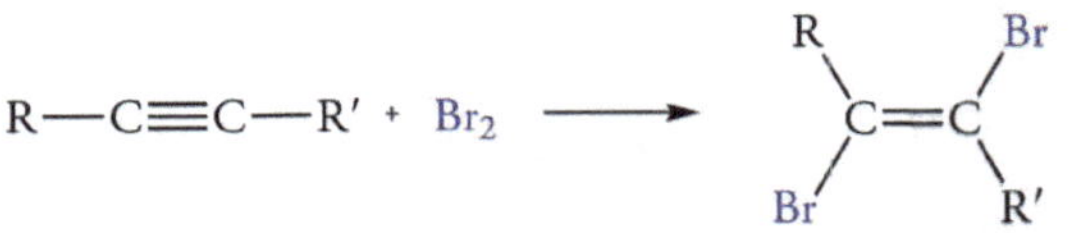

$$R—C≡C—R' + Br_2 \longrightarrow$$

$$\text{(alkene with R, Br, Br, R')} + Br_2 \longrightarrow R—\overset{\overset{Br}{|}}{\underset{\underset{Br}{|}}{C}}—\overset{\overset{Br}{|}}{\underset{\underset{Br}{|}}{C}}—R$$

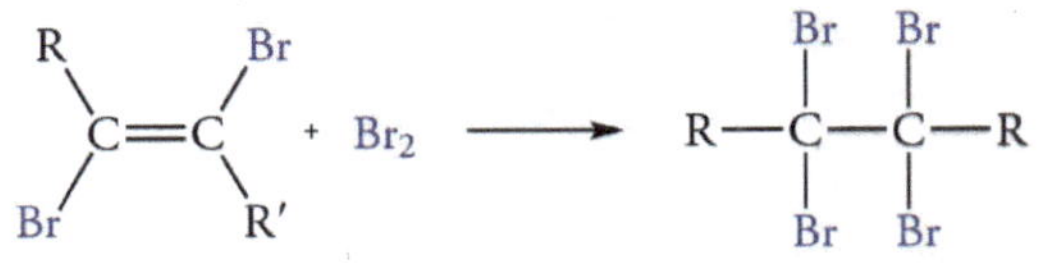

C. Hydration of Alkynes

Step 1. Formation of enol

$$R-C\equiv C-H \xrightarrow[\text{H}_2\text{SO}_4 \text{ / HgSO}_4]{\text{H}_2\text{O}}$$

enol

Step 2. Spontaneous conversion of enol to ketone

enol $\xrightarrow[\text{concerted}]{\text{1,3-proton shift}}$ ketone

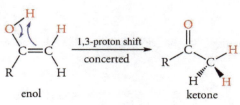

$$\bigcirc\!\!-\!C\equiv C-H \xrightarrow[\text{HgSO}_4]{\text{H}_2\text{O/ H}_2\text{SO}_4} \bigcirc\!\!-\!\overset{O}{\underset{}{C}}-CH_3$$

3. Synthesis of Alkynes
A. Synthesis of Alkynes by Dehydrohalogenation

$$R-\underset{\underset{Br}{|}}{\overset{\overset{Br}{|}}{C}}-\underset{\underset{H}{|}}{\overset{\overset{H}{|}}{C}}-R' \;+\; 2\,NaNH_2 \xrightarrow{NH_3\,(l)} R-C\equiv C-R' \;+\; 2\,NH_3 \;+\; 2\,NaBr$$

geminal dibromide

$$R-\underset{\underset{H}{|}}{\overset{\overset{Br}{|}}{C}}-\underset{\underset{Br}{|}}{\overset{\overset{H}{|}}{C}}-R' \;+\; 2\,NaNH_2 \xrightarrow{NH_3\,(l)} R-C\equiv C-R' \;+\; 2\,NH_3 \;+\; 2\,NaBr$$

vicinal dibromide

$$CH_3(CH_2)_3-\underset{\underset{H}{|}}{\overset{\overset{Cl}{|}}{C}}-\underset{\underset{Cl}{|}}{\overset{\overset{H}{|}}{C}}-H \xrightarrow[\text{2. H}_2\text{O}]{\text{1. 3NaNH}_2\text{ / NH}_3\,(l)} CH_3(CH_2)_3-C\equiv C-H$$

1,2-dichlorohexane   1-hexyne

B. Synthesis of Alkynes by Alkylation

$$R-C\equiv C-H \xrightarrow{NaNH_2\text{ / NH}_3\text{ (l)}} R-C\equiv C\!:^-\,Na^+ \;+\; NH_3$$

terminal alkyne   alkynide

$$R-C\equiv C\!:^- \;+\; R'CH_2Br \longrightarrow R-C\equiv C-CH_2R'$$

primary alkyl halide

112

End of Chapter Exercises

Structures of Alkynes

7.1   What is the molecular formula of each of the following compounds that contain carbon–carbon triple bonds?
       (a) mycomycin, an antibiotic

Answer: $C_{13}H_{10}O$ H—C≡C—C≡C—CH=C=CH—CH=CH—CH₂CO₂H

mycomycin

       (b) capillin, a skin fungicide

Answer: $C_{12}H_8O$

capillin

       (c)) ichthyothereol, a convulsant

Answer: $C_{15}H_{18}O$

CH₃—C≡C—C≡C—C≡C—C=C

ichthyothereol

7.2   Classify the triple bond in each of the following drugs. MDL 18962 is a drug used in breast cancer therapy. RU 486 is a drug used to induce abortion and may be useful in cancer therapy.

Answers:
(a) MDL 1280 monosubstituted

(b) RU 486 disubstituted

MDL 18962   RU 486

7.3   Write the molecular formula for the compounds with each of the following structural features.
       (a) six carbon atoms and one double bond   (b) five carbon atoms and two double bonds
       (c) seven carbon atoms, a ring, and one double bond   (d) four carbon atoms and one triple bond

Answers: (a) C_6H_{12} (b) C_5H_8 (c) C_7H_{12} (d) C_4H_6

7.4   What is the molecular formula for the compounds with each of the following structural features?
       (a) four carbon atoms and two triple bonds   (b) four carbon atoms, a double bond, and a triple bond
       (c) ten carbon atoms and two rings   (d) ten carbon atoms, two rings, and five double bonds

Answers: (a) C_4H_2 (b) C_4H_4 (c) $C_{10}H_{18}$ (d) $C_{10}H_8$

Properties of Alkynes

7.5 The heats of formation of 1-pentyne and 2-pentyne are 144 and 128.6 kJ mole^{-1}, respectively. Which compound is more stable? Based on this information, which compound has the larger heat of combustion?

Answer: One sp-hybridized carbon atom decreases the carbon–carbon bond length relative to the sp^3–sp^3 bond of propane from 154 to 146 pm. Changing a second sp^3-hybridized carbon atom to sp should decrease the bond length to 138 pm.

7.6 The heats of formation of 1-pentyne and 1,4-pentadiene are 144 and 106 kJ mole^{-1}, respectively. What does this information indicate about the relative stability of a triple bond compared to two double bonds?

Answer: The heats of formation are positive, and both compounds are unstable with respect to the elements. Thus, the compound with the lower positive heat of formation is more stable assuming that the entropies of formation of the two compounds are approximately equal. In this case, 2-pentyne, which has the more substituted triple bond, is the more stable. The heat of combustion measures the heat energy released ($\Delta H° < 0$) when carbon dioxide and water are formed. The less stable (highest energy) isomer releases more energy in the combustion reaction. In this case, 1-pentyne has the more negative heat of combustion.

7.7 The heats of formation of 1-propyne and 1,2-propadiene (allene) are 185 and 190 kJ mole^{-1}, respectively. Assuming that an equilibrium can be established, which compound would be present in the larger amount?

Answer: The heats of formation are positive for both compounds and both are unstable with respect to the elements. Thus, the compound with the lower positive heat of formation is more stable assuming that the entropies of formation of the two compounds are approximately equal. In this case, 1,4-pentadiene is the more stable isomer, so two double bonds are more stable than one triple bond.

7.8 Predict the direction of the dipole moment of 1-propyne. Why is its dipole moment larger than that of 1-propene?

Answer: The positive end of the dipole is the methyl group because the sp-hybridized carbon atom in the middle has greater s character and attracts electron density from the sp^3-hybridized carbon of the methyl group. The dipole moment of propene is smaller than that of propyne because the electrons are less strongly attracted to the sp^3-hybridized carbon atom of propene.

7.9 The boiling points of 1-alkynes are higher than those that of the 1-alkenes with the same number of carbon atoms. Suggest reasons for this fact.

Answer: The 1-alkynes are slightly more polar than 1-alkenes. The electrons in the two π bonds of an alkyne are more polarizable than the electrons in the single π bond of an alkene.

7.10 The boiling points of 3,3-dimethyl-l-butyne and 1-hexyne are 39.5 and 71.3 °C, respectively. Explain why the values are so different for these two isomers.

Answer: 3,3-Dimethyl-1-butyne has a more compact and somewhat spherical structure. The London forces for such compounds are smaller than for cylindrical structures such as 1-hexyne.

7.11 The boiling points of terminal alkynes are lower than the boiling points of isomeric internal alkynes. Is this fact consistent with the dipole moments of the compounds? If not, what other structural factors might contribute to the difference in the boiling points?

Answer: Terminal alkynes have a larger dipole moment than internal alkynes. Thus terminal alkynes should have higher boiling points than isomeric internal alkynes if polarity were the only structural feature determining this physical property. There is decreased freedom of motion of more carbon atoms of internal alkynes compared to terminal alkynes. Thus, the shape of the internal alkyne allow stronger London forces between the more linear chains.

Nomenclature

7.12 Name each of the following compounds.

(a) $CH_3CH_2CH_2C{\equiv}CH$ (b) $(CH_3)_3CC{\equiv}CCH_2CH_3$ (c) $CH_3{-}C{\equiv}C{-}CH{-}CH_3$
$$\overset{|}{CH_2CH_3}$$

Answers: (a) 1-pentyne   (b) 2,2-dimethyl-3-hexyne   (c) 4-methyl-2-hexyne

7.13   Name each of the following compounds.

Answers: (a) 4,5-dibromo-2-hexyne   (a) $CH_3CHBrCHBr$—$C\equiv C$—$CH_2(CH_2)_2CH_3$

(b) 1-chloro-3-octyne

(b) $ClCH_2$—CH_2—$C\equiv C$—CH—CH_3 with Cl below the CH

(c) 2-chloro-6-methyl-3-oc-tyne

(c) CH_3—$\overset{H}{\underset{CH_2CH_3}{C}}$—$CH_2$—$C\equiv C$—$\underset{Cl}{CH}$—$CH_3$

7.14   Write the structural formula for each of the following compounds.
        (a) 2-hexyne   (b) 3-methyl-l-pentyne   (c) 5-ethyl-3-octyne

Answers: (a) $CH_3CH_2CH_2$—$C\equiv C$—CH_3
                    2-hexyne

(b) CH_3—CH_2—$\underset{CH_3}{CH}$—$C\equiv C$—H
                    3-methyl-l-pentyne

(c) $CH_3CH_2CH_2CH$—$C\equiv C$—CH_2CH_3 with CH_2CH_3 below the CH
                    5-ethyl-3-octyne

7.15   Write the structural formula for each of the following compounds.
        (a) 3-heptyne   (b) 4-methyl-l-pentyne   (c) 5-methyl-3-heptyne

Answers: (a) $CH_3CH_2CH_2$—$C\equiv C$—CH_2CH_3 (b) CH_3—$\underset{CH_3}{CH}$—CH_2—$C\equiv C$—H
                    3-heptyne   4-methyl-l-pentyne

(c) CH_3—CH_2—$\underset{CH_3}{CH}$—$C\equiv C$—CH_2CH_3
                    5-methyl-3-heptyne

7.16   Write the structural formula for 4-ethynyl-l,5-nonadien-7-yne.

Answer: CH_3—$C\equiv C$—$CH=CH$—CH—CH_2—$CH=CH_2$
            with below the CH: $C \equiv C$—H (vertical triple bond to C, then H)
                    4-ethynyl-l,5-nonadien-7-yne

7.17   Write the structural formula for 1-ethyl-3-(2-propynyl)cyclopentene.

Answer: CH_2—$C\equiv C$—H

(cyclopentene ring with CH_2CH_3 substituent)

                    1-ethyl-3-(2-propynyl)cyclopentene

7.18 What is the IUPAC name for the group —C≡C—CH$_3$?

Answer: Propargyl

7.19 Which of the drugs listed in Exercise 7.2 contains a propargyl group?

Answer: (a) MDL 1280 contains a propargyl group, —CH$_2$—C≡C—H.

Acidity of Terminal Alkynes

7.20 Diisopropylamide ion [(CH$_3$)$_2$CH]$_2$N⁻ is a strong base commonly used in organic reactions. Is it expected to be a stronger or weaker base than the amide ion?

Answer: The two isopropyl groups are inductively electron donating relative to hydrogen, and they increase the electron density on the negatively charged nitrogen atom. Therefore, diisopropylamide ion is a stronger base than the amide anion.

7.21 Suggest an experimental procedure to prepare 1-deuterio-l-propyne from propene.
Answer: Prepare the conjugate base of 1-propyne using a strong base such as the amide ion. Then add D$_2$O to the reaction mixture. Since the conjugate base of propyne is a stronger base than DO⁻, it reacts with D$_2$O to give 1-deutero-1propyne.

$$CH_3—C≡C—H \xrightarrow{NaNH_2} CH_3—C≡C:^-$$

$$CH_3—C≡C:^- \ + \ \overset{\overset{..}{O}:}{\underset{D \quad D}{}} \longrightarrow CH_3—C≡C—D \ + \ D—\overset{..}{\underset{..}{O}}:^-$$

1-deutero-1-propyne

Hydrogenation Reactions

7.22 How many moles of hydrogen gas will react with each of the following compounds?
(a) CH$_3$=CH=CH—C≡CH   (b) HC≡C—C≡C—H

(c) CH$_2$=CH—C≡C—CH=CH$_2$ (d) HC≡C—C≡C—C≡CH

Answers: (a) three (b) four (c) four (d) six

7.23 How many moles of hydrogen gas will react with each of the compounds listed in Exercise 7.1?

Answers: (a) eight (b) four (c) seven

7.24 Which compound should have the larger heat of hydrogenation for the addition of two moles of hydrogen gas, 1-pentyne or 1,4-pentadiene? Why?

Answer: The heat of formation of 1-pentyne is more positive than the heat of formation of 1,4-pentadiene. In the hydrogenation reaction, pentane is the common product. Because heats of hydrogenation are negative, the energy difference between 1-pentyne and pentane is larger than the energy difference between 1,4-pentadiene and pentane.

7.25 Stearolic acid is converted to oleic acid by hydrogenation using the Lindlar catalyst. Elaidic acid is the product obtained by sodium/ammonia reduction of stearolic acid. Write the structures of oleic and elaidic acids.

$$CH_3(CH_2)_7C≡C(CH_2)_7CO_2H$$
stearolic acid

Answer:

oleic acid (Z isomer)   elaidic acid (Z isomer)

7.26 Disparlure, the pheromone of the gypsy moth, can be prepared by reduction of an alkyne followed by epoxidation of the alkene. What alkyne is required? What is the configuration of the alkene? What reagents are required for reduction of the alkyne?

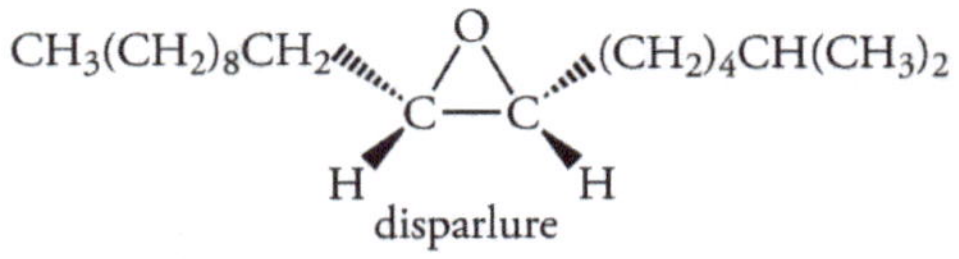

CH3(CH2)8CH2
O
(CH2)4CH(CH3)2
C—C
H
H

disparlure

Answer: The required alkene is (*Z*)-2-methyl-7-octadecene, which can be prepared by catalytic hydrogenation of 2-methyl-7-octadecyne using the Lindlar catalyst.

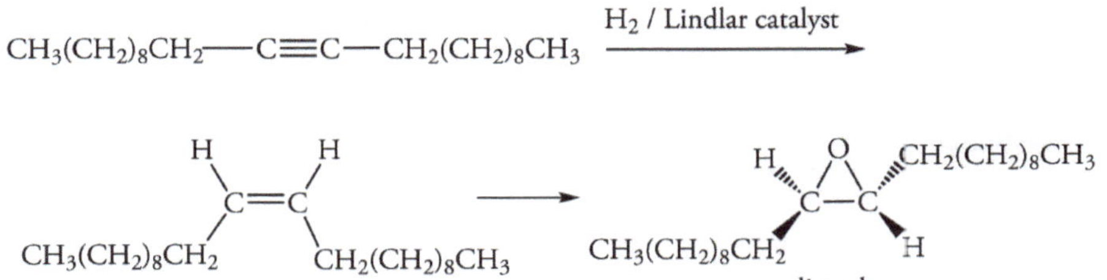

CH3(CH2)8CH2—C≡C—CH2(CH2)8CH3
H2 / Lindlar catalyst
H
H
C=C
CH3(CH2)8CH2
CH2(CH2)8CH3
H
O
CH2(CH2)8CH3
C—C
CH3(CH2)8CH2
H

disparlure

7.27 The pheromone of the grape berry moth is indicated below. How could this compound be prepared from a related alkyne. Would the ester functional group be affected by the reaction conditions?

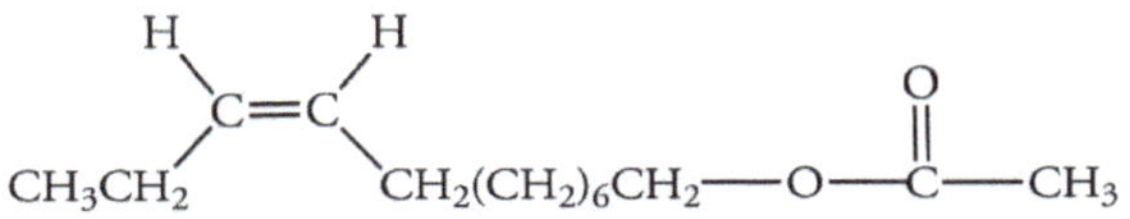

H
H
C=C
O
CH3CH2
CH2(CH2)6CH2—O—C—CH3

Answer: The (*Z*) configuration can be achieved by reduction of a structurally related alkyne using the sodium in liquid ammonia. The reaction conditions would result in abstraction of a proton from the alcohol, but it would be replaced in the workup of the reaction mixture. Its structure is shown below.

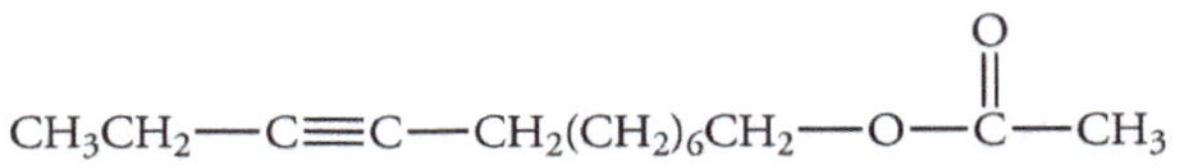

O
CH3CH2—C≡C—CH2(CH2)6CH2—O—C—CH3

7.28 (*E*)-11-Tetradecen-l-ol is one of the intermediate compounds required to synthesize the sex pheromone of the spruce budworm. How could this compound be prepared from an appropriate alkyne? Would the reaction conditions affect the hydroxyl group?

Answer: The (*E*) configuration can be achieved by reduction of a structurally related alkyne using the sodium in liquid ammonia. The reaction conditions would result in abstraction of a proton from the alcohol, but it would be replaced in the workup of the reaction mixture.

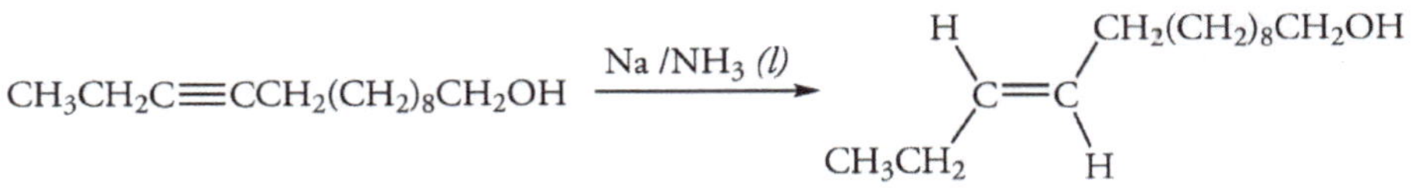

CH3CH2C≡CCH2(CH2)8CH2OH
Na /NH3 (l)
H
CH2(CH2)8CH2OH
C=C
CH3CH2
H

(*E*)-11-Tetradecen-l-ol

7.29 Draw the structure of the product of the reaction of the following compound with hydrogen using the Lindlar catalyst.

$$HO-CH_2CH_2-CH=\overset{\overset{\displaystyle CH_3}{|}}{C}-C\equiv C-H \xrightarrow{\text{H}_2\text{ / Lindlar Catalyst}} HO-CH_2CH_2-CH=\overset{\overset{\displaystyle CH_3}{|}}{C}-CH=CH_2$$

7.30 Draw the structure of the product of the reaction of the following compound with hydrogen using the Lindlar catalyst.

$$CH_3-C\equiv C-(CH_2)_3CH_2-\overset{\overset{\displaystyle O}{||}}{C}-OH \xrightarrow{\text{H}_2\text{ / Lindlar catalyst}}$$

Z-isomer

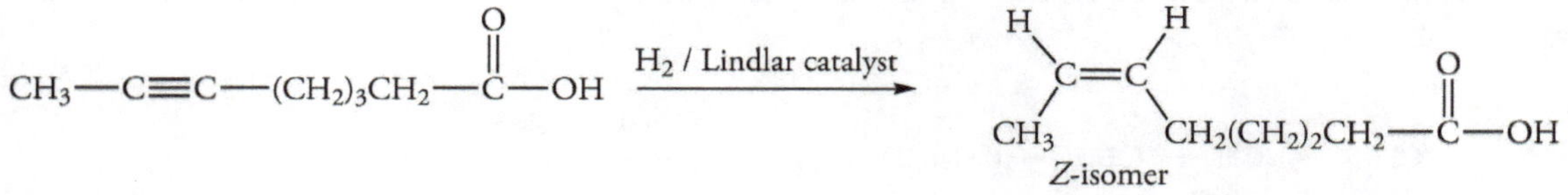

Electrophilic Addition Reactions

7.31 Addition of one mole of HCl to 2-hexyne gives a mixture of two products in approximately equal amounts. Draw their structures.

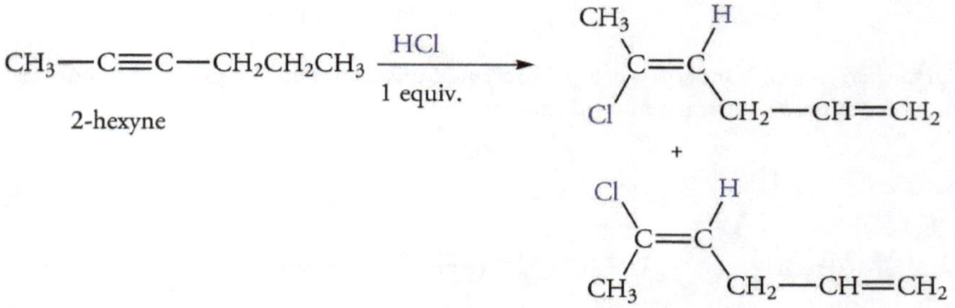

$$CH_3-C\equiv C-CH_2CH_2CH_3 \xrightarrow[\text{1 equiv.}]{\text{HCl}}$$

2-hexyne

+

7.32 Draw the structure of the addition of one mole of DBr to 1-propyne.

Answer: *Trans* addition of DBr gives the following (*E*) isomer.

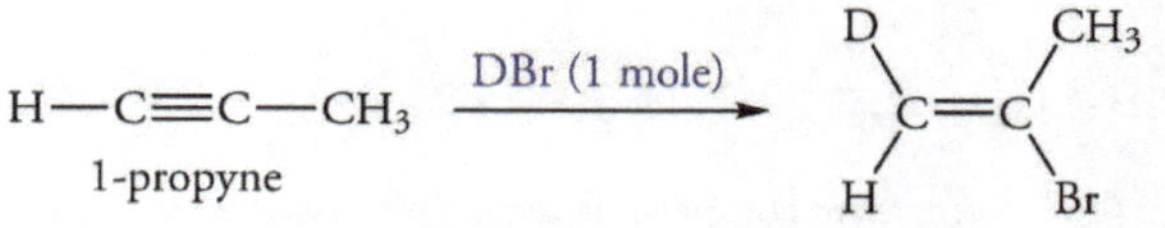

$$H-C\equiv C-CH_3 \xrightarrow{\text{DBr (1 mole)}}$$

1-propyne

7.33 Predict the product of the addition of one mole of Br$_2$ to 1-penten-4-yne.

Answer: The bromine adds *anti* to the triple bond to give the following product, which has no dipole moment.

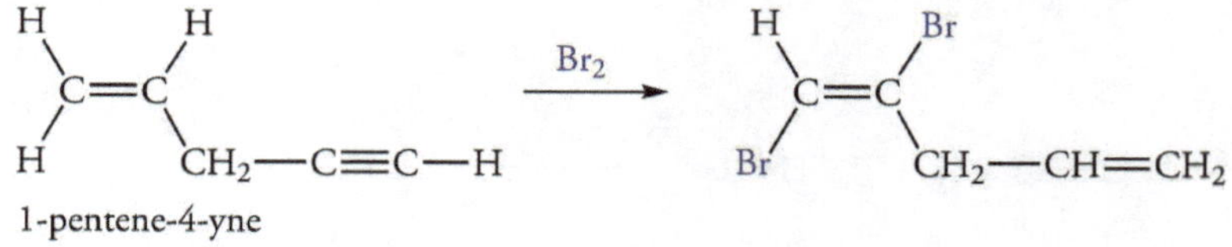

1-pentene-4-yne

7.34   Draw the structure of the compound resulting from the addition of one molar equivalent of bromine to acetylene dicarboxylic acid. What is the dipole moment of the product?

$$HO_2C—C≡C—CO_2H$$
acetylene dicarboxylic acid

Answer: The bromine adds *anti* to the triple bond to give the following product, which has no dipole moment.

7.35   Hydration of one of the following two compounds yields a single ketone product. The other compound yields a mixture of ketones. Which one yields only the single ketone product? Why?

$$CH_3CH_2C—C≡C—CH_2CH_3 \qquad\qquad CH_3C—C≡C—CH_2CH_2CH_3$$
$$\text{I} \qquad\qquad\qquad\qquad \text{II}$$

Answer: Compound I is a symmetrical alkyne and C-3 and C-4 are structurally equivalent. Hydration of this alkyne gives a single product with a carbonyl carbon atom located at C-3. Compound II forms two products. One has a carbonyl carbon atom located at C-3 and the other at C-2.

7.36   Hydration of 4-methyl-2-pentyne gives the following compounds in the indicated amounts. Suggest a reason for the observed product ratio.

Answer: The C-2 and C-3 carbon atoms each contain an alkyl group. Thus, based on inductive effects alone, there should be no regioselectivity, and either of them could react to give an intermediate enol in a hydration reaction. However, there is a small difference in the steric environments of each carbon atom. The slight regioselectivity in the reaction may be the result of this difference. The major product has the carbonyl oxygen atom at the least sterically hindered C-2 position.

Synthesis of Alkynes

7.37   Write the structure of all compounds that could yield the following alkyne upon dehydrohalogenation.

7.38   Which isomer, 2,2-dibromopentane or 3,3-dibromopentane, would give the better yield of 2-pentyne using sodium amide as the base?

Answer: 3,3-Dibromopentane can yield only 2-pentyne because C-2 and C-4 are equivalent in this symmetrical molecule. 2,2-Dibromopentane can yield 1-pentyne by elimination of hydrogen atoms at C-1 and 2-pentyne by elimination of hydrogen atoms at C-3.

7.39   Would the following reaction provide a good yield of the indicated product? Explain.

$$CH_3CH_2CH_2CBr_2CH_3 \xrightarrow{\ NaNH_2\ } CH_3CH_2—C\equiv C—CH_3$$

Answer: No, because elimination can result by abstraction of hydrogen atoms at either C-1 or C-3 to give a mixture of 1-pentyne and 2-pentyne.

7.40   Write the product of the reaction of 1,6-dibromohexane with excess sodium acetylide.

$$BrCH_2(CH_2)_4CH_2Br \xrightarrow[\text{(excess)}]{H—C\equiv C:^-} H—C\equiv C—CH_2(CH_2)_2CH_2—C\equiv C—H$$

7.41   Predict the product of the reaction of one equivalent of the alkynide of 1-propyne and 1-bromo-5-fluoropentane.

Answer: The carbon–fluorine bond is much stronger than the carbon–bromine bond, so bromide ion will be displaced to give 8-fluoro-2-octyne.

$$BrCH_2(CH_2)_3CH_2F \ + \ CH_3—C\equiv C:^- \longrightarrow CH_3—C\equiv C—(CH_2)_3CH_2F$$
$$\text{8-fluoro-2-octyne.}$$

7.42   Draw the structure of the final product of the following series of reactions.

$$H—C\equiv C—H \xrightarrow[\text{2. }CH_3CH_2CH_2I]{\text{1. }NaNH_2} \xrightarrow[\text{2. }CH_3CH_2I]{\text{1. }NaNH_2} ? \qquad CH_3CH_2—C\equiv C—(CH_2)_2CH_3$$
$$\text{3-heptyne}$$

7.43   Outline the steps of a synthesis of 2,2-dimethyl-3-octyne using reactants having no more than six carbon atoms.

Answer: Prepare the acetylide salt of 3,3-dimethyl-1-butyne and react it with 1-bromobutane. Note that reaction of the acetylide salt of 1-hexyne with 2-bromo-2-methylpropane would give only an elimination product because the alkyl halide is tertiary.

8 STEREOCHEMISTRY

KEYS TO THE CHAPTER

8.1 Configuration of Molecules

Stereoisomers have different configurations. The term "configuration" refers to the arrangement of atoms in space. Geometric isomers, which we previously studied in cycloalkanes and alkenes, are also stereoisomers.

8.2 Mirror Images and Chirality

Some molecules have mirror images that are not **superimposable**. Such molecules are **chiral**. Molecules that have a **plane of symmetry are achiral**; they are superimposable on their mirror image.

A **stereogenic center** in an organic molecule is a carbon atom bonded to four different atoms or groups of atoms. It is also called a **chiral center**. By inspecting the atoms or groups of atoms bonded to each carbon atom in a molecule, we can easily identify any chiral centers. If a carbon atom is bonded to two or more identical atoms or groups, such as two hydrogen atoms or two methyl groups, it is not a chiral center. *If a carbon atom is bonded to four different atoms or groups, it is a chiral center, and the molecule has a nonsuperimposable mirror image.* The two possible isomers having different configurations at a chiral center are **enantiomers.**

Another way to identify a molecule as chiral or achiral is to look for a plane of symmetry. A plane of symmetry can bisect atoms, groups of atoms, and bonds between atoms. In a molecule with a plane of symmetry, one side of the molecule is the mirror image of the other side. Thus, a molecule with a plane of symmetry is achiral. If a molecule contains two or more chiral centers and does not have a plane of symmetry, it is chiral. If is has a plane of symmetry, it is an achiral *meso* compound.

Pairs of enantiomers have the same physical properties but behave differently in a chiral environment such as a chiral binding site in an enzyme. Most of the molecules isolated from living organisms are chiral. They generate a chiral environment that allows distinctions to be made between enantiomers.

8.3 Optical Activity

Each member of a pair of enantiomers rotates the plane of polarized light in an instrument called a **polarimeter.** This phenomenon is called **optical activity.** The rotation observed for one enantiomer is equal in magnitude but opposite in direction for the other enantiomer. A chiral substance that rotates plane-polarized light clockwise is **dextrorotatory**; a chiral substance that rotates plane-polarized light counterclockwise is **levorotatory.** The amount of rotation under defined standard experimental conditions is the specific rotation. **Optical purity** is a measure of the excess of one enantiomer over another in a mixture.

8.4 Fischer Projection Formulas

Enantiomers in a Fischer projection are drawn according to the following conventions:
1. Arrange the carbon chain vertically with the most oxidized group (—CHO in glyceraldehyde) at the "top."
2. Place the carbon atom at the chiral center in the plane of the paper. It is C-2 in glyceraldehyde.
3. C-2 is bonded to four groups, the CHO group and the CH_2OH group extend behind the plane of the page, and the hydrogen atom and the hydroxyl group extend up and out of the plane.
4. Project these four groups onto a plane. The carbon atom at the chiral center usually not shown in this convention. It is located at the point where the bond lines cross. The vertical lines project away from the viewer. The horizontal lines project toward the viewer.

8.5 Absolute Configuration

The Kahn–Ingold–Prelog configurational nomenclature system, which is the same for both E,Z geometric isomers and chiral molecules, gives an unambiguous description of the absolute configuration of a molecule.

Priority is assigned to atoms based on the atomic numbers of directly bonded atoms. Atoms farther down the chain are ignored even though they may have still higher atomic numbers. Thus, a fluorine atom has a higher priority than a carbon atom even if that carbon atom is bonded to three chlorine atoms; for example, $F-> -CCl_3$. The chlorine atoms in this case are irrelevant because the comparison is between the atomic numbers of fluorine and carbon.

Once the priority order of the atoms or groups of atoms bonded to the chiral carbon atom has been determined, the molecule is viewed through the bond to the lowest priority group. The other three groups then lie on a circle. If the movement from priorities $1 \to 2 \to 3$ is clockwise, the molecule is R; if the motion is counterclockwise, the configuration is S.

The assignment of R or S configuration to a compound does not identify its optical rotation as being either (+) or (−). The direction of optical rotation is experimentally determined with a polarimeter. The absolute configuration is experimentally determined by X-ray crystallography.

8.6 Molecules with Two or More Stereogenic Centers

Some molecules have two or more stereogenic centers. The resulting stereochemistry depends on whether those centers are equivalent or nonequivalent. **Equivalent** sterogenic centers have identical sets of substituents. For n nonequivalent centers, there are 2^n stereoisomers. Some of those isomers are pairs of enantiomers. These stereoisomers have opposite configurations at every center and are thus mirror images. All other stereoisomers are termed **diastereomers.**

The configuration of each stereogenic center is determined independently. Then, the configuration of each center is written as R or S. For example, the enantiomer of a molecule with a sterogenic center $2S,3R$ is $2R,3S$. Any other combination—$2S,3S$ or $2R,3R$— is a diastereomer.

Compounds with two or more equivalent stereogenic centers have fewer stereoisomers than predicted by the 2^n formula. Some of the stereoisomers have a plane of symmetry and are not optically active; they are **meso compounds**. For two chiral centers, the configurations are R,S, which is the same as S,R because of the plane of symmetry. The isomers R,R and S,S are optically active and are enantiomers.

8.7 Cyclic Compounds with Stereogenic Centers

Cyclic compounds can have stereogenic centers. We apply the same rules to assign configuration to cyclic compounds and acyclic compounds. The only difference is that we eventually return to the stereogenic center as we move around the ring. However, in a chiral compound, the point of first difference is reached before that time.

Cyclic compounds having two nonequivalent stereogenic centers can exist in four stereoisomeric forms. An interesting feature of these molecules is seen when there are equivalent stereogenic centers. In those cases, there is at least one plane of symmetry. That plane, in some cases, bisects bonds, and in other cases bisects the atoms of the ring. In this latter case, it also bisects the atoms bonded to the stereogenic centers.

8.8 Separation of Enantiomers

Enantiomers have the same physical properties and, therefore, cannot be separated by physical methods. However, diastereomers have different physical properties and can be separated. Figure 8.18 illustrates the conversion of a mixture of enantiomers into a mixture of diastereomers. The diastereomers are separated, after which they are broken down to obtain one enantiomer from one diastereomer and the other enantiomer from the second diastereomer. Chiral chromatography provides a way to separate enantiomers based upon their diastereomeric interactions with a chiral column support.

8.9 Reactions at Stereogenic Centers

If a reaction at a stereogenic center does not change the bonds to the stereogenic center, then the configuration at that center is unchanged.

Reactions at the stereogenic center affect the configuration of the molecule. If the product has a configuration opposite that of the reactant, we postulate a transition state in which the nucleophile attacks opposite the bond to the leaving group and inverts the configuration as the reaction occurs.

Radical reactions proceed through a planar intermediate, which is achiral. Thus, subsequent reaction with another radical can occur with equal probability from either side of the plane of the molecule. The result is a racemic mixture.

8.10 Formation of Compounds with Stereogenic Centers

Formation of compounds with one stereogenic center from achiral compounds using achiral reagents cannot yield a single stereoisomer. However, in an enzyme catalyzed process, the reaction of an achiral compound generates a single stereoisomer. Such reactions are **stereospecific**.

In some reactions where two stereogenic centers are generated from an achiral substrate, some mechanistic information is obtained based on the diastereomers formed. For example, the formation of two equivalent centers might give a mixture of the *R,R* compound and the *S,S* compound. Although not optically active, that result is different than a process that gives the *R,S* (*meso*) compound.

8.11 Reactions that Form Diastereomers

If a new stereogenic center is generated in a reaction of a substrate that already has a stereogenic center, then a mixture of diastereomers results. The amounts of these isomers are not equal because the new center is generated in a chiral environment. The excess of one diastereomer over another is called the **stereoselectivity** of the reaction.

8.12 Prochiral Centers

In a chiral environment, two apparently equivalent groups can be distinguished, and the resulting product of a reaction involving those groups is chiral. The atomic center at which optical activity may result is prochiral. The equivalent groups bonded to the prochiral center are enantiotopic and are designated **pro-*R*** or **pro-*S*** to indicate the potential configuration, *R* or *S*, if the group is replaced.

Groups at a prochiral center in a molecule that contains a chiral center are **diastereotopic.** The "faces" of a planar site or functional group that contains a center that can be converted into a stereogenic center are designated as **re** or **si** depending on the priority ranking of the three groups and their arrangement using the *R,S* rules.

End of Chapter Exercises

Chirality

8.1 Which of the following isomeric methylheptanes has a chiral center?

(a) 2-methylheptane   (b) 3-methylheptane   (c) 4-methylheptane

Answers: (a) none (b) one   (c) none

$$CH_3-CH_2-\underset{\underset{H}{|}}{\overset{\overset{CH_3}{|}}{C}}-CH_2-CH_2-CH_2-CH_3$$

3-methylheptane

8.2 Which of the following isomeric bromohexanes has a chiral center?

(a) l-bromohexane   (b) 2-bromohexane   (c) 3-bromohexane

Answers: (a) none (b) one   (c) one

(b)
$$CH_3-\underset{\underset{H}{|}}{\overset{\overset{Br}{|}}{C}}-CH_2-CH_2-CH_2-CH_3$$

2-bromohexane

(c)
$$CH_3-CH_2-\underset{\underset{H}{|}}{\overset{\overset{Br}{|}}{C}}-CH_2-CH_2-CH_3$$

3-bromohexane

8.3 Which of the compounds with molecular formula $C_5H_{11}Cl$ has a chiral center?

Answer: 2-chloropentane and 2-chloro-3-methylbutane each have one chiral center.

$$CH_3-\underset{\underset{H}{|}}{\overset{\overset{Cl}{|}}{C}}-CH_2-CH_2-CH_3$$

2-chloropentane

$$CH_3-\underset{\underset{H}{|}}{\overset{\overset{Cl}{|}}{C}}-\underset{\underset{H}{|}}{\overset{\overset{CH_3}{|}}{C}}-CH_3$$

2-chloro-3-methylbutane

8.4 Which of the compounds with molecular formula $C_3H_5Cl_2$ has a chiral center?

Answer: 1,2-dichloropentane has one chiral center.

$$CH_3-\underset{\underset{H}{|}}{\overset{\overset{Cl}{|}}{C}}-\underset{\underset{H}{|}}{\overset{\overset{H}{|}}{C}}-Cl$$

1,2-dichloropropane

8.5 Which of the following isomeric methylheptanes has a chiral center?

Answers: (a) none (b) two (c) none
(d) one

(a)

(b)

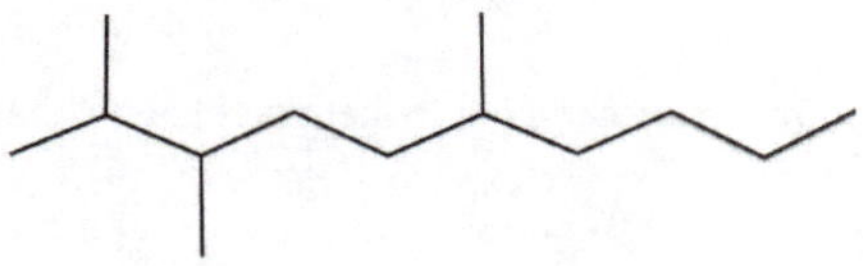

(c)

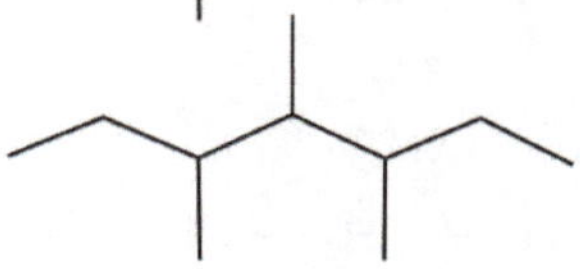

(d)

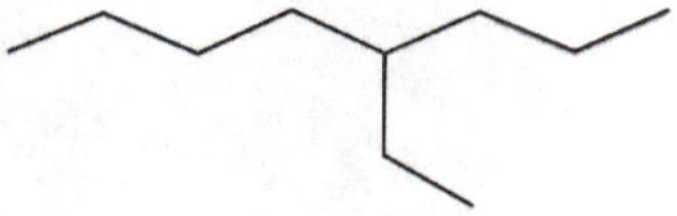

8.6 How many chiral centers does each of the following cyclic compounds have?

Answers: (a) two (b) one (c) none
(d) none

(a) 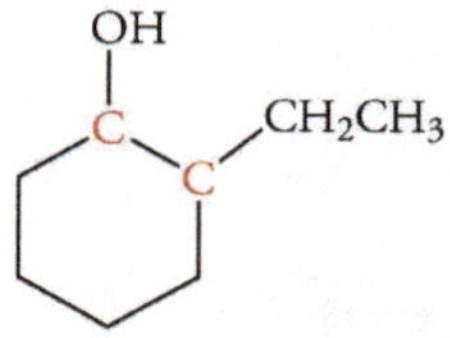 OH
C — CH$_2$CH$_3$
C

(b) 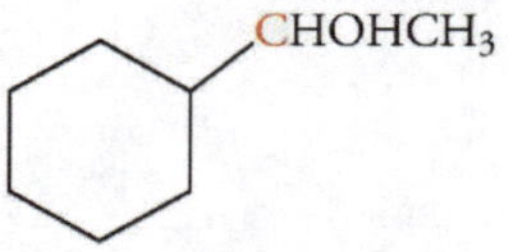 CHOHCH$_3$

(c) CH$_2$CH$_2$OH

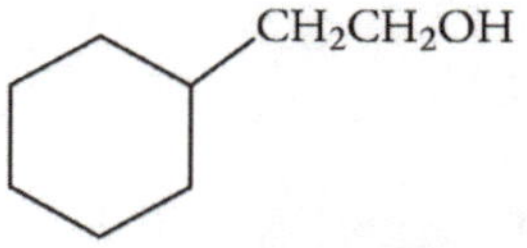

(d) 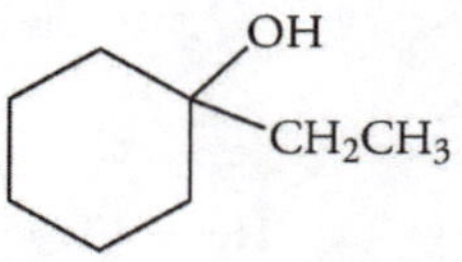 OH
CH$_2$CH$_3$

8.7 How many chiral centers does each of the following barbiturates have?

Answers: (a) one (b) two (c) one
(d) one

(a) phenobarbital

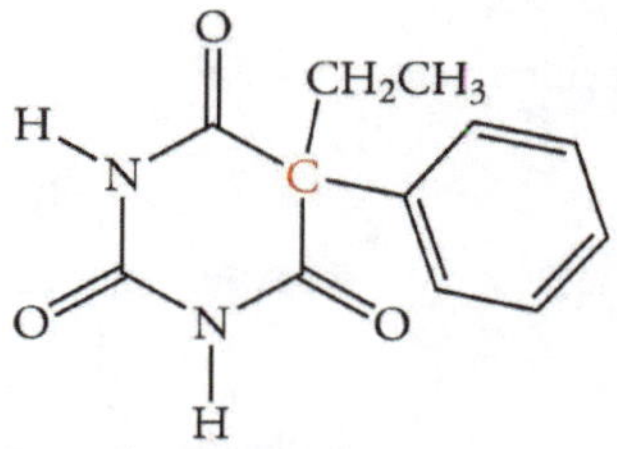

O
H–N
C — CH$_2$CH$_3$
O   N
O
H

(b) secobarbital

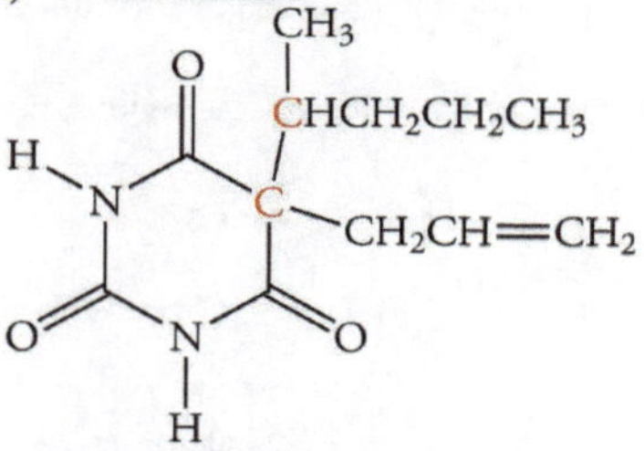

CH$_3$
O
H–N
C — CHCH$_2$CH$_2$CH$_3$
CH$_2$CH=CH$_2$
O   N
O
H

(c) hexobarbital

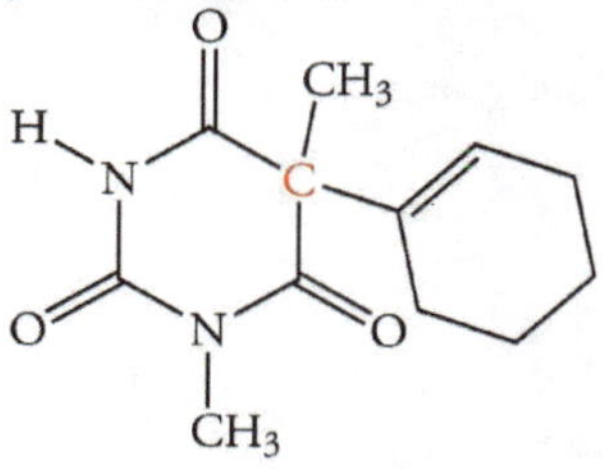

O
H–N
C — CH$_3$
O   N
O
CH$_3$

(d) amobarbital

O
H–N
C — CH$_2$CH$_3$
CH$_2$CH$_2$CH(CH$_3$)$_2$
O   N
O
H

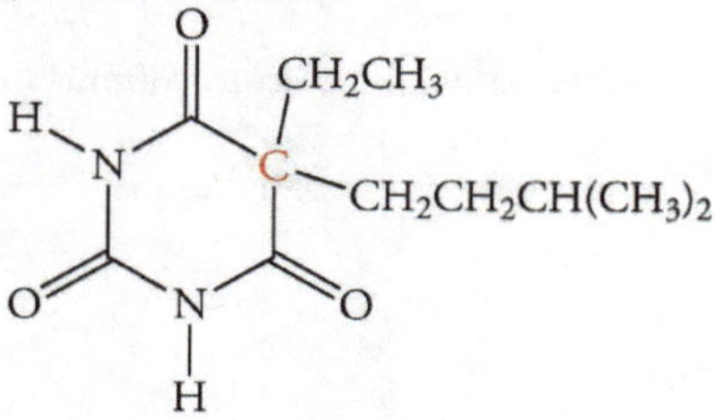

8.8 How many chiral centers does each of the following drugs have?

Answers: (a) none (b) one (c) none
(d) two

(a)   phenylbutazone, used to treat gout

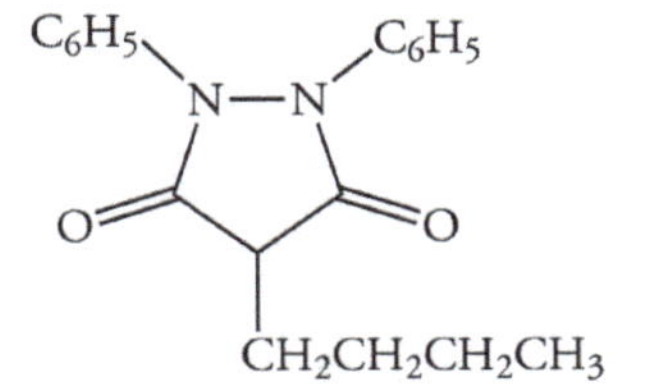

$CH_2CH_2CH_2CH_3$

(b)   ibuprofen, an analgesic

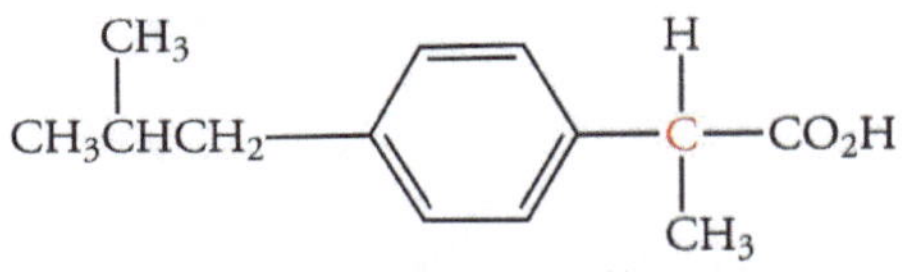

S enantiomer, most effective

(c)   chlorphentermine, a nervous system stimulent

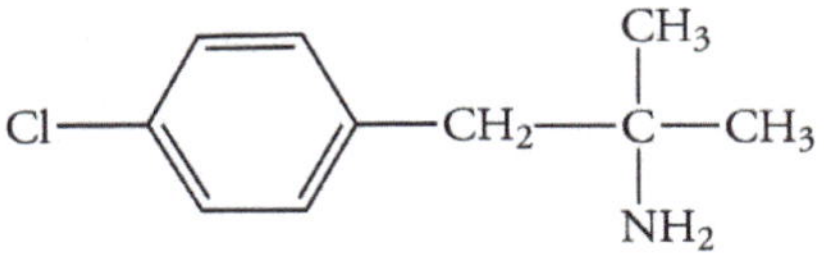

(d)   chloramphenicol, an antibiotic

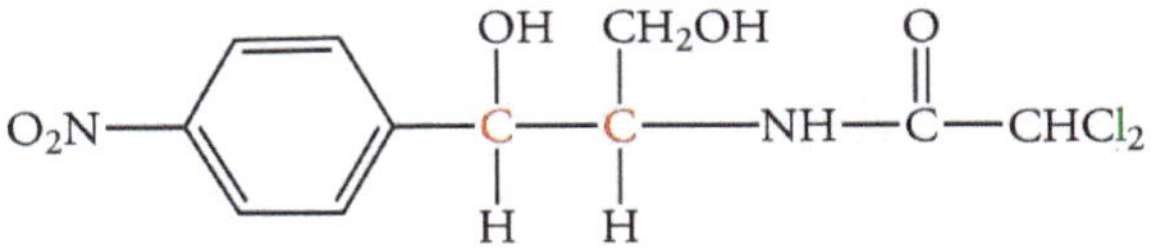

8.9 How many chiral carbon atoms are in each of the following synthetic anabolic steroids?

Answers: (a) six   (b) seven

(a)   Dianabol

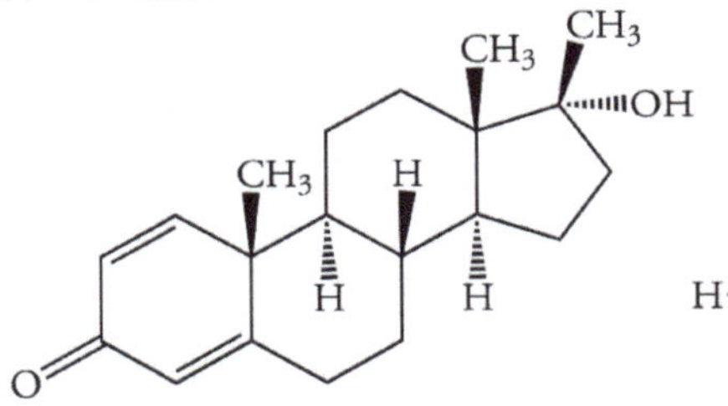

(b)   stanozolol

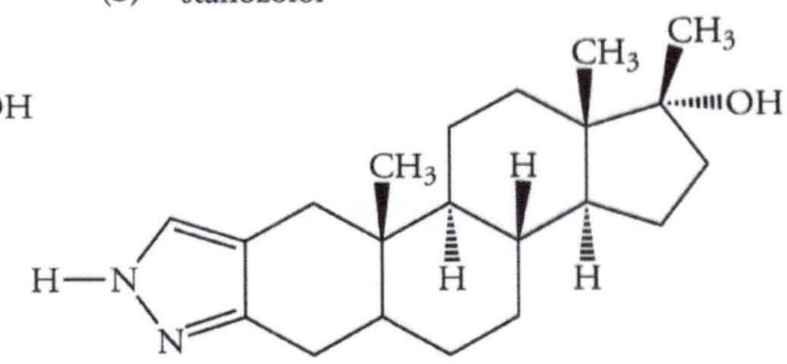

8.10 Determine the number of chiral centers in the male sex hormone testosterone and in the female sex hormone estradiol.

(a)   testosterone

Answers: (a) six   (b) five

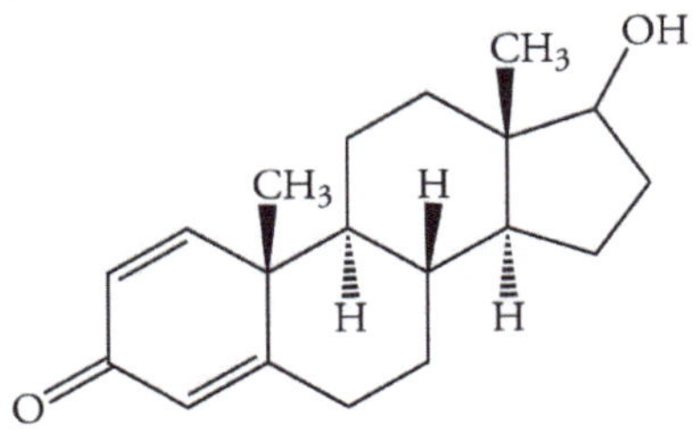

(b)   estradiol

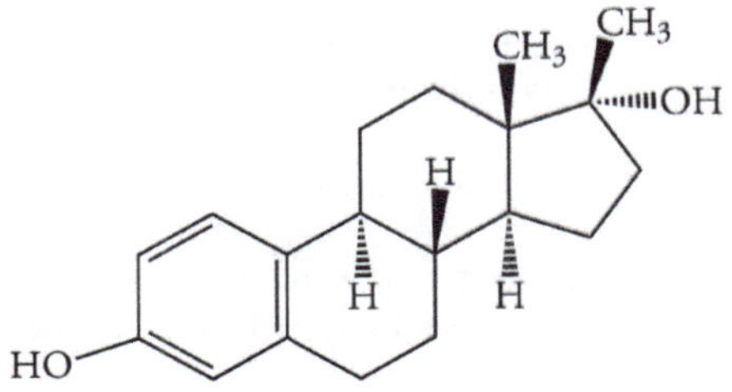

Plane of Symmetry

8.11 Determine whether each of the following compounds has a plane of symmetry.

Answer: Only (b) and (c) have a plane of symmetry.

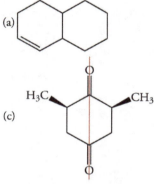

(a)

(b) —————————— plane of symmetry

(c)

(d)

plane of symmetry

8.12 Determine whether each of the following compounds has a plane of symmetry.

Answer: Only (a) and (d) have a plane of symmetry.

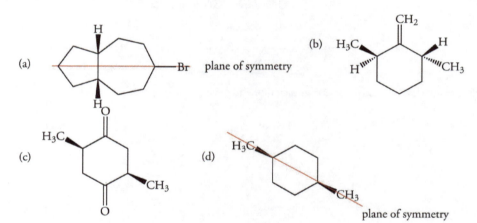

(a) ——————— Br   plane of symmetry

(b) H_3C

(c)

(d)

plane of symmetry

Optical Activity

8.13 Lactic acid in the blood has a specific rotation of +2.6°. A sample of lactic acid obtained from sour milk has a specific rotation of −2.6°. How do these compounds differ?

Answer: The compounds are enantiomers.

8.14 Optically pure (S)-(+)-citronellol from citronella oil has a specific rotation of +5.3°. An enantiomer of optically pure (S)-(+)-citronellol is obtained from geranium oil. What is its specific rotation?

Answer: The specific rotation of the enantiomer is −5.3°.

8.15 The configuration of naturally occurring MSG, which has a specific rotation of +24° is S. Is the assignment of configuration based upon the sign of the optical rotation correct?

Answer: No, it is incorrect. An assignment based upon the optical rotation of a compound is not related to its configuration. An S isomer can have a positive sign of rotation.

8.16 Carvone obtained from spearmint oil is the (R)-(−)-enantiomer. Explain the meaning of both terms within parentheses.

Answer: The R refers to the configuration at the chiral center. The (−) refers to the sign of the optical rotation of the compound.

8.17 A solution of 3 g of menthol in 50 mL of ethanol is prepared and a sample is placed in a 10-cm tube. The optical rotation is +3.0°. What is the specific rotation of menthol?

Answer: The concentration is 0.06 g/mL. The 10-cm tube is 1 dm long. Using the observed rotation and substituting into the equation to calculate specific rotation in Section 8.3, the specific rotation is 50°.

8.18 The specific rotation of (R)-2-bromobutane in ethanol is −23.1°. A solution of the compound in a 1-dm tube has $[\alpha]_D = 55°$.
What   is the concentration of the compound in grams per 100 mL?

Answer: Using the equation to calculate specific rotation, the concentration is calculated as 2.4 g/mL.

8.19 The specific rotation of (+)-2-butanol as a pure liquid is +13.9°. A synthetic sample of 2-butanol has an optical rotation of −4.5°. What is the composition of the sample?

Answer: The synthetic sample has a majority of the enantiomer of the opposite configuration of the reference (+) isomer. The optical purity of the synthetic sample is calculated as (4.5/13.9) × 100% = 32%, with the majority being the (−) isomer. If there is x% of the (+) isomer, there is (100 − x) % of the − isomer. Solving the equation 32% = (100−x) − x, there is 66% of the (−) isomer and 34% of the (+) isomer.

8.20 The specific rotation of the S enantiomer of MSG, a flavor enhancer, is +24°. What is the optical purity of a synthetic sample whose a is +6°? What are the percentages of the two enantiomers in the sample?

Answer: The synthetic sample has a majority of the reference (S) isomer. The optical purity of the synthetic sample is calculated as (6/24) × 100% = 25%, with the majority being the (S) isomer. If there is x% of the (S) isomer, there is (100 − x) % of the (R) isomer. Solving the equation 25% = x − (100−x), there is 62% of the (S) isomer and 38% of the (R) isomer.

Fischer Projection Formulas

8.21 Draw the Fischer projection formula of the following enantiomer of naturally occurring threonine obtained from proteins. Draw all diastereomers as well.

Answer: The structure must first be rotated around the C-2 to C-3 bond to obtain an eclipsed conformation of the CH_3 and CO_2H groups and turned to place the most oxidized group at the top. The Fischer projection of the enantiomer given is the first of the series of four stereoisomers depicted below.

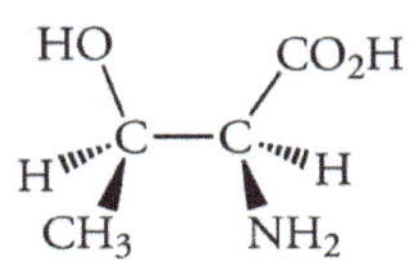

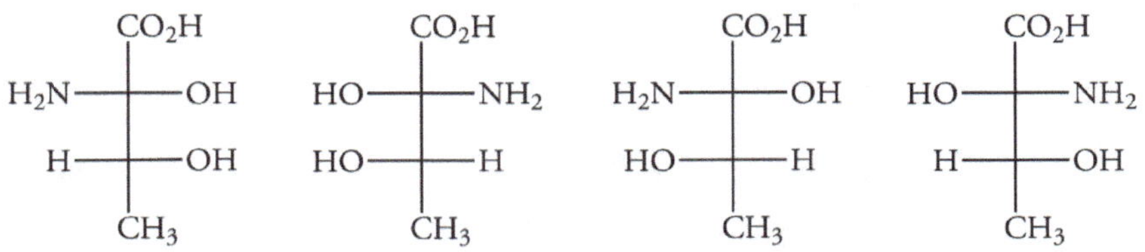

8.22 What stereochemical relationship exists between any and all pairs of the following structures of carbohydrates?

Answer: Compounds I and II are enantiomers, as are compounds III and IV. All other pairs of stereoisomers are diastereomers.

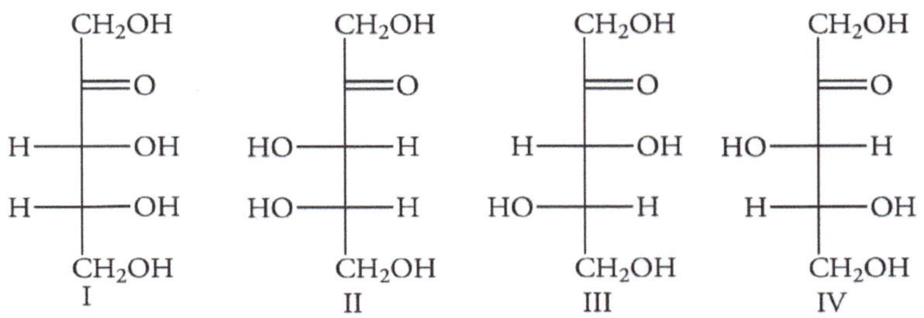

Priority Rules

8.23 Arrange the groups in each of the following sets in order of increasing priority.

(a) —OH, —SH, —SCH$_3$, —OCH$_3$
Answer: —OH < —OCH$_3$ < —SH < —SCH$_3$

(b) —CH$_2$Br, —CH$_2$Cl, —Cl, —Br
Answer: —CH$_2$Cl < —CH$_2$Br < —Cl < —Br

(c) —CH$_2$—CH=CH$_2$, —CH$_2$—O—CH$_3$, — CH$_2$—C≡CH, —C≡C—CH$_3$
Answer: —CH$_2$—CH=CH$_2$ < — CH$_2$—C≡CH < —C≡C—CH$_3$ < —CH$_2$—O—CH$_3$

(d) —CH$_2$CH$_3$, —CH$_2$OH, —CH$_2$CH$_2$Cl, —OCH$_3$
Answer: —CH$_2$CH$_3$ < —CH$_2$CH$_2$Cl < —CH$_2$OH < —OCH$_3$

8.24 Arrange the groups in each of the following sets in order of increasing priority.

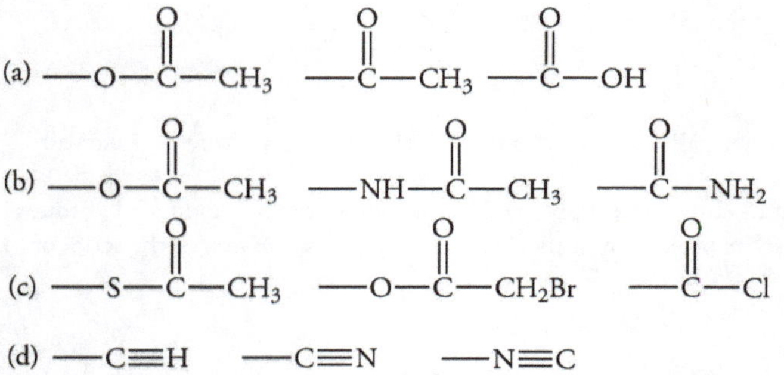

Answers:

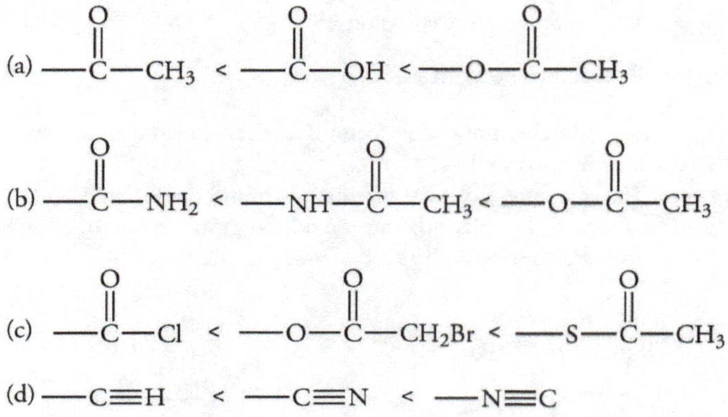

(a) (a)

(b) (b)

(c) (c)

(d) —C≡H   —C≡N   —N≡C (d) —C≡H < —C≡N < —N≡C

8.25 Examine the chiral carbon atom in each of the following drugs and arrange the groups from low to high priority.

(a) ethchlorvynol, a sedative-hypnotic

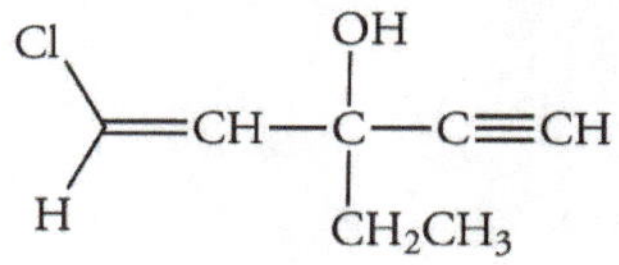

Answer: CH$_2$CH$_3$ < —CH=CHCl < —C≡CH < —OH

(b) chlorphenesin carbamate, a muscle relaxant

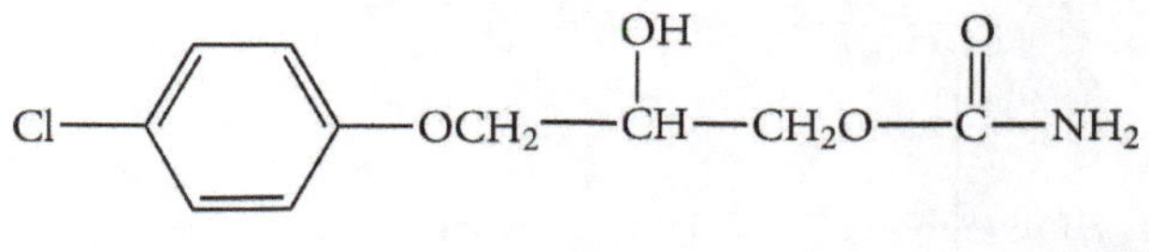

Answer: —H < —CH$_2$O—⟨benzene ring⟩—Cl < —CH$_2$O—C(=O)—NH$_2$ < —OH

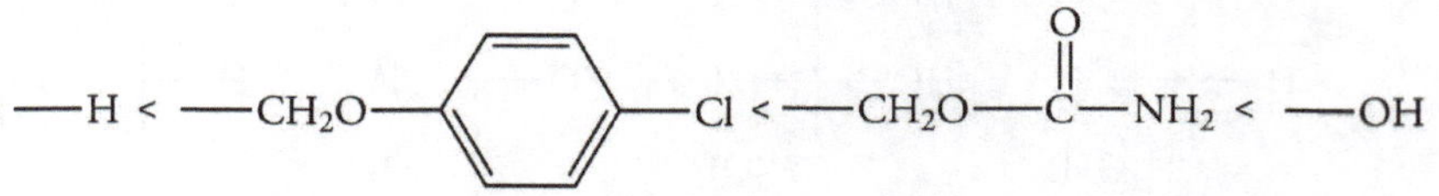

(c) mexiletine, an antiarrhythmic

Answer: —H < —CH₃ < —CH₂O— < —NH₂

8.26 Examine the chiral carbon atom in each of the following drugs and arrange the groups from low to high priority.

(a)   brompheniramine, an antihistamine

Answer: —H < —CH₂N(CH₃)₂ < <

(b)   fluoxetine, an antidepressant

Answer: —H < —CH₂CH₂NHCH₃ < < —O—

(c)   baclophen, an antispastic

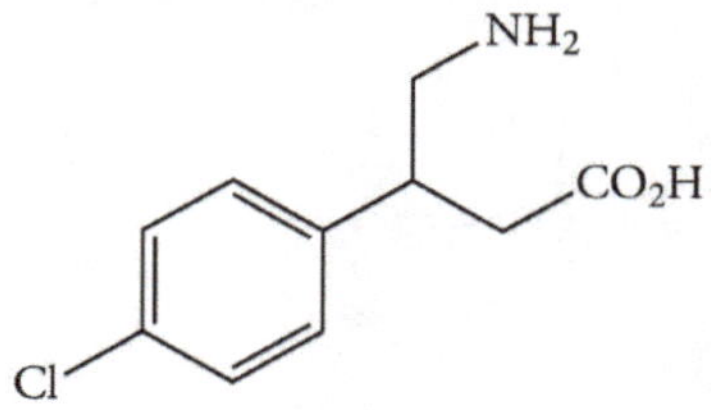

Answer: —H <   —CH$_2$CO$_2$H <   —Cl   <   —CH$_2$NH$_2$

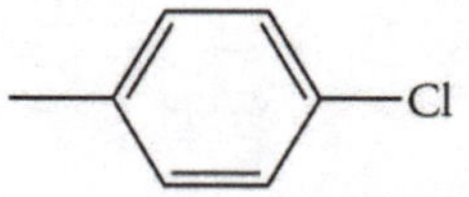

R,S Configuration

8.27 Draw the structure of each of the following compounds.

(a) (R)-2-chloropentane   (b) (R)-3-chloro-1-pentene   (c) (S)-3-chloro-2-methylpentane

Answers:

(a) 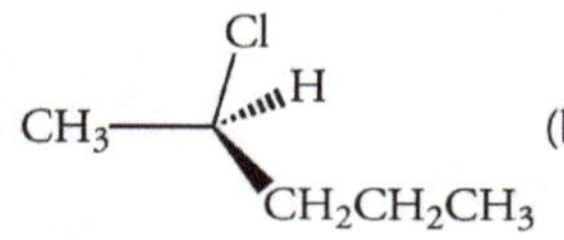 CH$_3$—   (b) 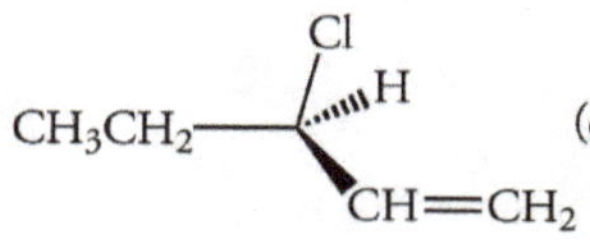 CH$_3$CH$_2$—   (c) 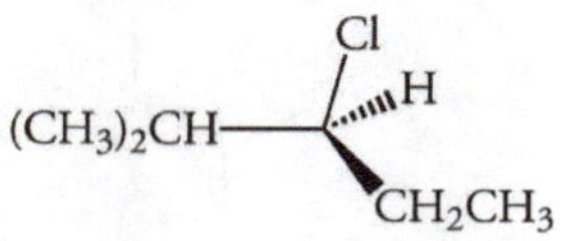 (CH$_3$)$_2$CH—

8.28 Draw the structure of each of the following compounds.

(a) (S)-2-bromo-2-phenylbutane   (b) (S)-3-bromo-1-hexyne   (c) (R)-2-bromo-2-chlorobutane

Answers:

(a) 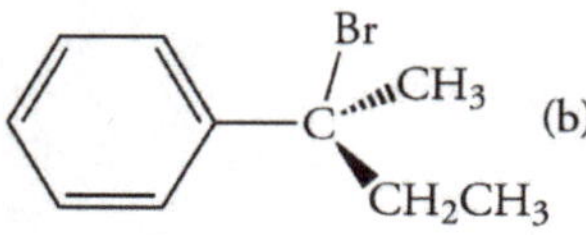   (b) 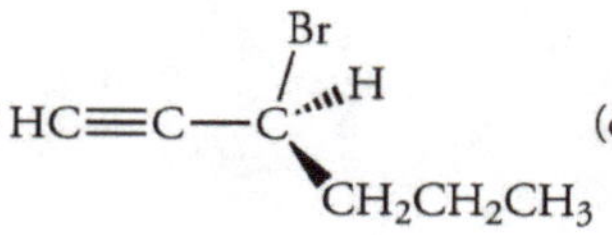 HC≡C—   (c) Br—

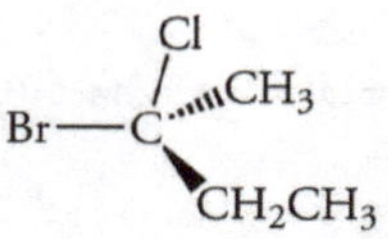

8.29 Assign the configuration of each of the following compounds.

Answers: (a) S (b) S (c) R (d) S

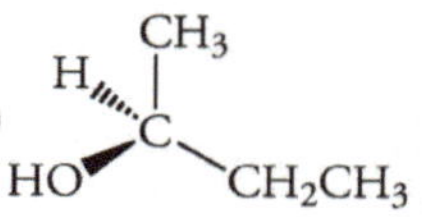

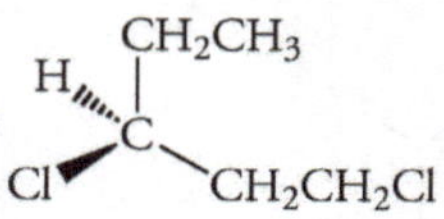

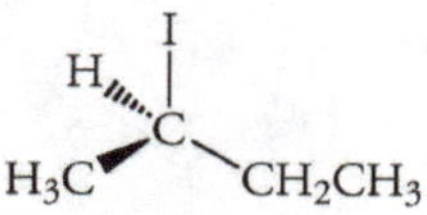

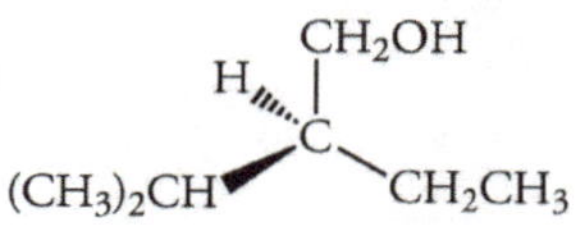

8.30 Assign the configuration of each of the following compounds.

Answers: (a) S (b) S (c) S (d) S

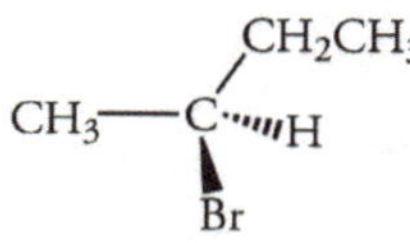

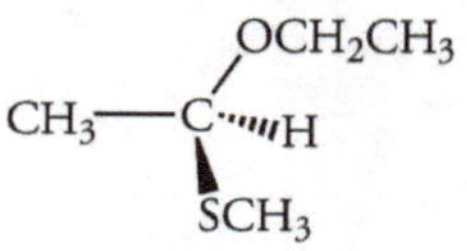

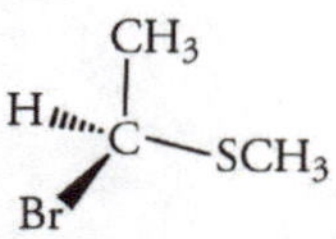

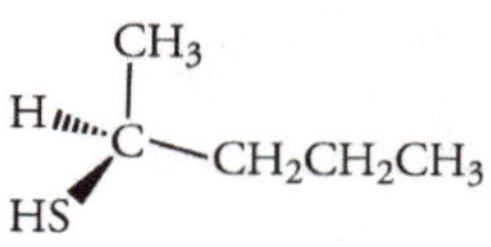

8.31 Assign the configuration of terbutaline, a drug used to treat bronchial asthma.

Answer:
(R)

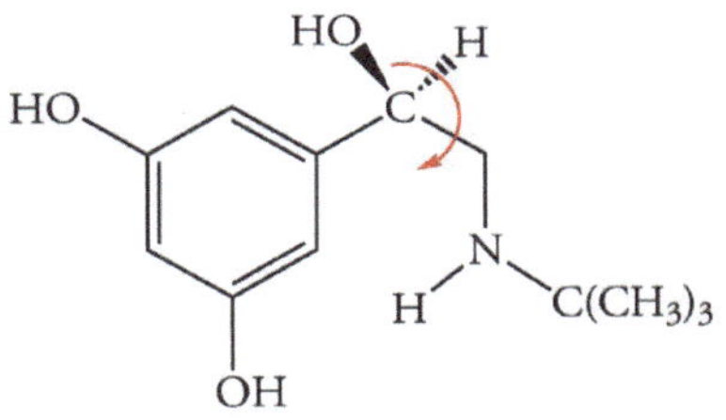

terbutaline

8.32 Assign the configuration of the following hydroxylated metabolite of diazepam, a sedative.

Answer: *S*. View from below the plane of the page!

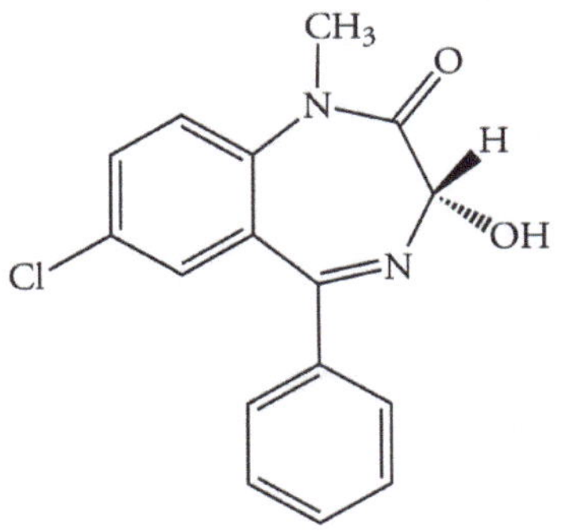

Diastereomers

8.33 Assign the configuration of each of the following compounds.

Answers: (a) *R,R* (b) *S,S* (c) *R,S* (d) *S,S*

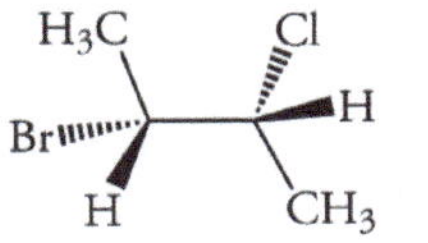

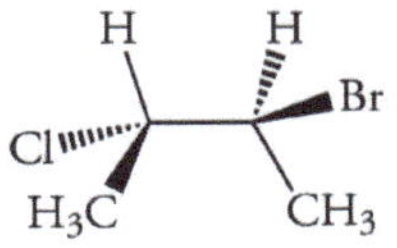

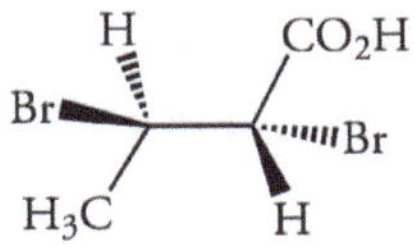

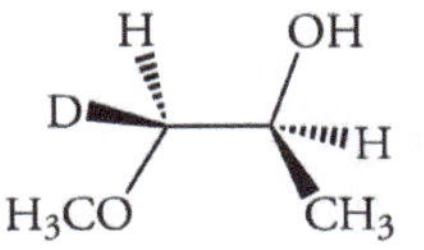

8.34 Assign the configuration of each of the following compounds.

Answers: (a) *R,R* (b) *R,S* (c) *R,R* (d) *R,S*

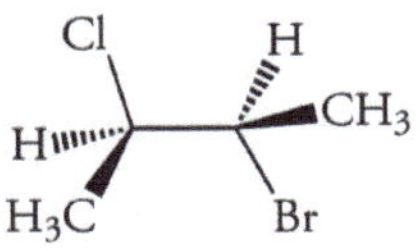

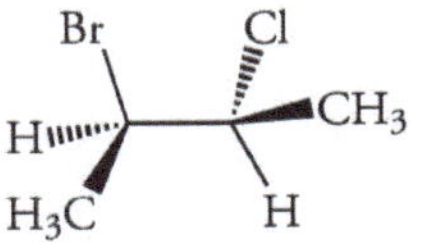

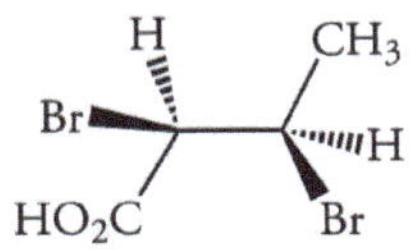

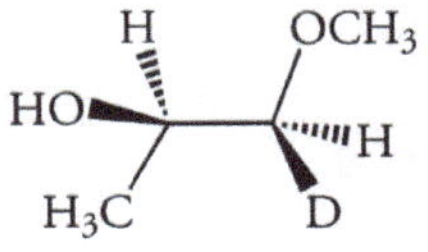

8.35 Assign the configuration of each of the following compounds.

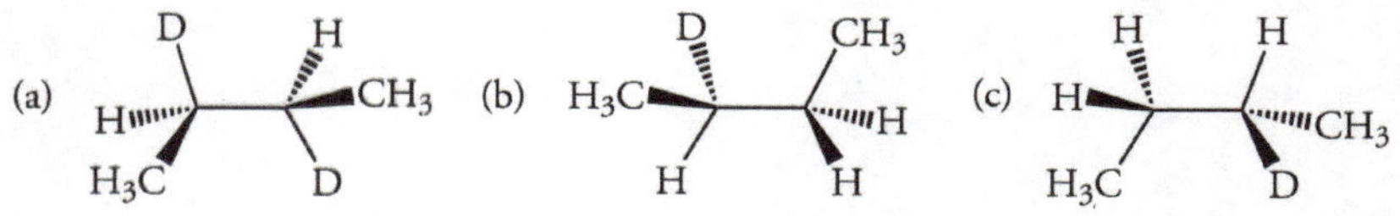

(a), (b), (c) structures

Answers: (a) *R,R* and not *meso* (b) *R,S* and is *meso*
(c) *S,S* and not *meso*

8.36 Assign the configuration of each stereogenic center in the following structures. Based on the assignment, determine if the structure is *meso*.

Answers: (a) *R,R* and not *meso* (b) *R,S* and is *meso*
(c) *S,S* and not *meso*

(a), (b), (c) structures

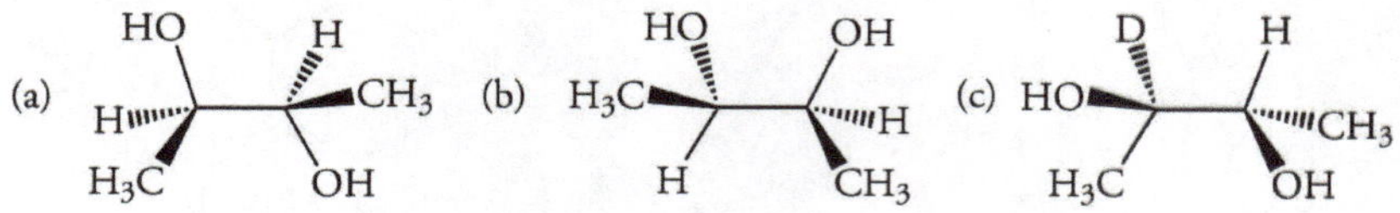

8.37 Ribose is optically active, but ribitol, its reduction product, is optically inactive. Why?

Answer: Ribitol has a plane of symmetry bisecting the Fischer projection through the C-3 atom and its attached C—H and C—OH bonds. Ribose does not have a plane of symmetry at that point because the "top" and "bottom" groups are not equivalent.

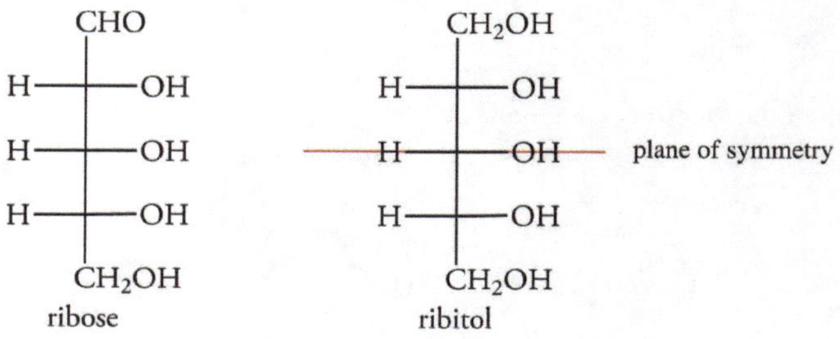

CHO
H——OH
H——OH
H——OH
CH₂OH
ribose

CH₂OH
H——OH
H——OH — plane of symmetry
H——OH
CH₂OH
ribitol

8.38 Which of the following carbohydrate derivatives are *meso* compounds?

Answer: (a) and (b) are meso compounds. Compound (a) has a plane of symmetry bisecting the Fischer projection through C-3 and its attached carbonyl oxygen atom. Compound (b) has a plane of symmetry bisecting the Fischer projection between C-3 and C-4.

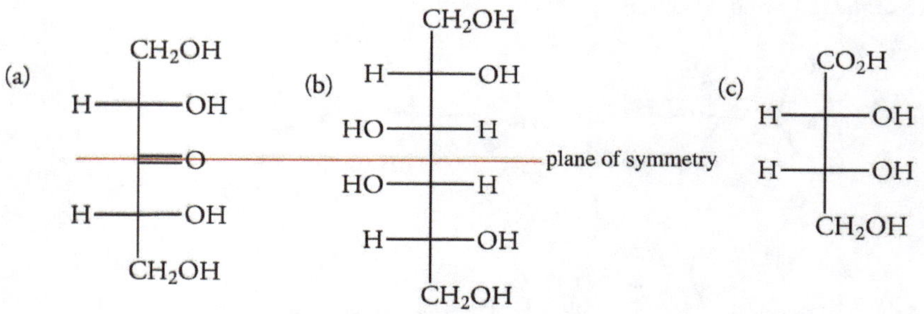

(a)
CH₂OH
H——OH
——O
H——OH
CH₂OH

(b)
CH₂OH
H——OH
HO——H
HO——H — plane of symmetry
H——OH
CH₂OH

(c)
CO₂H
H——OH
H——OH
CH₂OH

meso compounds

8.39   5-Hydroxylysine is an amino acid isolated from collagen. Determine the number of stereoisomers possible.

Answer: There are two nonequivalent stereogenic centers. One center, at C-2, bears an amino group. The second center, located at C-4, bears an hydroxyl group. There are four possible stereoisomers.

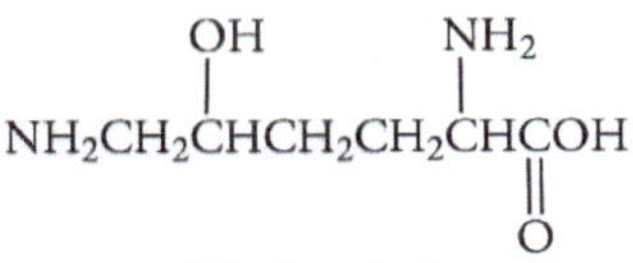

$$NH_2CH_2CHCH_2CH_2CHCOH$$

with OH on C, NH$_2$ on C, and =O below

5-hydroxylysine

Fischer projection 1: CO$_2$H / H—NH$_2$ / H—H / H—H / H—OH / CH$_2$OH

Fischer projection 2: CO$_2$H / H$_2$N—H / H—H / H—H / HO—H / CH$_2$OH

Fischer projection 3: CO$_2$H / H$_2$N—H / H—H / H—H / H—OH / CH$_2$OH

Fischer projection 4: CO$_2$H / H—NH$_2$ / H—H / H—H / HO—H / CH$_2$OH

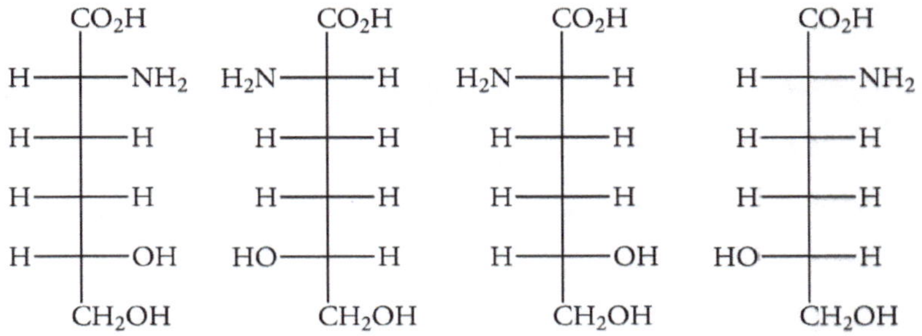

8.40 Consider the structure of pantothenic acid (vitamin B$_3$) and determine the number of stereoisomers possible.

$$HOCH_2—C—CHCNHCH_2CH_2CO_2H$$

with CH$_3$ and CH$_3$ on the C, and OH and O on the CHC group

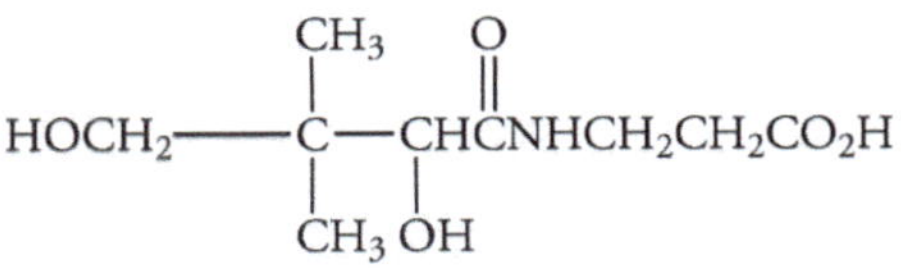

pantothenic acid

Answer: There is only one stereogenic center. There are two enantiomers.

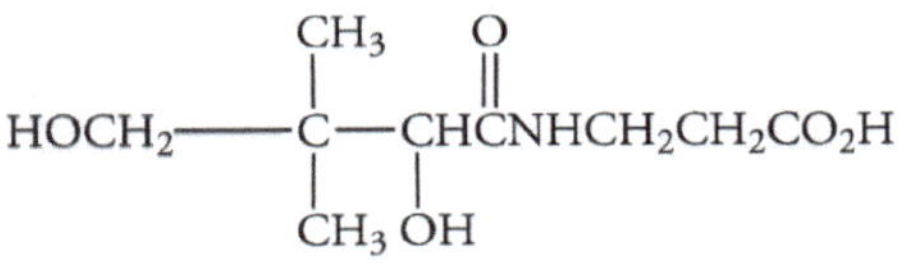

$$HOCH_2—C—CHCNHCH_2CH_2CO_2H$$

with CH$_3$ and CH$_3$ on the C, and OH and O on the CHC group

pantothenic acid

Fischer projection (left): CH$_2$OH / H$_3$C—CH$_3$ / HO—H / C—NHCH$_2$CH$_2$CO$_2$H with =O

Fischer projection (right): CH$_2$OH / H$_3$C—CH$_3$ / H—OH / C—NHCH$_2$CH$_2$CO$_2$H with =O

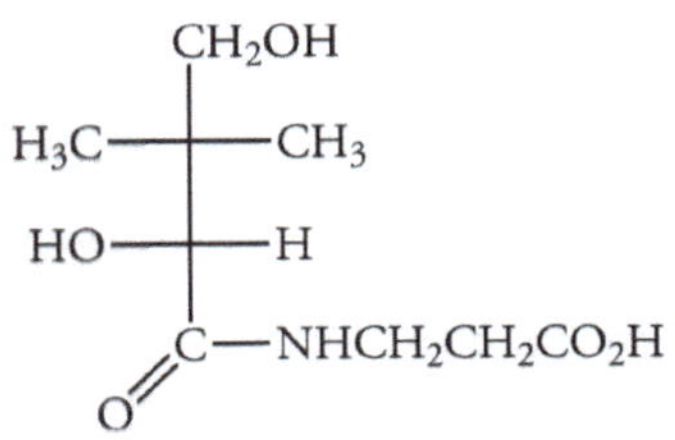

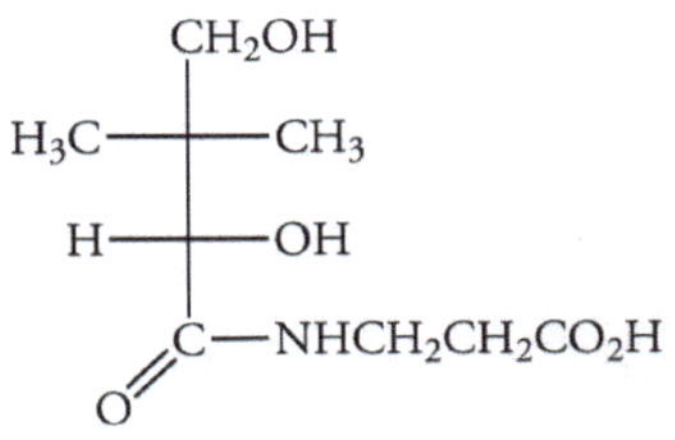

8.41 There are four isomeric 2,3-dichloropentanes, but only three isomeric 2,4-dichloropentanes. Explain why.

Answer: The two stereogenic centers in 2,3-dichloropentane are nonequivalent, so there are four possible stereoisomers. The two stereogenic centers in 2,4-dichloropentane are equivalent. A plane of symmetry can be placed through the C-3 atom. Thus, there are two enantiomers and one *meso* compound possible for this isomer.

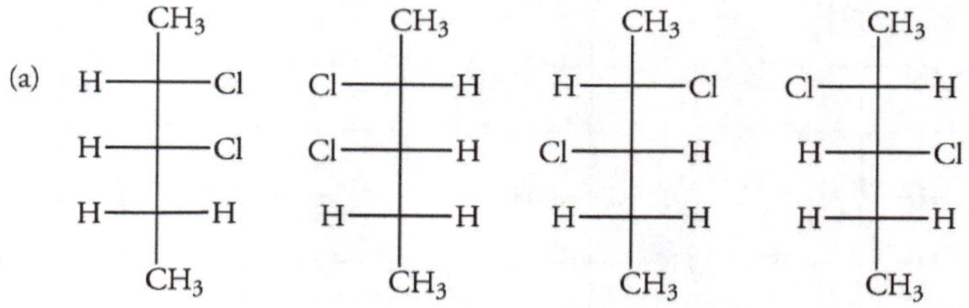

(a)

$$\begin{array}{cccc}
\text{CH}_3 & \text{CH}_3 & \text{CH}_3 & \text{CH}_3 \\
\text{H}-\!\!\!-\text{Cl} & \text{Cl}-\!\!\!-\text{H} & \text{H}-\!\!\!-\text{Cl} & \text{Cl}-\!\!\!-\text{H} \\
\text{H}-\!\!\!-\text{Cl} & \text{Cl}-\!\!\!-\text{H} & \text{Cl}-\!\!\!-\text{H} & \text{H}-\!\!\!-\text{Cl} \\
\text{H}-\!\!\!-\text{H} & \text{H}-\!\!\!-\text{H} & \text{H}-\!\!\!-\text{H} & \text{H}-\!\!\!-\text{H} \\
\text{CH}_3 & \text{CH}_3 & \text{CH}_3 & \text{CH}_3
\end{array}$$

stereoisomers of 2,3-dichloropentane

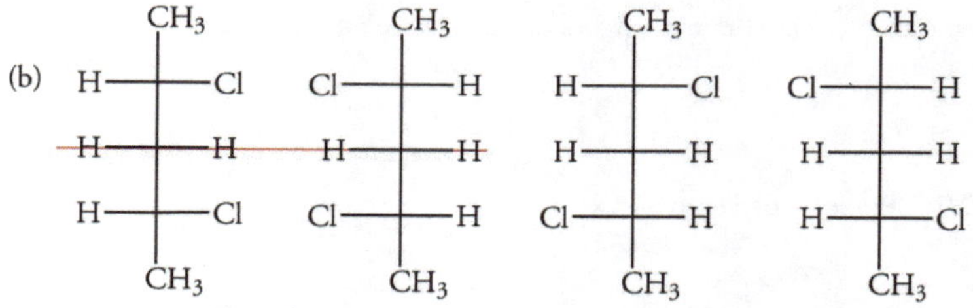

(b)

$$\begin{array}{cccc}
\text{CH}_3 & \text{CH}_3 & \text{CH}_3 & \text{CH}_3 \\
\text{H}-\!\!\!-\text{Cl} & \text{Cl}-\!\!\!-\text{H} & \text{H}-\!\!\!-\text{Cl} & \text{Cl}-\!\!\!-\text{H} \\
\text{H}-\!\!\!-\text{H} & \text{H}-\!\!\!-\text{H} & \text{H}-\!\!\!-\text{H} & \text{H}-\!\!\!-\text{H} \\
\text{H}-\!\!\!-\text{Cl} & \text{Cl}-\!\!\!-\text{H} & \text{Cl}-\!\!\!-\text{H} & \text{H}-\!\!\!-\text{Cl} \\
\text{CH}_3 & \text{CH}_3 & \text{CH}_3 & \text{CH}_3
\end{array}$$

meso compound   enantiomers

2,4-dichloropentane

8.42 Which of the following structures are *meso* compounds?

Answer: (a) and (c) are *meso* compounds. Compound (a) has a plane of symmetry bisecting the Fischer projection through C-3 and its attached C—H and C—Cl bonds. Compound (c) has a plane of symmetry bisecting the Fischer projection between C-2 and C-3. The symmetry planes are shown in red.

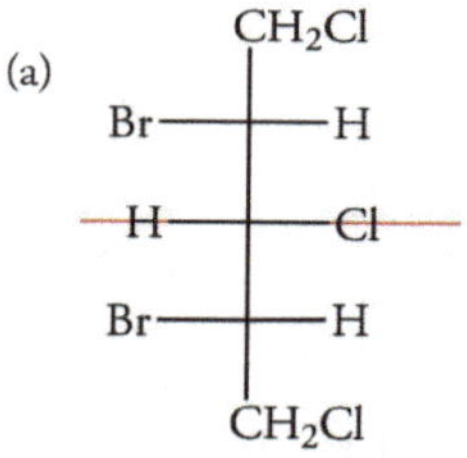

(a)
$$\begin{array}{c}
\text{CH}_2\text{Cl} \\
\text{Br}-\!\!\!-\text{H} \\
\text{H}-\!\!\!-\text{Cl} \\
\text{Br}-\!\!\!-\text{H} \\
\text{CH}_2\text{Cl}
\end{array}$$
meso compound

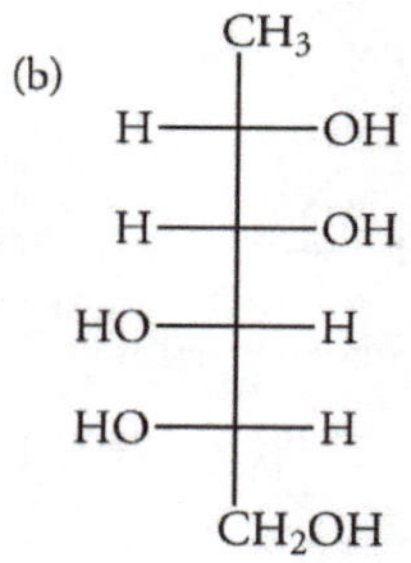

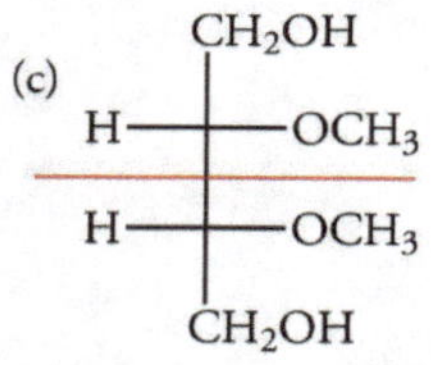

(b)
$$\begin{array}{c}
\text{CH}_3 \\
\text{H}-\!\!\!-\text{OH} \\
\text{H}-\!\!\!-\text{OH} \\
\text{HO}-\!\!\!-\text{H} \\
\text{HO}-\!\!\!-\text{H} \\
\text{CH}_2\text{OH}
\end{array}$$

(c)
$$\begin{array}{c}
\text{CH}_2\text{OH} \\
\text{H}-\!\!\!-\text{OCH}_3 \\
\text{H}-\!\!\!-\text{OCH}_3 \\
\text{CH}_2\text{OH}
\end{array}$$
meso compound

Cyclic Compounds

8.43 Which of the following compounds has a plane of symmetry?

(a) *cis*-1,2-dibromocyclobutane   (b) *trans*-1,2-dibromocyclobutane

(c) *cis*-1,3-dibromocyclobutane   (d) *trans*-1,3-dibromocyclobutane

Answer: (a) and (c) have a plane of symmetry and are *meso* compounds. (b) and (d) do not have a plane of symmetry.

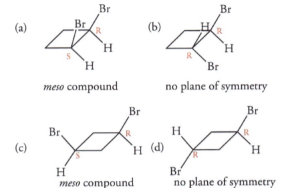

meso compound   no plane of symmetry

meso compound   no plane of symmetry

8.44 Which of the following structures has a plane of symmetry?
Answer: (a) and (b) have a plane of symmetry (shown in red) and are *meso* compounds. (c) does not have a plane of symmetry.

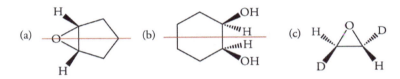

8.45 Assign the configuration of each stereogenic center in the following structures.

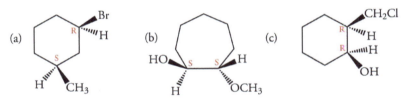

8.46 Assign the configuration of each stereogenic center in the following structures.

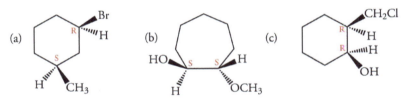

Resolution of Enantiomers

8.47   Reaction of a racemic mixture of A_R,A_S with a resolving agent X_R yields diastereomers. The A_R—X_R isomer is less soluble than A_R—A_S. Consequently, the A_S isomer is obtained optically pure. Describe the experimental results if X_S were available as a resolving agent.

Answer: The X_S compound would give a diastereomeric mixture of A_R—X_S and A_S—X_S compounds. The A_R—X_S compound, which is the enantiomer of the A_S—X_R compound, will be the less soluble. Consequently the A_R compound can be obtained optically pure.

8.48 Resolution of a racemic mixture yields one enantiomer with $[\alpha]_D = +44°$ and another enantiomer with $[\alpha]_D = -33°$. One enantiomer is optically pure. Which one? What is the optical purity of the other enantiomer?

Answer: The enantiomer with the larger specific rotation is the pure compound. Thus, the enantiomer with the +44 rotation is pure. The other sample is 75% optically pure.

Reactions of Chiral Compounds

8.49 (R)-(–)-Lactic acid is converted into a methyl ester when it reacts with methanol. What is the configuration of the ester? Can you predict its sign of rotation?

Answer: The methyl ester also has the R configuration. The bonds to the stereogenic center are *not* changed, nor is the priority series of the groups bonded to that center. The —CO_2H group has the second highest priority in the acid, as does the —CO_2CH_3 group of the methyl ester. The sign of rotation of the ester cannot be predicted.

(R)-(–)-Lactic acid

8.50 Free radical chlorination of (S)-2-bromobutane gives a mixture of compounds resulting from attack at any of the four nonequivalent carbon–hydrogen bonds. The products of reaction at C-l and C-4 are both optically active. Explain why.

Answer: The products have a chlorine atom bonded to carbon atoms that are not stereogenic centers, but the arrangement of the groups bonded to C-2 is unchanged by the reaction, so the products are still optically active.

(S)-2-bromobutane   (S)-1-chloro-3-bromobutane   (S)-1-chloro-2-bromobutane

8.51 Free radical chlorination of (S)-2-fluorobutane gives a 31% yield of 2-chloro-2-fluorobutane. What is the expected stereochemistry of the product?

Answer: The reaction occurs via a free radical that results from abstraction of hydrogen from C-2 atom, which is the stereogenic center. The radical is achiral and can react with chlorine from either face of this plane. Thus, a racemic mixture results.

(S)-2-fluorobutane   (R)-2-chloro-2-fluorobutane   (S)-2-chloro-2-fluorobutane

8.52 Free radical chlorination of (S)-2-bromobutane at the C-2 atom gives an optically inactive product, but reaction at C-3 gives an optically active product. Explain why.

Answer: The reaction occurs via a free radical that results from abstraction of hydrogen. C-2 is the stereogenic center, and loss of hydrogen from this center yields an achiral intermediate that can react with chlorine from either face of its plane. Thus, a racemic mixture results. The C-3 atom is not a stereogenic center but becomes one once substituted by chlorine. A mixture of 2S,3S and 2S,3R product results. It has a net optical activity because the optical rotations of the diastereomers do not cancel.

9
HALOALKANES AND ALCOHOLS
INTRODUCTION TO NUCLEOPHILIC SUBSTITUTION AND ELIMINATION REACTIONS

KEYS TO THE CHAPTER

9.1 Functionalized Hydrocarbons

In this chapter, we discuss the chemistry of haloalkanes and alcohols in which the halogen or hydroxyl group is bonded to an sp^3-hybridized carbon atom. Molecules with sp^2-hybridized carbon atoms bonded to a halogen or a hydroxyl group have different chemistry. We will discuss these compounds in later chapters. Haloalkanes and alcohols are classified as 1°, 2°, and 3° by the same method used to classify carbon atoms in alkanes.

9.2 Nomenclature of Haloalkanes

The rules for naming haloalkanes are very similar to the rules for naming alkanes (Section 4.3) and alkenes (Section 5.6). Halogen atoms and branching alkyl groups have equal priorities in terms of their positions in the parent chain. If no other functional groups are present, the chain is numbered from the end closest to the first substituent, whether it is a halogen or an alkyl group. However, the double bond of an alkene or the triple bond of an alkyne takes precedence in numbering a carbon chain, regardless of where the halogen is located or how many halogens there may be. The concept of the priority of one functional group over another is expanded with each new functional group we will study in later chapters. It is explicitly part of *R,S* configurational nomenclature.

9.3 Nomenclature of Alcohols

The common names of simple alcohols are based on the name of the alkyl group bonded to the hydroxyl group, as in methyl alcohol. The IUPAC method of naming alcohols is based on the longest continuous carbon chain that contains a hydroxyl group. The chain is numbered to give the carbon atom bonded to the hydroxyl group the lowest possible number. The suffix *-ol* is added to the stem of the name of the parent alkane. Note that in contrast to halogens, which have lower priority than multiple bonds in assigning names, the hydroxyl group has a higher priority than multiple bonds. Compounds with the hydroxyl group bonded to a ring are numbered from the carbon atom containing the hydroxyl group, and subsequent numbers are assigned in the direction to give the lowest possible numbers for any other structural features.

9.4 Structure and Properties of Haloalkanes

The electronegativities of the halogens decrease, and their polarizabilities increase from top to bottom in Group VII of the periodic table. The boiling points of homologous haloalkanes compounds increase in the same order, so the intermolecular attractive forces also increase from top to bottom. This indicates that polarizability is more important than bond polarity in determining the physical properties of haloalkanes. Other factors such as molecular shape and the extent of branching also influence the intermolecular forces and physical properties of haloalkanes.

9.5 Structure and Properties of Alcohols

Intermolecular hydrogen bonds between alcohols dominate their physical properties. Alcohols have higher boiling points than alkanes of similar molecular weight as a result of intermolecular hydrogen bonding. Hydrogen bonding between alcohol molecules and water also accounts for the solubility of alcohols. Alcohols serve as solvents for polar compounds, especially those that can also form hydrogen bonds with the solvent.

9.6 Organometallic Compounds

The Grignard reagent is a highly reactive organomagnesium compound formed by reacting a haloalkane with magnesium in an ether solvent. When a Grignard reagent is exposed to water, an alkane is produced. This reaction is useful in the synthesis of deuterium-substituted compounds. The carbon atom of a Grignard reagent has a partial negative charge and behaves as a nucleophile.

Organolithium compounds can also act as carbon nucleophiles. Organolithium compounds are very strong bases, and they are employed in many synthetic procedures. A Gilman reagent is used to "couple" two carbon groups by forming a new sigma bond. One group is provided by an organohalogen compound; the second is provided by the Gilman reagent. We will also discuss the Gilman reaction again in Chapter 17.

9.7 Reactions of Haloalkanes

We begin to explore the chemistry of haloalkanes in this section. We will continually encounter nucleophilic substitution and the competing elimination reaction. Nucleophiles can displace a halide ion as a leaving group in a substitution reaction. However, nucleophiles are often sufficiently basic to remove a proton from the carbon atom adjacent to the carbon atom bearing the halogen. The overall result is an elimination reaction.

9.8 Nucleophilic Substitution Reactions of Haloalkanes

Nucleophilicity refers to the ability of a nucleophile to displace a leaving group in a substitution reaction. We will describe trends in nucleophilicity in Chapter 10. Most common nucleophiles have a negative charge. However, it is the nonbonding electron pair that is important. For example, water, alcohols, ammonia, and amines are nucleophiles even though they are electrically neutral.

We must keep track of pairs of electrons in nucleophilic substitution reactions. In all cases, in this section, a nonbonded electron pair of a nucleophile forms a new bond to carbon, and the leaving group departs with a pair of electrons. If the nucleophile has a negative charge, and it reacts with a neutral substrate, then the leaving group also has a negative charge.

Although all of the examples of nucleophiles we discussed allow the synthesis of new compounds, two are especially interesting. Both cyanide ion and alkynide ions react with primary haloalkanes to give a product with a new carbon–carbon bond. The products, a nitrile or a terminal alkyne, can be converted to many other functional groups.

9.9 Mechanisms of Nucleophilic Substitution Reactions

In Chapter 6, we discussed electrophilic addition to unsaturated compounds. The essential features of S_N1 and S_N2 mechanisms are presented in this section. We will expand upon these reactions in Chapter 10.

The S_N2 reaction mechanism is based in part on kinetic experiments. The key points are

1. Both the substrate and the nucleophile are present in the transition state.
2. The nucleophile attacks along the same axis that the leaving group departs. (We will discuss the stereochemical consequences of this process in Chapter 10.)

The rate of the S_N2 reaction depends on the structure of the substrate. The order of reactivity is methyl > primary > secondary >> tertiary. Steric hindrance caused by the groups bonded to the reacting carbon center severely hinder approach of the nucleophile in tertiary compounds, and tertiary compounds do *not* react by an S_N2 mechanism. The ability of the nucleophile to displace the leaving group is improved in secondary compounds, and still further in primary compounds, so both can react with nucleophiles by an S_N2 mechanism. In chapter 10, we will discuss the stereochemical consequences of S_N2 reactions of chiral substrates.

A different mechanism accounts for substitution reactions at sterically hindered sites. Again, the mechanism is based on kinetic data. The rate determining step is the ionization of the haloalkane to give a carbocation that is subsequently captured by the nucleophile in a faster second step. Since only one molecule is present in the transition state for the reaction, it is unimolecular and is designated S_N1. Thus, the rate of the reaction depends only on the substrate concentration. The order of reactivity is tertiary > secondary > primary. In fact, it is unlikely that a primary compound would react by this mechanism when it has the S_N2 process as an available option. Why do tertiary compounds react so much faster by the S_N1 mechanism compared to secondary compounds, and why don't primary compounds react by this mechanism? The answer turns on carbocation stability: a highly unstable primary carbocation would form in an S_N1 reaction, so the reaction proceeds by an S_N2 mechanism instead. In chapter 10, we will explore the stereochemical consequences of S_N1 reactions of chiral substrates.

Substrates that react via the S_N1 mechanism may give rearranged products. This process is exactly like the rearrangements described in Section 6.4. The only difference is in the reaction that leads to the carbocation. In the case of alkenes, the carbocation resulted from addition of a proton to a π bond. We now find that the same types of carbocations form in an S_N1 reaction.

A carbocation can rearrange by a 1,2-hydride shift or 1,2-methide shift. The driving force for such rearrangements is the formation of a more stable carbocation. Hydride ion shifts occur when a secondary or primary carbocation is generated adjacent to a tertiary center. Shifts of alkyl groups most commonly occur when a carbocation is generated at a carbon atom adjacent to a quaternary center. Any of the alkyl groups bonded to that quaterary carbon can migrate. Thus, a mixture of products can result. As shown in Exercise 9.14, even ring bonds of cycloalkanes can migrate, resulting in rings of different size than the reactant.

9.10 Reactions of Alcohols

The reactions of alcohols can occur in several ways that differ in the number and type of bonds cleaved. These are:

1. Cleavage of the oxygen–hydrogen bond in an acid–base reaction.

2. Cleavage of the carbon–oxygen bond in a nucleophilic substitution reaction.

3. Cleavage of the carbon–oxygen bond as well as the carbon–hydrogen bond at the carbon atom adjacent to the carbon atom bearing the hydroxyl group in an elimination reaction.

4. Cleavage of the oxygen–hydrogen bond as well as the carbon–hydrogen bond at the carbon atom bearing the hydroxyl group in an oxidation reaction.

9.11 Acid–Base Reactions of Alcohols

In general, alcohols are somewhat weaker acids than water. Ethanol, 2-propanol, and 2-methyl-2-propanol have pK_a values of 15.9, 18.0, and 19.0, respectively. These values indicate a decrease in acidity from 1° to 2° to 3° alcohols. This agrees with the general principle that alkyl substituents are inductively electron-donating; this effect supplies more electron density to the oxygen atom, so it strengthens the O—H bond.

Also, electronegative substituents near the carbon atom bearing the hydroxyl group increase its acidity. Halogen substituents inductively withdraw electron density from the oxygen atom and weaken the O—H bond. The halogens also stabilize the negative charge of the conjugate base—an alkoxide ion.

Alcohols, like water, can be protonated. The product is a conjugate acid known as an alkyloxonium ion.

9.12 Substitution Reactions of Alcohols

The hydroxyl group of alcohols react with hydrogen halides such as HBr to give haloalkanes. The order of reactivity is tertiary > secondary > primary. Hydrogen bromide suffices to form bromoalkanes, but zinc chloride is required as a catalyst for the reaction with hydrogen chloride.

The substitution reactions of alcohols parallel that of haloalkanes. However, hydroxide ion is not the leaving group. Protonation of the hydroxyl group must occur to allow water to become the leaving group. In general, a weaker base is a better leaving group than a stronger base. Since hydroxide ion is a stronger base than water, it is a poor leaving group in both S_N1 and S_N2 reactions.

The order of reactivity of alcohols in S_N1 reactions is tertiary > secondary > primary. This order parallels the stability of the carbocation intermediates that form in the reaction. This order of reactivity is reversed for S_N2 reactions; however, tertiary alcohols do *not* react by an S_N2 mechanism.

9.13 Alternate Methods for the Synthesis of Alkyl Halides

Since undesirable rearrangements occur in acid-catalyzed reactions of alcohols, other methods have been developed to synthesize alcohols that do not require acid. In this section, we discussed two additional reagents that convert alcohols to haloalkanes. They are used for secondary and primary alcohols which react slowly with hydrogen halides.

Thionyl chloride is used to convert alcohols to chloroalkanes. The by-products are sulfur dioxide and hydrogen chloride, both of which escape from the solution as gases. Phosphorus tribromide is used to convert alcohols to bromoalkanes. The by-product, phosphorous acid, is soluble in water.

9.14 Elimination Reactions

The elimination reactions we consider result in loss of atoms from adjacent carbon atoms and is called a 1,2-elimination or a (β-elimination). β-Elimination reactions occur by either E1 or E2 mechanisms. An E1 mechanism is similar to an S_N1 mechanism in one key respect: it is a unimolecular reaction in which a carbocation intermediate forms in the rate determining step. And, as in S_N1 reactions, the carbocation can, and often does, rearrange to give several products. E1 reactions typically occur in the dehydration of tertiary alcohols. E1 reactions compete with S_N1 reactions. E2 reactions, like S_N2 reactions, are bimolecular processes in which there are two species in the transition state. E2 reactions are observed for primary and secondary alkyl halides and alcohols. The reactions of alcohols in both E1 and E2 reactions are acid catalyzed.

9.15 Regioselectivity in Dehydrohalogenation

Dehydrohalogenaton occurs to give a predominance of the most substituted alkene. This regioselectivity results in the so-called Zaitsev product. This product is the most stable alkene. The alkene with the most alkyl groups bonded to the carbon atoms of the double bond is favored. If geometric isomers are possible, the more stable *trans* isomer predominates.

9.16 Mechanisms of Dehydrohalogenation

There are two mechanisms for dehydrohalogenation. The E2 reaction occurs for primary haloalkanes and usually for secondary haloalkanes. The E1 process occurs with tertiary haloalkanes. In either case, the rate depends on the leaving groups. The reactivity order is iodo- > bromo- > chloroalkane.

The E2 mechanism is a concerted process in which a base abstracts a proton while a carbon–carbon double bond forms and simultaneously a halide ion leaves. Because a double bond starts to develop in the transition state, the energy of the transition state is lower for more highly substituted compounds. The stereochemistry of the E2 process involves an *anti* periplanar arrangement of the carbon–hydrogen bond and the bond from a carbon atom to the leaving group. This arrangement provides the necessary geometric alignment for the emerging orbitals required to form a π bond.

SUMMARY OF REACTIONS

1. Formation and Reactivity of Grignard Reagents

2. Coupling Reaction of Gilman Reagent and Halogen Compounds

$(CH_3)_2Cu^- \ Li^+ \ + \ CH_3(CH_2)_6CH_2I \ \longrightarrow \ CH_3(CH_2)_6CH_3$

                             1-iodooctane   nonane

trans-1-iodononene   *trans*-6-tetradecene

3. Nucleophilic Substitution of Haloalkanes

$CH_3CH_2CH_2CH_2CH_2Cl \ + \ CH_3O^- \ \longrightarrow \ CH_3CH_2CH_2CH_2CH_2OCH_3$

4. Synthesis of Haloalkanes from Alcohols

$CH_3CH_2CH_2CH_2CH_2OH \ + \ HCl \ \longrightarrow \ CH_3CH_2CH_2CH_2CH_2Cl$

5. Dehydrohalogenation of Haloalkanes

$$\text{(cyclodecane with Br, H)} \xrightarrow[\text{CH}_3\text{CH}_2\text{OH}]{\text{CH}_3\text{CH}_2\text{O}^-} \text{(cyclodecene)}$$

$$\text{H}_3\text{C-(ring with H, Br, H, H, D, H, D)} \xrightarrow[\text{(CH}_3)_3\text{COH}]{\text{(CH}_3)_3\text{COK}} \text{H}_3\text{C-(ring with H, H)}$$

6. Dehydration of Alcohols

$$\text{CH}_3\text{CH}_2\!-\!\underset{\underset{\text{CH}_2\text{CH}_3}{|}}{\overset{\overset{\text{OH}}{|}}{\text{C}}}\!-\!\text{CH}_2\text{CH}_3 \xrightarrow{\text{H}_2\text{SO}_4} \underset{\text{H}}{\overset{\text{H}_3\text{C}}{}}\text{C}=\text{C}\underset{\text{CH}_2\text{CH}_3}{\overset{\text{CH}_2\text{CH}_3}{}}$$

7. Synthesis of Haloalkanes from Alcohols

$$\text{CH}_3\text{CH}_2\text{CH}_2\text{CH}_2\text{CH}_2\text{OH} + \text{HCl} \longrightarrow \text{CH}_3\text{CH}_2\text{CH}_2\text{CH}_2\text{CH}_2\text{Cl}$$

$$\text{(tetralin-OH)} + \text{HBr} \longrightarrow \text{(tetralin-Br)}$$

$$\text{(cyclooctane-OH)} + \text{SOCl}_2 \longrightarrow \text{(cyclooctane-Cl)}$$

$$3 \text{ (cyclopentyl)}\!-\!\text{CH}_2\text{CH}_2\text{OH} + \text{PBr}_3 \longrightarrow \text{(cyclopentyl)}\!-\!\text{CH}_2\text{CH}_2\text{Br}$$

8. Dehydrohalogenation of Haloalkanes

$$\text{(cyclodecane with Br, H)} \xrightarrow[\text{CH}_3\text{CH}_2\text{OH}]{\text{CH}_3\text{CH}_2\text{O}^-} \text{(cyclodecene)}$$

$$\text{H}_3\text{C-(ring with H, Br, H, H, D, H, D)} \xrightarrow[\text{(CH}_3)_3\text{COH}]{\text{(CH}_3)_3\text{COK}} \text{H}_3\text{C-(ring with H, H)}$$

9. Dehydration of Alcohols

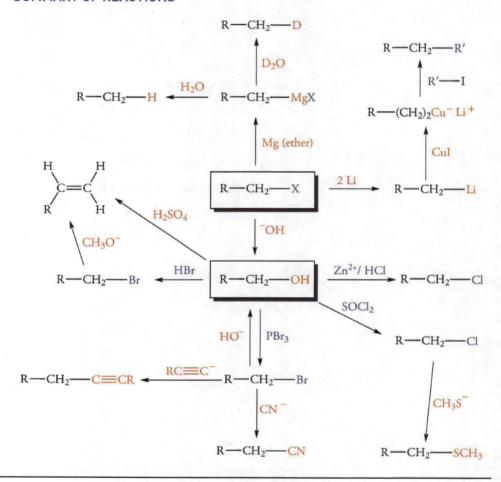

SUMMARY OF REACTIONS

End of Chapter Exercises

9.1 Classify each of the following haloalkanes.

Answers:
(a) secondary
(b) tertiary
(c) secondary
(d) tertiary

(a) CH_3—CH_2—CH—CH_2—CH_3
                       |
                       Br

(b)
             CH_3
             |
CH_3—C—CH_2—CH_3
        |
        Cl

(c) CH_3—CH_2—CH—CH_3
                   |
                   F

(d)
        CH_3
        |
CH_3—C—I
        |
        CH_3

9.2 Classify each of the following alcohols.

Answers:
(a) primary
(b) secondary
(c) secondary
(d) primary

(a)
        CH_3
        |
CH_3—C—CH_2—OH
        |
        CH_3

(b) CH_3—CH—CH_2—CH—CH_2—CH_3
            | |
            OH CH_3

(c) CH_3—CH—CH_2—CH_2—CH_3
            |
            OH

(d)
        CH_3
        |
CH_3—C—CH_2—CH_2—OH
        |
        CH_3

9.3 Classify each of the hydroxyl groups in the following vitamins.

Answers:
(a) phenol and primary
(b) primary
(c) primary at top, all others secondary

(a) pyridoxal (vitamin B_6)

(b) thiamine (vitamin B_1)

(c) riboflavin (vitamin B_2)

9.4 Classify each of the hydroxyl groups in the following steroids.

(a) digitoxigenin, a cardiac glycoside

(b) hydrocortisone, an antiinflammatory drug

(c) norethindrone, an oral contraceptive

Nomenclature of Haloalkanes

9.5 What is the IUPAC name for each of the following compounds?

(a) vinyl fluoride   (b) allyl chloride   (c) benzyl bromide

Answers: (a) fluoroethene   (b) 3-chloro-l-propene   (c) (bromomethyl)benzene

9.6 What is the IUPAC name for each of the following compounds?

(a) $(CH_3)_3CCH_2Cl$ (neopentyl chloride)   (b) $(CH_3)_2CHCH_2CH_2Br$ (isoamyl bromide)

(c) $C_6H_5CH_2CH_2F$ (phenethyl fluoride)

Answers: (a) 1-chloro-2,2-dimethylchloropropane   (b) 1-bromo-3-methylbutane   (c) 1-fluoro-2-phenylethane

9.7 Draw the structure of each of the following compounds.

(a) *cis*-l-bromo-2-methylcyclopentane   (b) 3-chlorocyclobutene

(c) (*E*)-1-fluoro-2-butene   (d) (*Z*)-l-bromo-1-propene

Answers:

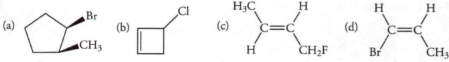

9.8 What is the IUPAC name for each of the following compounds?

(a)
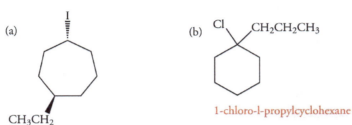
trans-1-ethyl-4-iodocycloheptane

(b)
CH₂CH₂CH₃ with Cl on cyclohexane

1-chloro-l-propylcyclohexane

(c)
trans-1-bromo-2-cyclobutylcyclodecane

(d)
2-bromo-3-cyclohexyl-1-propanol

Nomenclature of Alcohols

9.9 Write the structural formula of each of the following compounds

(a) 2-methyl-2-pentanol   (b) 2-methyl-1-butanol (c) 2,3-dimethyl-1-butanol (d)   cyclopentanol (e) *trans*-2-methylcyclohexanol

(f) 1,3-propandiol (g) 1,2,4-butanetriol

Answers:

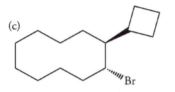

(a) 2-methyl-2-pentanol

$CH_3—CH_2—\overset{\overset{\displaystyle CH_3}{|}}{\underset{\underset{\displaystyle H}{|}}{C}}—CH_2—OH$

(b) 2-methyl-1-butanol

$CH_3—CH_2—\overset{\overset{\displaystyle CH_3}{|}}{\underset{\underset{\displaystyle CH_3}{|}}{C}}—CH_2—OH$

(c) 2,3-dimethyl-1-butanol

cyclopentane—OH

(d) cyclopentanol

cyclohexane with CH₃ and ""OH

(e) *trans*-2-methylcyclohexanol

$HO—CH_2—CH_2—CH_2—OH$

(f) 1,3-propanediol

$HO—CH_2—\overset{\overset{\displaystyle CH_3}{|}}{\underset{\underset{\displaystyle OH}{|}}{C}}—CH_2—OH$

(g) 1,2,4-butanetriol

9.10 What is the IUPAC name for each of the following compounds?

(a) 2-methyl-3-pentanol   (b) 3-ethyl-3-pentanol   (c) 4-methyl-2-pentanol (d) l-ethylcyclohexanol   (e) *cis*-3-ethylcyclopentanol

(f) 1,2-hexanediol   (g) 1,2,3,4,5,6-hexanehexol

Answers:

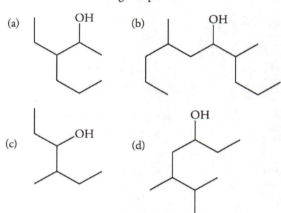

(a) 2-methyl-3-pentanol

(b) 3-ethyl-3-pentanol

(c) 4-methyl-2-pentanol

(d) 1-ethylcyclohexanol

(e) *cis*-3-ethylcyclopentanol

(f) 1,2-hexanediol

(g) 1,2,3,4,5,6-hexanehexol

9.11 What is the IUPAC name for each of the following compounds?

Answers:
(a) 3-ethyl-2-hexanol
(b) 4,7-dimethyl-5-decanol
(c) 4-methyl-3-hexanol
(d) 5,6-dimethyl-3-heptanol

(a)

(b)

(c)

(d)

9.12 What is the IUPAC name for each of the following compounds?

Answers:
(a) *trans*-4-methylcyclohexanol
(b) 4-cyclopentyl-1-butanol
(c) *trans*-2-bromocyclooctanol
(d) 2-bromo-3-cyclohexyl-1-propanol

(a)

(b)

(c)

(d)

9.13 Name the sex attractant of the Mediterranean fruit fly.

Answer:
(*E*)-6-nonen-1-ol

9.14 Name the following compound, which used as a mosquito repellent.

Answer:
2-ethyl-1,3-hexanediol

$$CH_3 — CH_2 — CH_2 — CH — CH — CH_2 — CH_3$$

OH   CH$_2$OH

Properties of Haloalkanes

9.15 Which compound is more polar, methylene chloride (CH_2Cl_2) or carbon tetrachloride (CCl_4)?
Answer: Methylene chloride is more polar because it has a dipole moment. Carbon tetrachloride has no dipole moment.

9.16 Tribromomethane is more polar than tetrabromomethane, but their boiling points are 150 and 189 °C, respectively. Explain why the more polar compound has the lower boiling point.
Answer: The polar tribromomethane has the lower molecular weight and is less polarizable than the higher-molecular-weight tetrabromomethane.

9.17 The dipole moment of (*Z*)-1,2-dichloroethene is 1.90 D. Predict the dipole moment of the *E* isomer.
Answer: The (*E*) isomer has no dipole moment because the bond moments of the two C—Cl groups cancel.

9.18 The dipole moment of 1,2-dichloroethane is 1.19 D. What does this value indicate about the conformational equilibrium of this compound?
Answer: The *anti* conformation has no dipole moment because the bond moments of the two C—Cl groups cancel one another. Therefore, there must be a substantial amount of the gauche conformation in the conformational equilibrium mixture to give a dipole moment of 1.19 D.

Physical Properties of Alcohols

9.19 1,2-Hexanediol is very soluble in water but 1-heptanol is not. Explain why these two compounds with similar molecular weights have different solubilities.
Answer: The two hydroxyl groups in 1,2-hexanediol can form more hydrogen bonds with water, thus greatly increasing its solubility.

9.20 Ethylene glycol and 1-propanol boil at 198 and 97 °C, respectively. Explain why these two compounds with similar molecular weights have different boiling points.
Answer: Ethylene glycol has two hydroxyl groups and can form more intermolecular hydrogen bonds. Therefore, its boiling point is higher than that of 1-propanol.

9.21 Explain why 1-butanol is less soluble than 1-propanol in water.
Answer: The nonpolar hydrocarbon portion of the molecule is larger in 1-butanol than in 1-propanol.

9.22 Suggest a reason why 2-methyl-1-propanol is much more soluble than 1-butanol in water.

Answer: The 2-methyl-1-propanol molecule has a more spherical shape, so it interferes less with the network of hydrogen-bonded water molecules.

Organometallic Reagents

9.23 Devise a synthesis of 1-deutero-1-methylcyclohexane starting from 1-methylcyclohexene.

Answer:

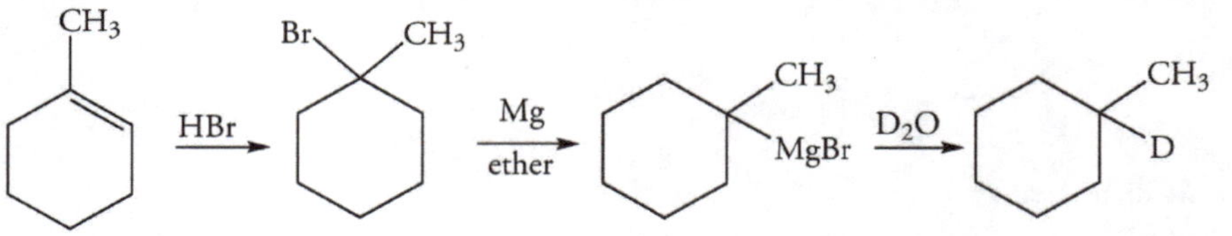

9.24 Devise a synthesis of 1,4-dideuterobutane starting from any organic compound that does not contain deuterium.

Answer:

$$BrCH_2CH_2CH_2CH_2Br \xrightarrow[\text{ether}]{Mg} BrMgCH_2CH_2CH_2CH_2BrMg$$

$$\downarrow D_2O$$

$$D—CH_2CH_2CH_2CH_2—D$$

9.25 Devise two syntheses to prepare 2-methyloctane using reagents containing alkyl groups with five or fewer carbon atoms.

Answer: Using a Gilman reagent and an alkyl halide, two possible combinations are (1) the Gilman reagent lithium di(3-methyl-1-iodobutyl) cuprate and 1-iodobutane and (2) the Gilman reagent lithium dibutyl cuprate and 3-methyl-l-iodobutane. The reaction with lithium dibutyl cuprate is shown below.

$$CH_3—CH(CH_3)—CH_2—CH_2I + (CH_3CH_2CH_2CH_2)_2CuLi$$

3-methyl-1-iodobutane; lithium dibutyl cuprate

$$\downarrow \text{ether}$$

$$CH_3—CH(CH_3)—CH_2—CH_2(CH_2)_3CH_3$$

2-methyloctane

9.26 Write the products of the following reactions for Gilman reagents that contain primary alkyl groups.

Answers:

(a)

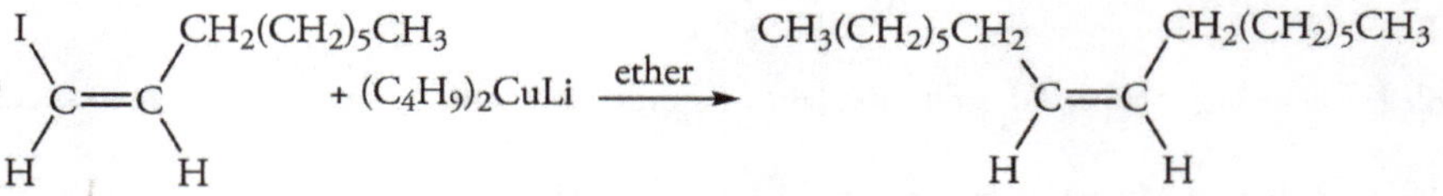

(b)

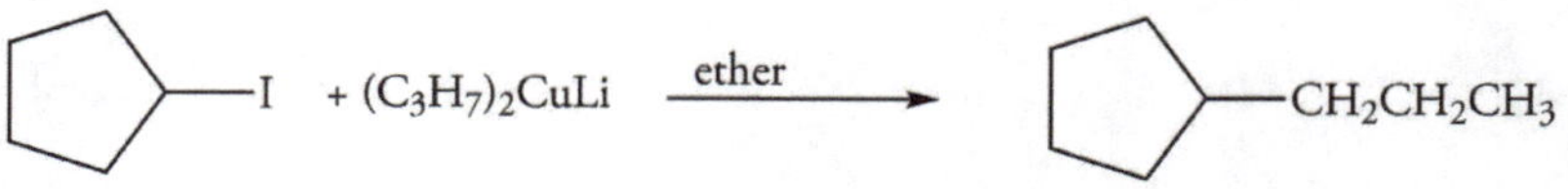

(c)

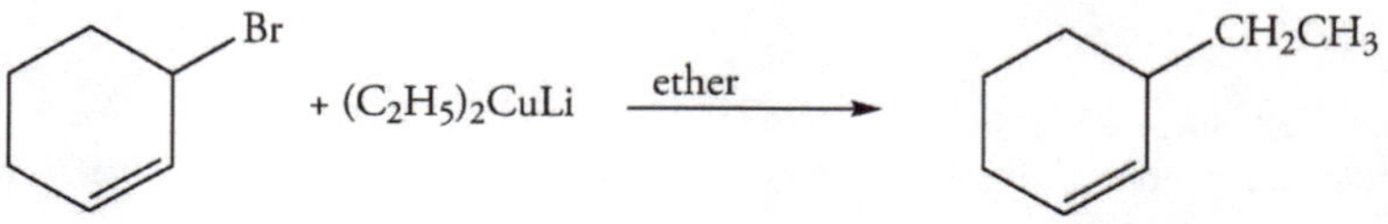

Nomenclature of Alcohols

9.27 Write the structure of the product obtained for each of the following combinations of reactants.

(a) 1-chloropentane and sodium iodide (b) 1,3-dibromopropane and excess sodium cyanide (c) benzyl chloride and sodium acetylide

(d) 2-bromobutane and sodium hydrosulfide (NaSH)

Answers:

(a) $CH_3CH_2CH_2CH_2CH_2Cl \xrightarrow{\text{NaI}} CH_3CH_2CH_2CH_2CH_2-I$

(b) $BrCH_2CH_2CH_2Br \xrightarrow[\text{excess}]{\text{NaCN}} N\equiv C-CH_2CH_2CH_2-C\equiv N$

(c)

(d)

9.28 What haloalkane and nucleophile are required to produce each of the following compounds?

(a) $CH_3CH_2CH_2C\equiv CH$ (b) $(CH_3)_2CHCH_2CN$ (c) $CH_3CH_2SCH_2CH_3$ (d) $C_6H_5CH_2OH$

Answers:

(a) $CH_3CH_2CH_2Br \xrightarrow{HC\equiv C^-Na^+} CH_3CH_2CH_2-C\equiv CH$

(b)

(c) $CH_3CH_2Br \xrightarrow{NaSCH_2CH_3} CH_3CH_2-S-CH_2CH_3$

(d)

Mechanism of Nucleophilic Substitution Reactions

9.29 Which compound in each of the following pairs reacts at the faster rate with sodium iodide in an S_N2 process to yield an alkyl iodide?

(a) 1-chlorohexane or 2-chlorohexane (b) bromocyclohexane or 1-bromo-1-methylcyclohexane

(c) 2-bromo-4-methylpentane or 2-bromo-2-methylpentane

Answers: (a) 1-Chlorohexane reacts fastest because it is a primary haloalkane; 2-chlorohexane is a secondary haloalkane.
(b) Bromocyclohexane reacts fastest because it is a secondary haloalkane; 1-bromo-1-methylcyclohexane is a tertiary haloalkane.
(c) 2-Bromo-4-methylpentane reacts fastest because it is a secondary haloalkane; 2-bromo-2-methylpentane is a tertiary haloalkane.

9.30 Rank the following compounds in order of increasing S_N2 reactivity with a common nucleophile.

     I: 1-bromohexane   II: 1-bromo-2-methylpentane   III: l-bromo-3-methylpentane

Answer: All are primary halogen compounds. The differences in rates are due to steric hindrance of the methyl groups in II and III. The order of reactivity is II < III < I based on the distance of the methyl group from the site of the reaction.

9.31 Which compound in each of the following pairs reacts at the faster rate in an S_N1 process under the same reaction conditions?

   (a) bromocyclohexane or 1-bromo-1-methylcyclohexane (b) 2-bromobutane or l-bromo-2-methylpropane

   (c) 2-bromobutane or 2-methyl-2-bromobutane

Answers: (a) 1-Bromo-l-methylcyclohexane is a tertiary haloalkane and is more reactive in an S_N1 reaction than the secondary haloalkane.
(b) 2-Bromobutane is a secondary haloalkane and is more reactive under S_N1 conditions than l-bromo-2-methylpropane, which is a primary haloalkane.
(c) 2-Bromo-2-methylbutane is a tertiary haloalkane and is more reactive under S_N1 conditions; 2-bromobutane is a secondary haloalkane.

9.32 Which compound in each of the following pairs reacts at the faster rate in an S_N1 process under the same reaction conditions?

   I: 2-bromohexane   II: 2-bromo-2-methylpentane   III: 1-bromo-2-methylpentane

Answer: The order of reactivity is III < I < II, which is in the order primary < secondary < tertiary.

9.33 Predict the product of the reaction of one molar equivalent of sodium iodide with 1,3-dichlorohexane.

Answer: The product is 3-chloro-l-iodohexane, which results from S_N2 displacement of the primary halogen by iodide ion rather than displacement of the secondary halogen at C-3.

1,3-dichlorohexane

9.34 Treatment of the following compound with sodium sulfide yields C_4H_8S. What is the structure of the product? How is it formed?

$$Br-CH_2-CH_2-CH_2-CH_2-CH_2-Br + NaS \rightarrow C_4H_8S$$

Answer: Displacement of one chloride ion by sulfide gives an intermediate that can displace a second halide ion in an intramolecular S_N2 reaction.

9.35 Reaction of the following compound with water under S_N1 conditions yields a mixture of two alcohols. Explain why.

Answer: The compound has a tertiary C—Cl bond which reacts under S_N1 conditions to give a tertiary carbocation. Capture of the carbocation can occur from either face of the planar carbocation, resulting in two geometric isomers.

9.36 Reaction of either 3-bromo-1-butene or (Z)-1-bromo-2-butene with water under S_N1 conditions yields the same product. Explain why.

Answer: The same resonance-stabilized allyl carbocation results from either of the two compounds under S_N1 conditions. Therefore, only one product forms.

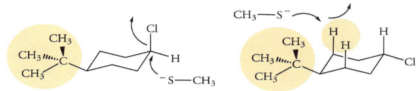

9.37 The rate of reaction of *cis*-1-bromo-4-*tert*-butylcyclohexane with methylthiolate (CH_3S^-) is faster than for the *trans* isomer. Suggest a reason for this difference.

Answer: The reaction of the *trans* isomer requires attack of the nucleophile by a path over the cyclohexane ring to displace the equatorial halogen. This process is more sterically hindered than attack of the nucleophile to displace the axial halogen in the *cis* isomer in a path in the equatorial plane of the ring.

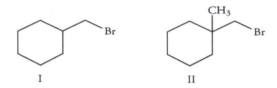

cis-1-bromo-4-*tert*-butylcyclohexane   *trans*-1-bromo-4-*tert*-butylcyclohexane
(nucleophile not sterically hindered)   (nucleophile sterically hindered)

9.38 Which of the following two compounds reacts at the faster rate with sodium cyanide?
Answer: Compound I is less sterically hindered and reacts at the faster rate.

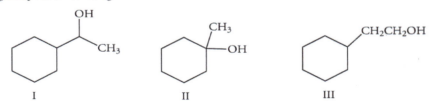

Acid–Base Properties of Alcohols

9.39 1,1,1-Trichloro-2-methyl-2-propanol is used as a bacteriostatic agent. Compare its pK_a to that of 2-methylpropanol.
Answer: The trichloro compound is more acidic as a result of inductive electron withdrawal by the three chlorine atoms. Its pK_a is 12.87, much less the pK_a value of 2-methyl-2-propanol, which is 19.

9.40 Which base is the stronger, methoxide ion or *tert*-butoxide ion? Explain your reasoning.
Answer: The *tert*-butoxide ion is a stronger base than methoxide ion, because *tert*-butyl alcohol is a weaker acid than methanol. The electron-releasing alkyl groups of *tert*-butyl alcohol destabilize the anion.

Formation of Alkyl Halides from Alcohols

9.41 Rank the following compounds according to their rates of reaction with HBr.

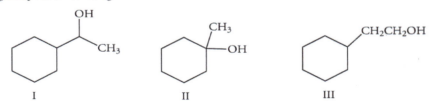

Answer: The reaction will occur by an S_N1 mechanism, so the order of reactivity parallels the order of carbocation stability: III < I < II.

9.42 Rank the following compounds according to their rates of reaction with HCl and $ZnCl_2$.

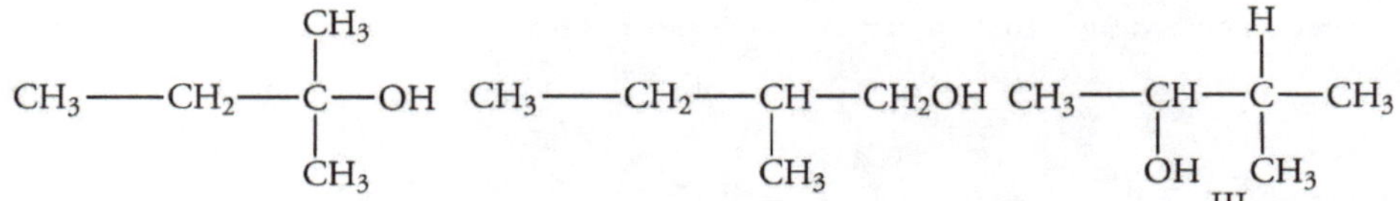

$$CH_3-CH_2-\underset{\underset{CH_3}{|}}{\overset{\overset{CH_3}{|}}{C}}-OH \qquad CH_3-CH_2-\underset{\underset{CH_3}{|}}{CH}-CH_2OH \qquad CH_3-\underset{\underset{OH}{|}}{CH}-\underset{\underset{CH_3}{|}}{\overset{\overset{H}{|}}{C}}-CH_3$$

$$\text{I} \qquad\qquad\qquad \text{II} \qquad\qquad\qquad \text{III}$$

Answer: The reaction will occur by an S_N1 mechanism, so the order of reactivity parallels the order of carbocation stability: II < III < I.

9.43 Write the structure of the product of reaction for each of the following compounds with PBr_3.

Answers:

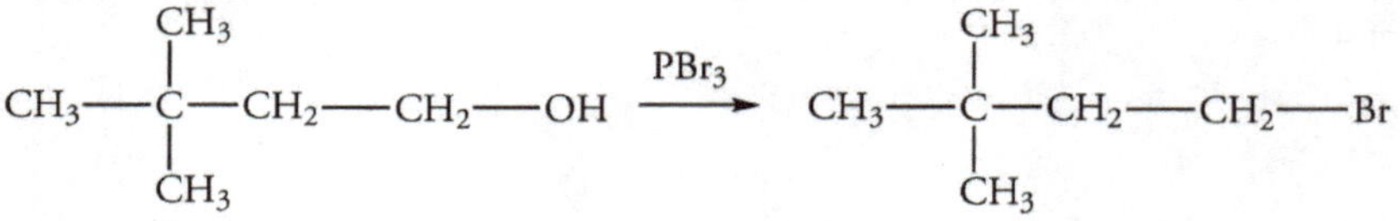

$$CH_3-\underset{\underset{CH_3}{|}}{\overset{\overset{CH_3}{|}}{C}}-CH_2-CH_2-OH \xrightarrow{PBr_3} CH_3-\underset{\underset{CH_3}{|}}{\overset{\overset{CH_3}{|}}{C}}-CH_2-CH_2-Br$$

$$\text{I}$$

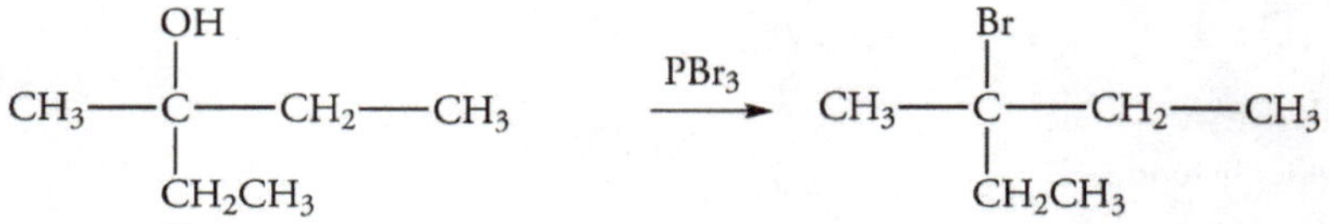

$$CH_3-\underset{\underset{CH_2CH_3}{|}}{\overset{\overset{OH}{|}}{C}}-CH_2-CH_3 \xrightarrow{PBr_3} CH_3-\underset{\underset{CH_2CH_3}{|}}{\overset{\overset{Br}{|}}{C}}-CH_2-CH_3$$

$$\text{II}$$

$$CH_3-\underset{\underset{CH_3}{|}}{\overset{\overset{H}{|}}{C}}-CH_2-\underset{\underset{OH}{|}}{CH}-CH_3 \xrightarrow{PBr_3} CH_3-\underset{\underset{CH_3}{|}}{\overset{\overset{H}{|}}{C}}-CH_2-\underset{\underset{Br}{|}}{CH}-CH_3$$

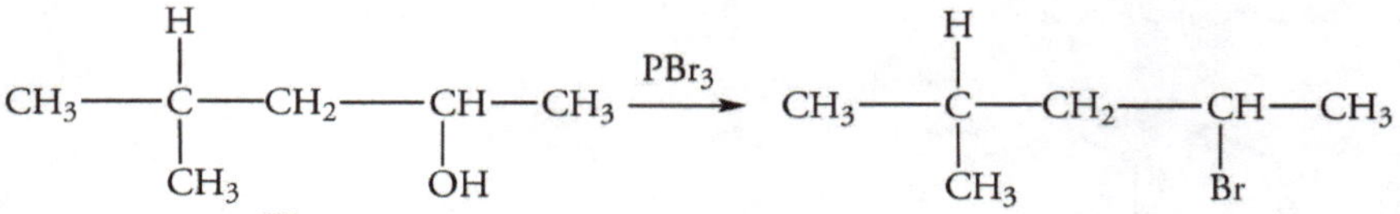

$$\text{III}$$

9.44 Write the structure of the product of reaction for each of the following compounds with $SOCl_2$.

Answers:

(a) $\xrightarrow{SOCl_2}$

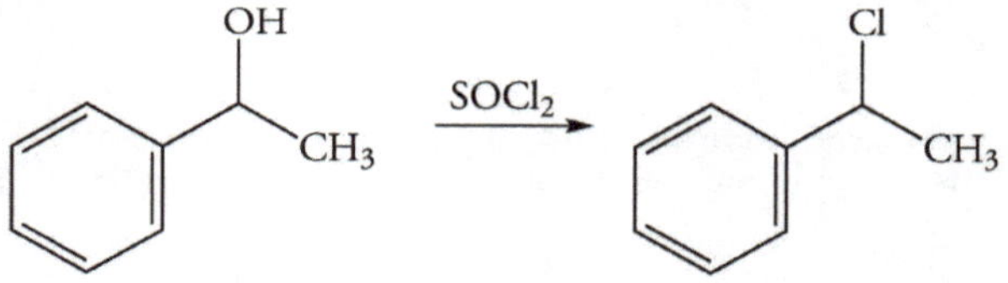

(b) $\xrightarrow{SOCl_2}$

(c) $\xrightarrow{SOCl_2}$

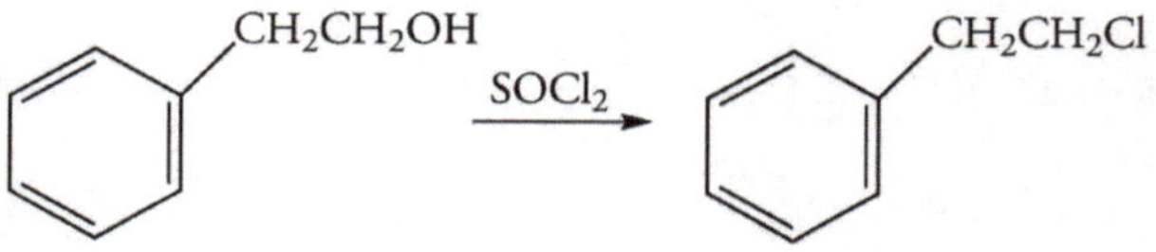

9.45 Reaction of 3-buten-1-ol with HBr yields a mixture of two products: 3-bromo-l-butene and 1-bromo-2-butene. Explain why. (Hint: The reaction of this allyl alcohol occurs via an S_N1 process.)

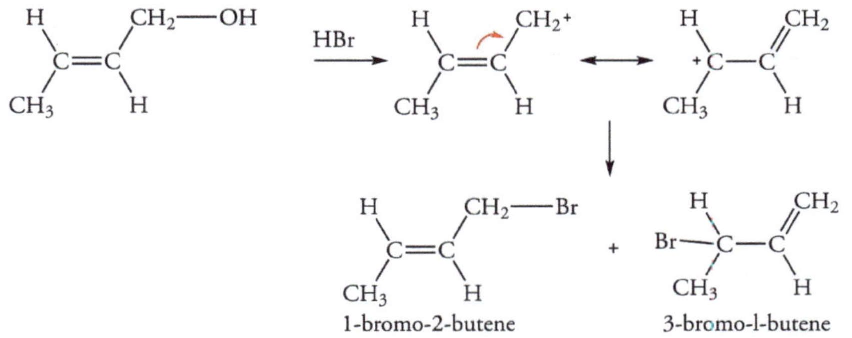

1-bromo-2-butene

3-bromo-l-butene

Answer: A resonance-stabilized allyl carbocation forms as an intermediate under these S_N1 conditions, and it can react with Br⁻ to give two different products.

9.46 The rate of reaction of the following unsaturated alcohol with HBr is faster than the rate of reaction of the saturated alcohol. Explain why.

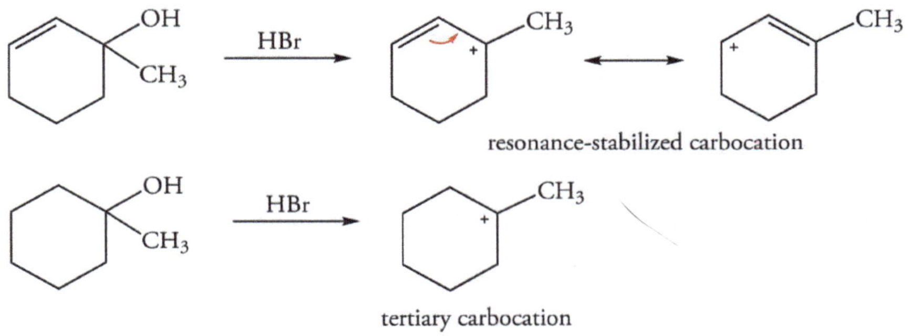

resonance-stabilized carbocation

tertiary carbocation

Answer: Both compounds are tertiary alcohols. However, the first compound yields a tertiary allylic carbocation in an S_N1 reaction. As a result, the transition state for formation of this intermediate has a lower activation energy, and the reaction occurs at a faster rate.

9.47 Which of the compounds in Exercises 9.41 and 9.42 may yield rearranged products?

Answer: Compound I in Exercise 9.41 can rearrange from a secondary carbocation intermediate to give a tertiary carbocation. Compound II in Exercise 9.42 can rearrange from a primary carbocation intermediate to give a tertiary carbocation. Compound III in Exercise 9.42 can rearrange from a secondary carbocation intermediate to give a tertiary carbocation.

I Question 9.41   2° carbocation   3° carbocation

CH_3—CH_2—CH—CH_2OH → CH_3—CH_2—C—CH_2^+

II Question 9.42   1° carbocation

1,2-hydride shift

CH_3—CH_2—C—CH_3

3° carbocation

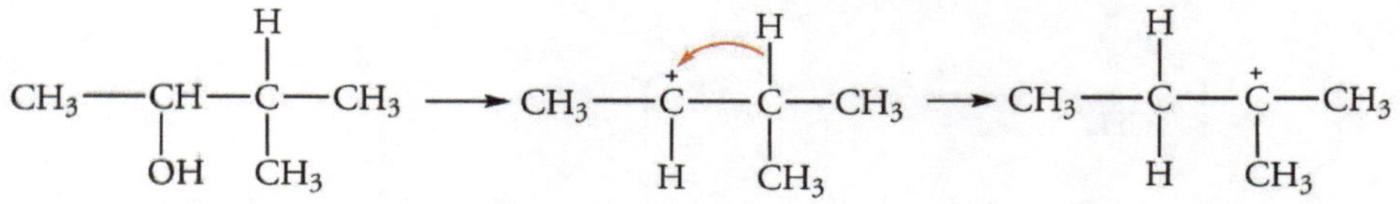

CH_3—CH—C—CH_3 → CH_3—C—C—CH_3 → CH_3—C—C—CH_3

III Question 9.42   1° carbocation   3° carbocation

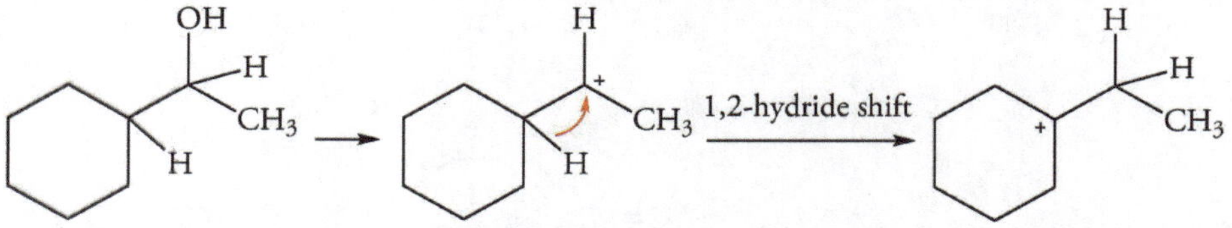

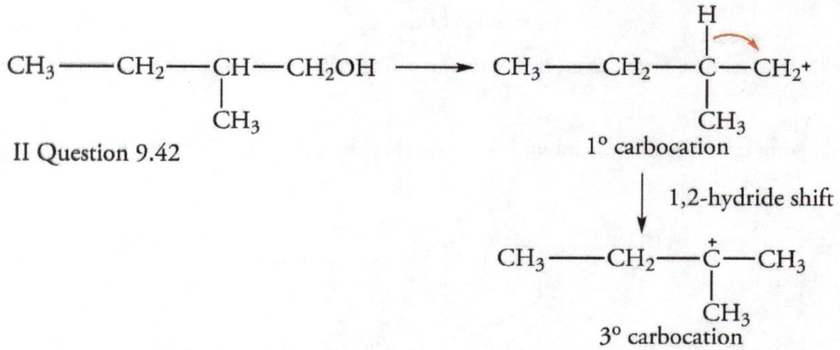

9.48 The reaction of 2-octanol with HBr gives 2-bromooctane and 3-bromooctane in a 13:1 ratio. Explain how 3-bromooctane forms in this reaction.

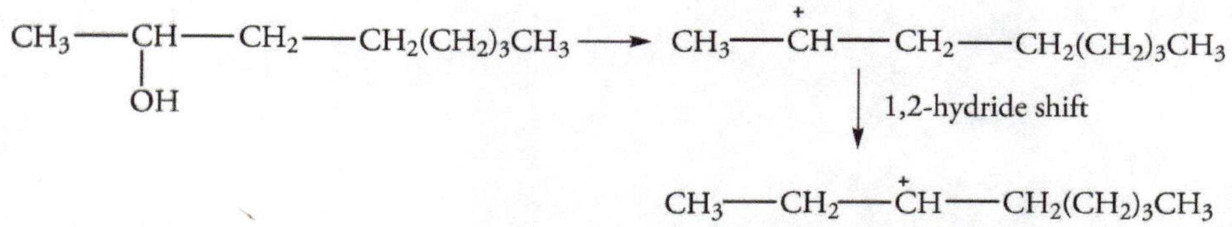

CH_3—CH—CH_2—$CH_2(CH_2)_3CH_3$ → CH_3—$\overset{+}{CH}$—CH_2—$CH_2(CH_2)_3CH_3$

1,2-hydride shift

CH_3—CH_2—$\overset{+}{CH}$—$CH_2(CH_2)_3CH_3$

Answer: The carbocation with positive charge at C-2 can rearrange by a hydride shift of hydrogen at C-3 to give a carbocation with positive charge at C-3. Each carbocation can capture Br⁻.

Regioselectivity in Dehydrohalogenation

9.49 Consider each of the following isomeric compounds with the molecular formula $C_6H_{13}Br$. Which ones will give only a terminal monosubstituted alkene when they undergo dehydrobromination by an E2 process?

I: $CH_3CH_2CH_2CH_2CH_2CH_2Br$ II: $CH_3CHBrCH_2CH_2CH_2CH_3$

III: $CH_3CH_2CHBrCH_2CH_2CH_3$ IV: $(CH_3)_2CHCH_2CH_2CH_2Br$

V: $(CH_3)_2CHCH_2CHBrCH_3$ VI: $(CH_3)_2CHCHBrCH_2CH_3$

$$VII: CH_3CH_2\overset{\overset{\displaystyle CH_3}{|}}{C}HCH_2CH_2Br$$ $$VIII: CH_3CH_2\overset{\overset{\displaystyle CH_3}{|}}{C}HCHBrCH_3$$

$$IX: CH_3CH_2\overset{\overset{\displaystyle CH_3}{|}}{C}BrCH_2CH_3$$ $$X: \quad BrCH_2\overset{\overset{\displaystyle CH_3}{|}}{\underset{\underset{\displaystyle CH_3}{|}}{C}}CH_2CH_3$$

XI: $(CH_3)_3CCH_2CH_2Br$ XII: $(CH_3)_3CCHBrCH_3$

$$XIII: (CH_3)_2CH\overset{\overset{\displaystyle CH_3}{|}}{C}HCH_2Br$$ $$XIV: (CH_3)_2CH\overset{\overset{\displaystyle CH_3}{|}}{C}BrCH_3$$

Answer: Only compounds I, IV, VII, and XI can lose HBr in an E2 reaction to give a single product. Each is a primary bromoalkane.

9.50 Consider each of the compounds in Exercise 9.49. Which ones can undergo dehydrobromination by an E2 process to give only a terminal disubstituted alkene?

Answer: Only compound XIII can yield a terminal disubstituted alkene.

XIII

9.51 Consider each of the compounds in Exercise 9.49. Which ones can undergo dehydrobromination by an E1 process?

Answer: Only compounds IX and XIV are tertiary haloalkanes, and they will undergo dehydrobromination by an E1 process.

9.52 Consider each of the compounds in Exercise 9.49. Which ones cannot undergo dehydrobromination?

Answer: Compound XIII cannot undergo dehydrobromination because the β carbon does not have a hydrogen.

9.53 Consider each of the compounds in Exercise 9.49. Which ones can undergo dehydrobromination to give at least one set of E, Z stereoisomers among the products?

Answer: Compounds II, III, V, VI, VIII, and IX can undergo dehydrobromination to give at least one set of E, Z stereoisomers among the products?

9.54 Consider each of the compounds in Exercise 9.49. Which ones can undergo dehydrobromination to give a trisubstituted alkene among the products? Which ones can undergo dehydrobromination to give a tetrasubstituted alkene among the products?

Answer: Compounds VI, VIII, and IX give trisubstituted products. XIV can give a tetrasubstituted alkene.

9.55 How many alkenes can form from each of the following compounds via an E2 process? Write the structure of each alkene.

(a) l-bromopentane (b) 2-chlorohexane (c) 3-iodoheptane (d) S-bromononane

Answers:

(a)

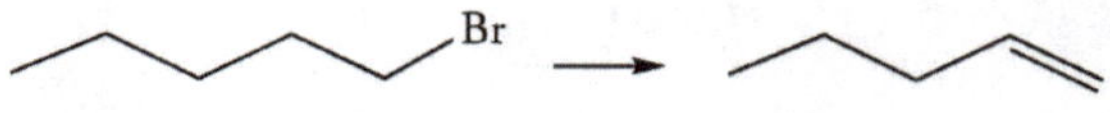

(b)

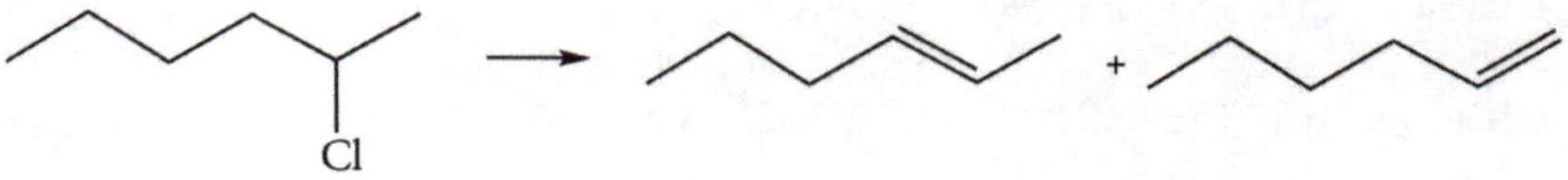

(c)

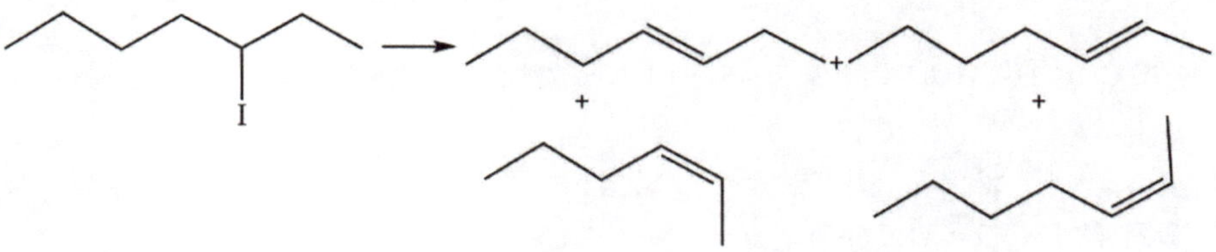

(d)

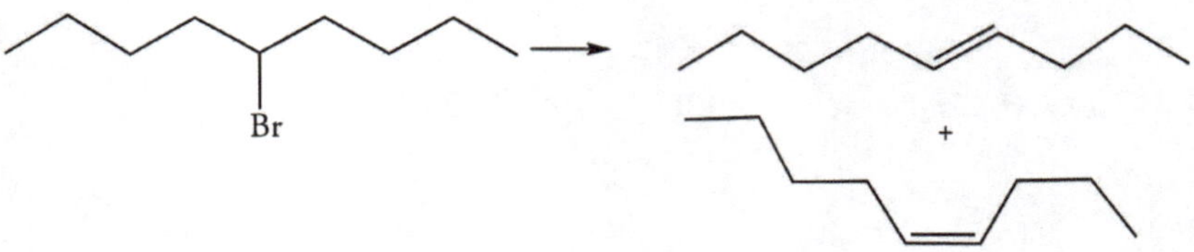

9.56 How many alkenes can form from each of the following compounds via an E2 process? Write the structure of each alkene.

(a) 3-bromo-2-methylhexane (b) 2-chloro-3-methylhexane (c) 3-iodo-4-ethylhexane (d) 4-bromo-4-methylheptane

Answers:

9.57 What bromocycloalkane can give each of the following unsaturated compounds in the best yield by an E2 process?

9.58 Which of the following unsaturated compounds can be obtained in good yield by an E2 process from a bromocycloalkane?

Answer: Only (a) and (c) can be converted to a single product. No bromocycloalkane for (b) or (d) can be made that leads to a single product.

(a)

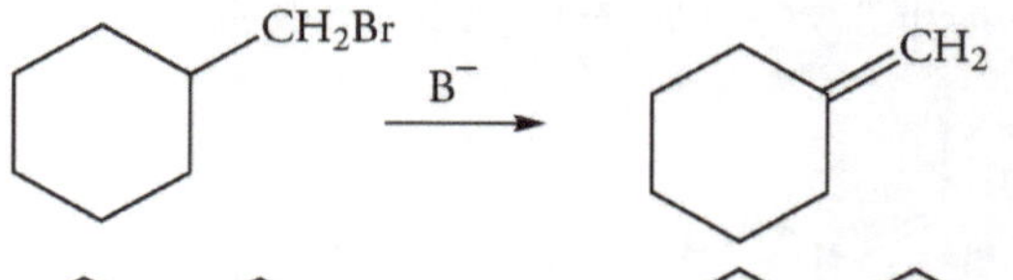

(b)

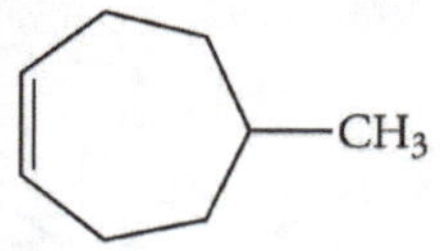

(c)

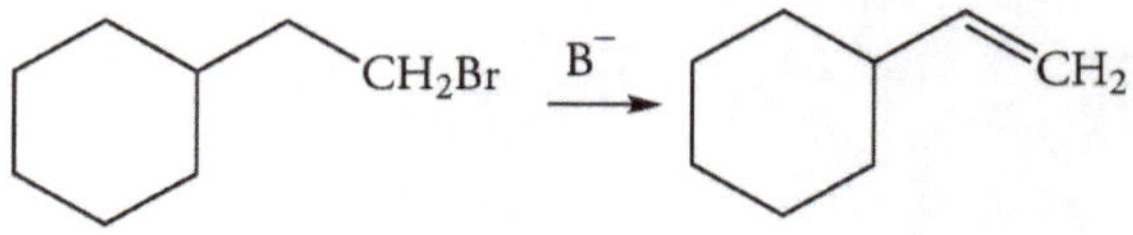

(d)

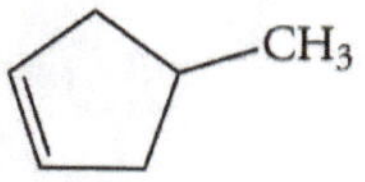

Stereoelectronic Effects in Dehydrohalogenation

9.59 The following isomer undergoes an E2 reaction about 1000 times slower than any of the other stereoisomers of 1,2,3,4,5,6-hexachlorocyclohexane. Why?

Answer: The required *trans*-diaxial arrangement of C—H and C—Cl cannot be achieved even by a ring flip. In any chair conformation, all chlorine atoms are in axial positions and all hydrogen atoms are in equatorial positions with a dihedral angle of 60°.

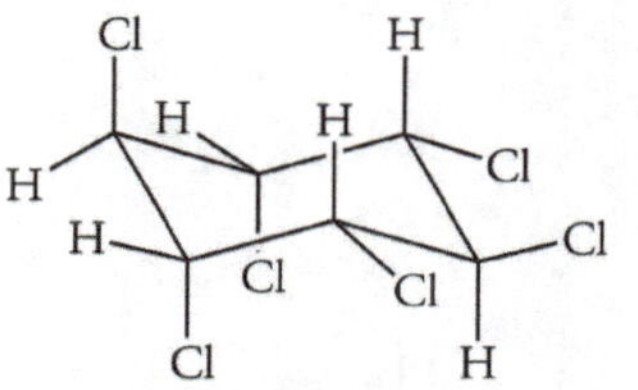

lindane

9.60 One of the following two isomeric bicyclic compounds undergoes an E2 elimination much faster than the other. Identify the compound that reacts at the faster rate and explain why.

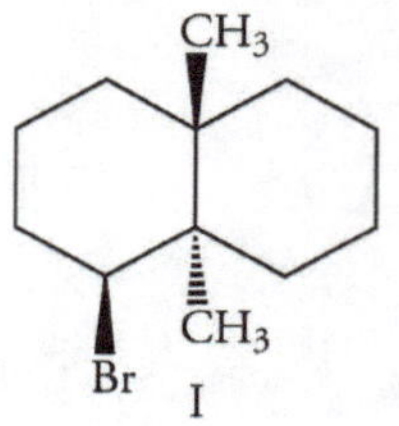

I

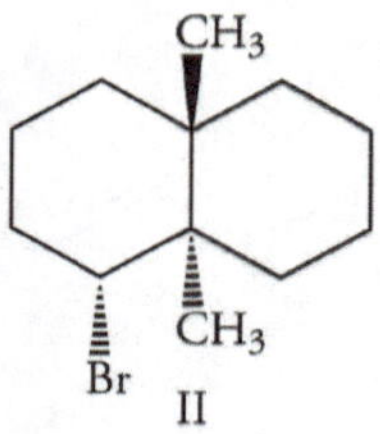

II

Answer: The fused rings of *trans*-decalin are conformationally rigid. Compound I has an axial C—Br bond that is in a position to undergo *trans*-diaxial elimination. Compound II does not have the required *trans*-diaxial arrangement, so the reaction is much slower.

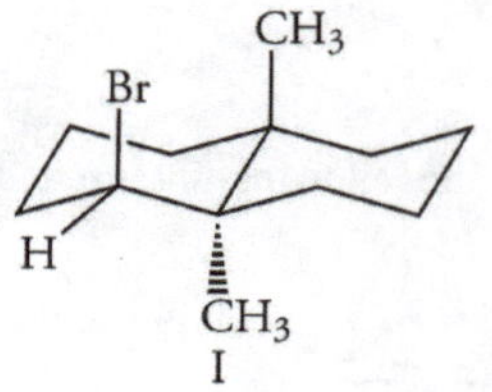

I

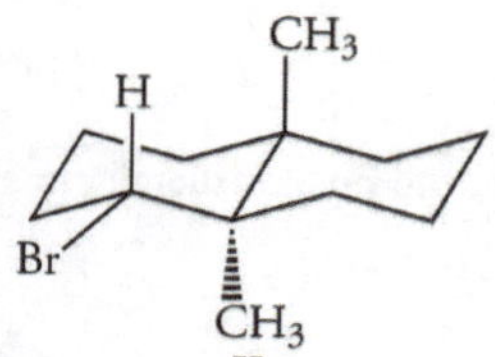

II

9.61 What is the configuration of the alkene formed by the elimination of one molar equivalent of HBr from the following compound?

Answer: The *anti* periplanar arrangement of the C—H and C—Br bonds in the conformation shown gives the (*E*) isomer shown to the right.

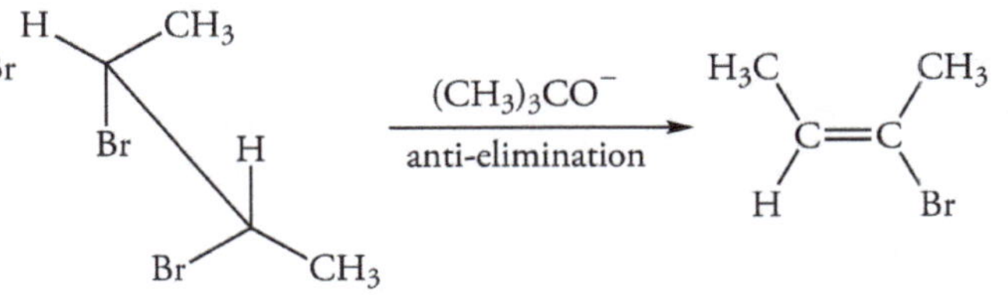

9.62 An E2 elimination of 1-chloro-1,2-diphenylethane can yield a mixture of (*E*)- and (*Z*)-1,2-diphenylethene. How would the *E/Z* ratio of isomers for this reaction compare to the *E/Z* ratio for the E2 elimination of 2-bromopentane?

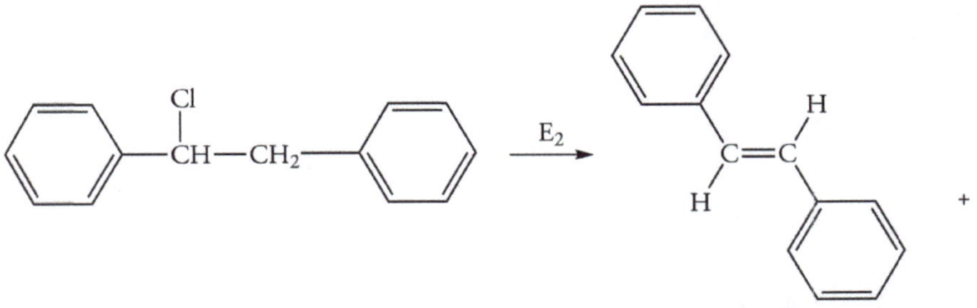

(*E*)-1,2-diphenylethene
(major product)

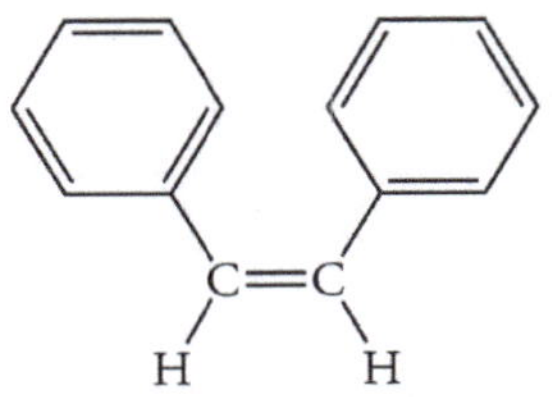

+

(*Z*)-1,2-diphenylethene
(minor product)

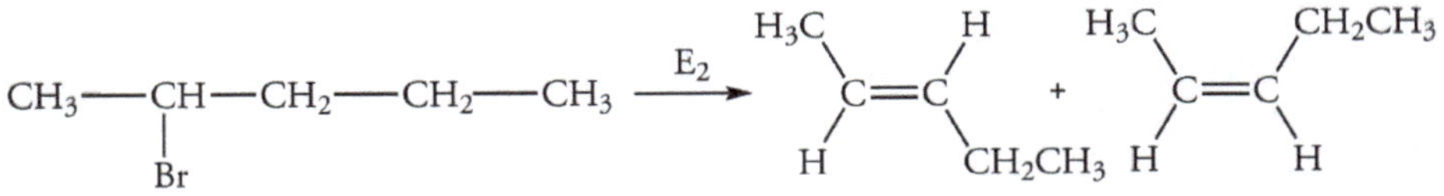

(*E*)-2-pentene   (*Z*)-2-pentene
(major product)   (minor product)

Answer: There would be a larger *E/Z* ratio of isomers for 1,2-diphenyiethene because the two phenyl groups present a larger steric hindrance in the formation of the (*Z*) isomer than the methyl and ethyl groups do.

9.63 When menthyl chloride reacts with sodium ethoxide in ethanol, the only alkene product is 2-menthene. Explain why.

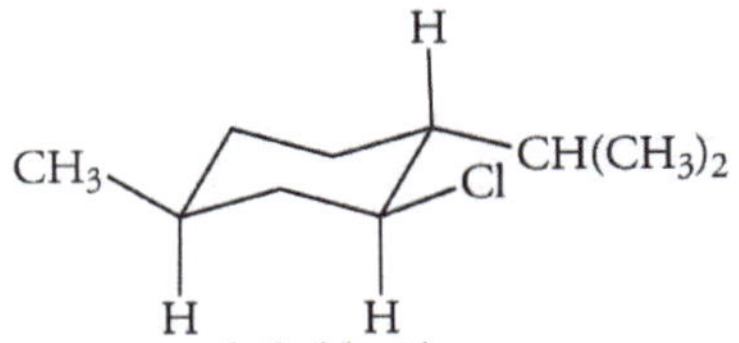

menthyl chloride 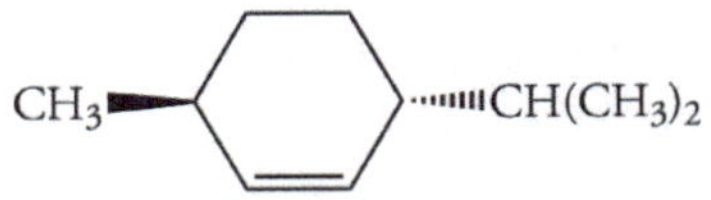   2-menthene

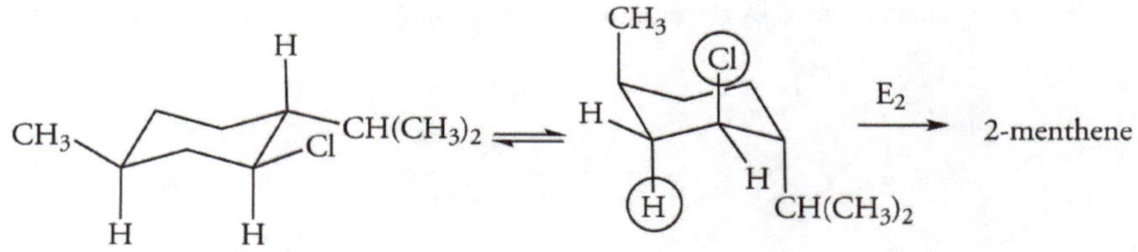

menthyl chloride

Answer: In the conformation obtained by a ring flip, there is only one axial C—H bond situated to undergo an *anti* periplanar E2 elimination. It gives 2-menthene.

9.64 When neomenthyl chloride reacts with sodium ethoxide in ethanol, the only alkene product is 3-menthene. Explain why.

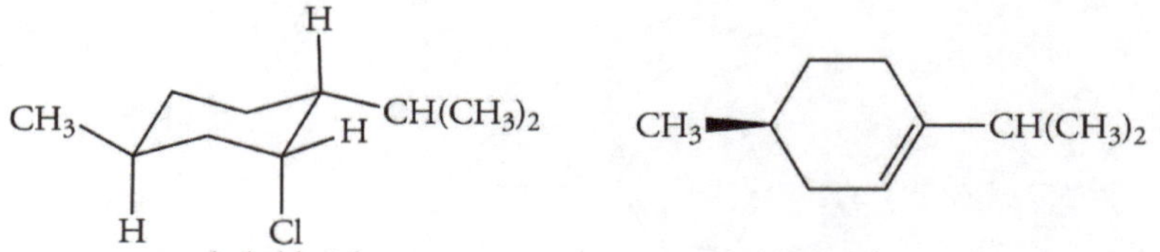

neomenthyl chloride   3-menthene

Answer: There are two C—H bonds situated to undergo *anti* periplanar E2 eliminations. The ratio of products is determined by the degree of substitution of the double bonds. 3-Menthene is the more stable compound because it has a trisubstituted double bond.

9.65 Draw the structure of the alkene formed in an E2 elimination of the following compound.

Answer: The *anti* periplanar arrangement of the C—H and C—Cl bonds gives the alkene shown to the right. It is the more stable, tetrasubstituted double bond.

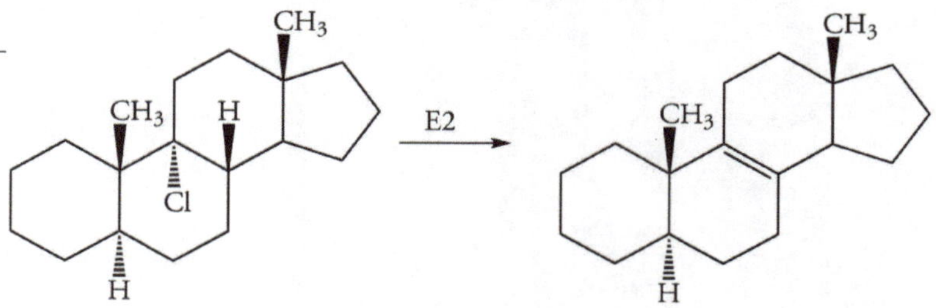

9.66 Which of the following compounds reacts at the faster rate in an E2 elimination reaction?

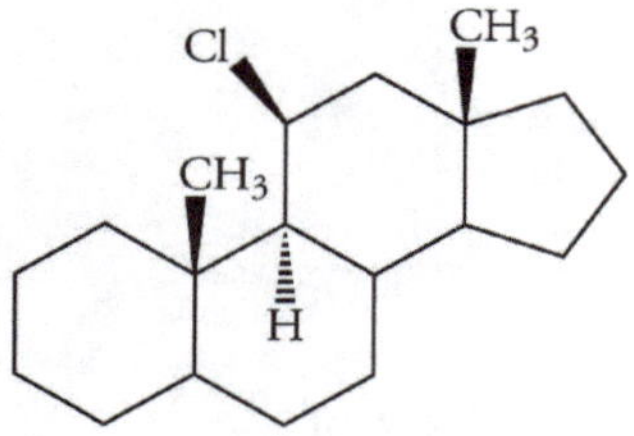

I

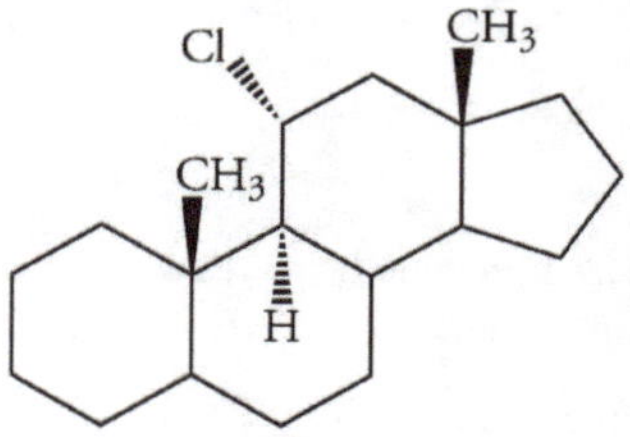

II

Answer: Compound I has axial C—H and C—Cl bonds as shown in the structure and can undergo an *anti* periplanar E2 elimination reaction. Compound II has an equatorial C—Cl bond, so it would react at a slower rate.

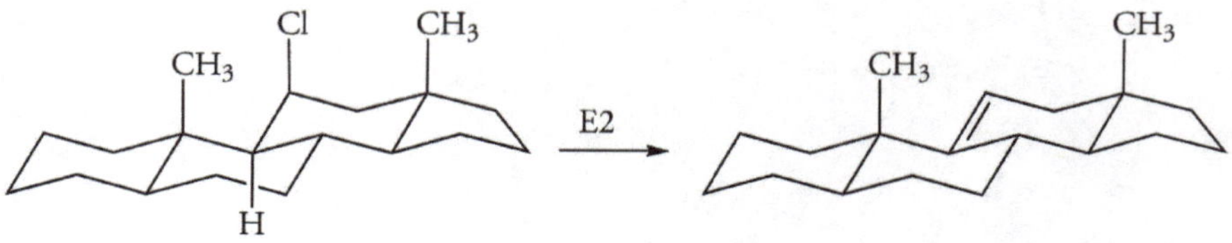

9.67 The following compound cannot undergo dehydrobromination under either E1 or E2 conditions. Explain why.

Answer: There is no *anti* periplanar C—H bond as required for an E2 elimination reaction. The bridgehead carbon cannot form the required planar arrangement for the carbocation that would have to be generated in an E1 reaction.

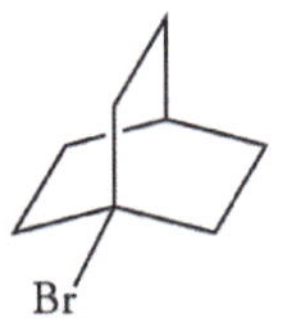

1-bromobicyclo[2.2.2]octane

9.68 Explain why 1-*tert*-butylcyclopropene is difficult to synthesize by a dehydrohalogenation reaction.

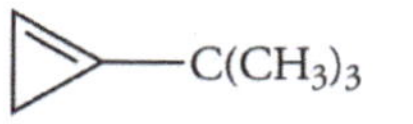

1-*tert*-butylcyclopropene

Answer: The reaction of either the required 1-halo or *trans*-2-halo compound could undergo an E2 elimination because there is a *syn* periplanar arrangement of the required bonds. However, the resulting compound is very strained. Therefore, a high energy transition state would be required for elimination.

9.69 Reaction of 1-bromo-2-deutero-2-phenylethane with *tert*-butoxide in *tert*-butyl alcohol gives a 7:1 ratio of deuterated and nondeuterated phenylethenes. Write the structures of the products. What does the data suggest about the ease of abstraction of deuterium versus hydrogen?

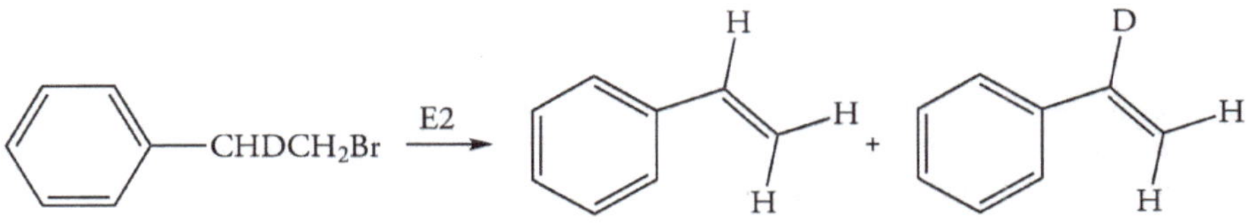

Answer: The energy required to break a C—D bond is larger than for a C—H bond. This energy difference is reflected in the transition state for the E2 reaction where these bonds are partially broken.

9.70 Dehydrobromination of each of the following compounds gives a single product. One compound yields a cycloalkene containing deuterium, and the other yields a cycloalkene that does not contain deuterium. Which compound is which?

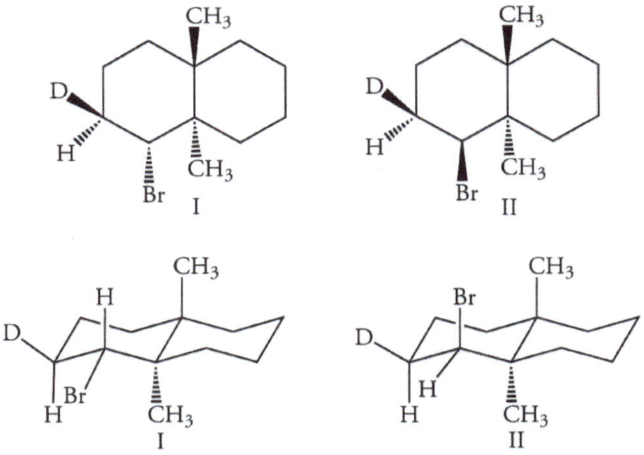

Answer: Compound II has axial C—H and C—Br bonds and can undergo an *anti* periplanar E2 elimination reaction in which the deuterium is retained. Compound I cannot easily undergo an E2 elimination reaction. However, it can eliminate the deuterium atom situated *trans* to the bromine atom in a twisted conformation, which improves the relationship between the two bonds required for elimination of DBr.

9.71 Although the following reaction of the deuterated bicyclic compound with a strong base occurs at a somewhat slow rate, it gives the indicated product by an E2 mechanism. Explain why the product forms.

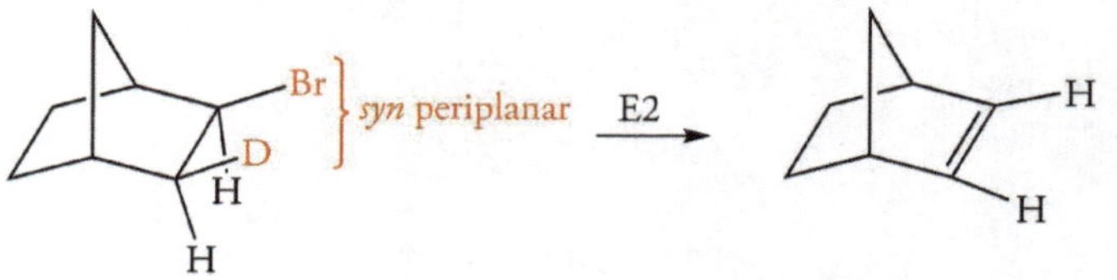

Answer: The E2 reaction occurs from a *syn* periplanar arrangement of the C—D and C—Br bonds, even though breaking a C—D bond requires more energy than breaking a C—H bond. The only C—H bond available for an elimination reaction is at a 120° dihedral angle to the C—Br bond, which is unfavorable for an E2 reaction.

9.72 One of the following 2,3-dichlorobicyclo[2.2.1]heptanes undergoes an E2 elimination using potassium *tert*-butoxide in *tert*-butyl alcohol about 100 times as fast as the other. Which compound reacts at the faster rate? The same product, 2-chlorobicyclo[2.2.1] hept-l-ene, forms in both reactions.

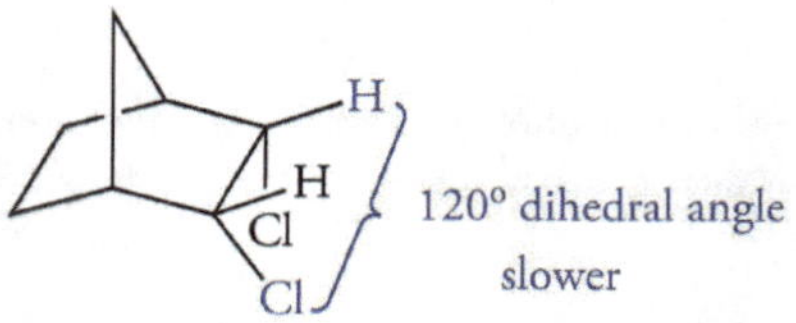

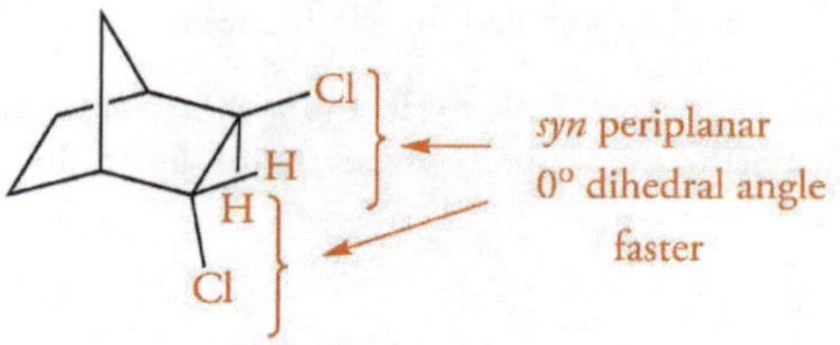

Answer: The E2 reaction of the compound on the right occurs from a *syn* periplanar arrangement of the C—H and C—Cl bonds in the *exo* positions. The only C—H bond available for an elimination reaction in the first compound is at a 120° dihedral angle to the C—Cl bond, which is unfavorable for an E2 reaction.

Dehydration of Alcohols

9.73 Draw the structure of the dehydration product(s) when each of the following compounds reacts with sulfuric acid. If more than one product forms, predict the major isomer assuming that no rearrangement reactions occur.

(a)
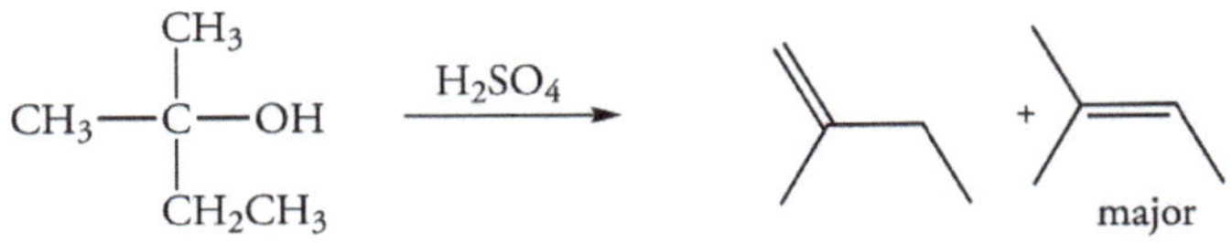
CH3
CH3—C—OH
CH2CH3
H2SO4
+
major

(b)
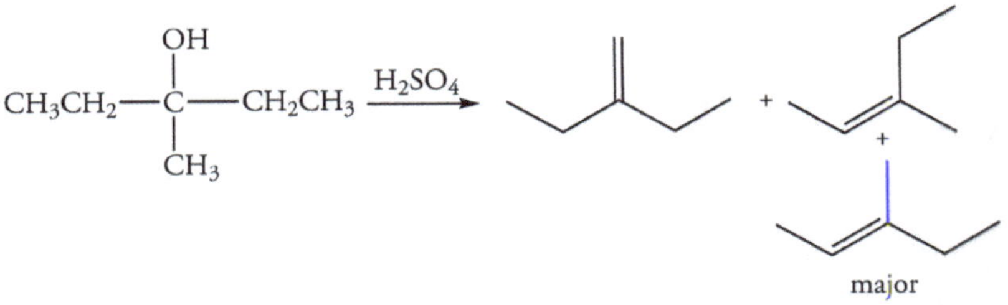
OH
CH3CH2—C—CH2CH3
CH3
H2SO4
+
+
major

(c)
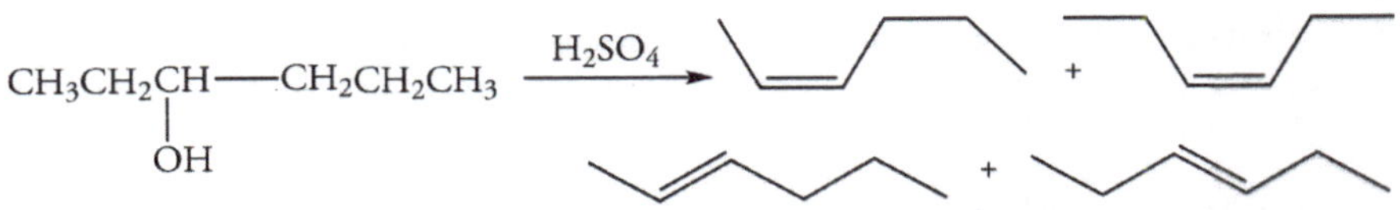
CH3CH2CH—CH2CH2CH3
OH
H2SO4
+
+

These two isomers have comparable stability.

(d)
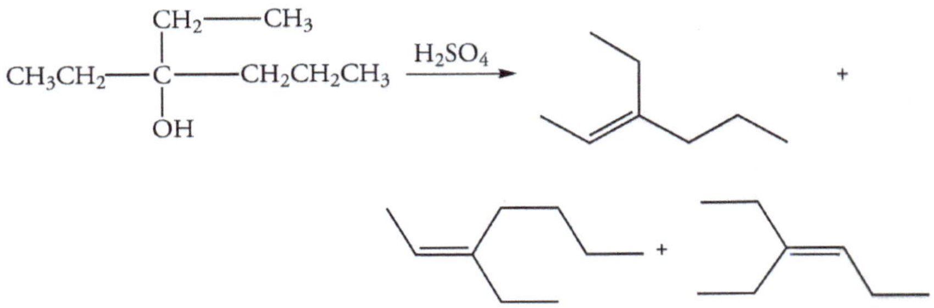
CH2—CH3
CH3CH2—C—CH2CH2CH3
OH
H2SO4
+
+

These three isomers have comparable stability.

9.74 Draw the structure of the dehydration product(s) when each of the following compounds reacts with sulfuric acid. If more than one product forms, predict the major isomer assuming that no rearrangement reactions occur.

(a)
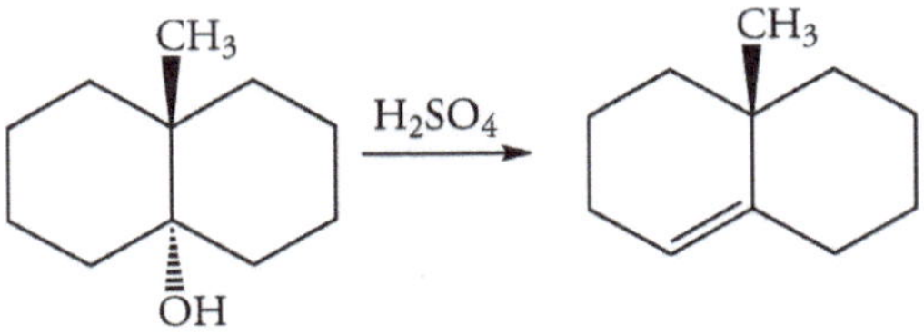
CH3
OH
H2SO4
CH3

(b)
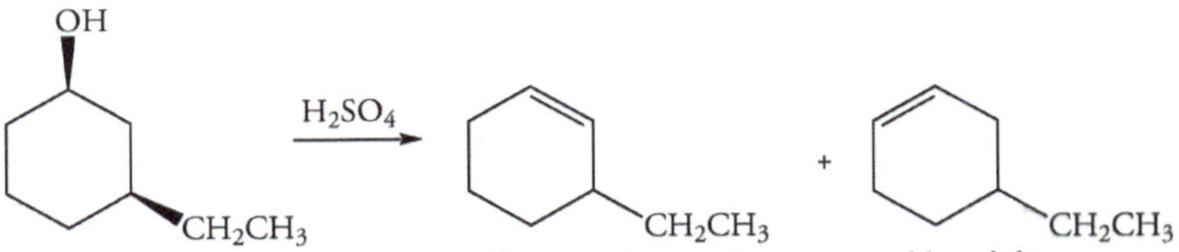
OH
CH2CH3
H2SO4
CH2CH3
+
CH2CH3

These two isomers have comparable stability

(c)
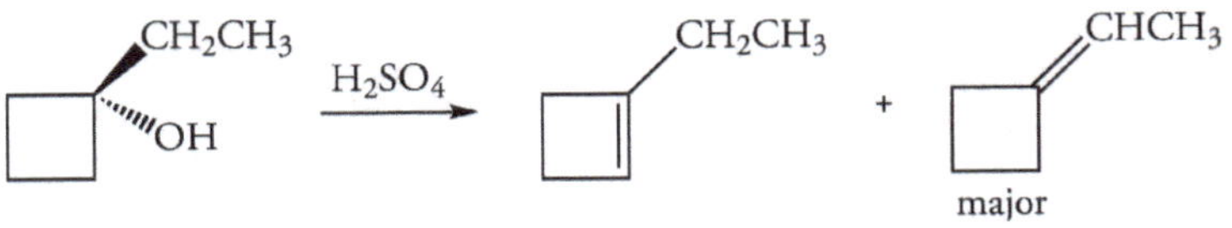
CH2CH3
OH
H2SO4
CH2CH3
+
CHCH3
major

9.75 Write the expected product of the acid-catalyzed dehydration of 1-phenyl-2-propanol. The reaction is more rapid than the dehydration of 2-propanol. Explain why.

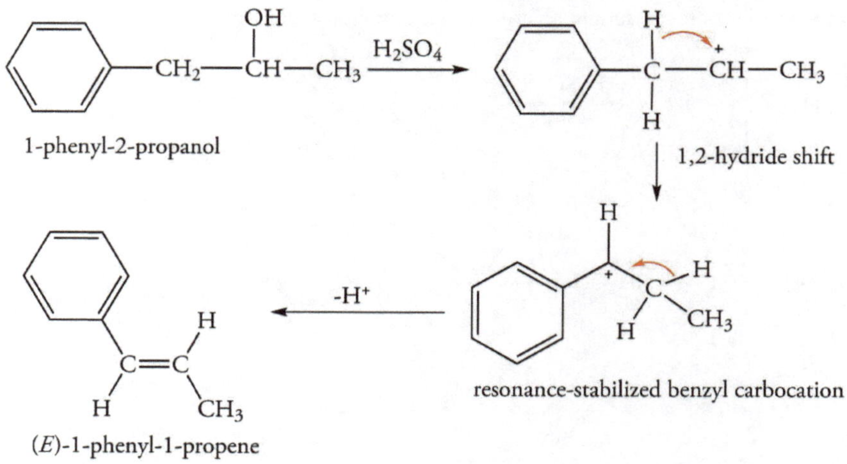

1-phenyl-2-propanol

1,2-hydride shift

resonance-stabilized benzyl carbocation

-H⁺

(E)-1-phenyl-1-propene

Answer: The major product is the more highly substituted alkene. The reaction proceeds through a resonance-stabilized, benzyl carbocation intermediate, and the product has a double bond that can interact with the benzene ring giving a resonance-stabilized structure. Therefore, the dehydration is more rapid than for 2-propanol.

9.76 1,2-Diphenylethanol dehydrates extremely easily. Explain why.

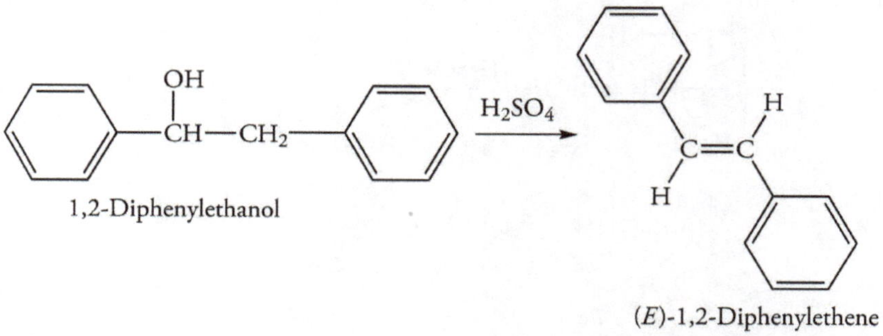

1,2-Diphenylethanol

(E)-1,2-Diphenylethene

Answer: The alkene has a double bond that can interact with the two benzene rings giving a resonance-stabilized structure.

9.77 Dehydration of *cis*-2-methylcyclohexanol yields two products in a 5:1 ratio. What are the structures of the two products?

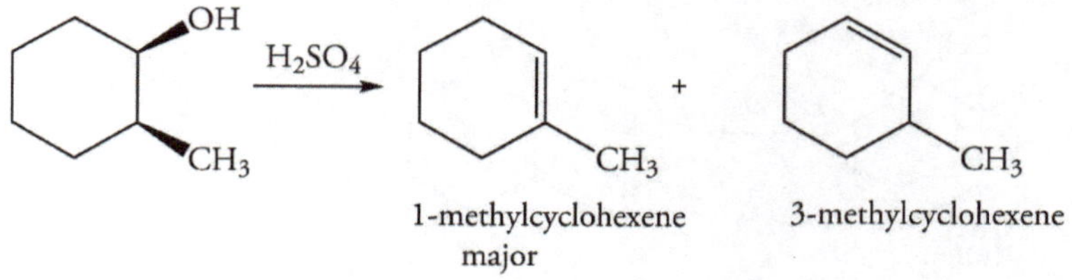

1-methylcyclohexene
major

3-methylcyclohexene

Answer: 1-Methylcyclohexene, which has a trisubstituted double bond, predominates over 3-methylcyclohexene, which has a disubstituted double bond.

9.78 Dehydration of cyclododecanol yields two isomeric products in approximately equal amounts. Catalytic hydrogenation of either compound yields cyclododecane. What are the structures of the two products?

Answer: The two compounds formed in the dehydration reaction are geometric isomers (*E*)-cyclodecene and (*Z*)-cyclodecene.

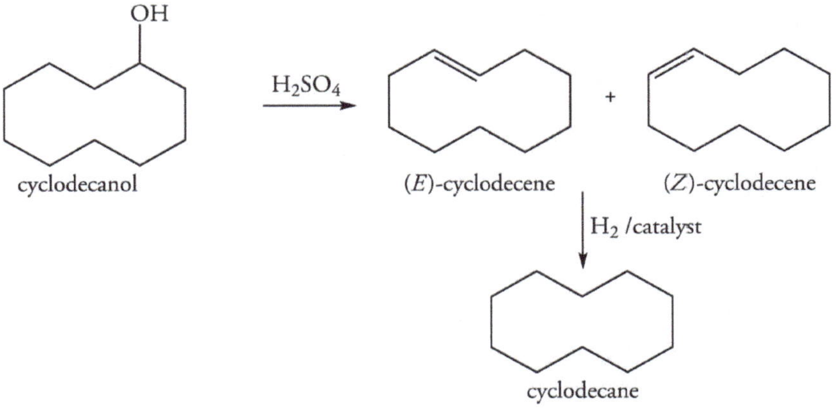

Carbocation Rearrangement in S$_N$1 and E1 Reactions

9.79 The following isomerization reactions occur in some industrial processes. Write a mechanism that accounts for each step. Indicate whether each reaction is energetically favorable or unfavorable.

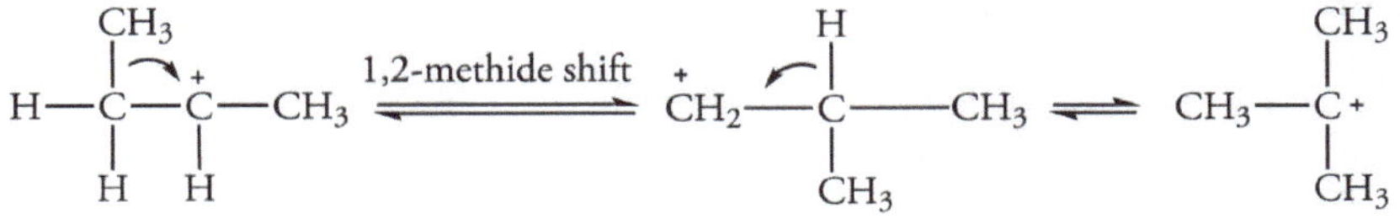

Answer: The first step, a 1,2-methide shift, converts a 2° carbocation into a 1° carbocation. Therefore, it is unfavorable. However, the second step, a 1,2-hydride shift, converts a 1° carbocation into a 3° carbocation. Thus, the pathway from the secondary to the tertiary carbocation is favorable overall.

9.80 Write a mechanism that accounts for each step of the rearrangement of the carbocation shown below. Indicate whether each reaction is energetically favorable or unfavorable.

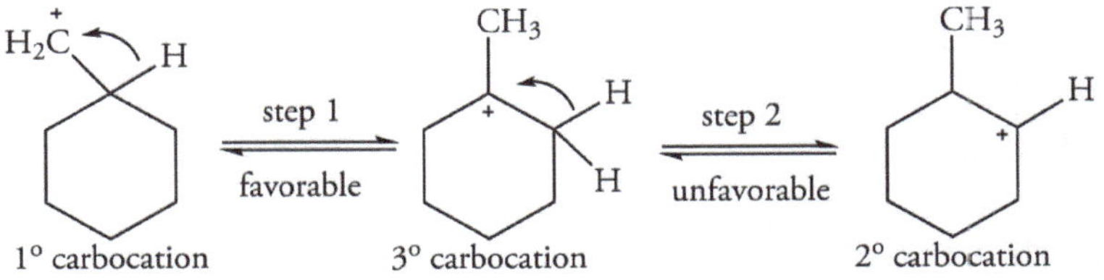

Answer: The first step, a 1,2-hydride shift, converts a primary carbocation into a tertiary carbocation and is favorable. The second step, another 1,2-hydride shift, converts a tertiary carbocation into a secondary carbocation and is unfavorable.

9.81 Ethylidenecyclohexane and l-ethylcyclohexene can be equilibrated using an acid catalyst. Write a mechanism that accounts for this conversion.

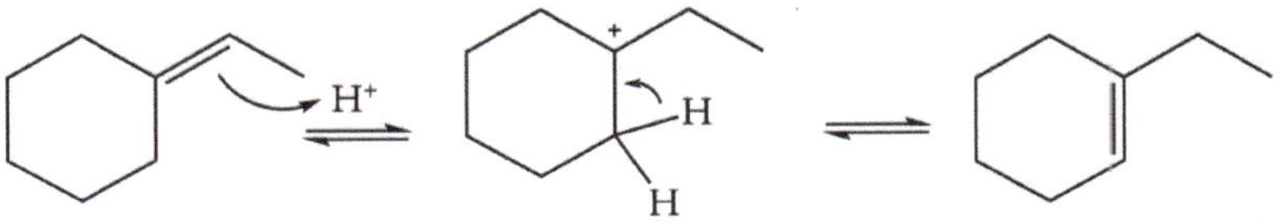

Answer: The rearrangement occurs by protonation of the double bond to give a tertiary carbocation, which is followed by loss of a proton from C-2 of the cyclohexane ring.

9.82 4-Methylcyclohexene isomerizes to 1-methylcyclohexene over alumina (an acidic substance). Write a mechanism that accounts for this conversion.

Answer: First, protonation of the double bond at C-4 gives a secondary carbocation at C-3. Second, a hydride shift from C-2 to C-3 gives a secondary carbocation at C-2. Third, loss of a proton from C-1 of the cyclohexane ring forms the 1-methylcyclohexene isomer.

Rearrangement in Dehydration Reactions

9.83 Dehydration of 2,2,4-trimethyl-3-pentanol with acid gives a complex mixture of the alkenes in the indicated percentages. Write a mechanism that accounts for each product.

I: 2,3,4-trimethyl-1-pentene   29%   II: 2,4,4-trimethyl-1-pentene   24%

III: 3,3,4-trimethyl-1-pentene 2%   IV: 2,4,4-trimethyl-2-pentene   24%

V: 3,3,4-trimethyl-2-pentene 18%   VI: 2-isopropyl-3-methyl-1-butene 3%

Answer: The initial secondary carbocation at C-3 can rearrange either by a hydride ion shift from C-4 to give a tertiary carbocation (blue arrow) or by a methide ion shift from C-2 to give a different tertiary carbocation (red arrow). These three carbocations account for most of the alkene products. Compounds III and VI are formed from loss of a proton by carbocations that result from further rearrangement of the one of the tertiary carbocations. Compounds III and VI are formed in significantly smaller amounts than the other four.

9.84 Dehydration of 2,2-dimethylcyclohexanol with acid gives the following isomeric alkenes. Write a mechanism that accounts for each product.

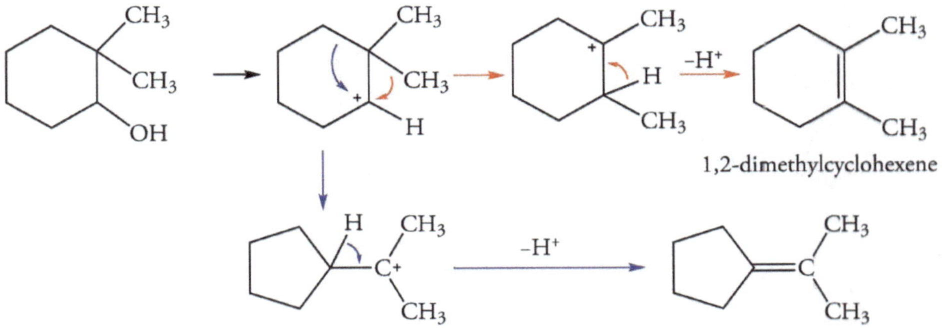

1,2-dimethylcyclohexene

isopropylidinecyclopentane

Answer: The initial secondary carbocation can rearrange by either a 1,2-methide shift (red arrow) or a methylene group shift (blue arrow) of the cyclohexane ring. Each process forms a tertiary carbocation. Subsequent loss of a proton from each of these carbocations gives the observed products.

9.85 1-Methylcyclopentene is one of the dehydration products obtained from l-cyclobutyl-l-ethanol. Write a mechanism that accounts for this reaction.

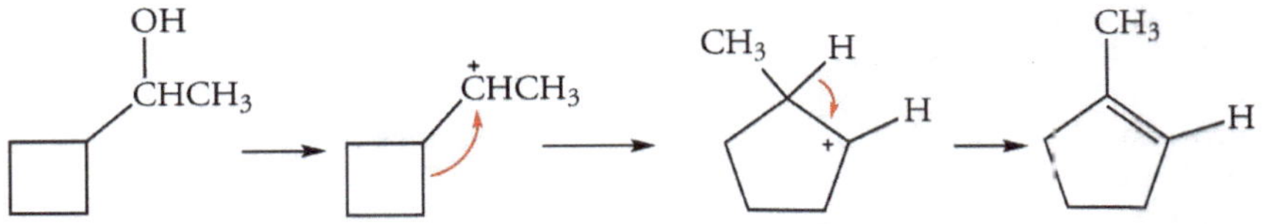

Answer: The initial secondary carbocation can rearrange by migration of a methylene group of the cyclobutane ring to give another secondary carbocation. The driving force of the reaction is the decrease in ring strain. Subsequent loss of a proton from the cyclopentyl carbocation gives 1-methylcyclopentene.

9.86 3,3-Dimethylcyclopentene is one of the dehydration products obtained from 2-cyclobutyl-2-propanol. Write a mechanism that accounts for this reaction.

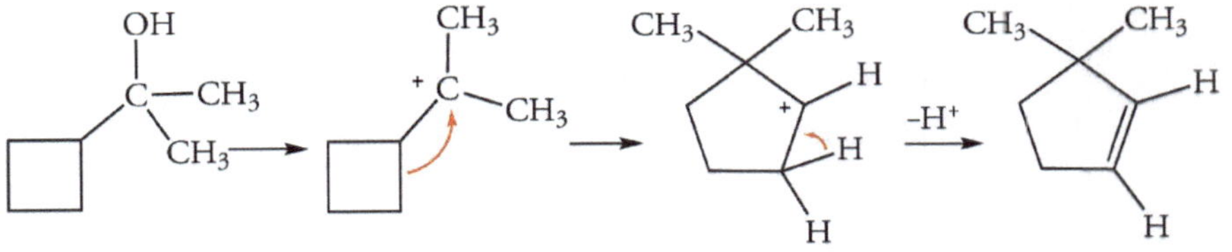

Answer: The initial tertiary carbocation can rearrange by migration of a methylene group of the cyclobutane ring to give a secondary carbocation. Although the carbocation is less stable, the driving force of the reaction is the decrease in ring strain. Subsequent loss of a proton from the cyclopentyl carbocation gives 3,3-dimethylcyclopentene.

9.87 1-*tert*-Butylcyclohexene is one of several dehydration products obtained from 1,2,2-trimethylcycloheptanol. Two rearrangements are required for this transformation. Write a mechanism accounting for these reactions.

Answer: The initial tertiary carbocation can rearrange by migration of a methylene group at C-3. Another tertiary carbocation results, but there is a small reduction of ring strain since a cycloheptane ring is converted into a more stable cyclohexane ring. A subsequent 1,2-methide shift forms a tertiary butyl group and generates a tertiary carbocation with the charge on the six-membered ring. This carbocation loses a proton to give a trisubstituted double bond.

9.88 Dehydration of 2-methyl-2-spiro[4.4]nonanol gives a mixture containing 1-methyl-6-bicyclo[4.3.0]nonene. Write a mechanism that accounts for formation of this product.

2-methyl-2-spiro[4.4]nonanol

1-methyl-6-bicyclo[4.3.0]nonene

Answer: The initial tertiary carbocation can rearrange by migration of a methylene group at the adjacent carbon atom that is part of the other cyclopentane ring. Another tertiary carbocation results, but there is a reduction of ring strain as a cyclopentane ring is converted into a cyclohexane ring. This carbocation loses a proton to give a trisubstituted double bond.

10 NUCLEOPHILIC SUBSTITUTION AND ELIMINATION REACTIONS

Keys to the Chapter

10.1 Nucleophilicity and Basicity

Nucleophilicity refers to the ability of a nucleophile to displace a leaving group in a substitution reaction. We describe various trends in nucleophilicity in this section. There is a trend within a period, a trend within a group, a trend based on the charge of the nucleophile, and a steric effect of the nucleophile. Nucleophiles are also bases, and they can abstract protons in elimination reactions. However, although nucleophilicity and basicity are related to the availability of the same electron pair, the reactions of a series of nucleophiles do not necessarily parallel those of the same species as bases.

Within a period, nucleophilicity parallels basicity and decreases from left to right in the periodic table for elements in similarly structured species with the same charge. For example, hydroxide ion is a better nucleophile than fluoride ion. Nucleophilicity is decreased by hydrogen bonding of protic solvents such as alcohols.

Within a group, the order of nucleophilicity is opposite to the order of basicity. This order of nucleophilicity is related to the polarizability of the nucleophile. The order $I^- > Br^- > Cl^-$ is one that we encounter many times in the study of reaction mechanisms. Another important relationship is $RS^- > RO^-$.

Charge has a large effect on nucleophilicity. A species with a negative charge is more nucleophilic than a neutral species with a similar structure. For example, alkoxide ion, RO^-, is a better nucleophile than ROH.

A nucleophile must approach a carbon reaction center to form a bond. Therefore, steric hindrance affects the rate of reaction. Sterically hindered nucleophiles react at a slower rate than similarly charged, smaller nucleophiles containing the same nucleophilic element. For example, *tert*-butoxide reacts more slowly than ethoxide in S_N2 reactions.

10.2 Stereochemistry of Substitution Reactions

The stereochemistry of a nucleophilic substitution reaction is also a powerful probe of the reaction mechanism. In an S_N2 reaction, a chiral substrate undergoes inversion of configuration because the nucleophile, the reactive center of the substrate, and the leaving group are colinear. The substrate turns "inside out," like an umbrella in the wind, during the reaction. In the S_N1 mechanism, a planar carbocation forms. It is achiral. Thus, the stereochemical result of an S_N1 reaction is the formation of a racemic product. The degree of racemization depends on the degree to which the leaving group leaves the site of the carbocation prior to its capture by the nucleophile. If both sides of the plane of the carbocation are symmetrically solvated, then complete racemization occurs. If one side is still shielded by the leaving group, then there is some residual net inversion.

10.3 S_N2 Versus S_N1 Reactions

The prediction of which mechanism will prevail for a specific reaction is not always an absolute process. Four factors determine whether an S_N1 or S_N2 reaction prevails.

1. The structure of the substrate
2. The nucleophile
3. The leaving group
4. The solvent

1. The Structure of the Substrate

The order of reactivity in S_N1 reactions is tertiary > secondary > primary. The order of reactivity in S_N2 reactions is primary > secondary > tertiary. At the extremes, tertiary substrates usually react by an S_N1 mechanism, and primary compounds usually react by an S_N1 mechanism. The reaction of secondary substrates may proceed by either mechanism depending on the nucleophile, leaving group, and solvent.

The effect of branching at the β carbon atom and conjugation of allyl and benzyl carbocations also have an important effect on S_N1 and S_N2 reactions. Branching at the β carbon atom decreases the rate of an S_N2 reaction due to steric hindrance. Conjugation of the allyl and benzyl carbocations increases their stability. Primary allyl and benzyl carbocations are about as stable as unconjugated secondary carbocations. Secondary allyl and benzyl carbocations are about as stable as unconjugated tertiary carbocations.

2. The Nucleophile

The nucleophile can affect which mechanism prevails in those borderline substrates such as secondary compounds. A charged nucleophile tends to favor the S_N2 mechanism. The structurally related neutral nucleophile tends to favor an S_N1 mechanism.

3. The Leaving Group

The leaving group strongly affects the rate of both S_N1 and S_N2 reactions. Any feature that stabilizes the negative charge of the leaving group increases the rate of the reaction. The same polarizability order of the halide ions as nucleophiles is seen in the order of their reactivity as leaving groups: $I^- > Br^- > Cl^-$. The range of rates is larger for S_N1 reactions than for the S_N2 reactions because the differences in the stabilities of carbocations are larger than the energy differences between different transition states for S_N2 reactions. For leaving groups containing the same element, such as oxygen, the order of leaving group abilities parallels their basicity. Weak bases are better leaving groups. Thus, leaving groups that are derived from strong acids are the best leaving groups.

4. The Solvent

Polar solvents stabilize charged intermediates such as carbocations better than nonpolar solvents. Thus, the rate of an S_N1 reaction increases with increasing solvent polarity. The effect of solvent polarity on S_N2 reactions is much smaller because there is less charge separation in the transition state. Aprotic solvents increase the effective nucleophilicity of the nucleophile because decreased solvation of the nonbonding electron pair makes that pair more available for reaction with a substrate.

10.4 Mechanisms of Elimination Reactions

An E2 reaction is a concerted, *bimolecular* process in which a base extracts a proton from the β-carbon of the substrate and a bond to the leaving group departs simultaneously. Both the substrate and the base are present in the transition state for the reaction. An El reaction occurs in two steps, the first step is rate determining, and does not involve the base. Thus, the formation of a carbocation in an El reaction is the same as the first step in the S_N1 process. The preferred conformation for an E2 reaction is *anti* periplanar. Based on the *anti* periplanar transition state for an E2 mechanism, the stereochemistry of the alkene can be predicted based on the stereochemistry of both stereogenic centers of the substrate. First, write the structure with the correct configuration at the centers containing the hydrogen atom to be eliminated and the leaving group. Then, rotate about the carbon–carbon bond to the staggered conformation with the hydrogen atom and the leaving group *anti* to one another. This conformation has the correct geometry for the bonded groups that will remain in the alkene.

The rate of an E2 reaction is slower when C—D bond breaks than when a C—H bond breaks. This is called a deuterium isotope effect. If a C—H (or C—D) bond is not broken in the rate determining step, as in an El reaction, there is no deuterium isotope effect.

The basicity of the base increases the rate of an E2 reaction. Thus, the amount of E2 product increases even for substrates that can react by an El reaction. The size of the base affects the regiochemistry. More hindered bases abstract protons from less sterically hindered sites and increase the amount of the less substituted alkene.

10.5 Effect of Substrate Structure on Competing Substitution and Elimination Reactions

Many factors control the competition of S_N1, S_N2, E2, and E1 reactions. In general, tertiary haloalkanes react by S_N1 and E1 mechanisms. The ratio of substitution and elimination products depends on the balance between the basicity and the nucleophilicity of the reagent. Primary haloalkanes react by S_N2 and E2 mechanisms. The amount of elimination increases with the basicity of the nucleophile and with its steric size. Secondary haloalkanes are much more sensitive to the conditions of the reaction, and S_N1, S_N2, E2, and E1 processes occur to varying degrees. Highly polarizable nucleophiles favor an S_N2 reaction; neutral nucleophiles are more apt to be seen in an S_N1 reaction. Aprotic solvents favor S_N2 reactions; polar solvents enhance the rate of S_N1 reactions. Effective nucleophiles that are weak bases, such as thiolates, favor substitution over elimination. Strong bases that are less effective nucleophiles, such as *tert*-butoxide, favor elimination.

End of Chapter Exercises

Nucleophilicity

10.1 Hydroxylamine (NH_2OH) is a nucleophile. Write its Lewis structure. Which atom supplies the electrons in nucleophilic substitution reactions?

$$H-\overset{\cdot\cdot}{N}-\overset{\cdot\cdot}{O}:$$
$$\underset{H}{\diagup} \qquad \diagdown H$$

hydroxylamine

Answer: The nucleophilicity for the atoms within a period decreases from left to right. Thus, the nitrogen atom supplies the electrons in nucleophilic substitution reactions.

10.2 Thiocyanate ion (SCN^-) reacts with alkyl halides to give thiocyanate products (R—SCN). The cyanate ion (OCN^-) reacts to form isocyanate products (R—NCO). Write the Lewis structures of the ions. Explain the difference in the sites of reactivity for the two ions.

Answer: The nucleophilicity for the atoms within a family increases from top to bottom. Thus, the sulfur atom of the thiocyanate ion is more nucleophilic than the oxygen atom of the cyanate ion. In the case of the cyanate ion, the nitrogen atom is more basic and a better nucleophile than the oxygen atom, so it forms isocyanate products, R—NCO.

$$:\overset{\cdot\cdot}{\underset{\cdot\cdot}{O}}-C\equiv N: \quad \longleftrightarrow \quad :\overset{\cdot\cdot}{O}=C=\overset{\cdot\cdot}{N}:^- \qquad :\overset{\cdot\cdot}{\underset{\cdot\cdot}{S}}-C\equiv N: \quad \longleftrightarrow \quad :\overset{\cdot\cdot}{S}=C=\overset{\cdot\cdot}{N}:^-$$

isocyanate   thiocyanate

10.3 Reaction of methoxide ion with an alkyl halide to give dimethyl ether is about 100 times faster than the reaction of acetate ion with an alkyl halide to give an ester, methyl acetate. Explain this observation

$$CH_3-\overset{\cdot\cdot}{\underset{\cdot\cdot}{O}}:^- + CH_3-\overset{\cdot\cdot}{\underset{\cdot\cdot}{I}}: \longrightarrow CH_3-\overset{\cdot\cdot}{\underset{\cdot\cdot}{O}}-CH_3$$

methoxide   dimethyl ether

$$CH_3-\overset{\overset{:O:}{\|}}{C}-\overset{\cdot\cdot}{\underset{\cdot\cdot}{O}}:^- \longleftrightarrow CH_3-\overset{:\overset{\cdot\cdot}{O}:^-}{\underset{|}{C}}=\overset{\cdot\cdot}{\underset{\cdot\cdot}{O}}: + CH_3-\overset{\cdot\cdot}{\underset{\cdot\cdot}{I}}: \longrightarrow CH_3-\overset{\overset{:O:}{\|}}{C}-\overset{\cdot\cdot}{\underset{\cdot\cdot}{O}}-CH_3$$

acetate   methyl acetate

Answer: The negative charge is localized in the methoxide ion, which makes the ion an effective nucleophile. The negative charge is delocalized in the acetate ion, and the decreased charge on the oxygen atom decreases its nucleophilicity.

10.4 Diethylphenylphosphine reacts with iodoethane about 10^3 times faster than the nitrogen analog, diethylaniline. Explain this result.

$$\text{(reaction scheme: diethylphenylphosphine} + CH_3CH_2I \xrightarrow[\text{10\% water}]{\text{90\% acetone}} \text{product} + I^- \quad k_{rel} = 10^3)$$

$$\text{(reaction scheme: diethylaniline} + CH_3CH_2I \xrightarrow[\text{10\% water}]{\text{90\% acetone}} \text{product} + I^- \quad k_{rel} = 1)$$

Answer: The nucleophilicity for the atoms within a family increases from top to bottom as the atoms become more polarizable. Thus, the phosphorus atom of the phosphine is more nucleophilic than the nitrogen atom of the aniline.

10.5 Dimethyl sulfide, $(CH_3)_2S$, reacts with iodomethane to displace iodide ion twice as fast as diethyl sulfide, $(CH_3CH_2)_2S$. Explain why.

Answer: Dimethyl sulfide is a smaller nucleophile than diethyl sulfide. Nucleophilicity decreases with increasing size of the nucleophile because the larger nucleophile experiences steric hindrance in the transition.

10.6 Triethylarsine, $(CH_3CH_2)_3As$, reacts with iodomethane only four times as fast as dimethyl selenide, $(CH_3)_2Se$. Compare this difference in rate with the rate difference of ammonia relative to water, about 3×10^5.

Answer: The nucleophilicity for the atoms within a period decreases from left to right. Thus, if the structures are similar, the nucleophilicity of an arsenic compound will be greater than that of a selenium compound, and the nucleophilicity of a nitrogen compound will be greater than that of an oxygen compound. The sizes of ammonia and water are similar. However, triethylarsine is larger than dimethyl selenide. The larger size of triethylarsine decreases its nucleophilicity and decreases its rate of reaction.

Stereochemistry of Substitution Reactions

10.7 Reaction of (R)-$(-)$-2-butanol with HBr yields a mixture of 87% (S)-$(+)$-2-bromobutane and 13% (R)-$(-)$-2-bromobutane. What is the optical purity of the product? What is the mechanism for this substitution reaction?

Answer: The optical purity is 87% – 13% = 74%. Because a mixture of enantiomers results, the mechanism must be S_N1. However, the leaving group (H_2O) substantially shields one face of the carbocation as it leaves, and the bromide ion reacts with substantial inversion of configuration.

10.8 Reaction of (R)-2-methyl-1-butanol with HBr yields 1-bromo-2-methylbutane. Predict the configuration of the product. What is the mechanism for this substitution reaction?

Answer: The mechanism for a substitution reaction at a primary center is S_N2. The stereogenic center at C-2 does not change in this reaction. In the process, the highest priority group —CH_2OH is changed into the highest priority group —CH_2Br. Thus, the configuration is unchanged and is R.

10.9 The rate of incorporation of radioactive iodide into optically active 2-iodooctane in acetone as solvent leads to racemization at twice the rate of incorporation of radioactive iodine. Explain how these data support an S_N2 mechanism.

Answer: Each time a radioactive iodide ion reacts by an inversion process in the S_N2 mechanism, the resulting molecule of inverted product cancels the optical rotation of one molecule of the reactant. Thus, two molecules of optically active 2-iodobutane are effectively "lost" (in terms of their optical rotation) for every one molecule that reacts with radioactive iodide.

10.10 The reaction of (S)-2-bromooctane with cyanide ion gives a cyano compound with an R configuration. However, reaction of (S)-2-bromooctane with iodide ion followed by reaction of the alkyl iodide with cyanide ion gives a cyano compound with the S configuration. Explain these data.

Answer: In the substitution reaction with cyanide ion, the highest priority bromine group is replaced by a high priority cyanide group. The reaction occurs by inversion, and the net result is a product with the R configuration. In the two step reaction sequence, two inversion steps occur, resulting in net retention of configuration. First, replacing bromide by iodide ion yields the R product, which subsequently is converted into the S product when cyanide replaces iodide in the second step.

10.11 *trans*-1-Chloro-3-methylcyclopentane reacts with sodium iodide in acetone to give *cis*-1-iodo-3-methylcyclopentane. What is the mechanism of this reaction?

Answer: Since the stereochemistry at C-1 is inverted when iodide replaces chloride, the reaction must occur by an S_N2 mechanism.

10.12 Write the product expected from the reaction of *cis*-1-bromo-2-methylcyclopentane with cyanide ion.

Answer: The stereochemistry at C-1 is inverted when cyanide ion replaces bromide ion. Thus, the product has the *trans* configuration.

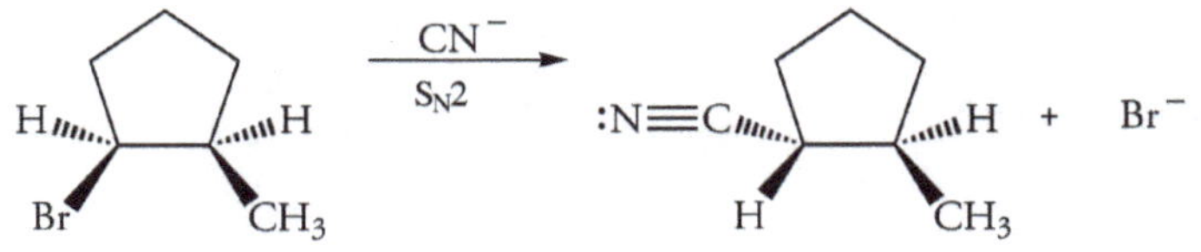

10.13 The following compound has the *R* configuration. Draw the product expected from the reaction of this compound in ethanol, indicating the stereochemistry.

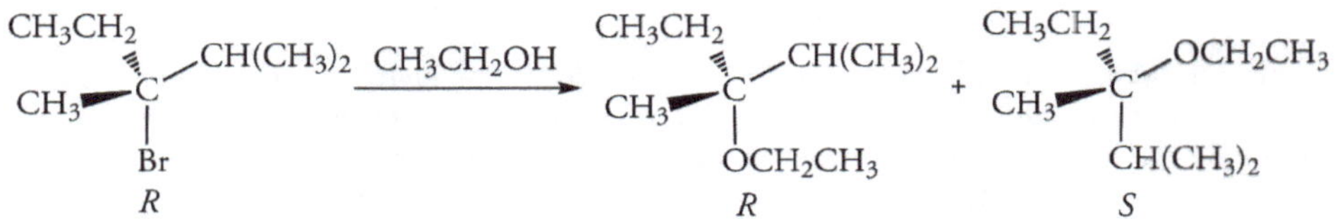

Answer: The compound is a tertiary halide, which reacts by an S_N1 mechanism, so the reaction gives a racemic mixture of ethyl ethers.

10.14 (*S*)-1-Chloro-1-phenylethane reacts in a 20% water–80% acetone solution to give a 51:49 ratio of (*R*)- and (*S*)-1-phenyl-1-ethanols. Explain why the product is highly racemic even though the reactant is a secondary alkyl halide.

Answer: The benzene ring stabilizes the secondary carbocation, which is a benzylic carbocation, by resonance. The benzylic carbocation can be attacked from either side of the plane to give an almost completely racemized product.

10.15 The reactant in the following reaction has the *S* configuration. Based on the composition of the product mixture, what is the mechanism of the reaction?

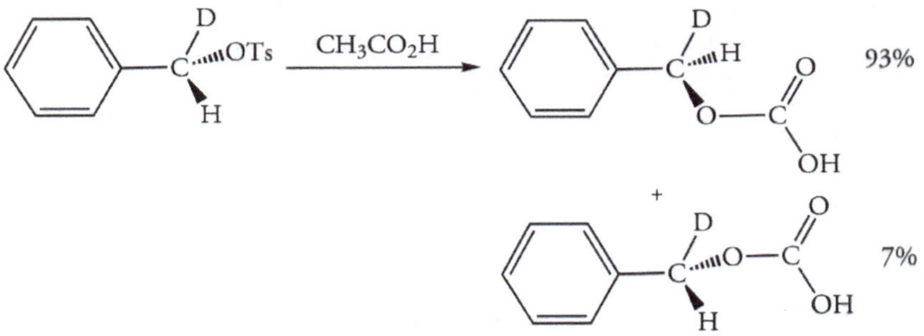

Answer: A primary tosylate tends to react by the S_N2 mechanism. However, in this case, the benzene ring can stabilize a primary benzyl carbocation by resonance and allow the reaction to occur by the S_N1 mechanism. However, the leaving tosylate group, which is quite large, substantially shields one face of the carbocation as it leaves, so the acetate ion reacts with substantial inversion of configuration.

10.16 The reactant in the following reaction has the *S* configuration. Based on the composition of the product mixture, what is the mechanism of the reaction?

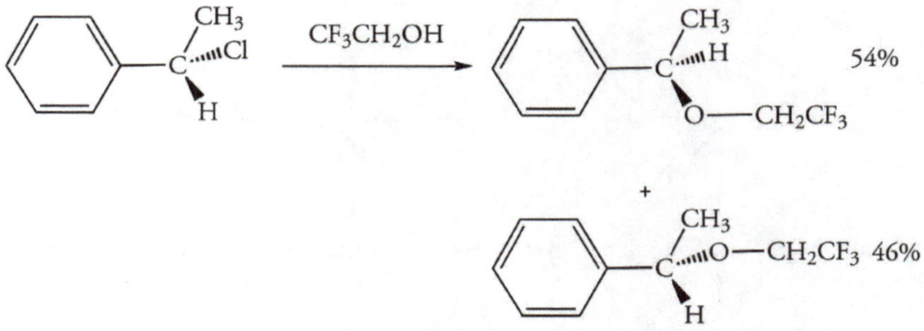

Answer: A secondary chloroalkane can react by an S_N1 or S_N2 mechanism depending on reaction conditions and other structural features of the compound. In this case, the secondary carbon is benzylic. Thus, the benzene ring can stabilize a secondary carbocation—it is a benzyl carbocation—by resonance, and the reaction then occurs by the S_N1 mechanism. The product is an almost equal mixture of enantiomers. The slight excess of inverted product forms because the leaving chloride shields one face of the carbocation, so trifluoroethoxide ion reacts from the back side.

10.17 What is the configuration of the tosylate prepared from (*R*)-2-butanol? What is the configuration of the iodide obtained by reacting that tosylate with iodide ion in acetone?

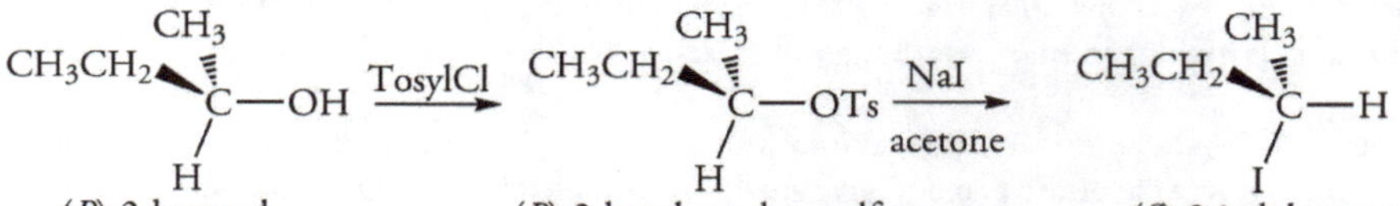

(*R*)-2-butanol (*R*)-2-butyl-p-tolunesulfonate (*S*)-2-iodobutane

Answer: A formation of the tosylate ester occurs by displacement of a chloride ion from the sulfur atom of *p*-toluenesulfonyl chloride (TosylCl) by the oxygen atom of the alcohol. The reaction occurs with retention of configuration because no bonds at the C-2 stereogenic center are affected. The highest priority hydroxyl group in the reactant is transformed into the highest priority tosylate group in the product. Thus, the configuration is still *R*. Subsequent displacement of the tosylate group by iodide occurs with inversion of configuration. The highest priority tosylate group in the reactant is replaced by a high priority iodide group in the product. Thus, the configuration of the inverted product is *S*.

10.18 Write the product expected from the reaction of each of the following compounds in aqueous acetone.

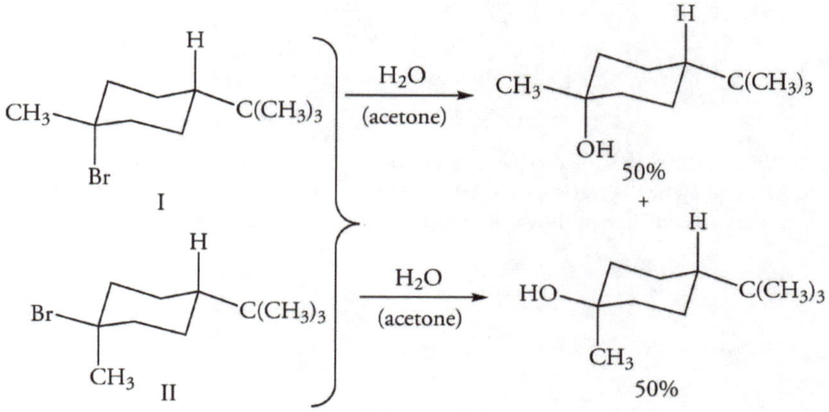

Answer: Both compounds are tertiary bromides that react with nucleophiles such as water in aqueous acetone by an S_N1 mechanism. A mixture of compounds with equatorial and axial hydroxyl groups results. The composition of the mixture is the same for both compounds I and II.

Reactivity in Substitution Reactions

10.19 1-Bromo-1,1-diphenylethane reacts very rapidly in ethanol. Explain why.

Answer: The bromide ion leaves readily because the resulting carbocation is both tertiary and benzylic. The carbocation is resonance stabilized by the two benzene rings.

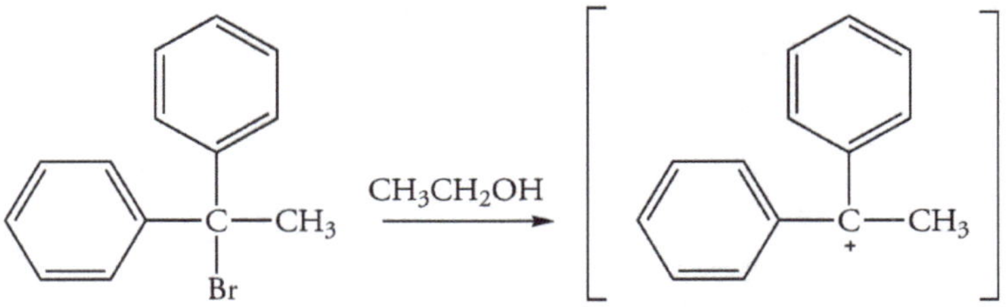

1-bromo-1,1-diphenylethane   tertiary, benzylic carbocation

10.20 4-Chloro-2,2,4,6,6-pentamethylheptane reacts in aqueous acetone about 500 times faster than *tert*-butyl chloride does. Explain this observation.

Answer: The tertiary carbocation has two *tert*-butyl groups and a methyl group bonded to the positively charged carbon atom compared to three methyl groups in the *tert*-butyl carbocation. Steric crowding around the C-4 atom of the reactant is decreased in the carbocation compared to the reactant as the chloride ion leaves.

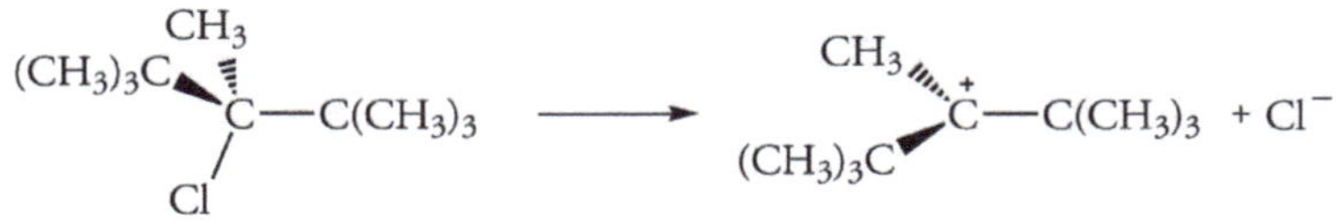

4-Chloro-2,2,4,6,6-pentamethylheptane

10.21 3-Bromo-1-butene and (*E*)-1-bromo-2-butene react at the same rate in aqueous acetone. Explain why. An identical mixture of two substitution products is obtained from either compound. What are the structures of the products?

Answer: The same resonance-stabilized carbocation results from both compounds. Attack of water at either C-1 or C-3 gives the same mixture of alcohols.

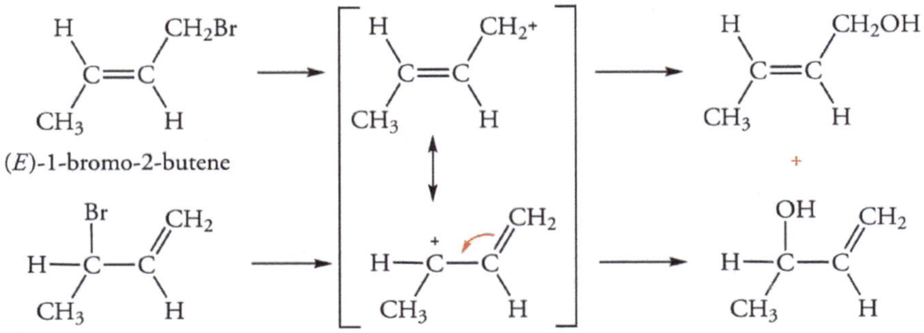

(*E*)-1-bromo-2-butene

10.22 The following compound reacts in methanol to rapidly replace one of the two bromine atoms by a methoxy group. Which bromine atom is replaced?

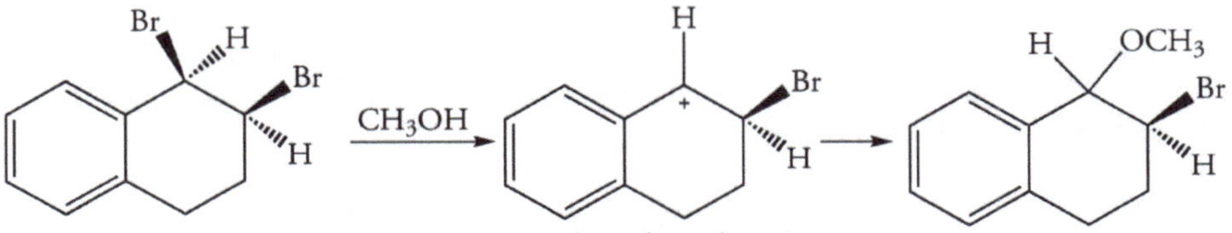

benzylic carbocation

Answer: The bromine atom nearer the benzene ring is replaced because the resulting secondary carbocation is benzylic, and therefore it is resonance stabilized.

10.23 The following sulfonium ion reacts in 80% ethanol–20% water to give 36% 2-methyl-1-propene. The remaining 64% of the product is a mixture of two substitution products. What are the substitution products? *tert*-Butyl chloride reacts under the same conditions to give the identical mixture of products. Explain this observation.

$$CH_3-\overset{\overset{\displaystyle CH_3}{|}}{\underset{\underset{\displaystyle CH_3}{|}}{C}}-\overset{+}{S}\overset{\displaystyle CH_3}{\underset{\displaystyle CH_3}{<}} \quad\xrightarrow[\text{20\% H}_2\text{O}]{\text{80\% CH}_3\text{CH}_2\text{OH}}\quad CH_3-\overset{\overset{\displaystyle CH_3}{|}}{\underset{\underset{\displaystyle CH_3}{|}}{C}}{+} \quad + \quad CH_3-\overset{..}{\underset{..}{S}}-CH_3$$

$$\downarrow$$

$$CH_3-\overset{\overset{\displaystyle CH_3}{|}}{\underset{\underset{\displaystyle CH_3}{|}}{C}}-OH \;+\; CH_3-\overset{\overset{\displaystyle CH_3}{|}}{\underset{\underset{\displaystyle CH_3}{|}}{C}}-OCH_2CH_3$$

Answer: Dimethyl sulfide is the leaving group and a *tert*-butyl carbocation forms. The substitution products are an alcohol resulting from nucleophilic attack of water on the tertiary carbocation and an ethyl ether resulting from nucleophilic attack of ethanol. The same products are formed from *tert*-butyl chloride because the same carbocation is formed when the chloride ion leaves.

10.24 The relative rates of substitution of bromide by ethoxide in ethanol for methyl, ethyl, propyl and butyl bromides are 1, 0.057, 0.018, and 0.013, respectively. Explain these data.

Answer: The methyl compound reacts fastest because the three hydrogen atoms do not present any steric hindrance to the nucleophile in the S_N2 reaction. The other three compounds all have one alkyl group and two hydrogen atoms in the path of the nucleophile. Each alkyl group is primary, and their steric sizes are similar. The slight decrease in rate suggests that as the length of the chain increases there is some increase in steric hindrance. Note that the change is larger when comparing a methyl group to an ethyl group. Extending the chain to a propyl group does not decrease the rate of reaction by as large a factor.

10.25 Trifluoromethanesulfonyl chloride reacts with alcohols to form sulfonate esters. Would you expect the "triflates" to be more or less reactive than the methanesulfonate esters?

Answer: The fluorine atoms withdraw electron density from the sulfonyl group and also stabilize the triflate ion compared to the methanesulfonate ion. Thus, the trifluoromethanesulfonate ion is a better leaving group, and therefore the triflate is more reactive.

10.26 A nitro group is electron withdrawing. Would you expect the sulfonate esters of *p*-nitrobenzenesulfonic acid to be more or less reactive than the tosylates?

Answer: The nitro group inductively withdraws electron density from the sulfonyl group, and it also stabilizes the sulfonate ion compared to the tosylate ion. Thus, the nitro-substituted sulfonate ion is a better leaving group, so its esters are more reactive.

$$O_2N-\!\!\left\langle\text{benzene ring}\right\rangle\!\!-\overset{\overset{\displaystyle O}{\|}}{\underset{\underset{\displaystyle O}{\|}}{S}}-OH$$

p-nitrobenzenesulfonic acid

10.27 Explain why ethanol (CH_3CH_2OH) can solvate both cations and anions. Explain why dimethyl ether, $(CH_3)_2O$, is a poorer solvent for ionic compounds. Discuss the solvation characteristics of both solvents for both cations and anions.

Answer: The lone pair electrons of the oxygen atom of both the alcohol and the ether coordinate with cations. The partially positive hydrogen atom of the hydroxyl group of alcohols can help solvate anions. Because the ether cannot solvate anions, the solubility of ionic compounds in ether is much smaller than in alcohols.

10.28 *trans*-1-Iodo-3-methylcyclopentane reacts with KF in DMF to give a fluoro compound. What is its configuration? The iodo compound reacts with KF in ethanol to give products that do not contain fluorine. Explain these data.

Answer: In an aprotic solvent, the fluoride ion is sufficiently nucleophilic to react with the iodo compound in an S_N2 mechanism that occurs with inversion of configuration. The product of this reaction is *cis*-1-fluoro-3-methylcyclopentane. In ethanol, the fluoride is solvated and is not an effective nucleophile. The substitution product is an ethyl ether.

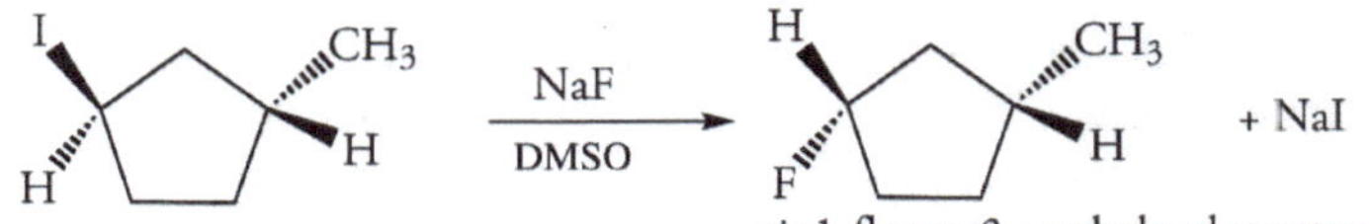

cis-1-fluoro-3-methylcyclopentane

Solvent Effect in Substitution Reactions

10.29 The structure of hexamethylphosphoramide is shown below. Its dielectric constant is 30. How do you expect HMPA to affect the rates of S_N1 and S_N2 reactions?

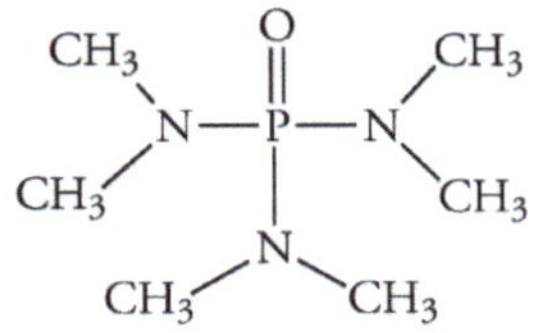

hexamethylphosphoramide
(HMPA)

Answer: Hexamethylphosphoramide (HMPA) is very polar, as indicated by its dielectric constant. It has no protic sites and is thus aprotic. HMPA can be used for substitution reactions. Because it has a high dielectric constant, it will favor S_N1 processes. Because it is aprotic, it will accelerate S_N2 processes.

10.30 The rate constant for the displacement of iodide from iodomethane by fluoride ion is 10^6 times faster in dimethyl formamide than in methanol. Explain why.

Answer: The fluoride ion is solvated by methanol, and its nucleophilicity is much less than it is in an aprotic solvent such as dimethylformamide.

10.31 Methyl tosylate reacts with halide ions in water. The rate constants for the reaction in water decrease in the order $k_I > k_{Br} > k_{Cl}$. The rate constants for the reactions in acetone stand in the order $k_I < k_{Br} < k_{Cl}$. Explain these data.

Answer: The reaction rates in water reflect the nucleophilicity of the solvated anion. The degree of solvation decreases as ionic size increases. Thus, iodide ion is the best nucleophile of the series. In the aprotic acetone solvent, the order of nucleophilicity is related to the strength of the ions as bases. Chloride ion is the strongest base of the series.

10.32 The equilibrium constant for the reaction of bromomethane with iodide ion is 15 in water and 0.6 in acetone. Explain these data.

$$CH_3Br + I^- \rightleftharpoons CH_3I + Br^-$$

Answer: The position of the equilibrium in water is affected by the degree of solvation of the anion. Bromide ion is more strongly solvated than iodide ion, and the reaction tends to proceed to the right as written. In acetone as solvent, the position of the equilibrium is controlled by factors other than the stability of the anions, neither of which is solvated.

Elimination Reactions

10.33 Attempted displacement of iodide ion by fluoride in acetone usually fails because elimination products result. Explain why elimination is favored over substitution.

Answer: Fluoride ion is the strongest base of the halides because HF is the weakest acid of the hydrogen halides. In an aprotic solvent the fluoride ion is not solvated, and it is an even stronger base than in protic solvents. Elimination is favored in reactions with strong bases.

10.34 The product mixture obtained in the reaction of isobutyl bromide with sodium ethoxide in ethanol contains 62% 2-methyl-1-propene. The reaction using potassium *tert*-butoxide in *tert*-butyl alcohol contains 92% 2-methyl-1-propene. Explain why.

Answer: The *tert*-butoxide ion is not as nucleophilic as ethoxide ion because it is sterically larger. Thus, substitution reactions occur less readily with *tert*-butoxide ion. In addition, the *tert*-butoxide ion is a stronger base than the ethoxide ion, and a larger fraction of elimination product results with a stronger base.

10.35 The product mixture obtained in the reaction of *sec*-butyl bromide with 1 M sodium ethoxide in ethanol contains 78% unsaturated material. What are the products and which of them should predominate? Using 4 M sodium ethoxide in ethanol, the product mixture is 91% unsaturated material. Why?

Answer: The unsaturated products are 1-butene, *cis*-2-butene, and *trans*-2-butene. The 2-butenes are more substituted alkenes and are the major products. The alkenes result in part from an El mechanism derived from loss of a proton from the carbon atom adjacent to the carbocation center, as well as an E2 mechanism in which either ethanol or ethoxide ion abstracts a proton from the reactant. Increasing the concentration of ethoxide ion increases the amount of product in the E2 mechanism.

10.36 The unsaturated compounds obtained in the reaction of 2-bromo-2,3-dimethylbutane with the alkoxide of 3-ethyl-3-pentanol are 92% 2,3-dimethyl-l-butene and 8% 2,3-dimethyl-2-butene. Compare these data with the data for reaction of *tert*-butoxide with the same compound (Section 10.5).

Answer: The alkoxide derived from 3-ethyl-3-pentanol is a more sterically hindered base than *tert*-butoxide ion. Their base strengths are similar. The increased amount of the 1-butene indicates that the more sterically hindered base does not abstract the tertiary hydrogen atom at C-3 as readily and is far more likely to abstract the primary hydrogen atom. Thus, the less substituted alkene predominates.

10.37 E2 reactions of tosylates occur using alkoxide ions as bases in the related alcohol solvent. Determine the stereochemistry of the 2-phenyl-2-butene formed from reaction of the tosylate of (2*R*,3*R*)-3-phenyl-2-butanol.

Answer: Arrange the structure in a conformation so that the hydrogen atom at C-3 and the tosylate group are *anti* periplanar. The structure of the product has the (*E*) configuration.

10.38 The tosylate of *cis*-2-phenylcyclohexanol undergoes an elimination reaction much more rapidly with *tert*-butoxide in *tert*-butyl alcohol than does the *trans* isomer. The product is exclusively 1-phenylcyclohexene. Explain these data.

Answer: In the *cis* isomer, the tosylate group and the hydrogen at the C-2 atom are *anti* periplanar and elimination can readily occur.

Answer: In the *trans* isomer, elimination cannot occur unless the chair undergoes a ring flip to place the tosylate in an axial position. However, in this conformation, the product will not be 1-phenylcyclohexene because the phenyl ring will be in an axial position. Elimination will occur by abstraction of the hydrogen atom at C-6 to give the isomeric 3-phenylcyclohexene.

10.39 The following products are obtained from the E2 reaction of (2S,3R)-2-bromo-3-deuterio-butane using sodium ethoxide in ethanol. Explain why. Predict the products from the E2 reaction of (2S,3S)-2-bromo-3-deuteriobutane.

trans-2-butene

(Z)-2-deutero-2-butene

Answer: Arrange the structure in a conformation so that the hydrogen atom at C-3 and the bromide ion are *anti* periplanar. This conformation gives (Z)-2-deutero-2-butene. An alternate conformation has the deuterium atom at C-3 and the bromide ion in an *anti* periplanar arrangement. This conformation gives trans-2-butene.

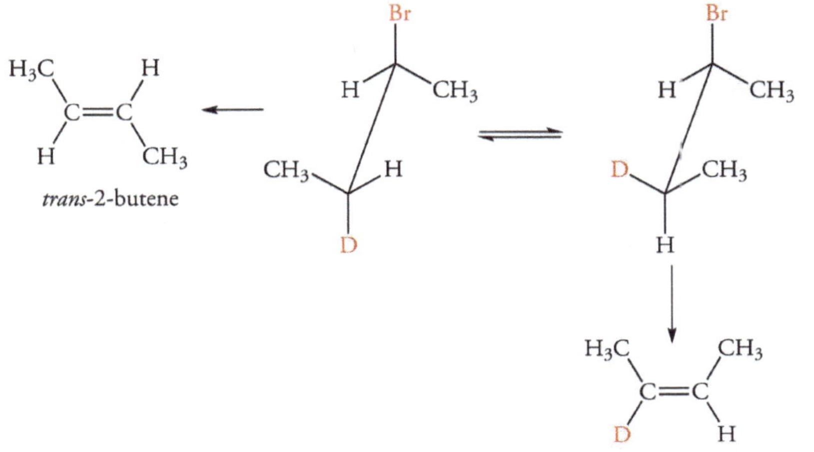

trans-2-butene

(Z)-2-deutero-2-butene

Answer: Reversing the configuration at C-3 from R to S gives a (Z) compound without deuterium and an (E) compound containing deuterium.

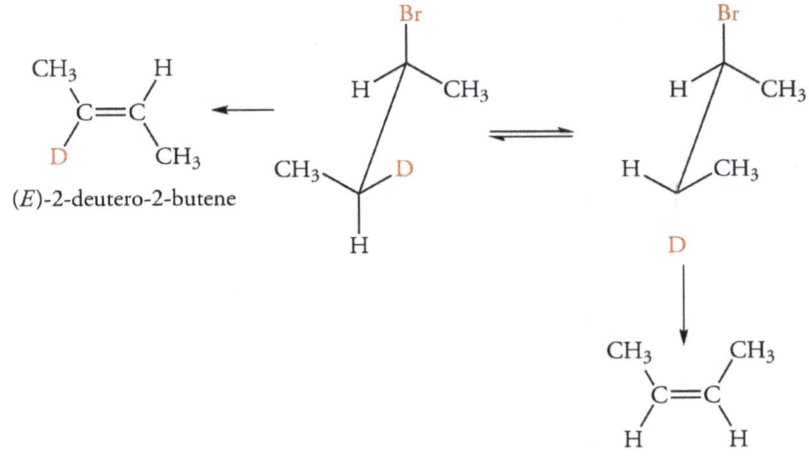

(E)-2-deutero-2-butene

cis-2-butene

10.40 The E2 reaction of l-bromo-2-deutero-2-phenylethane gives the following compounds. Explain why the indicated percentage of each compound is formed.

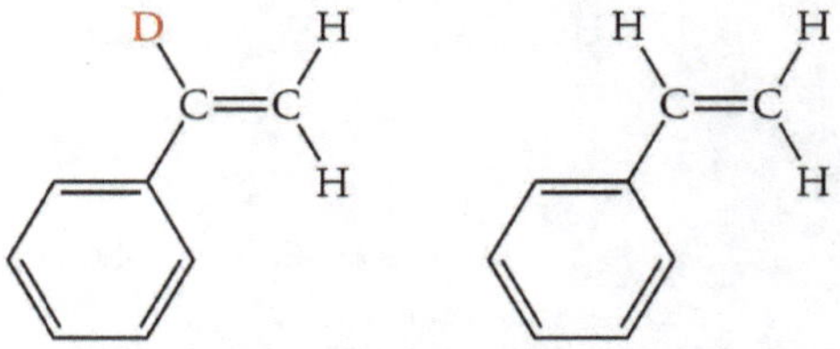

87%   13%

Answer: The carbon–deuterium bond is not as easily broken in an E2 reaction as the carbon–hydrogen bond. Thus, the major product contains the deuterium atom. It is formed by abstraction of the hydrogen atom from C-2.

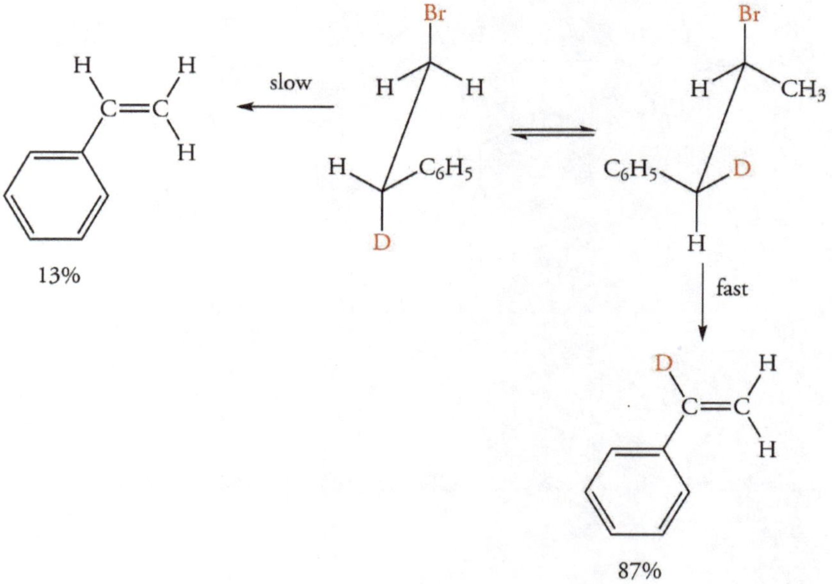

13%

87%

10.41 The E2 reaction of each of the following compounds with sodium methoxide in methanol proceeds regiospecifically to give different compounds. What is the structure of the compounds derived from each stereoisomer?

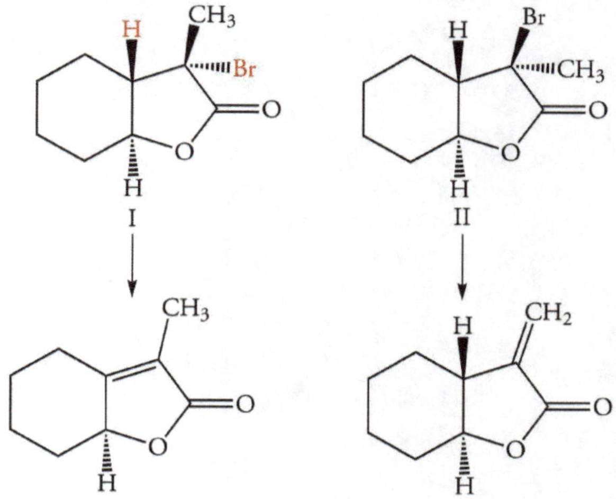

I   II

Answer: Compound I has a hydrogen atom at the bridgehead position that may be abstracted to give a tetrasubstituted alkene. The elimination reaction of compound II can only occur by abstraction of a proton from the methyl group.

10.42 Predict the E2 product formed in the reaction of each of the following compounds. Which compound reacts at the faster rate?

Answer: Compound I has an axial bromine atom that is *anti* periplanar to the hydrogen atom. This compound reacts at a faster rate than compound II, which has an equatorial bromine atom. The elimination product of compound II does not contain deuterium.

11

CONJUGATED ALKENES
AND
ALLYLIC SYSTEMS

KEYS TO THE CHAPTER

In this chapter, we have focused on compounds and intermediates that have conjugated double bonds. Conjugated alkenes are resonance stabilized, and this property affects both the structures of the conjugated molecules and the stabilities of reaction intermediates derived from them. Although we can describe these properties with conventional Lewis structures, molecular orbital theory provides far deeper insights into their properties and reactions. We will use the ideas developed in this chapter again in the next chapter when we discuss the properties and reactions of aromatic compounds.

11.1 Classes of Dienes

Conjugated dienes and higher polyenes contain a series of alternating single and double bonds. Isolated dienes have more than one single bond separating the two double bonds and are regarded as two separate alkenes. Cumulated dienes have two double bonds sharing a common carbon atom.
Terpenes are naturally occurring compounds that are synthesized in cells from isoprene units. Isoprene is a conjugated butadiene with a methyl branch. The terpenes are named according to the number of five-carbon units that make up their skeleton.

11.2 Stability of Conjugated Dienes

The effect of the interaction of two double bonds is the first case of the interaction of functional groups that we have discussed. Others such as the interaction of the carbon–oxygen double bond of a carbonyl group with a carbon–carbon double bond will be considered in later chapters. The π electrons in a conjugated system are *delocalized*.

Conjugated dienes are stabilized by resonance interaction of the double bonds. Therefore, these dienes are of lower energy than dienes in which the double bonds do not interact. Thus, the energy released upon hydrogenation is smaller than predicted based on the heats of hydrogenation of the component double bonds. The difference between the experimental value and the predicted value based on isolated double bonds is the resonance energy.

11.3 Molecular Orbital Models of Conjugated Systems

Molecular orbitals are made from **linear combinations of atomic orbitals**. We obtain these orbitals by adding or subtracting the equations that describe the energies of atomic orbitals. Adding orbitals with the proper sign of the wave function corresponds to constructive overlap of atomic orbitals. This combination gives rise to bonding **molecular orbitals**. Subtracting the equations for the wave functions of atomic orbitals with opposite signs gives destructive overlap and produces an **antibonding molecular orbital**.

Molecular orbitals may be **symmetric** or **antisymmetric** based on the sign of the molecular orbital at one point compared to a related point on the other side of a **nodal plane**. The electron density at this nodal plane is zero. The bonding molecular orbital of ethene is symmetric; the antibonding molecular orbital of ethene is antisymmetric with respect to a nodal plane. The energies of the molecular orbitals increase as the number of nodal planes increases.

A group of molecular orbitals for a polyene can be separated into a group of bonding molecular orbitals (corresponding to the number of double bonds) and an equal number of antibonding molecular orbitals. In a neutral polyene, all of the π electrons are in the bonding molecular orbitals. Our picture of the degree of double bond character between adjacent carbon atoms in a polyene is based on whether the various contributing molecular orbitals are bonding or antibonding. For the C-2 to C-3 bond in butadiene, the π_2 provides no electron density, but π_1 does provide electron density, so there is some double bond character in this bond.

11.4 Effects of Conjugation on Diene Structure

Conjugation accounts for the partial double bond character of the C-2 to C-3 bond of butadiene. The bond length is shorter than predicted based on the hybridization of the component sp^2 orbitals that form the sigma bond. This shortening is attributed to the double bond character as a result of the contribution of the π_1 molecular orbital.

To maintain the overlap of the 2p orbitals that make up the molecular orbitals, the atoms of the double bonds must be coplanar. Two conformations are observed, consistent with this requirement. The *s-trans* conformation is more stable than the *s-cis* conformation due to steric hindrance in the *s-cis* conformation. Rotation about the C-2 to C-3 bond in butadiene requires more energy than rotation about a single bond without double bond character. The rotational energy barrier is due to both a torsional component in which there is some steric interference and the loss of resonance energy that occurs when the π orbitals are orthogonal in the transition state of the process.

11.5 Allylic Systems

Allylic halides react to give allylic carbocations under S_N1 conditions. The positive charge of an allyl carbocation or any substituted allylic carbocation is distributed between two "end" carbon atoms. There is no charge on the "center" carbon atom. However, the charge distribution is equal only in the allyl carbocation itself. Substituted allylic carbocations necessarily have unequal charge distributions based on the identity of the attached groups. This effect is seen in the product distribution of the S_N1 reactions of substituted allyl chlorides. The major product corresponds to capture of a nucleophile at the more highly substituted center because the positive charge is better stabilized at that center.

Allylic radicals have an electron deficiency at either end of the radical. (There is no radical character at the "center" carbon atom.) Delocalization in the allyl radical is reflected in the lower bond energy of the C—H bond that gives rise to the radical. *N*-bromosuccinimide (NBS) reacts with allyl systems to give selective halogenation in which a bromine atom replaces an allylic hydrogen.

11.6 Molecular Orbitals of Allylic Systems

The molecular orbitals formed from an odd number of p orbitals in an allylic system are arranged somewhat differently from the arrangement of molecular orbitals made from an even number of p orbitals. First, there is one molecular orbital, called a **nonbonding orbital**, in which the nodal plane contains the central carbon atom. This means that there is no π electron density at the central carbon atom. Second, the energy of the nonbonding orbital is the same as the contributing atomic orbital. Hence, there is no net stabilization as a result. The symmetry of the molecular orbitals is defined in the same way as for polyenes. The lowest energy molecular orbital is symmetric, and the symmetry alternates with each higher energy molecular orbital. And, as the number of nodal planes increases, the energy increases. The nodal plane in an allylic system, or any π system made from an odd number of 2p orbitals, may be at an atom or between atoms.

The distribution of electrons among the molecular orbitals follows Hund's rule. The allyl cation has electrons only one electron in π_1, and the resulting positive charge is felt at the terminal carbon atoms, corresponding to the π_2 orbital. This orbital contains one electron in the allyl radical and two electrons in the allyl anion.

11.7 Electrophilic Conjugate Addition Reactions

Conjugates dienes undergo electrophilic addition. However, rather than forming a localized carbocation in the first step of the reaction, as in alkenes, an allylic carbocation is formed. The nucleophile can then add to either of two carbon atoms that bear the positive charge of the allylic carbocation. Therefore, either 1,2- or 1,4 addition can occur. The amounts of 1,2- and 1,4-addition products depend on the groups bonded to the two "end" carbon atoms of the allylic system.

If the 1,2-addition product of the reaction is stable, and does not have sufficient energy to revert to an allyl cation by ionization of the original nucleophile, then the result is termed **kinetic control**. This means that the product that forms fastest is the major product. However, if the energy of the system is high enough, the 1,2-product ionizes. Repeating this process eventually gives the most thermodynamically stable product, hence the term thermodynamic control.

The product of kinetic control may be influenced by the stability of the intermediate and the partial positive charge at the two "ends" of the allylic carbocation. However, the product of thermodynamic control is influenced only by structural features of the product, such as the stability of the remaining double bond. Products with the more highly substituted double bond are favored.

11.8 The Diels–Alder Reaction

The Dields–Alder reaction is a powerful synthetic method for the formation of compounds that contain six-membered rings. The addition of a diene to a dienophile produces a cyclohexene ring. The reaction is concerted, and the stereochemistry of the products is the same as the stereochemistry of the reactants. In bicylic ring systems, the *endo* isomer predominates.

11.9 Spectroscopy

The energy of light is directly proportional to its frequency. The energy of light is given by the relation $E = h\nu$. The wavelength of light is inversely proportional to the frequency, as given by $\lambda = c/\nu$. Spectroscopy is used to probe the physical changes in a molecule as the result of absorption of energy.

11.10 Ultraviolet–Visible Spectroscopy

The portions of the electromagnetic spectrum known as the ultraviolet and the visible regions are associated with energies sufficient to cause electronic transitions of conjugated systems. The position of the absorption peak is referred to as λ_{max}. Each electronic transition corresponds to promotion of an electron from the highest occupied molecular orbital (HOMO) to the lowest energy unoccupied molecular orbital (LUMO).

The λ_{max} of a compound increases as the number of conjugated π bonds increases. Some conjugated systems absorb light in the visible region of the spectrum.

Summary of Reactions

1. Allylic Bromination

2. Electrophilic Conjugate Addition

1,2-addition   1,4-addition

3. Diels–Alder Reaction

Diene   Dienophile

Classes of Polyenes

11.1   Which of the following compounds has conjugated double bonds?

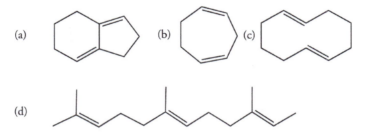

(a)   (b)   (c)

(d)

Answer: Conjugated compounds have only one single bond separating the double bonds. Only compound (a) has conjugated double bonds. In (b), the double bonds are separated by two single bonds. In (c), four single bonds separate the two double bonds. In (d), each of the three double bonds is separated by three single bonds.

11.2   Which of the following compounds has conjugated double bonds?

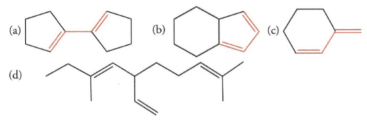

(a)   (b)   (c)

(d)

Answer: The compounds in (a), (b), and (c) all have conjugated double bonds. In (d), the closest double bonds are the one on the left and the vinyl group, shown as a branch. However, there are two intervening single bonds, so the double bonds are not conjugated.

11.3   How many compounds in each of the following sets of isomeric compounds contain conjugated double bonds?

(a)

(b)

Answer: (a) Only the middle compound has conjugated double bonds. (b) The first and second compounds have conjugated double bonds.

11.4   Classify the double bonds in each of the following compounds.
(a) mycomycin, an antibiotic.

$$H-C\equiv C-C\equiv C-CH=C=CH-CH=CH-CH=CH-CO_2H$$
         cumulated   conjugated

Answer: The double bonds shown in read are cumulated; those shown in blue are conjugated. What about the alkynes? The triple bonds are separated by single bonds. We know that one pair of 2p orbitals makes a π bond that is perpendicular to a second pair of 2p orbitals making the other π bond. Thus, one pair of two conjugated π bonds is perpendicular to a second pair of two conjugated π bonds.

(b) vitamin A$_2$, contained in freshwater fish

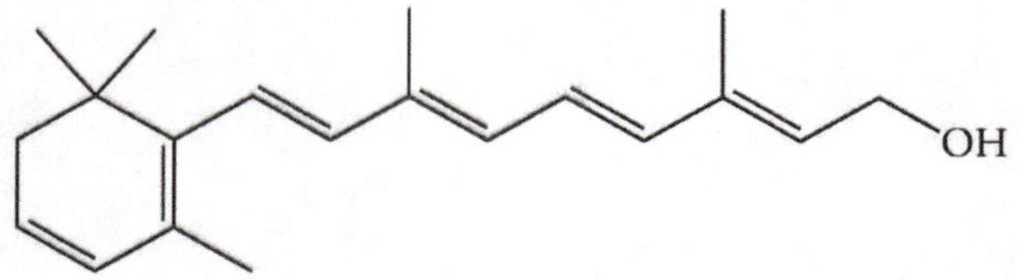

(c) humulene, a compound found in hops

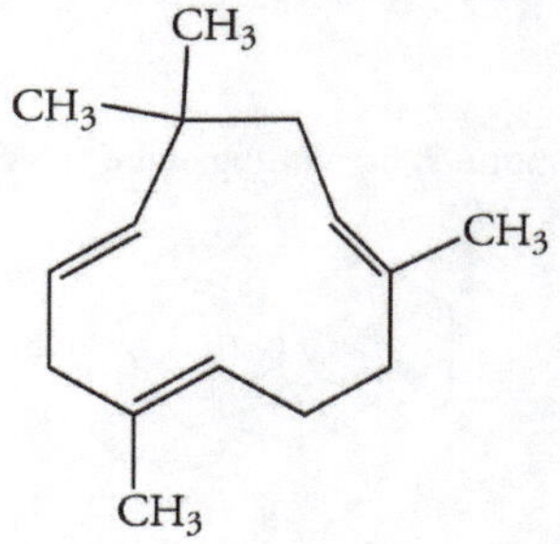

Answers: (a) All of the double bonds in vitamin A$_2$ are conjugated; (b) humulene has no conjugated double bonds.

11.5   Cyanodecapentayne has been identified in intergalactic space by radio astronomers. How many conjugated π bonds are in this compound?

H—C≡C—C≡C—C≡C—C≡C—C≡C—C≡N

Answer: The series of five carbon–carbon triple bonds and the carbon–nitrogen triple bond are all separated by single bonds. One pair of 2p orbitals makes a π bond that is perpendicular to a second pair of 2p orbitals making the other π bond. Thus, there are six conjugated π bonds perpendicular to a second set of six conjugated π bonds.

11.6   How many conjugated π bonds are in lycopene, the red pigment in tomatoes?

Answer: Lycopene has 16 conjugated double bonds.

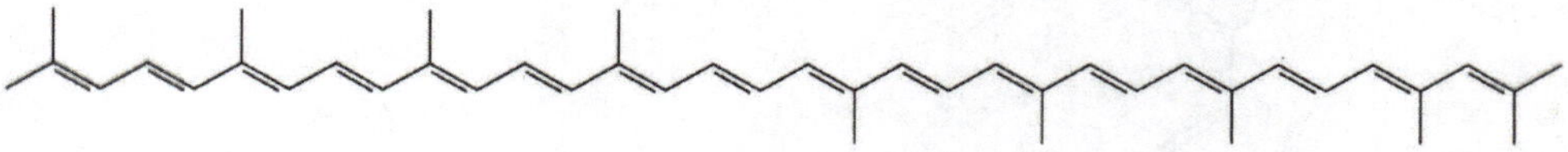

Terpenes

11.7   Classify each of the following terpenes and divide it into isoprene units.

Answers:
(a) monoterpene
(b) sesquiterpene
(c) sesquiterpene

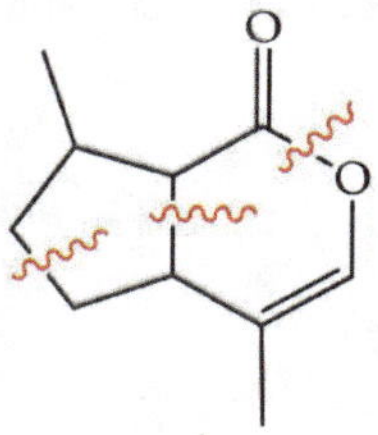

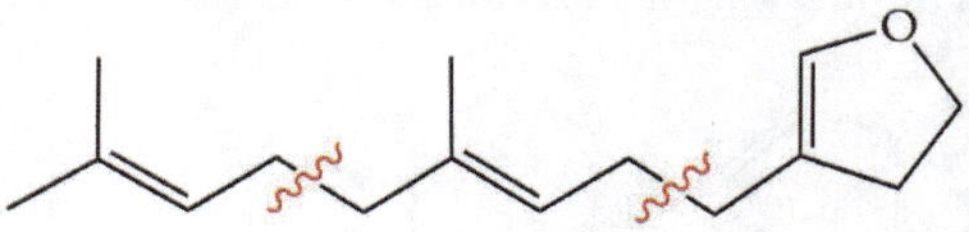

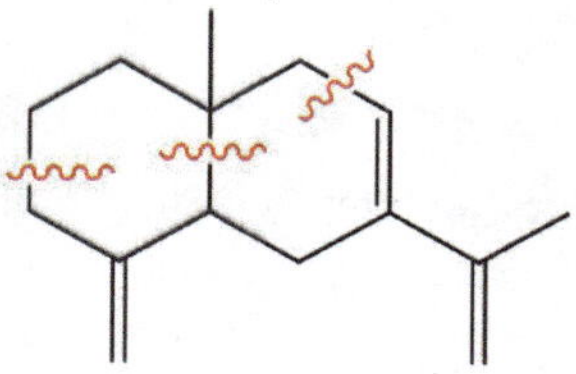

12 ARENES AND AROMATICITY

KEYS TO THE CHAPTER

12.1 Aromatic Compounds

The most common aromatic compounds contain a benzene ring that may have one or more of its hydrogen atoms replaced by substituents of varying complexity. Although many of these compounds are termed "aromatic" because of their odor, that property is a human physiological response and is not the criterion used to classify them.

Several benzene rings can be fused to include two common carbon atoms between rings. The possibilities are endless, as more and more rings can be fused. The common feature in all such compounds is the alternating series of single and double bonds within the rings. The structures of benzene, naphthalene, anthracene, and phenanthrene are the simplest arenes and the ones most often encountered.

12.2 Covalent Structure of Benzene

Benzene consists of a six-membered ring in which all carbon and hydrogen atoms are identical. We now see that the structures that Kekulé proposed for benzene are contributing structures to a resonance hybrid that we can draw with equivalent Lewis structures.

12.3 Aromaticity and the Resonance Stabilization of Benzene

The chemical criterion for aromaticity is the lack of reactivity toward reagents, such as bromine, that normally react readily with carbon–carbon double bonds. Benzene owes its stability to the resonance stabilization of its cyclic, conjugated π bonds.

The resonance energy of benzene is determined by comparing its heat of hydrogenation to that expected by the heat hydration of three isolated double bonds. The resonance energy for benzene is substantially larger than for cyclohexadiene.

12.4 Molecular Orbitals of Benzene

Benzene has six molecular orbitals that form a delocalized π system; three of benzene's molecular orbitals are bonding and each has two electrons; three are antibonding, and they are vacant.

One way to determine the number of molecular orbitals and their relative energies in benzene is to inscribe a hexagon in a circle. Place one vertex at the six o'clock position. The points of contact of the hexagon with the circle give the relative energies of the molecular orbitals. A horizontal line through the center of the circle bisects the polygon and corresponds to an energy level of zero; that is, any orbitals having this energy are nonbonding. Vertices below this level correspond to bonding orbitals; vertices above this level are antibonding.

12.5 Aromaticity and the Hückel Rule

To be aromatic, a molecule must be cyclic, planar, and it must contain only sp^2-hybridized atoms. The number of π electrons in the delocalized system must equal $4n + 2$, where n is an integer. (The value of n is not necessarily the number of carbon atoms in the ring, or the number of π molecular orbitals in a ring.)

The "$4n + 2$ rule" is known as the Hückel rule. It predicts that cyclic π systems having 2 ($n = 0$), 6 ($n = 1$), 10 ($n = 2$), and 14 ($n = 3$) electrons will be unusually stable, that is, they will be aromatic. For example, benzene has six π electrons, $4n + 2 = 6$ when $n = 1$.

12.6 Heterocyclic Aromatic Compounds

The criteria described above for establishing aromaticity also apply to heterocyclic compounds. Heteroatoms in the ring must be sp^2 hybridized and the ring must contain $4n+2$ π electrons. The most common heteroatoms are nitrogen, oxygen, and sulfur. The nitrogen atom in pyridine contributes one 2p electron to the π system. A lone pair of electrons on the sp^2-hybridized nitrogen atom of pyridine lies in the plane of the ring and does not contribute to the π system. In pyrrole, however, and in furan, and thiophene, the heteroatom contributes two electrons to the π system.

12.7 Polycyclic Aromatic Compounds

Polycyclic aromatic compounds that contain fused benzene rings contain $4n+2$ π electrons. Since they obey the Hückel rule, they are aromatic. For example, naphthalene contains 10 π electrons ($n = 2$) and anthracene has 14 π electrons ($n = 3$).

End of Chapter Exercises

Criteria for Aromaticity

12.1   Determine whether each of the following is an aromatic compound.

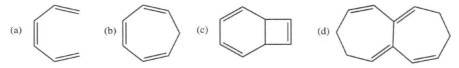

(a)   (b)   (c)   (d)

Answers:
(a) Although there are six π electrons in the triene, it is not aromatic because the π bonds are not in a ring.

(b) There are six π electrons in the cyclic triene. However, there is an sp^3-hybridized carbon atom interrupting the conjugation of the π bonds, so the compound is not aromatic.

(c) There are six p electrons in the bicyclic triene. However, only two of the π bonds are conjugated and each end of that system is separated from the third π bond by an sp^3-hybridized carbon atom. The compound is not aromatic.

(d) There are eight π electrons, so the compound does not have $4n+2$ π electrons. Two other facts tell us that it is not aromatic: (1) the π bonds are not in a continuous cyclic arrangement, and (2) there are intervening sp^3-hybridized carbon atoms.

12.2   Determine whether each of the following is an aromatic compound.

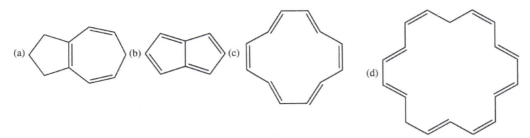

(a)   (b)   (c)   (d)

Answers:
(a) There are six π electrons in the bicyclic compound. However, there is an sp^3-hybridized carbon in the seven-membered ring. The compound is *not* aromatic.

(b) The π bonds are in a cyclic arrangement without any intervening sp^3-hybridized carbon atoms. However, there are 8 π electrons, a number that is not consistent with the Hückel rule. The compound is not aromatic.

(c) The π bonds are in a cyclic arrangement without any intervening sp^3-hybridized carbon atoms. However, there are 12 π electrons, a number that is not consistent with the Hückel rule. The compound is not aromatic.

(d) There are 14 π electrons, a number that is consistent with the Hückel rule for $n = 3$. However, there are two intervening sp^3-hybridized carbon atoms. Therefore, compound is not aromatic.

Hückel Rule

12.3   Borazole is an aromatic compound. Explain why.

H
|
H—N—H
\ B   B /
H—N   N—H
 \ B /
   |
   H

borazole

Answer:

Boron is a member of group III and has only six bonding electrons in its trivalent compounds. Thus, it is sp² hybridized and has a vacant 2p orbital. Nitrogen has an unshared pair of electrons in its trivalent compounds. The six electrons of nitrogen can be delocalized over the six atom ring using the 2p orbitals of both nitrogen and boron. Borazole is isoelectronic with benzene and is aromatic.

12.4   Is the following compound aromatic? Explain your answer.

B
|
CH₃

Answer:

Boron is sp² hybridized and has a vacant 2p orbital. There is a network of 2p orbitals of carbon and boron that can form a delocalized π system. However, there are only four π electrons, indicated by the two carbon–carbon π bonds. This number does not fit the Hückel rule, and the compound is not aromatic.

12.5   Are the following compounds aromatic according to the Hückel rule?

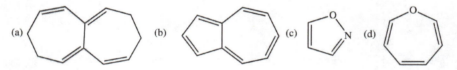

(a)   (b)   (c)   (d)

Answers:

(a) No, because 8 π electrons does not fit the Hückel rule of 4n+2; second, there are four sp³-hybridzied carbon atoms.

(b) Yes, because 10 π electrons fits the Hückel rule.

(c) Yes, because oxygen contributes one of its two unshared electron pairs to the π system. The four electrons of the carbon–carbon π bond and the carbon–nitrogen π bond, and one unshared electron pair of electrons from oxygen give a total of 6 π electrons, which fits the Huckel rule. Note that the other electron pair of oxygen and that of nitrogen are in sp² hybrid orbitals that are perpendicular to the π system.

(d) No, because oxygen would contribute one of its two unshared electron pairs to the π system. The six electrons of the three carbon–carbon π bonds and one unshared electron pair from oxygen give a total of 8 π electrons, which does not fit the Hückel rule. Note that the other electron pair of oxygen cannot be included to give a total of 10 electrons because it is in an sp²-hybrid orbital perpendicular to the π system.

12.6 Are the following compounds aromatic according to the Hückel rule?

Answers:
(a) Yes, because the nitrogen atom bonded to a hydrogen atom contributes one of its unshared electron pairs to the π system. The four electrons of the two carbon–carbon π bonds and this one unshared electron pair from nitrogen give a total of 6 π electrons, which does fit the Hückel rule. Note that each of the other two nitrogen atoms contribute one electron to the π system as part of their double bonds. The remaining unshared electron pair of each nitrogen atom is in an sp² orbital that is perpendicular to the π system.

(b) No, because each nitrogen atom would contribute its unshared electron pair to the π system. The four electrons of the two carbon–carbon π bonds and the two unshared electron pairs give a total of 8 π electrons, which does not fit the Hückel rule.

Aromatic Ions

12.7 1,2,3,3-Tetrachlorocyclopropene reacts with one mole of $SbCl_5$, to give the $C_3Cl_3^+$ ion. Draw the structure of the ion and explain why it forms.

Answer:
The carbocation that results from abstraction of a chloride ion by the Lewis acid $SbCl_5$ is a cyclopropenium ion. It has two π electrons and fits the Hückel rule (with $n = 0$). The electrons can be delocalized over the three carbon atoms of the ring.

12.8 The dipole moment of dipropylcyclopropenone is 5 D. This value is significantly higher than that of acetone, which is 3 D. Write a resonance form that accounts for the larger dipole moment of the cyclic ketone.

Answer:
The dipolar resonance form of the carbonyl group places a negative charge on oxygen and a positive charge on one of the carbon atoms of the cyclopropene ring. The resulting cyclopropenium ion has two π electrons and fits the Hückel rule. The positive charge can be delocalized over the three carbon atoms of the ring. The increased polarity of the carbonyl group as a result of resonance increases the dipole moment.

12.9 Cyclooctatetraene reacts with potassium to give a stable dianion. Explain why. Inscribe the dianion in a circle, with one vertex pointed down, and draw a molecular orbital energy diagram for it.

Answer:

The cycloöctatetraenyl dianion has 10 π electrons, which fits the Hückel rule, so the dianion is aromatic.

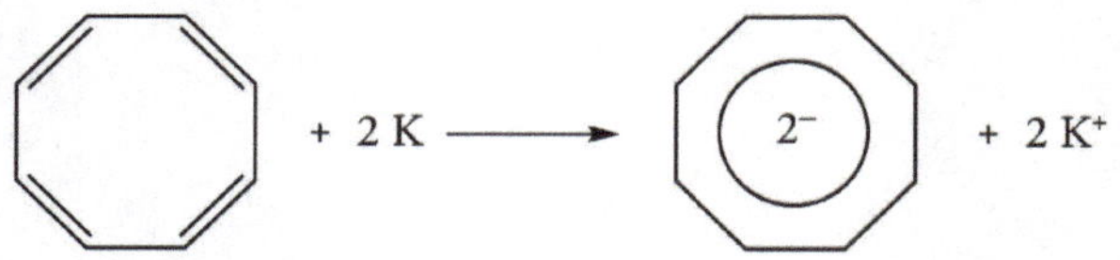

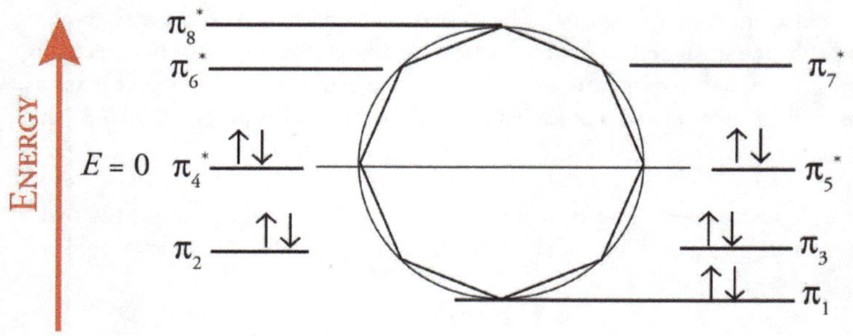

12.10 The following hydrocarbon reacts with two moles of butyllithium to form the stable ion $C_8H_6^{2-}$. Draw the structure of this ion. Explain why it is stable.

Answer:

Loss of one proton from each of the methylene groups gives a dianion that has a total of 10 π electrons, which fits the Hückel rule. Therefore, the dianion is aromatic.

10 π electrons, aromatic

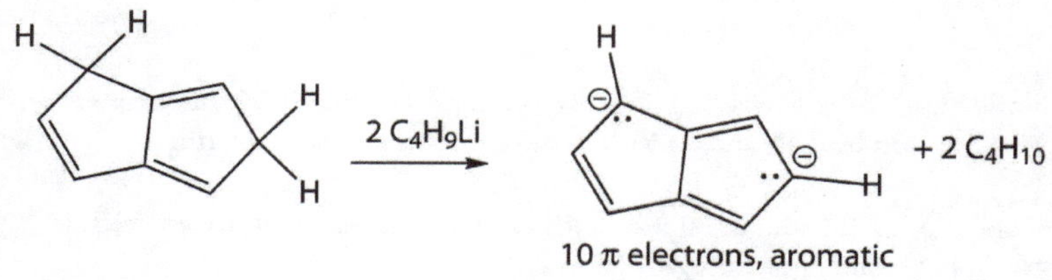

12.11 Is the following hydride ion transfer reaction favorable in the direction written?

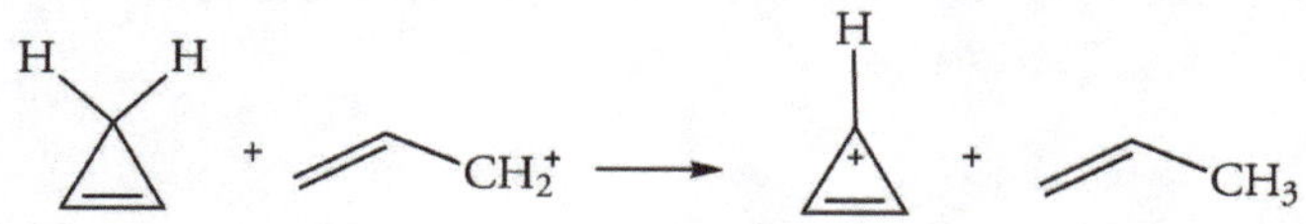

Answer:

The cyclopropenium cation product is aromatic. Thus, its resonance stabilization is larger than the resonance stabilization of the allyl cation. The reaction is favorable in the direction written.

ELECTROPHILIC AROMATIC SUBSTITUTION

13.1 Names of Benzene Derivatives

The derivatives of benzene have both common names and IUPAC names. Some monosubstituted benzene compounds have common names such as toluene, phenol, and aniline. Disubstituted compounds may be named using a numbering system or the prefixes o-, m-, and p- for ortho, meta, and para, respectively. However, compounds with three or more substituents must use numbers to locate the substituents. Substituents are named in alphabetical order.

13.2 Electrophilic Substitution Reactions

Aromatic rings are attacked by electrophiles, E^+, to give substituted aromatic compounds represented by Ar—E. The common reactions of aromatic compounds are designated by the type of group substituted on the aromatic ring. Thus, the term halogenation means that a halogen has substituted for hydrogen on the aromatic ring to give a product represented by Ar—X. In the case of chlorination, the product is Ar—CI.

The mechanism of aromatic substitution consists of two steps. The first is electrophilic attack to give a carbocation intermediate. The second step is loss of a proton to regenerate the aromatic system. We discussed five types of electrophilic substitution reactions. They are summarized in the following table:

Reaction Type	Reagents
1. Halogenation	$FeBr_3/Br_2$ or $FeCl_3/Cl_2$
2. Nitration	Nitric acid/sulfuric acid
3. Sulfonation	Fuming sulfuric acid
4. Friedel–Crafts alkylation	$AlCl_3/CH_3Cl$
5. Friedel–Crafts acylation	Acyl chloride/ $AlCl_3$

13.3 Limitations of Friedel–Crafts Reactions

In a Friedel–Crafts alkylation, the electrophile is a carbocation that tends to rearrange to the most stable carbocation by either a hydride or methide shift. An alternate approach to obtaining the desired alkyl group on the aromatic ring is to acylate first and then reduce the carbonyl group.

A second limitation reflects the effects of substituents already on the aromatic ring. Deactivating, meta-directing substituents preclude Friedel–Crafts alkylation and acylation.

13.4 Effects of Substituents on the Reactivity of Aromatic Compounds

Substituents on the aromatic ring affect the rate of electrophilic substitution. Substituents that increase reactivity are activating groups; substituents that decrease the reactivity are deactivating groups. The degree to which the groups affect the reactivity is qualified by the adjectives "strongly" and "weakly" (see Table 13.1). The substituents already on the aromatic ring determine where the electrophile attacks. Substituents are either ortho, para, or meta directors. All activating groups are ortho, para directors, except halogens, which are weakly deactivating, but ortho, para directing. Other deactivating groups are meta directors.

13.5 Interpreting Rate Effects

Any group increases the electron density of the aromatic ring makes it more inviting to an electrophile, so the rate of reaction is faster. Conversely, groups that reduce the electron density decrease the rate of reaction. Groups can affect the electron density by an inductive effect, a resonance effect, or a combination of both.

Except for alkyl groups, all common substituents withdraw electron density from the ring by an inductive effect. This effects is greatest for groups with a formal positive charge (such as the nitro group), followed by groups with a partial positive charge (such as the carbonyl group).

Substituents that either have lone pair electrons or have multiple bonds to an electronegative atom directly bonded to the ring have a resonance interaction with the ring. Groups with lone pair electrons (such as hydroxyl and amino) increase the electron density of the ring. Groups with multiple bonds to an electronegative atom (such as the carbonyl group) withdraw electron density from the ring.

Halogens deactivate the ring toward electrophilic substitution but are *ortho–para* directors. The halogens withdraw electron density by an inductively effect but can donate electrons to the ring by resonance. Substituents such as the hydroxyl and amino groups donate electrons to the ring by resonance and overcome inductive electron withdrawal by their electronegative atoms.

The nitro group and cyano group withdraw electrons by a combination of both an inductive effect and a resonance effect, so they are strongly deactivating.

13.6 Interpreting Directing Effects

The location of the positive charge in the carbocation intermediate in electrophilic substituent determines whether *ortho*, *para*, or *meta* substitution occurs. The charge is always *ortho* and *para* to the position at which the electrophile enters. If the substituent can stabilize a positive charge by resonance, then substitution at the *ortho*, para sites is favored. *Meta* directors destabilize positive charge when an electrophile attacks *ortho* and *para* to them. As a consequence, the electrophile enters the *meta* position by default.

The location of the positive charge in the carbocation intermediate in electrophilic substituent determines whether *ortho*, *para*, or *meta* substitution occurs. The charge is always *ortho* and *para* to the position at which the electrophile enters. If the substituent can stabilize a positive charge by resonance, then substitution at the *ortho*, *para* sites is favored. *Meta* directors destabilize positive charge when an electrophile attacks *ortho* and *para* to them. As a result, the electrophile enters the meta position by default.

13.7 Functional Group Modification

Some functional groups can be introduced on the benzene ring indirectly when a functional group already on the ring is transformed by a chemical reaction. The major examples are
1.   Oxidizing an alkyl group to a carboxylic acid
2.   Reducing an acyl group to an alkyl group
3.   Reducing a nitro group to an amino group. The amino group can be converted to other functional groups by converting the amino group to a diazonium ion. The groups that can replace the diazonium ion are:
1.   halogens, using copper(I) halide
2.   nitrile, using copper(I) cyanide
3.   hydroxyl, using aqueous acid
4.   hydrogen, using hypophosphorous acid

13.8 Synthesis of Substituted Aromatic Compounds

The *ortho*, *para*, or *meta*-directing properties of the each group added to the ring, and the chemical modifications that can be performed on these groups after they have been added dictate the order in which groups are added to an aromatic ring.

If two groups can be introduced by direct reaction, and their location is consistent with the directing characteristic of one of them, then the synthesis is straightforward. If the effect of two groups are cooperative—both *ortho*, *para* directing or both *meta*-directing—they act in concert. If the effects of the two groups are opposed, then the one with the greater *ortho*, *para* directing activity determines the distribution of products.

Summary of Reactions

1. Halogenation

2. Nitration

3. Sulfonation

4. Friedel–Crafts Alkylation

5. Friedel–Crafts Acylation

6. Side Chain Oxidation

7. Acyl Side Chain Reduction

8. Nitro Group Side Chain Reduction

(a)

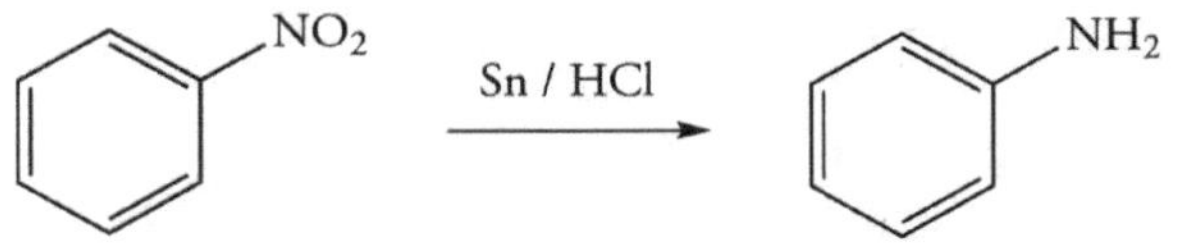

$$C_6H_5NO_2 \xrightarrow{\text{Sn / HCl}} C_6H_5NH_2$$

(b)

$$C_6H_5NO_2 \xrightarrow{\text{H}_2\text{ / Pd}} C_6H_5NH_2$$

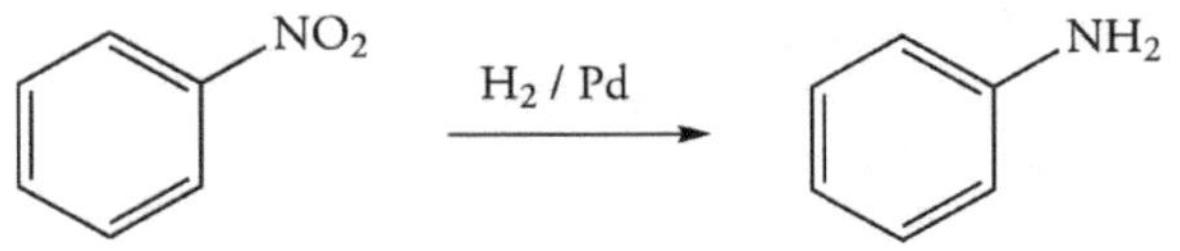

9. Amino Group Side Chain Reactions via Diazonium Ions

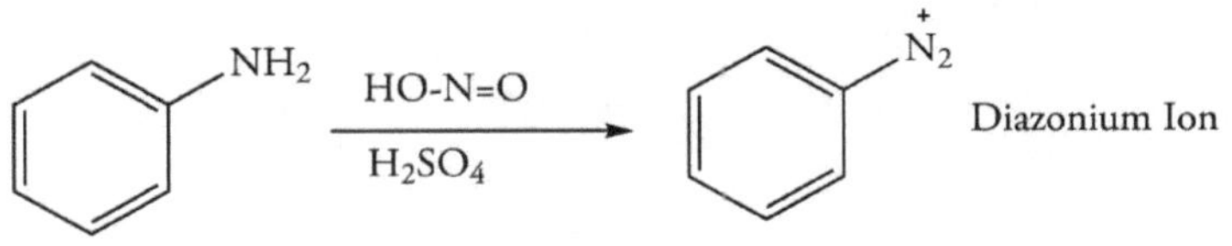

$$C_6H_5NH_2 \xrightarrow[\text{H}_2\text{SO}_4]{\text{HO-N=O}} C_6H_5N_2^+ \quad \text{Diazonium Ion}$$

(a)

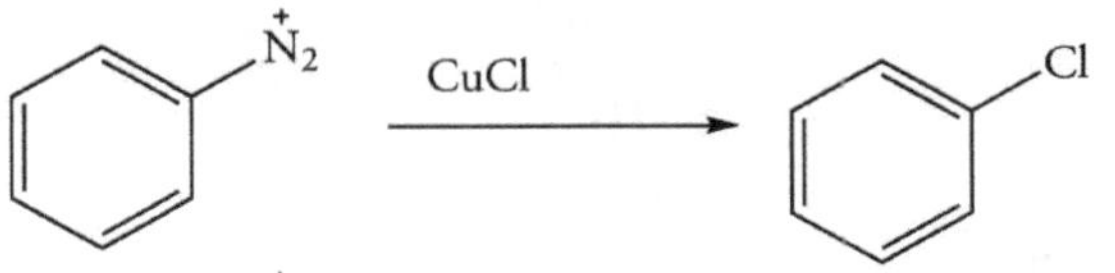

$$C_6H_5N_2^+ \xrightarrow{\text{CuCl}} C_6H_5Cl$$

(b)

$$C_6H_5N_2^+ \xrightarrow{\text{CuBr}} C_6H_5Br$$

(c)

$$C_6H_5N_2^+ \xrightarrow{\text{CuCN}} C_6H_5CN$$

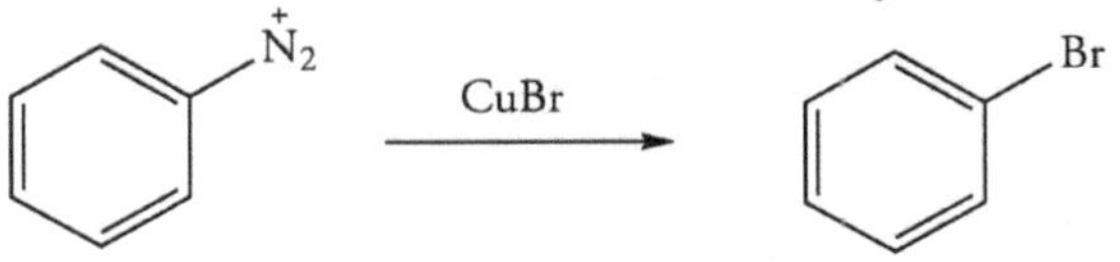

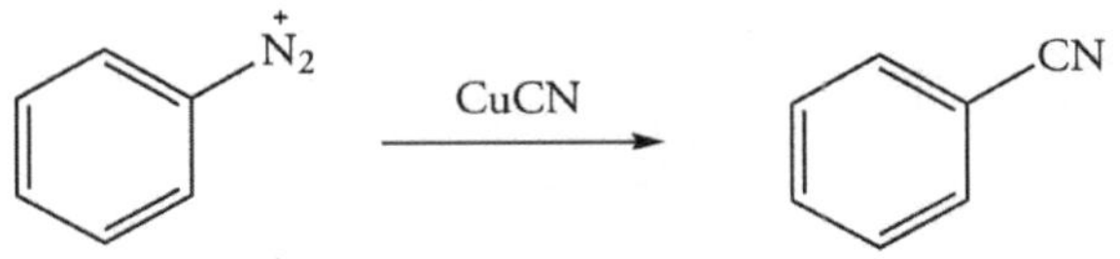

(d)

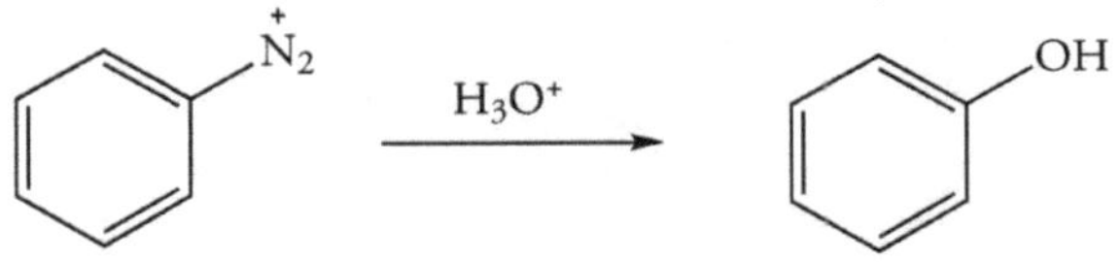

$$C_6H_5N_2^+ \xrightarrow{\text{H}_3\text{O}^+} C_6H_5OH$$

(e)

$$C_6H_5N_2^+ \xrightarrow{\text{H}_3\text{PO}_2} C_6H_5H$$

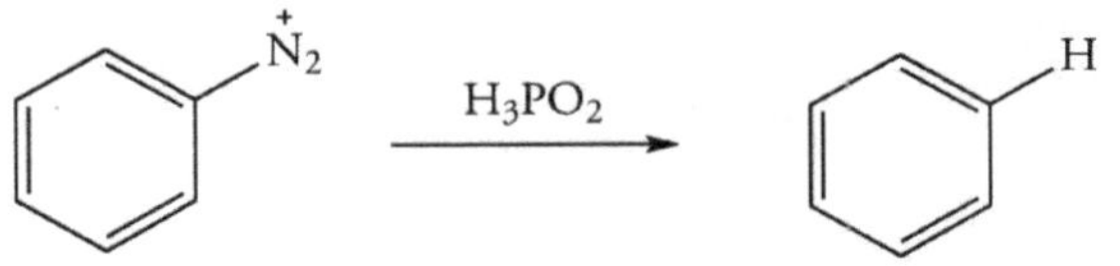

10. Benzylic Bromination with *N*-Bromosuccinimide (NBS)

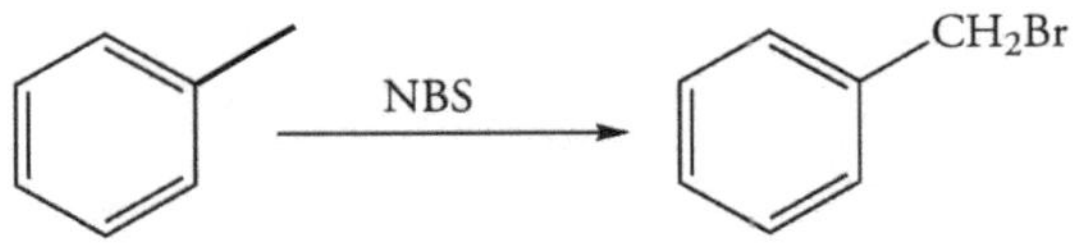

$$C_6H_5CH_3 \xrightarrow{\text{NBS}} C_6H_5CH_2Br$$

End of Chapter Exercises

Nomenclature of Aromatic Compounds

13.1 Identify each of the following as an *ortho-*, *meta-*, or *para*-substituted compound.

Answers:

(a) *para*

(b) *ortho*

(c) *meta*

13.2 Identify each of the following as an *ortho-*, *meta-*, or *para*-substituted compound.

Answers:

(a) *meta*

(b) *para*

(c) *ortho*

13.3 Name each of the following compounds.

Answers:
(a) ethylbenzene
(b) isopropylbenzene
(c) 1,4-diethylbenzene
(d) 1,3,5-trimethylbenzene

(a) CH_2CH_3

(b)

(c) CH_3CH_2—CH_2CH_3

(d)

13.4 Name each of the following compounds.

Answers:
(a) 1,2,4-trichlorobenzene
(b) 2,4-dibromophenol
(c) 3-bromoaniline
(d) 3,4-dichlorotoluene

(a)

(b)

(c)

(d)

13.5 Name each of the following compounds.

Answers:
(a) 4-chloro-3,5-dimethylphenol
(b) 2,6-dimethylaniline
(c) 4-chloro-2-phenylbenzene

(a)

(b)

(c)

13.6 Draw the structure of each of the following compounds.
  (a) 5-isopropyl-2-methylphenol, found in oil of marjoram
  (b) 2-isopropyl-5-methylphenol, found in oil of thyme
  (c) 2-hydroxybenzyl alcohol, found in the bark of the willow tree

Answers:

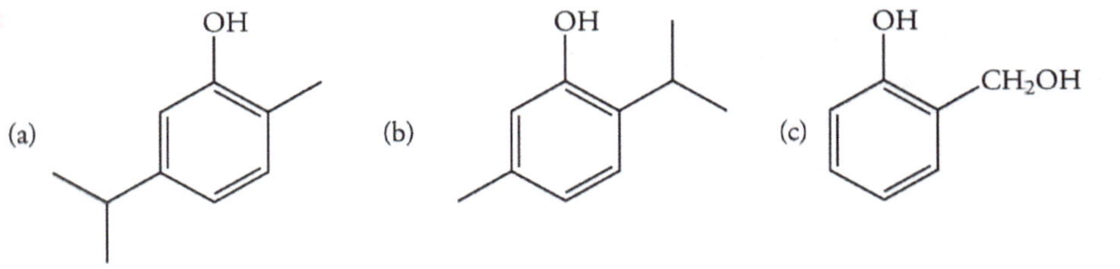

(a)   (b)   (c)

13.7 Draw the structure of 3,4,6-trichloro-2-nitrophenol, a lampricide used to control sea lampreys in the Great Lakes.

Answer:

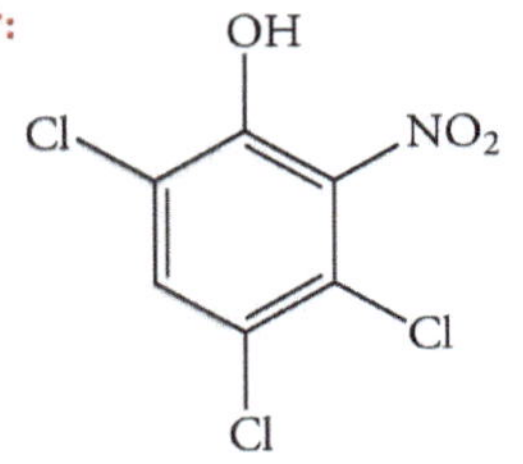

13.8 *N,N*-Dipropyl-2,6-dinitro-4-trinuoromethylaniline is the IUPAC name for Treflan, a herbicide. Draw its structure. (The prefix *N* signifies the location of a substituent replacing hydrogen on a nitrogen atom.)

Answer:

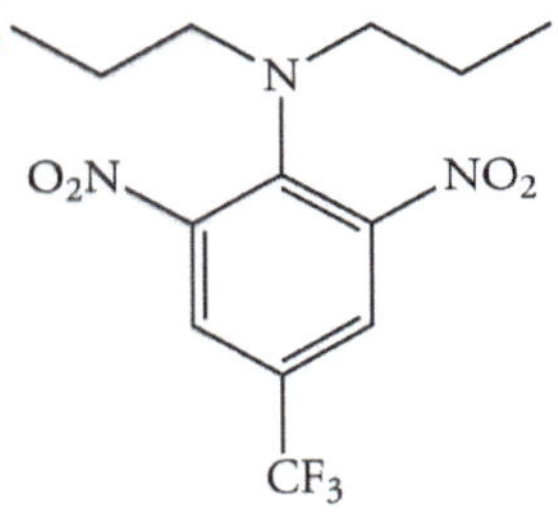

Electrophiles

13.9 Some activated rings may be hydroxylated by reacting hydrogen peroxide (H_2O_2) with acid. What is the formula of the electrophile? How does it form?

Answer: The oxygen–oxygen bond of the conjugate acid of hydrogen peroxide cleaves heterolytically to give water and a hydroxyl cation, which is the electrophile.

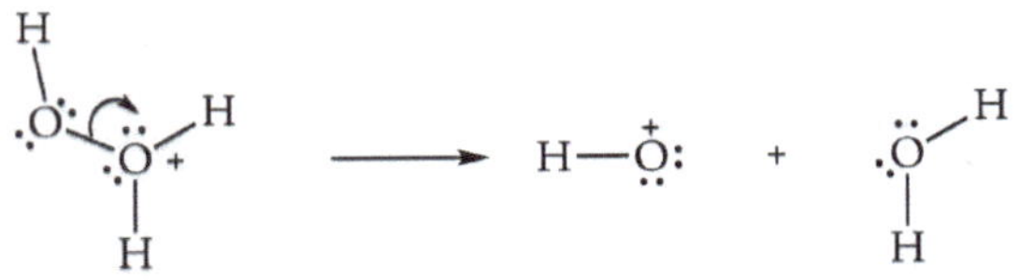

13.10 Reactive aromatic rings can be iodinated using iodine monochloride (ICl). What is the electrophile?

Answer: The electrophile is I^+, which is produced by the heterolytic cleavage of the I—Cl bond.

13.11 Benzene reacts with mercuric acetate to give phenylmercuric acetate using perchloric acid ($HClO_4$) as a catalyst. What is the electrophile? How does it form?

Answer: The electrophile is $AcOHg^+$, which is formed by protonation of one of the acetate groups of the covalent $Hg(OAc)_2$ followed by heterolytic cleavage of the Hg—O bond. Acetic acid is the other product.

13.12 Treating an aromatic rings with *tert*-butyl alcohol, $(CH_3)_3COH$, in acid solution places a tertiary butyl group on the ring. What is the formula of the electrophile? How does it form?

Answer: The electrophile is $AcOHg^+$, which is formed by protonation of one of the acetate groups of the covalent $Hg(OAc)_2$ followed by protonation of the oxygen atom followed by heterolytic cleavage of the C—O bond. Water is the leaving group.

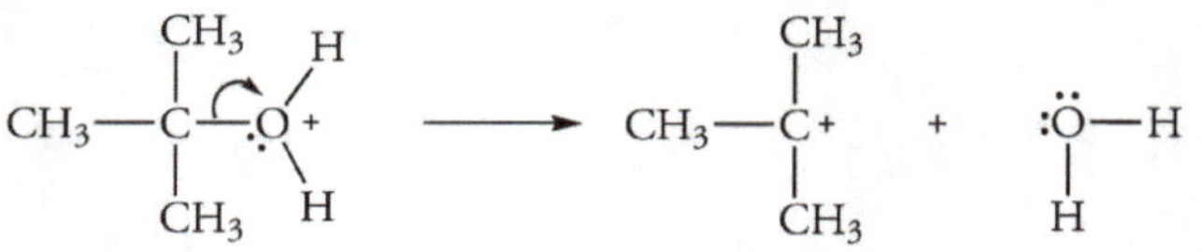

Properties of Ring Substituents

13.13 Some activated rings may be hydroxylated by reacting hydrogen peroxide (H_2O_2) with acid. What is the formula of the electrophile? How does it form?

Answer: The oxygen–oxygen bond of the conjugate acid of hydrogen peroxide cleaves heterolytically to give water and a hydroxyl cation, which is the electrophile.

13.14 Is the thiomethyl group, —S—CH_3, an activating or deactivating group. Will it be *ortho, para* directing or *meta* directing?

Answer: Both chlorine and sulfur are third row elements. However, since sulfur is less electronegative than chlorine, the ring is less deactivated by inductive electron withdrawal. Although third row elements are not effective electron donors by resonance, sulfur is a better donor of electrons than a chloro group because it is less electronegative. Therefore, the —SCH_3 group is an *ortho, para*-directing group.

13.15 The sulfonamide group is found in sulfa drugs. Is it an activating or deactivating group. Will it be *ortho, para*-directing or *meta* directing?

Answer: The sulfur atom is bonded to three electronegative atoms. Therefore, the sulfonamide group will inductively withdraw electron density from the aromatic ring and deactivate it toward electrophilic substitution. There are no lone pair electrons on sulfur to be donated by resonance. Thus, the sulfonamide group is a *meta* director.

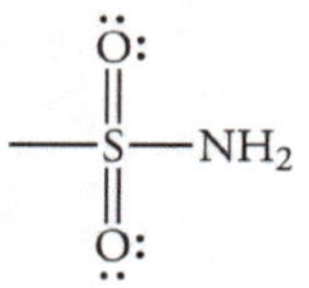

13.16 Nitration of *N,N*-dimethylaniline, $C_6H_5N(CH_3)_2$, in 85% sulfuric acid gives a *meta* nitro compound as the major product. What is the structure of the ring substituent responsible for the orientation of the nitro product?

Answer: An ammonium ion is formed in acidic solution because the nitrogen atom is protonated. The conjugate acid of *N,N*-dimethylaniline has a positive charge on the nitrogen atom. Therefore, the group withdraws electron density from the aromatic ring and deactivates benzene toward electrophilic substitution. Since there are no lone pair electrons on the protonated nitrogen atom to be donated by resonance, the group is a meta director.

13.17 The percentages of *meta* nitro product formed in the nitration of benzene compounds containing CH_3—, CH_2Cl—, $CHCl_2$—, and CCl_3— groups are 5%, 16%, 34%, and 64%, respectively. Explain this trend in the data.

Answer: As the number of chlorine atoms bonded to the carbon atom increases, the inductive withdrawal of electron density from that carbon atom and the adjacent benzene ring increases. Since there are no lone pair electrons on the carbon atom to be donated by resonance, the groups become increasingly better *meta* directors by increased electron withdrawal.

Reagents for Electrophilic Substitution

13.18 What reagent is required for each of the following reactions? Write the structure of the major product(s) expected from each reaction.
(a) bromination of anisole   (b) sulfonation of toluene
(c) nitration of benzoic acid   (d) acetylation of bromobenzene

Answers: The reagents are as follows. The structures of the products are shown below.
(a) bromine with iron(III) bromide
(b) fuming sulfuric acid (sulfur trioxide and sulfuric acid)
(c) nitric acid with sulfuric acid
(d) acetyl chloride with aluminum trichloride

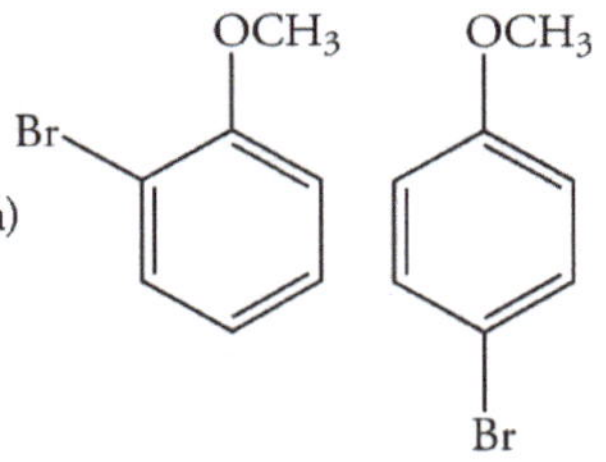

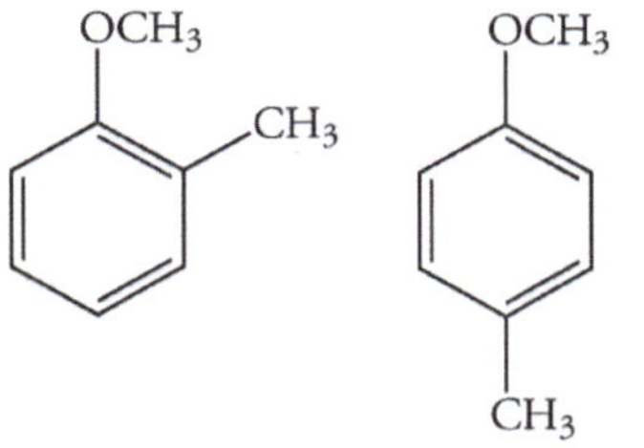

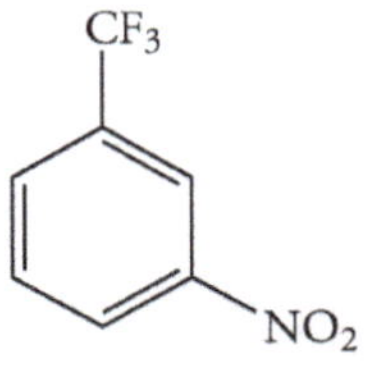

13.19 What reagent is required for each of the following reactions? Write the structure of the principal product(s) expected from each reaction.

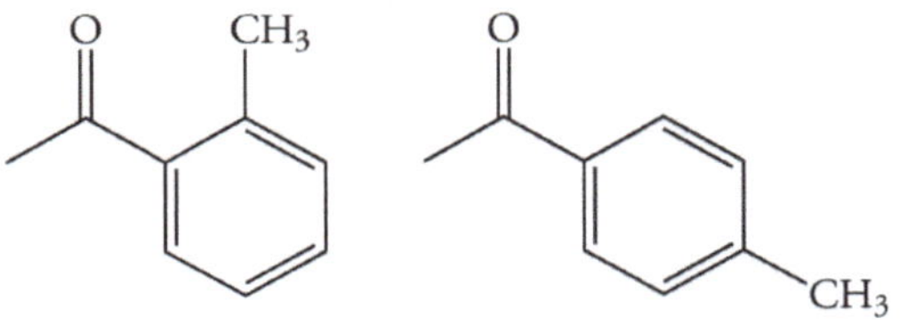

(a) chlorination of bromobenzene
(b) Friedel–Crafts methylation of anisole
(c) Friedel–Crafts acetylation of toluene
(d) nitration of trifluoromethylbenzene

Answers: The reagents are as follows. The structures of the products are shown below.
(a) chlorine with iron(III) chloride
(b) methyl chloride with aluminum trichloride
(c) acetyl chloride with aluminum trichloride
(d) nitric acid with sulfuric acid

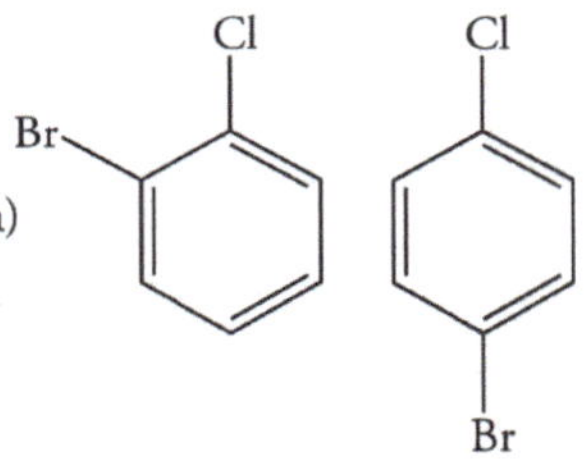

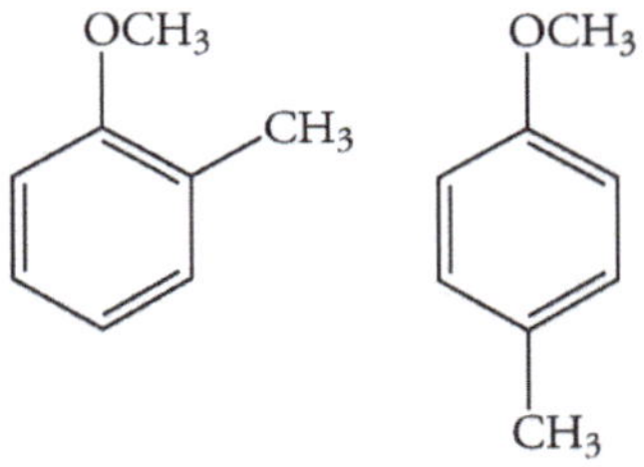

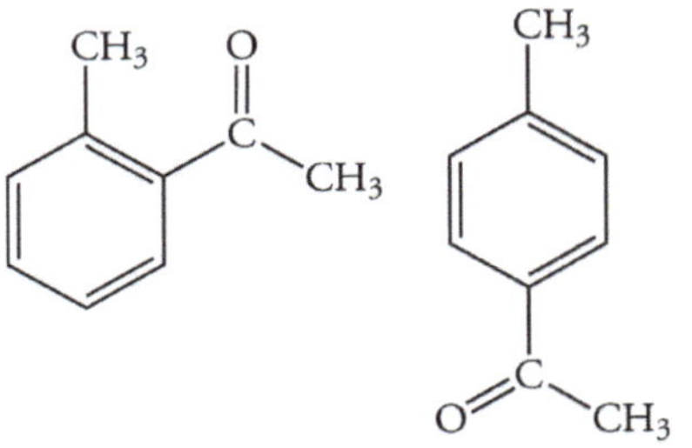

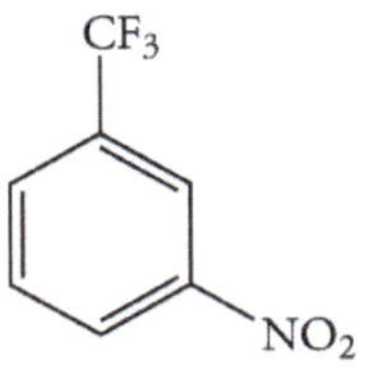

Friedel–Crafts Alkylation and Acylation

13.20 Write the structure of the product resulting from the Friedel–Crafts alkylation of benzene using chlorocyclohexane and aluminum trichloride.

Answer: Reaction of chlorocyclohexane with aluminum trichloride gives the cyclohexyl carbocation, which reacts with benzene to give cyclohexylbenzene.

13.21 What product results from the Friedel–Crafts alkylation of benzene using 1-chloro-2-methylpropane and aluminum trichloride?

Answer: Reaction of 1-chloro-2-methylpropane with aluminum trichloride gives a primary isobutyl carbocation, which rearranges to a *tert*-butyl carbocation by a 1,2-hydride shift. Thus, the product of the reaction is *tert*-butylbenzene.

13.22 Alkylation of benzene can be accomplished using an alkene such as propene and an acid catalyst. Identify the electrophile and the product.

Answer: Protonating C-1 of propene gives a secondary, isopropyl carbocation, which is the electrophile. The product is isopropylbenzene.

13.23 Write the structure of the product formed by alkylation of *p*-methylanisole using 2-methyl-1-propene and sulfuric acid.

Answer: The electrophile is the *tert*-butyl carbocation, $(CH_3)_3C^+$, which is formed by protonating the double bond. The *tert*-butyl carbocation adds *ortho* or *para* to the methyl group. The major product is the *para* isomer because steric hindrance diminishes attack at the *ortho* position.

13.24 Reaction of toluene with isopropyl alcohol, $(CH_3)_2CHOH$, using sulfuric acid gives a mixture of two isomers with the molecular formula $C_{10}H_{13}$. Write the structures of these compounds. How does the electrophile form?

Answer: The electrophile is the isopropyl carbocation, $(CH_3)_2CH^+$, which is formed by protonation of the oxygen atom followed by heterolytic cleavage of the C—O bond. Water is the leaving group. The isopropyl carbocation alkylates *ortho* or *para* to the methyl group. The major product will be the *para* isomer due to steric hindrance in the attack at the *ortho* position.

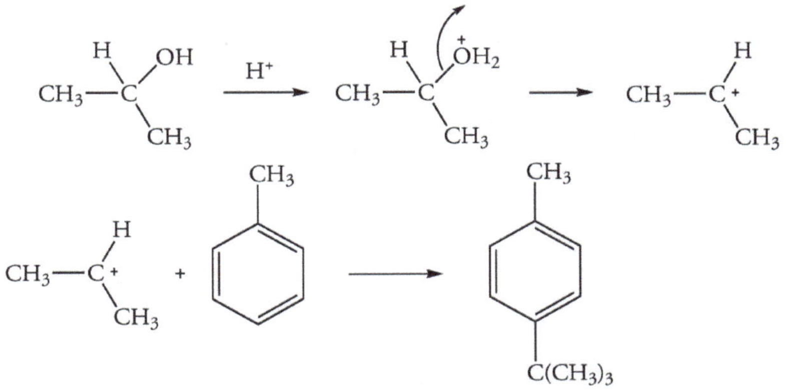

13.25 The following compound reacts with sulfuric acid to give a tricyclic hydrocarbon with molecular formula $C_{17}H_{18}$. Write its structure.

Answer:

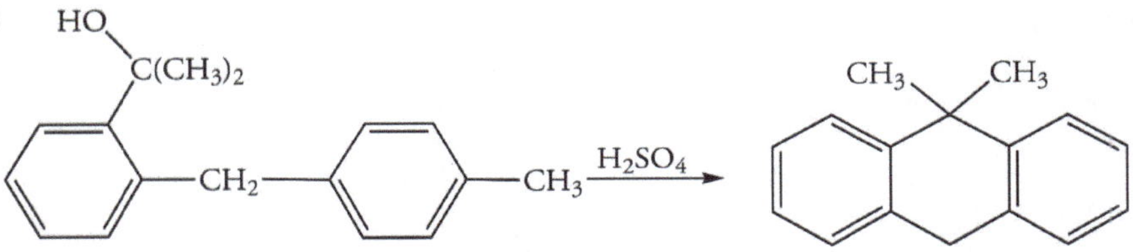

13.26 4-Phenylbutanoyl chloride reacts in carbon disulfide with aluminum trichloride to give a ketone with molecular formula $C_{10}H_{10}O$. Write the structure of the product.

Answer: An intramolecular Friedel–Crafts acylation reaction occurs *ortho* to the butanoyl chain to give a cyclic ketone.

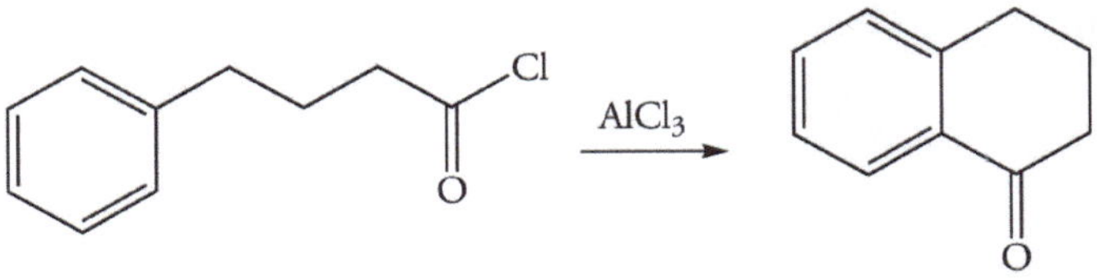

13.27 The following compound undergoes an intramolecular Friedel–Crafts acylation to give a cyclic ketone. Write the structure of the product.

Answer: An intramolecular Friedel–Crafts acylation reaction occurs on the ring containing the methoxy group because it is the more activated aromatic ring.

Electrophilic Aromatic Substitution Reactions

13.28 Indicate on which ring and at what position bromination of each compound will occur.

Answer: (a) Reaction occurs in the ring bonded to the nitrogen atom because that ring is activated. The other ring is deactivated by the carbonyl group. Reaction will occur at the *ortho* or *para* position.

Answer: (b) Reaction occurs in the ring bonded to the methylene group because it is slightly activated. The other ring is deactivated by the carbonyl group. Reaction will occur at the *ortho* or *para* position.

13.29 Indicate on which ring and at what position nitration of each compound will occur.

Answer: (a) Both rings are activated by the oxygen atom. However, the ring on the right is deactivated by the nitro group. Nitration will occur in the other ring at the *ortho* or *para* position.

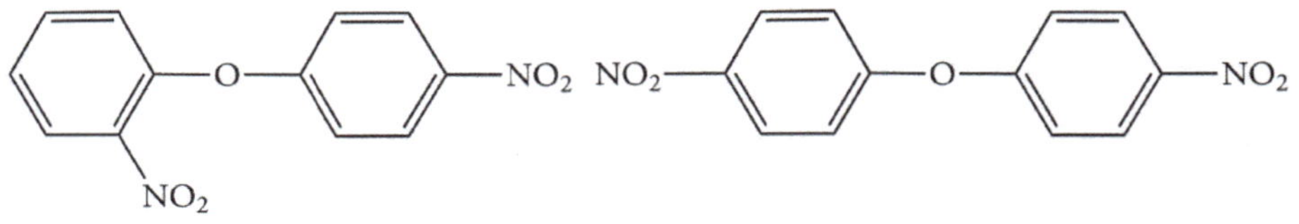

Answer: (b) Both rings are slightly activated by the methylene group. However, the ring on the right is also activated by the hydroxyl group. Nitration will tend to occur in that ring *ortho* to the activating group. A smaller amount of product with the nitro group *ortho* to the methylene group also forms.

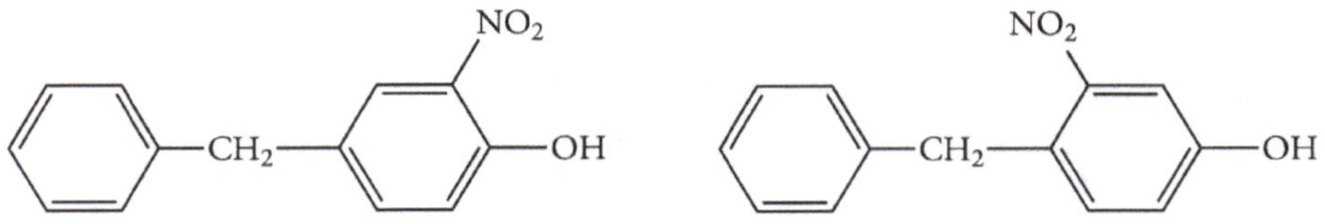

13.30 Write the structure of the major product of each of the following reactions, assuming that only monosubstitution occurs.

Answers: (a) The alkyl group is slightly activating and the bromo group is slightly deactivating. Bromination occurs *ortho* to the alkyl group.
(b) The bromo groups are deactivating but are *ortho, para*-directing groups. Nitration at C-2 between the two bromine atoms is sterically hindered. The other *ortho* positions at C-4- and C-6 are structurally equivalent, and only one product results.
(c) The dimethylamino group is strongly activating and the isopropyl group is slightly activating. Acetylation occurs *ortho* to the dimethylamino group.

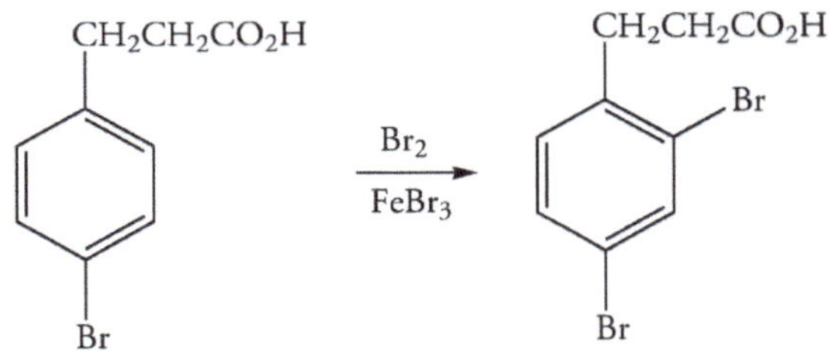

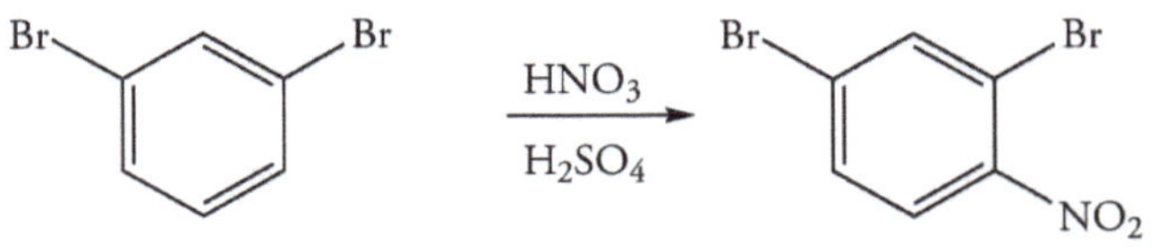

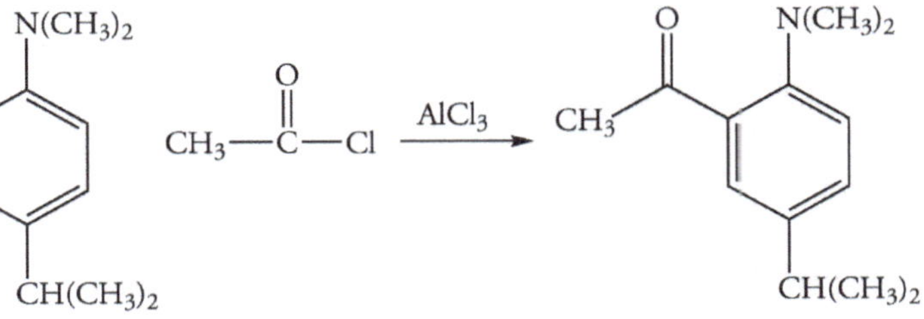

13.31 Write the structure of the major product of each of the following reactions, assuming that only monosubstitution occurs.

Answers: (a) The amino group is strongly activating and the nitro group is strongly deactivating. Bromination occurs *ortho* to the amino group.
(b) The two carboxylic acid groups are deactivating and are meta directing groups. Nitration occurs at the C-5 position, which is meta to both groups.
(c) The isopropyl group is slightly activating and the bromo group is slightly deactivating. Acetylation may occur *ortho* or para to the diisopropyl group. The *ortho* position between the isopropyl and bromo group is sterically hindered. The other *ortho* position is substantially hindered by the isopropyl group. The *para* position with respect to the isopropyl group is also *ortho* with respect to the bromine. This is the most likely site for attack.

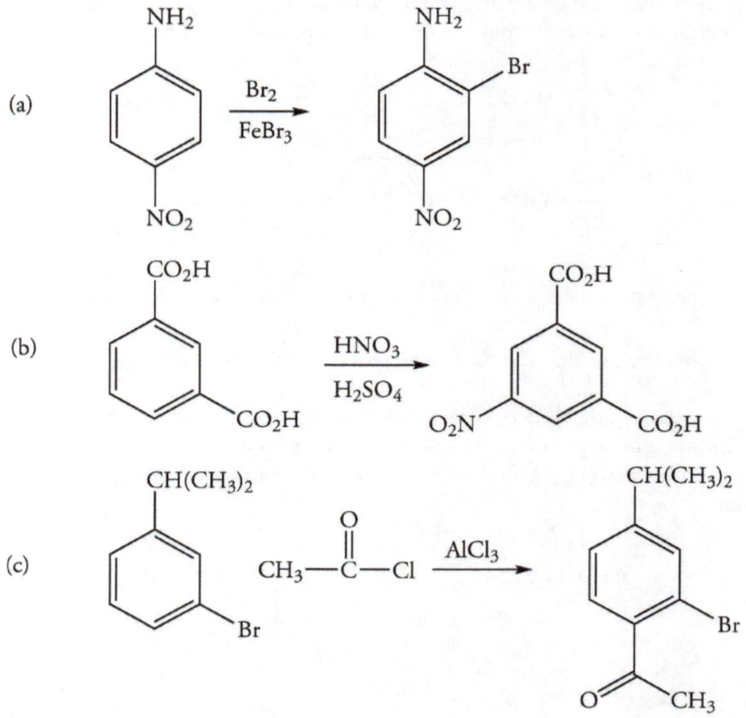

(a)

NH_2 benzene ring with NO_2 para $\xrightarrow[\text{FeBr}_3]{\text{Br}_2}$ NH_2 benzene ring with Br ortho and NO_2

(b)

CO_2H benzene ring with CO_2H meta $\xrightarrow[\text{H}_2\text{SO}_4]{\text{HNO}_3}$ benzene ring with two CO_2H groups and O_2N at C-5

(c)

$CH(CH_3)_2$ benzene ring with Br meta + $CH_3-\overset{\displaystyle O}{\overset{\|}{C}}-Cl \xrightarrow{\text{AlCl}_3}$ $CH(CH_3)_2$ benzene ring with Br and $\overset{O}{\underset{\|}{C}}CH_3$

13.32 Write the product of the reaction of each of the following compounds with zinc-mercury amalgam and HCl.

Answer: Zinc-mercury amalgam and HCl convert carbonyl groups to methylene groups.

(a)

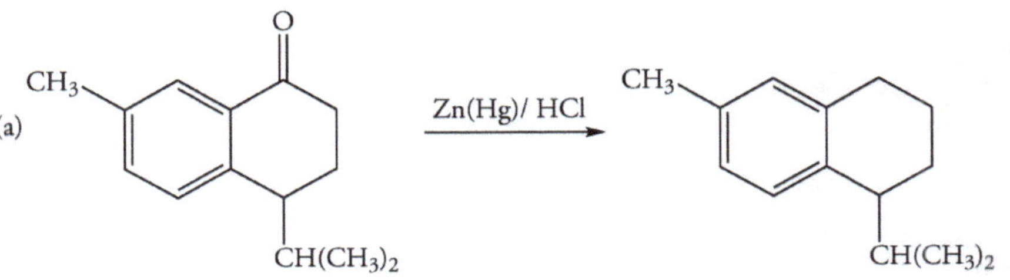

(b)

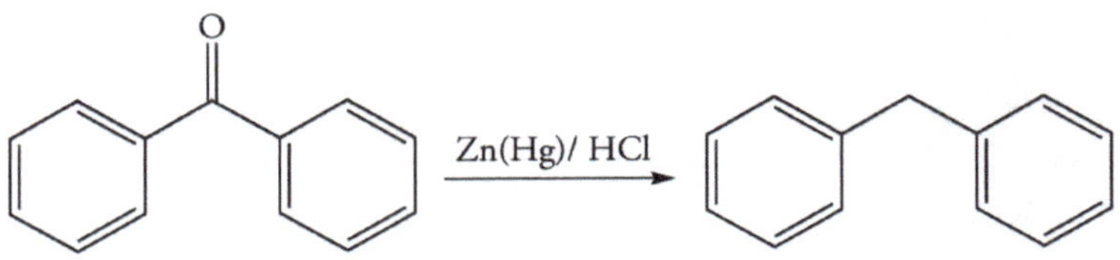

(c)

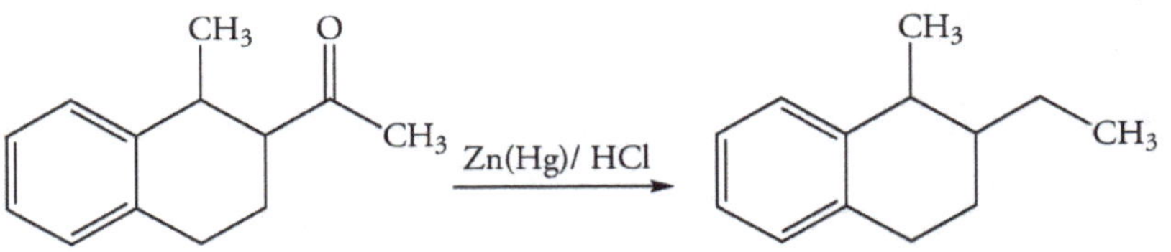

13.33 Write the product of the reaction of each of the following compounds with tin and HCl.

Answer: Tin and HCl reduces nitro groups to produce amino groups.

(a)

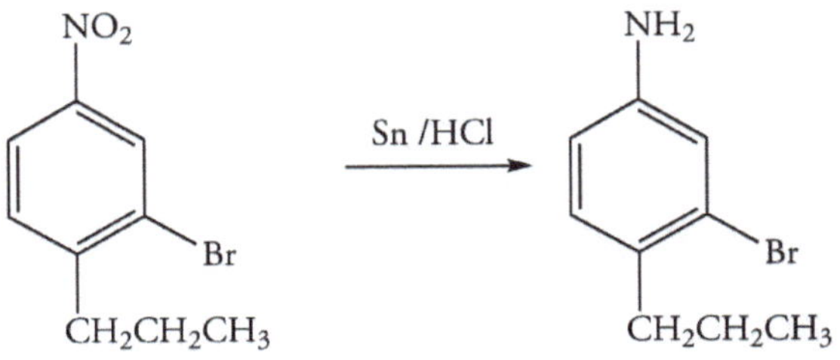

(b)

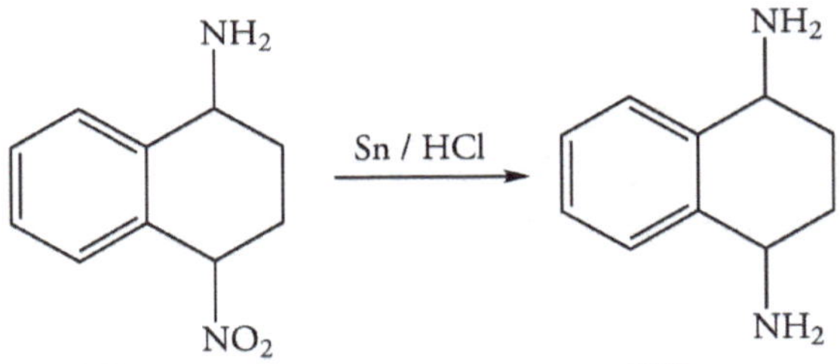

(c)

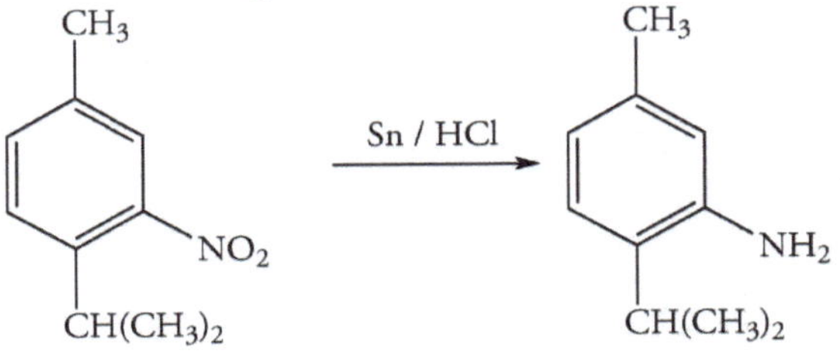

14

METHODS FOR STRUCTURE DETERMINATION

NUCLEAR MAGNETIC RESONANCE

AND

MASS SPECTROMETRY

KEYS TO THE CHAPTER

14.1 Structure Determination

Although the identity of a molecule and its structure can be determined indirectly based on its chemical reactions, such methods destroy some portion of the sample of the compound. Spectroscopic methods determine molecular structure by physical methods. Therefore, the sample is not destroyed.

Even a relatively simple molecule can exist in many isomeric forms, and spectroscopic methods rapidly narrow the range of possibilities, greatly shortening the time required to determine a molecular structure.

14.2 Nuclear Magnetic Resonance

Many nuclei have a nuclear spin. A spinning nucleus generates a magnetic field, whose energy depends on the direction of spin in the presence of an applied magnetic field. The NMR method depends on detecting the absorption of energy required to change the direction of the spin of a nucleus.

Two nuclei that are important in the determination of the structures of organic compounds are ^1H and ^{13}C. The magnetic field strength required to "flip" the spin of various hydrogen atoms (or carbon atoms) within a molecule differs. The local magnetic fields differ throughout a molecule because the bonding characteristics differ. Thus, each hydrogen (or carbon) nucleus is unique, and distinct resonances are obtained for each structurally nonequivalent atom in a molecule.

14.3 The Chemical Shift

A spinning nucleus induces a small local magnetic field that opposes the applied magnetic field. This local field shields the nucleus from the applied field. A relative scale called the delta scale, in which one delta unit (δ) is 1 ppm of the applied magnetic field, is used to measure the chemical shift of hydrogen atoms. The resonance for the hydrogen atoms of tetramethylsilane, $(CH_3)_4Si$, is defined as $0\ \delta$. The delta scale is independent of the applied magnetic field. Shielded nuclei are found at high field and have small δ values. The chemical shifts for the hydrogens in organic compounds ranges from 0 to 10 δ.

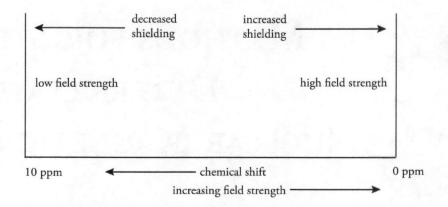

decreased shielding ← → increased shielding

low field strength | high field strength

10 ppm ← chemical shift → 0 ppm

increasing field strength →

14.4 Detecting Sets of Nonequivalent Hydrogen Atoms

To understand the relationship between an NMR spectrum and the structure of a compound, we have to be able to recognize the equivalence of nuclei in the structure. The simplest examples, such as the six equivalent protons on the two methyl groups of 2-bromopropane, are straightforward. However, some nuclei that might look equivalent at first glance are actually nonequivalent. For example, both hydrogen atoms in 1-bromo-1-chloroethene are bonded to the same carbon atom, but they are not equivalent. One hydrogen atom is *cis* to the chlorine atom and the other is *cis* to the bromine atom. Replacing one hydrogen atom or the other by deuterium gives a set of diastereomers, and the hydrogen atoms are **diastereotopic**. Such hydrogen atoms usually have different chemical shifts.

Hydrogen atoms that are in mirror image environments are **enantiotopic** (Section 8.12). They have same chemical shifts. Replacing enantiotopic hydrogen atoms by deuterium atoms gives enantiomers, and their physical properties are identical, including their chemical shifts.

14.5 Structural Effects on Chemical Shifts

The chemical shifts of hydrogen atoms depend on the local electron density, which in turn affects the local magnetic field. Electronegative atoms deshield hydrogen atoms by an effect analogous to the inductive effect we have discussed many times. Deshielding results in a shift to lower field and larger δ values.

Electrons in π bonds are easily polarized, and they induce substantial local magnetic fields. These effects extend across three or four chemical bonds. Of particular importance is the deshielding of hydrogen atoms bonded to an aromatic ring. The chemical shifts of aromatic hydrogen atoms are large, meaning that they appear at low field with δ values of 7-8 ppm.

Table 14.5 gives the chemical shifts of hydrogen atoms in various structural environments.

14.6 Relative Peak Areas and Proton Counting

The NMR spectrum tells us how many sets of structurally nonequivalent hydrogen atoms are present in a molecule. Each set causes resonance absorptions in its own characteristic region. The area of each resonance peak is proportional to the relative number of hydrogen atoms of each kind. Therefore, NMR also tells us the ratios of the nonequivalent hydrogen atoms in a molecule. These ratios are proportional to the relative peaks heights of the resonances of the nonequivalent hydrogen atoms.

14.7 Spin-Spin Splitting

Multiple lines, called **multiplets**, are often observed for the absorptions of equivalent hydrogen atoms. This phenomenon is called **spin-spin splitting**. It results from the interaction of the nuclear spin of the hydrogen atoms on an adjacent carbon atoms. In general, sets of hydrogen atoms on nonequivalent neighboring carbon atoms couple with each other. If hydrogen atom A couples and causes splitting of the resonance for hydrogen atom B, then the resonance for hydrogen atom B is also split by hydrogen atom A. The separation of the components of the multiplet is the **coupling constant**, which is designated as *J*.

A set of one or more hydrogen atoms that has n equivalent neighboring hydrogen atoms has $n+1$ peaks in the NMR spectrum. Common multiplets include **doublets**, **triplets**, and **quartets**. The appearance of several sets of multiplets resulting from $n = 1$ to $n = 4$ is shown in Figure 14.15. The areas of the component peaks of a doublet are equal; the areas of the component peaks of other multiplets are not equal and are summarized in Table 14.16.

14.8 Structural Effects on Coupling Constants

The conformation of a molecule atom contributes to the magnitude of the coupling constant of vicinal hydrogen atoms. The coupling constant is largest for *anti* periplanar arrangements in saturated H—C—C—H compounds. The coupling constant for vinyl hydrogen atoms in an E arrangement is larger than for hydrogen atoms in an isomer with a Z configuration.
Coupling constants over more than three bonds are termed **long range**. These coupling constants can be used to determine the structures of isomeric aromatic compounds.

14.9 Ion Impact Mass Spectrometry

In **electron impact mass spectrometry**, the collision of a high energy electron with a sample molecule produces a **radical cation**, M^+. The first ion that forms in this process is the **parent ion.** Since the charge of the ion is $+1$, and since the mass of an electron is much smaller than the mass of a proton or neutron, the ratio of the mass of the parent ion, M^+, to its charge, m/z, equals the molecular mass of the compound.

The parent ion has very high energy, and it fragments in the instrument before it reaches the detector. The peak for the most abundant ion is assigned an arbitrary intensity of 100; this is the **base peak**. The parent ion, M^+, fragments two give two products: one is a cation or a radical cation, the other is neutral. We can identify the mass of the neutral particle that forms in a fragmentation reaction by subtracting the mass of the base peak from the mass of the parent ion.

One per cent of carbon atoms exist as the stable isotope ^{13}C. If the parent ion is sufficiently intense, then a **P+1 ion** will be present in the mass spectrum. The probability that a molecule will contain ^{13}C increases with the number of carbons in the compound.

Each type of functional group fragments with a characteristic pattern. The parent ion of a normal alkane fragments by forming a neutral methyl group, $CH_3\cdot$, and a primary radical cation. This species continues to fragment by losing successive $CH_2\cdot$ groups. The base peak for a normal alkane typically has a mass of $CH_3(CH_2)_n\cdot^+$. In contrast, the parent ion of a branched alkane usually does not have a parent ion. Instead, it fragments at a branch point to give a neutral fragment and a secondary or tertiary radical cation as the base peak.

Mass spectrometry allows us to chlorine and bromine by their isotopic abundances. Chlorine has two isotopes, ^{35}Cl and ^{37}Cl. The atomic mass of chlorine corresponds to an isotopic ratio $^{35}Cl/^{37}Cl$ of 3:1. The parent ion of a chloro compound has two peaks whose mass differ by two units having intensities are 3:1. It is also easy to identify bromine since the $^{79}Br/^{81}Br$ ratio is 1:1.

An alcohol ionizes in a mass spectrometer to produce a radical cation in which the oxygen atom bears the positive charge. The fragmentation of the base peak results from cleavage of the C—H bond rather than the C—O bond because the C—O bond is stronger.

The mass spectrum of a compound provides clues about the presence or absence of nitrogen. If a compound contains a single nitrogen atom, or any odd number of nitrogen atoms, its mass is an odd number. If the mass spectrum of a compound is an even number, then it has either an even number of nitrogen atoms, or none.

14.10 Effect of Dynamic Processes

Dynamic processes, such as rotation about a single bond, occur on such a rapid time scale that two seemingly nonequivalent hydrogen atoms have the same chemical shift. One such circumstance is found in cyclohexane because chair-chair interconversion exchanges the axial and equatorial hydrogen atoms so rapidly that they cannot be distinguished at room temperature. A second example is found in alcohols, in which the rapid exchange of protons between oxygen atoms of various alcohol molecules in a sample.

14.11 Carbon-13 NMR Spectroscopy

Carbon-13 (^{13}C) NMR spectroscopy permits the direct determination of the number of nonequivalent carbon atoms in a structure. If the carbon atoms are nonequivalent, each has a distinct resonance. More importantly, the count of the number of resonances gives the number of sets of equivalent carbon atoms that must be in the structure. The list of chemical shifts in Table 14.7 gives some idea about the identity of each carbon atom responsible for a resonance.

The intensity of the resonances of ^{13}C atoms is *not* proportional to the number of equivalent carbon atoms.

 **End of Chapter Exercises**

Infrared Spectroscopy

14.1 How can infrared spectroscopy be used to distinguish between propanone and 2-propen-1-ol?

$$CH_3-\overset{\overset{\textstyle O}{\|}}{C}-CH_3 \qquad\qquad CH_2{=}CH-CH_2OH$$
$$\text{propanone} \qquad\qquad\qquad \text{2-propene-1-ol}$$

Answer: The carbonyl group of propanone (acetone) has a strong absorption at 1749 cm⁻¹. 2-Propen-1-ol (allyl alcohol) has an absorption for the carbon-carbon double bond at 1645 cm⁻¹ and an absorption for the oxygen-hydrogen bond at 3400 cm⁻¹.

14.2 How can infrared spectroscopy be used to distinguish between 1-pentyne and 2-pentyne?

Answer: 1-Pentyne is a terminal alkyne, so its sp-hybridized C—H bond has an absorption in the 3450 cm⁻¹, and another strong C≡C absorption at 2120 cm⁻¹. 2-Pentyne, which is an internal alkyne, does not have a C—H absorption at 3450 cm⁻¹. Also, the C≡C absorption is so weak that it is barely visible.

14.3 The carbonyl stretching vibration of ketones is at a longer wavelength than the carbonyl stretching vibration of aldehydes. Suggest a reason for this observation.

Answer: The longer wavelength absorption (smaller wavenumber) corresponds to a lower energy vibration. The dipolar resonance form of a ketone is more stable than that of an aldehyde because the extra alkyl group donates electron density. The increased contribution of the resonance form with a carbon-oxygen single bond means that the ketone carbonyl bond absorption requires less energy.

14.4 The carbonyl stretching vibrations of esters and amides occur at 1735 and 1670 cm⁻¹, respectively. Suggest a reason for this difference.

Answer: Both oxygen and nitrogen are inductively electron withdrawing, and they destabilize the dipolar resonance form of the carbonyl group. Since oxygen is more electronegative than nitrogen, this effect is larger for oxygen, so the dipolar resonance form of an ester is less stable that of an amide. The relative ability of the two atoms to donate electrons by resonance is also important. Because nitrogen donates electrons by resonance more effectively than oxygen, there is an increased contribution of a dipolar resonance form for the amide.

14.5 An infrared spectrum of a compound with molecular formula $C_4H_8O_2$ has an intense, broad band between 3500 and 3000 cm⁻¹ and an intense peak at 1710 cm⁻¹. Which of the following compounds best fits this data?

I: $CH_3CH_2CO_2CH_3$ II: $CH_3CO_2CH_2CH_3$ III: $CH_3CH_2CH_2CO_2H$

Answer: The absorptions correspond to an O—H and a carbonyl group, respectively. Only the carboxylic acid group of III has both structural features. The other two compounds are esters that would have an absorption corresponding to a carbonyl group but, because esters do not have an O—H group, would have no absorption in the 3500–3000 cm⁻¹ region.

14.6   Explain why the carbonyl stretching vibrations of the following two esters differ.

$$CH_2\!\!=\!\!CH\!\!-\!\!CH_2\!\!-\!\!\overset{\displaystyle O}{\overset{\|}{C}}\!\!-\!\!O\!\!-\!\!CH_3 \qquad CH_3\!\!-\!\!CH\!\!=\!\!CH\!\!-\!\!\overset{\displaystyle O}{\overset{\|}{C}}\!\!-\!\!O\!\!-\!\!CH_3$$
$$\text{1735 cm}^{-1} \qquad\qquad\qquad \text{1720 cm}^{-1}$$

Answer: The carbonyl group of the second compound is conjugated with a double bond. As a result, there is some contribution of a resonance form in which the carbon-oxygen bond has single bond character. The increased contribution of the resonance form with a carbon-oxygen single bond means that the carbonyl bond absorption requires less energy.

14.7   Explain how the two isomeric nitration products of isopropylbenzene can be distinguished using infrared spectroscopy.

Answer: The *ortho* nitro isomer has four adjacent C—H bonds, and the out-of plane bending of these bonds occurs at 748 cm^{-1}. The para nitro isomer has two sets of two adjacent C—H bonds, and the out-of plane bending occurs at 866 cm^{-1}.

14.8   Explain how the structures of the three isomeric trimethylbenzenes can be established using infrared spectroscopy.

Answer: The 1,2,3 isomer has three adjacent C—H bonds, and the out-of plane bending absorptions of these bonds occur in the 810–750 cm^{-1} region. There is another absorption in the 745–690 cm^{-1} region. The 1,2,4 isomer has two adjacent C—H bonds and one lone C—H bond. The absorptions for these bonds occur in the 860–800 and 900–860 cm^{-1} regions, respectively. The 1,3,5 isomer has three C—H bonds with no neighbors. The absorptions for these bonds occur in the 900–860 cm^{-1} region.

Calculation of Chemical Shift

14.9   The hydrogen NMR spectrum of CHCl$_3$, measured with a 360 MHz spectrometer, is a singlet that is 2622 Hz downfield from TMS. Calculate δ.

Answer: Divide 2622 Hz by (60×10^6) and multiply by 10^6 to obtain 7.28 δ.

14.10   The hydrogen NMR spectrum of CHI$_3$, measured with a 360 MHz spectrometer, is a singlet at 5.37 δ. Calculate the chemical shift in Hz relative to TMS.

Answer: Multiply 5.37 δ by 360 Hz because each ppm or δ unit equals 360 Hz. The chemical shift is 1933 Hz downfield from TMS.

Chemical Shifts and Structure

14.11   How many NMR signals should be observed for the hydrogen atoms in each of the following compounds?
(a) 2,2-dimethylpropane   (b) 2-methyl-l-propene   (c) 1,3,5-trimethylbenzene

Answers: (a) Only one, because all four methyl groups are equivalent. C-2 atom does not have a C—H.
(b) Two, because there are two equivalent methyl groups and two equivalent sp^2-hybridized C—H bonds.
(c) Two, because there are three equivalent methyl groups and three equivalent sp^2-hybridized C—H bonds on the benzene ring.
(d) There are four. The C-1 methyl group and the branching methyl group at C-2 are not equivalent! There are also resonances for sp^2-hybridized C—H bond at C-3, and the C-4 methyl group.

14.12   How many NMR signals should be observed for the hydrogen atoms in each of the following compounds?
         (a) 1,1-dichloroethene   (b) vinyl chloride (c) allyl bromide (d) 1-bromo-l-chloroethene

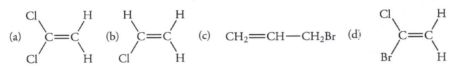

Answers: (a) Only one, because there are two equivalent sp²-hybridized C—H bonds at C-2.
         (b) The sp²-hybridized C—H bonds at C-2 are not equivalent. Thus, these two hydrogen atoms and the hydrogen at C-1 give
             three resonances.
         (c) The sp²-hybridized C—H bonds at C-3 are not equivalent. Thus, these two hydrogen atoms, the hydrogen atom of the sp²-
             hybridized C—H bond at C-2, and the C-1 methylene hydrogen atoms give four resonances.
         (d) The sp²-hybridized C—H bonds at C-2 are not equivalent, so each hydrogen atom has a different resonance.

14.13   How can the compounds of each pair be distinguished using hydrogen NMR spectroscopy?
         (a) isopropyl ethyl ether and *tert*-butyl methyl ether (b) cyclohexane and *cis*-3-hexene
         (c) 2,2-dimethyloxirane and *cis*-2,3-dimethyloxirane

Answers: (a) Each compound has resonances integrating as three hydrogen atoms in the 3.3-4.0 δ region associated with ethers. However,
             this signal for *tert*-butyl methyl ether is a singlet due to the three hydrogen atoms of the methyl group. Isopropyl ethyl
             ether has two resonances in this region. The resonance arising from the two hydrogen atoms of the methylene group is a
             quartet due to coupling of the methyl group. The other hydrogen resonance in this region is due to the methine hydrogen of
             the isopropyl group and is a septet because it is split by the two methyl groups.
         (b) All 12 of the hydrogen atoms of cyclohexane are equivalent, and the resonance appears as a singlet at high field. The
             isomeric alkene has a multiplet in the 5.0-6.5 δ region due to the hydrogen atoms of the sp²-hybridized C—H bonds, as
             well as peaks due to the methylene and methyl hydrogen atoms.
         (c) The resonance due to the hydrogen atoms of the equivalent methyl groups of the 2,2-dimethyloxirane is a singlet. The
             hydrogen atoms of the equivalent methyl groups of the isomeric compound are split by the hydrogen atom of the ring.

14.14   How can the compounds of each pair be distinguished using hydrogen NMR spectroscopy?
         (a) 1,3-dibromopropane and 2,2-dibromopropane
         (b) 1,1-dichlorobutane and 1,4-dichlorobutane
         (c) *cis*-2-butene and 2-methyl-l-propene

Answers: (a) The equivalent C-1 and C-3 methylene groups of 1,3-dibromopropane give a triplet due to splitting by the hydrogen atoms
             at C-2. The methylene hydrogen atoms at C-2 give a quintet. The equivalent methyl groups of 2,2-dibromopropane give one
             peak, a singlet.
         (b) The low-field portion of the spectrum of 1,1-dichlorobutane has an absorption due to the C-1 hydrogen atom. It is a triplet.
             The low-field portion of the spectrum of the isomeric 1,4-dichlorobutane has an absorption with an integrated intensity of
             four hydrogen atoms due to the C-1 and C-4 methylene groups. The signal is also a triplet.
         (c) Both compounds have resonances in the 5.0-6.5 δ region due to the two hydrogen atoms of the sp²-hybridized C—H
             bonds. However, those of 2-methyl-1-propene have no nearest neighbor hydrogen atoms to split the signal. The signal for
             the isomeric *cis*-2-butene is split by the hydrogen atoms of the methyl groups.

14.15   Draw the structure of each of the following hydrocarbons whose hydrogen NMR spectrum consists of a singlet with the indicated
         chemical shift.
         (a) C_5H_{10}; $\delta = 1.5$ (b) C_8H_{18}; $\delta = 0.9$ (c) $C_{12}H_{18}$; $\delta = 2.2$ (d) C_8H_8; $\delta = 5.8$

Answers: (a) [pentagon structure]   (b) $CH_3-\underset{\underset{CH_3}{|}}{\overset{\overset{CH_3}{|}}{C}}-\underset{\underset{CH_3}{|}}{\overset{\overset{CH_3}{|}}{C}}-CH_3$

          (c) $CH_3-\underset{\underset{CH_3}{|}}{\overset{\overset{CH_3}{|}}{C}}-C\equiv C-C\equiv C-\underset{\underset{CH_3}{|}}{\overset{\overset{CH_3}{|}}{C}}-\underset{\underset{CH_3}{|}}{\overset{\overset{CH_3}{|}}{C}}-CH_3$ (d) [cyclooctatetraene structure]

14.16 Draw the structure of each of the following halogen compounds whose hydrogen NMR spectrum consists of a singlet with the indicated chemical shift.

(a) $C_2H_3Cl_3$; $\delta = 2.7$ (b) $C_2H_4Cl_2$; $\delta = 3.7$ (c) C_4H_9Br; $\delta = 1.8$ (d) $C_3H_6Br_2$; $\delta = 2.6$

Answers:

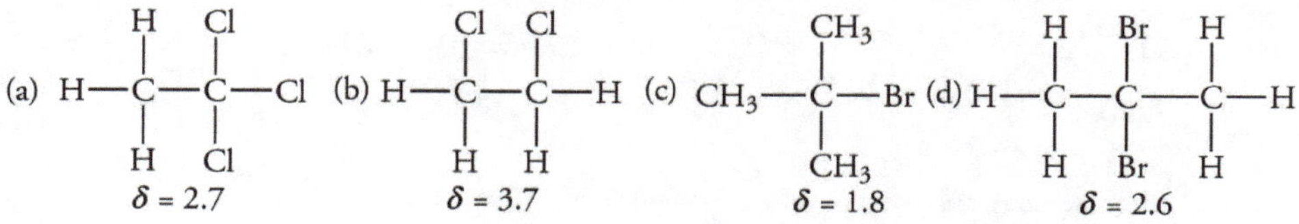

(a)
$$H-\overset{\overset{\displaystyle H}{|}}{\underset{\underset{\displaystyle H}{|}}{C}}-\overset{\overset{\displaystyle Cl}{|}}{\underset{\underset{\displaystyle Cl}{|}}{C}}-Cl$$
$\delta = 2.7$

(b)
$$H-\overset{\overset{\displaystyle Cl}{|}}{\underset{\underset{\displaystyle H}{|}}{C}}-\overset{\overset{\displaystyle Cl}{|}}{\underset{\underset{\displaystyle H}{|}}{C}}-H$$
$\delta = 3.7$

(c)
$$CH_3-\overset{\overset{\displaystyle CH_3}{|}}{\underset{\underset{\displaystyle CH_3}{|}}{C}}-Br$$
$\delta = 1.8$

(d)
$$H-\overset{\overset{\displaystyle H}{|}}{\underset{\underset{\displaystyle H}{|}}{C}}-\overset{\overset{\displaystyle Br}{|}}{\underset{\underset{\displaystyle Br}{|}}{C}}-\overset{\overset{\displaystyle H}{|}}{\underset{\underset{\displaystyle H}{|}}{C}}-H$$
$\delta = 2.6$

14.17 The hydrogen NMR spectrum of [18]annulene consists of signals at $\delta = 8.8$ ppm and $\delta = -1.9$ ppm. The negative value of δ corresponds to an "unusual" chemical shift that is upfield from TMS. The ratio of intensities of the 8.8 ppm to -1.9 ppm resonances is 2:1. Explain these data.

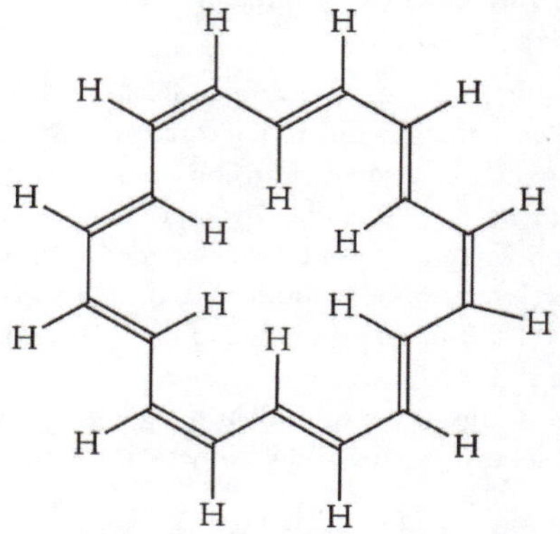

[18]annulene

Answer: The local magnetic field generated by the electrons of the ring is similar to that shown for benzene in Figure 14.12. The 12 hydrogen atoms extending away from the ring are deshielded, and their resonance occurs at 8.8 δ. There are also 6 hydrogen atoms inside the ring. The magnetic field experienced by these hydrogen atoms is reversed, and they are shielded. Thus, the resonance occur at high field. The negative value indicates that the resonance is at higher field than TMS.

14.18 The hydrogen NMR spectrum of [14]annulene consists of signals at $\delta = 7.8$ ppm and $\delta = -0.6$ ppm. Assign the resonances and predict the relative intensities of each.

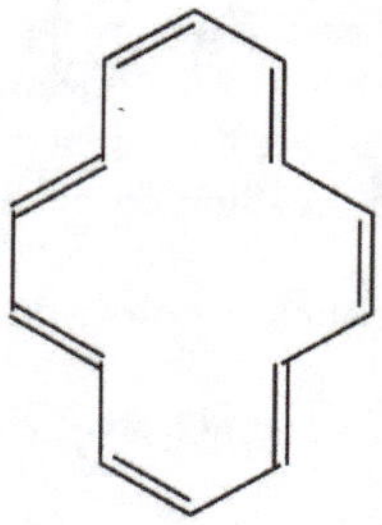

[14]annulene

Answer: The 10 hydrogen atoms extending away from the ring are deshielded. Their resonance occurs at 7.8 δ. There are also 4 hydrogen atoms inside of the ring. The magnetic field experienced by these hydrogen atoms is reversed, and they are shielded. They are shielded, and their resonance occurs at -0.6 δ. The ratio of the low-field to high-field resonances is 5:2.

Multiplicity and Structure

14.19 Describe the multiplicity of each of the signals corresponding to a set of equivalent hydrogen atoms in each of the following ethers.

(a) $CH_3CH_2OCH_2CH_3$ (b) $CH_3OCH(CH_3)_2$ (c) $ClCH_2OCHClCH_3$ (d) $C_{\text{-}2}CHOCHClCHCl_2$

Answers: (a) The resonance of the six hydrogen atoms of the two equivalent methyl groups is a triplet, and the resonance of the four hydrogen atoms of the two equivalent methylene groups is a quartet.

(b) The resonance of the hydrogen atoms of the methyl group bonded to the oxygen atom is a singlet. The resonance of the hydrogen atom of the other carbon atom bonded to the oxygen atom is a heptet. The resonance of the hydrogen atoms of the two equivalent methyl groups is a doublet.

(c) The resonance of the hydrogen atoms of the methylene group bonded to a chlorine atom is a singlet. The resonance of the hydrogen atom of the other carbon atom bearing a chlorine atom is a quartet. The resonance of the hydrogen atoms of the methyl group is a doublet.

(d) The resonance of the hydrogen atom at the carbon atom bearing two chlorine atoms and an oxygen atom is a singlet. The resonance of the hydrogen atom of the carbon atom bearing an oxygen atom and one chlorine atom is a doublet. The resonance of the third hydrogen atom is a doublet.

14.20 Describe the multiplicity of the lowest field resonance of each of the following alkyl halides.

(a) 1-chloropentane (b) 1-chloro-2,2-dimethylpropane (c) 3-chloropentane (d) 1-chloro-2-methyl-2-butene

Answers: The lowest field resonance in each case is for hydrogen atom(s) at the carbon atom that is bonded to the chlorine atom.

(a) triplet, split by the two hydrogen atoms on C-2
(b) singlet, because there are no hydrogen atoms on C-2
(c) quintet, split by the four equivalent hydrogen atoms at C-2 and C-4
(d) singlet, because there are no hydrogen atoms on C-2

14.21 The chemical shifts of the C-1, C-2, and C-3 hydrogen atoms of 1,1,2-trichloropropane are 5.50, 4.22, and 1.20 ppm. The coupling constant of the C-2 and C-3 hydrogen atoms is 6.5 Hz and that of the C-2 and C-1 hydrogen atoms is 4.5 Hz. Draw the splitting diagram for the C-2 hydrogen atom.

Answer: The resonance for the hydrogen atom at C-2 is split into a quartet by the three hydrogen atoms at C-3 and is further split into doublets by the hydrogen atom at C-1.

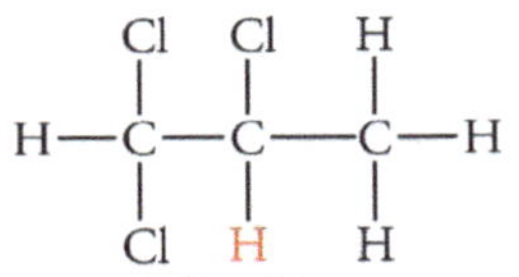

$\delta = 4.2$

14.22   Assume that the coupling constants for three nonequivalent hydrogen identified as H_a, H_b, and $H_{a,b}$ are $J_{a,b}$ = 6 Hz, $J_{a,c}$ =2 Hz, and $J_{b,c}$ = 6 Hz. Draw the splitting diagram for H_b. What is the appearance of this resonance?

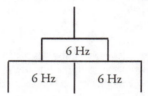

Answer: Because H_b is split by both H_a and H_c, its resonance should be a doublet of doublets. However, the coupling constants $J_{a,b}$ and $J_{b,c}$ are both 6 Hz. As a result, the inner lines of each doublet in the doublet of doublets overlap, so the resonance of H_b appears as a triplet.

14.23   Hydrogen bromide adds to 3-bromopropene under certain experimental conditions to give a compound whose NMR spectrum is a quintet at 2.10 δ and a triplet at 3.60 δ. The ratio of the total intensity of the quintet to that of the triplet is 1:2. What is the structure of the compound?

quintet

triplet - - - - - H   H   H

Br—C—C—C—Br

H   H   H

1,3-dibromopropane

Answer: The product is 1,3-dibromopropane. The resonance of the C-2 hydrogen atoms appears as a quintet. The resonance of the hydrogen atoms at C-1 and C-3 is a triplet.

14.24   The spectrum of a compound with molecular formula $C_3H_3Cl_5$ consists of a triplet at 4.5 δ and a doublet at 6.0 δ. The intensity ratio of the high-field to low-field signal is 1:2. What is the structure of the compound?

1H triplet

Cl   H   Cl

Cl—C—C—C—Cl

2H doublet - - - - - H   Cl   H

1,1,2,3,3-pentachloropropane

Answer: 1,1,2,3,3-Pentachloropropane has equivalent hydrogen atoms at C-1 and C-3 that give the low-field resonance which is split into a doublet by the hydrogen atom at C-2. The higher field resonance is due to the hydrogen atom at C-2, which is split into a triplet by the hydrogen atoms at C-1 and C-3. Note that 1,1,1,2,3-pentachloropropane also has two equivalent hydrogen atoms at C-3 another nonequivalent hydrogen atom at C-2. However, the low-field resonance would be for a single hydrogen atom at C-2, and it would be a triplet. The high-field resonance would be due to two hydrogen atoms at C-1, and it would be a doublet. These resonances are just the opposite of those in the spectrum described.

1H triplet

Cl   H   H

Cl—C—C—C—Cl

Cl   Cl   H - - - - - 2H doublet

1,1,1,2,3-pentachloropropane

Analysis of Spectra

14.25   Determine the structure of the compound corresponding to each of the following hydrogen NMR spectra.

(a) $C_3H_6Cl_2$

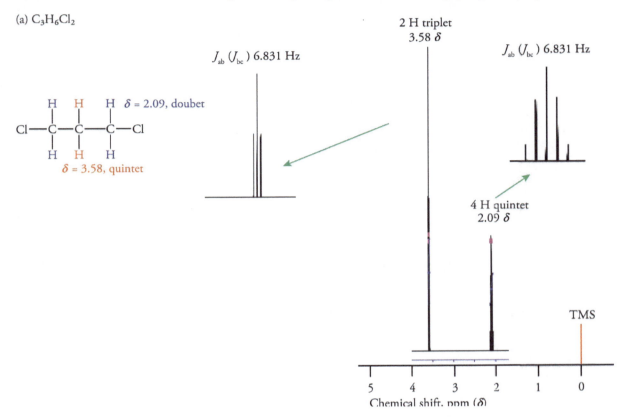

J_{ab} (J_{bc}) 6.831 Hz

2 H triplet
3.58 δ

J_{ab} (J_{bc}) 6.831 Hz

H   H   H δ = 2.09, doubet

Cl—C—C—C—Cl

H   H   H

δ = 3.58, quintet

4 H quintet
2.09 δ

TMS

5   4   3   2   1   0
Chemical shift, ppm (δ)

(b) C_4H_9Cl

δ = 3.52, 2H doubet H   CH₃ δ = 0.981, 6 H doublet

Cl—C—C—CH₃

H   H

δ = 1.88,
9 line multiplicity

J_{ab} = 6.598 Hz

1 H 9 line multiplicity
1.88 δ

1.00

The low-field resonance corresponds to the two hydrogen atoms at C-1. The nine line multiplet at 1.88 δ corresponds to the hydrogen atom at C-2, which is split by eight hydrogen atoms. However, the intensity of two of the "outer" lines is so small relative to the other seven lines that they are barely visible. The high-field doublet corresponds to the six hydrogen atoms in the two methyl groups.

J_{bc} = 6.351 Hz

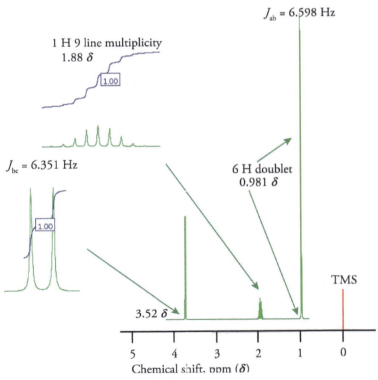

6 H doublet
0.981 δ

1.00

TMS

3.52 δ

5   4   3   2   1   0
Chemical shift, ppm (δ)

227

ALCOHOLS: REACTIONS AND SYNTHESIS

15.1 Overview Alcohol Reactions

Alcohols can react in several ways that differ in the number and type of bonds cleaved. These are:

1. Cleavage of the oxygen–hydrogen bond
2. Cleavage of the carbon–oxygen bond
3. Cleavage of the carbon–oxygen bond as well as the carbon–hydrogen bond at the carbon atom adjacent to the carbon atom bearing the hydroxyl group.
4. Cleavage of the oxygen–hydrogen bond as well as the carbon–hydrogen bond at the carbon atom bearing the hydroxyl group.

15.2 Converting Alcohols to Esters

Esters formed by the reaction of an alcohol with either an inorganic acid or a carboxylic acid have one feature in common. In both reactions, the oxygen atom bridging the alcohol and acid fragments is derived from the alcohol. The esterification reaction occurs by a substitution reaction mechanism in which a nucleophile attacks the carbonyl carbon atom to give a tetrahedral intermediate that subsequently ejects a leaving group. This process, known as nucleophilic acyl substitution, is another of the limited number of important mechanisms that dominate organic chemistry. We will meet this mechanism again in later chapters when we discuss the chemistry of acids and acid derivatives.

Hydroxide ion is a poor leaving group, and ester synthesis proceeds much more easily with a better leaving group such as the chloride ion. Thus, the formation of esters by reaction of alcohols with acid chlorides is a more favorable process than reaction of alcohols with the acids themselves.

15.3 Converting Alcohols to Haloalkanes

The conversion of an alcohol to a haloalkane can be done with thionyl chloride or phosphorus tribromide. In both cases, an intermediate is formed that converts the hydroxyl group into a better leaving group.

The reaction mechanism of the reaction of an alcohol with thionyl chloride occurs through a chlorosulfite intermediate. Subsequent displacement by chloride ion can occur with retention or inversion of configuration depending on the solvent. In pyridine as solvent, inversion occurs. In dioxane, retention occurs via a solvated ion pair, in which the chloride attacks by an internal return mechanism.

The reaction mechanism of the reaction of an alcohol with phosphorus tribromide occurs through a phosphite ester. The oxygen atom is now bound to phosphorus and as such is a better leaving group than the hydroxide ion. Nucleophilic substitution by bromide gives an alkyl bromide.

15.4 Oxidation of Alcohols

Alcohols are oxidized to carbonyl compounds by chromium(VI) compounds. The products depend on the structure of the substrate and the specific chromium reagent.

Alcohols are oxidized by the Jones reagent (CrO_3 in H_2SO_4). Primary alcohols are oxidized to aldehydes, which are further oxidized to carboxylic acids under the reaction conditions. The aldehyde usually cannot be isolated. Secondary alcohols are oxidized to ketones. Tertiary alcohols are not oxidized. Pyridinium chlorochromate (PCC), a reagent generated from CrO_3 and pyridine in CH_2Cl_2 as solvent, oxidizes alcohols only to carbonyl compounds. Secondary alcohols are oxidized to ketones, but primary alcohols are converted into aldehydes without being further oxidized to carboxylic acids.

15.5 Reactions of Vicinal Diols

Vicinal diols are oxidized by periodic acid, HIO_4, to give aldehydes if the two hydroxyl groups are secondary and ketones if they are tertiary. If one of the -OH groups is primary, the products are formic acid and an aldehyde.

15.6 Synthesis of Alcohols

Alcohols can be prepared by reduction of carbonyl compounds. Aldehydes yield primary alcohols; ketones yield secondary alcohols. The reduction of alcohols by hydrogen gas with a transition metal catalyst occurs more slowly than the reduction of alkenes, so both carbonyl groups and double bonds are reduced in compounds that contain both functional groups. Lithium aluminum hydride and sodium borohydride both reduce carbonyl compounds to alcohols without affecting carbon–carbon double bonds.

Alcohols with more complex hydrocarbon structures can be made by alkylation methods using a carbonyl compound and a Grignard reagent, which is carbanion-like reagent. This material is discussed in Section 15.10 and we will consider it again in Chapter 17.

15.7 Synthesis of Alcohols From Haloalkanes

The substitution of a halide ion by a hydroxide ion occurs in competition with an elimination reaction. We can circumvent this problem by using an oxygen-containing derivative that is a weak base as the nucleophile. The acetate ion is such a species. The product of the reaction of a haloalkane with acetate is an ester. Subsequent hydrolysis of the ester gives the alcohol and acetate ion. The hydrolysis occurs by nucleophilic acyl substitution.

15.8 Indirect Hydration Methods: Oxymercuration–Demercuration and Hydroboration–Oxidation

Direct hydration of an alkene to give an alcohol is limited by both rearrangement and the reversibility of the reaction. Two indirect methods for hydration are given in this section, each with different regioselectivity.

The first step in the conversion of an alkene to an alcohol by oxymercuration–demercuration resembles the addition of bromine in a mechanistic sense. Oxymercuration proceeds through a cyclic mercurinium ion. It does not rearrange because much of the positive charge is on the mercury atom. Subsequent attack by a nucleophile, water in this case, gives an oxymercury addition product. Treating the oxymercury adduct with sodium borohydride leads to the alcohol. The entire sequence, known as oxymercuration–demercuration, results in addition of water with the same regiospecificity as the direct hydration reaction. Thus, the process is a Markovnikov addition reaction.

Hydroboration–oxidation of an alkene gives an alcohol in a two-step sequence that is equivalent to an anti-Markovnikov addition of water. Hydroboration occurs by the addition of an H—B bond across the double bond by a cyclic four-center mechanism. The hydrogen atom that adds is not a proton but has the character of a hydride anion, and the boron is the "positive" part of the reagent. Thus, in a sense, the boron adds to the same carbon atom that adds a proton in the addition of reagents such as HBr.

The stereochemistry of the addition of an H—B bond across the double bond is *syn*. The subsequent replacement of the boron atom by oxygen using basic hydrogen peroxide occurs with retention of configuration.

15.9 Reduction of Carbonyl Compounds with Metal Hydrides

Both lithium aluminum hydride and sodium borohydride can reduce carbonyl compounds to alcohols without affecting carbon–carbon double bonds.

The mechanism of reduction of carbonyl compounds with hydride reagents occurs by attack of a nucleophilic hydride anion whose source is either borohydride or aluminum hydride on the carbonyl carbon. Lithium aluminum hydride must be used in an aprotic solvent such as ether to avoid its reactions with acidic hydrogens. Thus, this reagent shouldn't be used for reduction of carbonyl compounds that have hydroxyl groups or other acidic protons. Sodium borohydride can be used in protic solvents such as alcohols, and in fact, it is necessary to do so. The proton of the alcohol solvent is transferred to the oxygen atom, giving the alcohol product.

Lithium aluminum hydride and sodium borohydride have different reactivities: Lithium aluminum hydride reduces esters to alcohols; sodium borohydride does not.

15.10 Grignard Reagent

The Grignard reagent is a highly reactive organomagnesium compound formed by reacting a haloalkane with magnesium in an ether solvent. The carbon atom of a Grignard reagent has a partial negative charge. The Grignard reagent is a versatile material that can be used to form new carbon–carbon bonds. It acts as a nucleophile and attacks the carbonyl carbon atom to give an alkoxide which forms a salt with $(MgBr)^+$. Hydrolysis of the magnesium bromide salt gives the alcohol.

To determine how to combine two molecules to give an alcohol in a Grignard synthesis, examine the substituents bonded to the carbon bearing the hydroxyl group. One component of the new compound can come from the Grignard reagent. The other component must have been present in a carbonyl compound.

1. Primary alcohols are made from a Grignard reagent and formaldehyde.
2. Secondary alcohols can be made by reacting a Grignard reagent with an aldehyde.
3. Tertiary alcohols can be made by reacting a Grignard reagent with a ketone.

The use of the Grignard reagent is precluded if there is an acidic hydrogen in the substrate selected to react with the Grignard reagent. This acidic hydrogen destroys the Grignard reagent before it adds to the carbonyl group. If the substrate has a hydroxyl group, it can be protected by forming a trimethylsilyl ether.

15.11 Thiols

Thiols contain a sulfhydryl group (—SH). Thiols, also called mercaptans, have significantly different physical and chemical properties than alcohols. Thiols are lower boiling compounds because the —SH group does not form intramolecular hydrogen bonds. Thiols are stronger acids than alcohols. Like alcohols, thiols can be synthesized by displacement of a halide ion from haloalkanes. The SH^- ion is an excellent nucleophile. Oxidation of thiols produces disulfide bonds rather than analogs of aldehydes and ketones.

Summary of Reactions

1. Formation of Esters

$$CH_3-\overset{\overset{\displaystyle O}{\|}}{C}-Cl \quad + \quad H-O-CH_2CH_3 \longrightarrow CH_3-\overset{\overset{\displaystyle O}{\|}}{C}-O-CH_2CH_3 + \quad HCl$$

ethanoyl chloride   ethanol   ethyl ethanoate
(acetyl chloride)   (ethyl acetate)

2. Synthesis of Alkyl Halides

$$CH_3(CH_2)_4CH_2OH + SOCl_2 \longrightarrow CH_3(CH_2)_4CH_2Cl \; + \; HCl(g) \; + \; SO_2(g)$$

$$3\,R-OH \; + PBr_3 \longrightarrow R-Br \; + H_3PO_3$$

3. Oxidation of Alcohols

$$CH_3(CH_2)_4CH_2OH \xrightarrow{\text{Jones reagent}} CH_3(CH_2)_4CH_2CO_2H$$

4. Pinacol Rearrangement

2,3-dimethyl-2,3-butanediol
(pinnacol)

Summary of Reactions

1. Formation of Esters

CH₃—C(=O)—Cl + H—O—CH₂CH₃ ⟶ CH₃—C(=O)—O—CH₂CH₃ + HCl

ethanoyl chloride ethanol ethyl ethanoate
(acetyl chloride) (ethyl acetate)

+ $\xrightarrow{\text{pyridine}}$

2. Synthesis of Alkyl Halides

$CH_3(CH_2)_4CH_2OH + SOCl_2 \longrightarrow CH_3(CH_2)_4CH_2Cl + HCl(g) + SO_2(g)$

—CH₂OH + SOCl₂ ⟶ —CH₂Cl + HCl(g) + SO₂(g)

$3\,R—OH + PBr_3 \longrightarrow R—Br + H_3PO_3$

3 —CH₂OH + PBr₃ ⟶ 3 —CH₂Br + H₃PO₃

3. Oxidation of Alcohols

$CH_3(CH_2)_4CH_2OH \xrightarrow{\text{Jones reagent}} CH_3(CH_2)_4CH_2CO_2H$

—CH₂OH $\xrightarrow{\text{PCC}}$ —C(=O)H

—OH $\xrightarrow{\text{Jones reagent}}$ =O

4. Pinacol Rearrangement

CH₃—C(OH)(CH₃)—C(OH)(CH₃)—CH₃ $\xrightarrow{H_2SO_4}$ CH₃—C(CH₃)₂—C(=O)—CH₃ + H₂O

2,3-dimethyl-2,3-butanediol
(pinnacol)

5. Oxidative Cleavage of Vicinal Diols

2-methyl-1-phenyl-1,2-butanediol → benzaldehyde + 2-butanone

cis-1,2-cyclohexanediol → 6-oxohexanal

6. Synthesis of Alcohols from Haloalkanes

7. Synthesis of Alcohols From Alkenes

A. Oxymercuration–Demercuration

B. Hydroboration–Oxidation

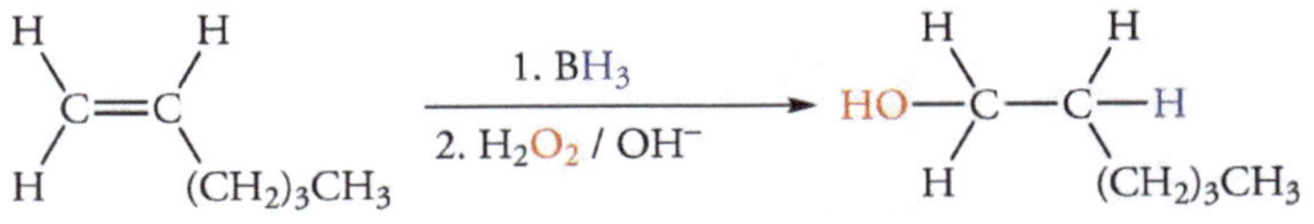

$$\text{H}_2\text{C}=\text{CH}-(\text{CH}_2)_3\text{CH}_3 \xrightarrow[\text{2. H}_2\text{O}_2 \,/\, \text{OH}^-]{\text{1. BH}_3} \text{HO}-\text{CH}_2-\text{CH}_2-(\text{CH}_2)_3\text{CH}_3$$

anti-Markovnikov product

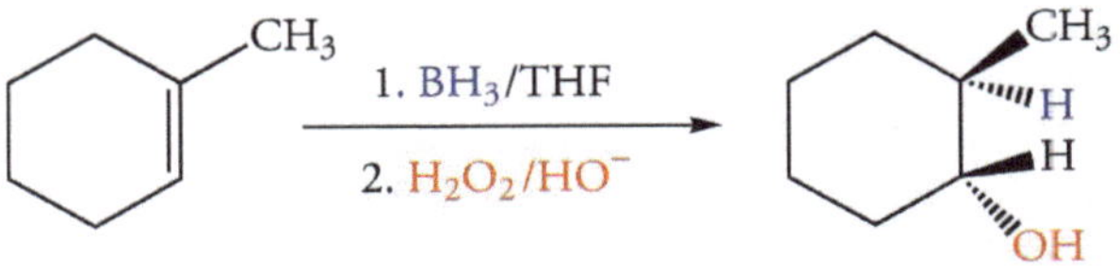

1-methylcyclohexene $\xrightarrow[\text{2. H}_2\text{O}_2/\text{HO}^-]{\text{1. BH}_3/\text{THF}}$ (trans-2-methylcyclohexanol)

8. Reduction of Carbonyl Compounds

A. Catalytic Hydrogenation

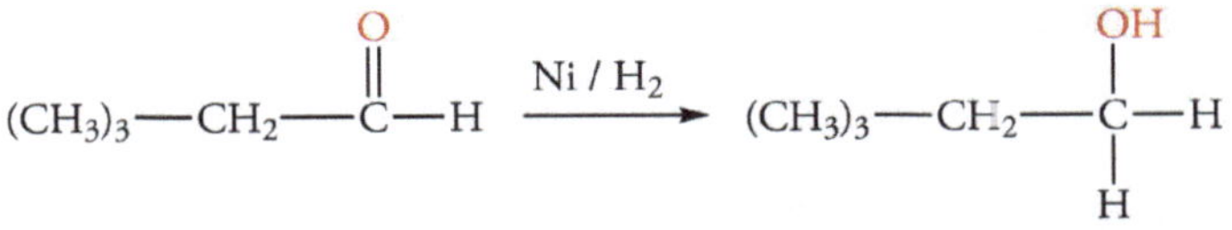

$$(\text{CH}_3)_3-\text{CH}_2-\overset{\text{O}}{\overset{\|}{\text{C}}}-\text{H} \xrightarrow{\text{Ni} \,/\, \text{H}_2} (\text{CH}_3)_3-\text{CH}_2-\overset{\text{OH}}{\underset{\text{H}}{\overset{|}{\text{C}}}}-\text{H}$$

3,3-dimethylbutanal — 3,3-dimethylbutanol

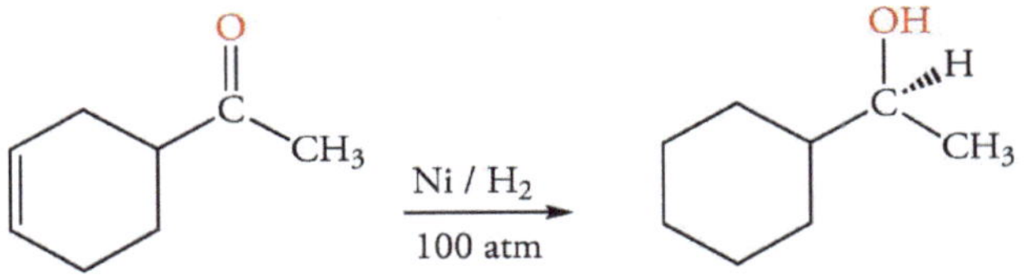

(cyclohexenyl methyl ketone) $\xrightarrow[\text{100 atm}]{\text{Ni} \,/\, \text{H}_2}$ (1-cyclohexylethanol)

At high pressure, both the vinyl and the carbonyl groups are reduced.

B. Sodium Borohydride Reduction

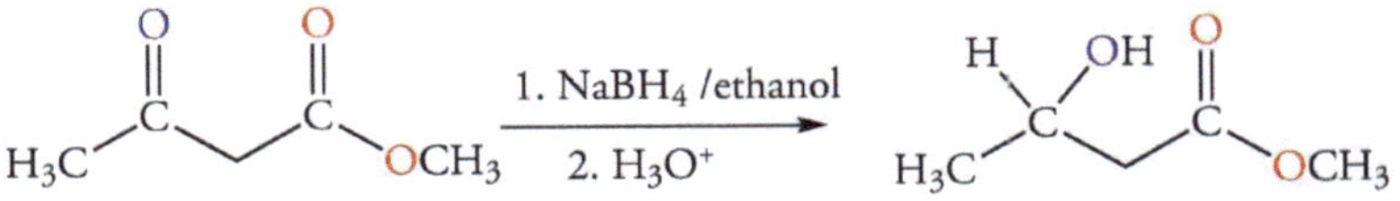

$$\text{H}_3\text{C}-\overset{\text{O}}{\overset{\|}{\text{C}}}-\text{CH}_2-\overset{\text{O}}{\overset{\|}{\text{C}}}-\text{OCH}_3 \xrightarrow[\text{2. H}_3\text{O}^+]{\text{1. NaBH}_4 \,/\text{ethanol}} \text{H}_3\text{C}-\overset{\text{OH}}{\underset{\text{H}}{\overset{|}{\text{C}}}}-\text{CH}_2-\overset{\text{O}}{\overset{\|}{\text{C}}}-\text{OCH}_3$$

Only the ketone is reduced.

C. Lithium Aluminum Hydride Reduction

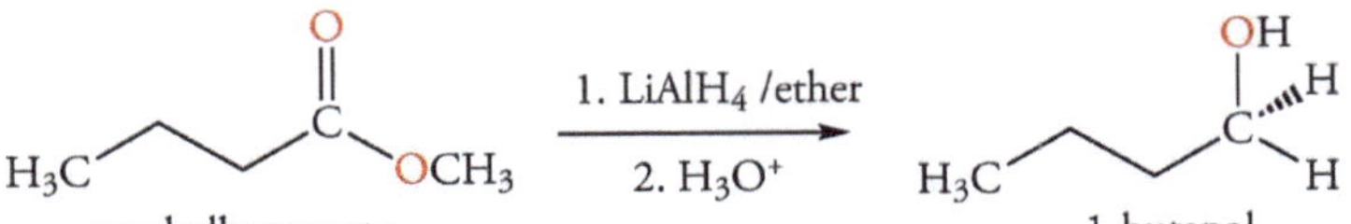

$$\text{H}_3\text{C}-\text{CH}_2-\text{CH}_2-\overset{\text{O}}{\overset{\|}{\text{C}}}-\text{OCH}_3 \xrightarrow[\text{2. H}_3\text{O}^+]{\text{1. LiAlH}_4 \,/\text{ether}} \text{H}_3\text{C}-\text{CH}_2-\text{CH}_2-\overset{\text{OH}}{\underset{\text{H}}{\overset{|}{\text{C}}}}-\text{H}$$

methylbutanoate — 1-butanol

9. Synthesis of Alcohols Using Grignard Reagents

$$\text{(structure with Br)} \xrightarrow[\substack{\text{2. HCHO} \\ \text{3. } H_3O^+}]{\text{1. Mg(s) / ether}} \text{(structure with } CH_2OH\text{)}$$

$$\text{(structure with Br)} \xrightarrow[\substack{\text{2. } C_6H_5CHO \\ \text{3. } H_3O^+}]{\text{1. Mg(s) / ether}} \text{(product structure with H, OH)}$$

 **End of Chapter Exercises**

Formation of Esters

15.1   Write the structural formula of each of the following esters.
    (a) ethyl sulfate   (b) dimethyl phosphate   (c) propyl nitrate (d) 2-propylmethanesulfonate

Answers:

(a) $CH_3CH_2-\overset{\cdot\cdot}{\underset{\cdot\cdot}{O}}-\overset{\overset{:O:}{\|}}{\underset{\underset{:O:}{\|}}{S}}-\overset{\cdot\cdot}{\underset{\cdot\cdot}{O}}-H$ (b) $CH_3-\overset{\cdot\cdot}{\underset{\cdot\cdot}{O}}-\overset{\overset{:O:}{\|}}{\underset{\underset{:O:}{\|}}{P}}-\overset{\cdot\cdot}{\underset{\cdot\cdot}{O}}-CH_3$

        ethyl sulfate   dimethyl phosphate

(c) $CH_3CH_2CH_2-\overset{\cdot\cdot}{\underset{\cdot\cdot}{O}}-\overset{\overset{:O:}{\|}}{N}-\overset{\cdot\cdot}{\underset{\cdot\cdot}{O}}:$ (d) $CH_3-\overset{\overset{:O:}{\|}}{\underset{\underset{:O:}{\|}}{S}}-\overset{\cdot\cdot}{\underset{\cdot\cdot}{O}}-CH(CH_3)_2$

         propyl nitrate   2-propylmethanesulfonate

15.2   Write the structural formula of each of the following esters.
    (a) trimethyl phosphate   (b) dipropyl sulfate   (c) 2-propyl nitrate   (d) l-butyl-*p*-toluenesulfonate

Answers:

(a) $CH_3-\overset{\cdot\cdot}{\underset{\cdot\cdot}{O}}-\overset{\overset{:O-CH_3}{|}}{\underset{\underset{:O:}{\|}}{P}}-\overset{\cdot\cdot}{\underset{\cdot\cdot}{O}}-CH_3$ (b) $CH_3CH_2CH_2-\overset{\cdot\cdot}{\underset{\cdot\cdot}{O}}-\overset{\overset{:O:}{\|}}{\underset{\underset{:O:}{\|}}{S}}-\overset{\cdot\cdot}{\underset{\cdot\cdot}{O}}-CH_2CH_2CH_3$

       trimethyl phosphate   dipropyl sulfate

(c) $(CH_3)_2CH-\overset{\cdot\cdot}{\underset{\cdot\cdot}{O}}-\overset{\overset{:O:}{\|}}{N}-\overset{\cdot\cdot}{\underset{\cdot\cdot}{O}}:$ (d) $CH_3-\langle\text{benzene ring}\rangle-\overset{\overset{:O:}{\|}}{\underset{\underset{:O:}{\|}}{S}}-\overset{\cdot\cdot}{\underset{\cdot\cdot}{O}}-(CH_2)_3CH_3$

        2-propyl nitrate   1-butyl-*p*-toluenesulfonate

15.3   Oxalyl chloride is a diacid chloride having the following structure. Draw the structure of the related diacid. Draw the structure of the product from reaction of one equivalent of benzyl alcohol with oxalyl chloride. Draw the structure of the product from the reaction of two equivalents of methyl alcohol with oxalyl chloride.

Answers:

$Cl-\overset{\overset{O}{\|}}{C}-\overset{\overset{O}{\|}}{C}-Cl$ (a) $HO-\overset{\overset{O}{\|}}{C}-\overset{\overset{O}{\|}}{C}-OH$ (b) $\langle\text{phenyl}\rangle-CH_2-\overset{\overset{O}{\|}}{C}-\overset{\overset{O}{\|}}{C}-Cl$

    oxalyl chloride   oxalic acid

(c) $CH_3-O-\overset{\overset{O}{\|}}{C}-\overset{\overset{O}{\|}}{C}-O-CH_3$

15.4   What acid is contained in the following diester? Write the structure of an acid chloride that could be used to synthesize the diester.

Answer: The compound is a dimethyl ester of carbonic acid, which is an unstable add. The acid chloride is phosgene.

$CH_3-O-\overset{\overset{O}{\|}}{C}-O-CH_3$ $\overset{\overset{O}{\|}}{\underset{\underset{Cl\quad Cl}{}}{C}}$ $\xrightarrow{CH_3OH}$ $\overset{\overset{O}{\|}}{\underset{\underset{CH_3-O\quad\quad O-CH_3}{}}{C}}$

                             phosgene

15.5 The following diol reacts with one equivalent of tosyl chloride to give a single ester in good yield. Write the structure of the ester. Explain why the reaction is regioselective.

Answer: The ester is formed with the secondary alcohol, because its oxygen atom is a better nucleophile than the oxygen atom of the tertiary alcohol, which is more sterically hindered. The tertiary hydroxyl group does not react because only one equivalent of tosyl chloride is used.

15.6 The following diol reacts with one equivalent of tosyl chloride to give a single ester in good yield. Write the structure of the ester. Explain why the reaction is regioselective.

Answer: The equatorial hydroxyl group in the A ring of the steroid is less sterically hindered than the axial hydroxyl group of the C ring. Esterification occurs at the oxygen of the equatorial hydroxyl group because it is a more effective nucleophile.

Reactivity of Esters

15.7 Are alkyl esters of trifluoromethanesulfonic acid expected to be more or less reactive in S_N1 reactions than alkyl esters of methanesulfonic acid?

methanesulfonic acid   trifluoromethanesulfonic acid

Answer: The leaving group of the trifluoromethanesulfonic acid is a weak conjugate base of the more acidic trifluoromethanesulfonic acid. Weak bases are better leaving groups. Thus, alkyl esters of trifluoromethanesulfonic acid are more reactive than those of methanesulfonic acid because trifluoromethanesulfonic acid is the stronger acid and its conjugate base is the weaker base.

15.8   Describe the expected reactivity of the following compounds in S_N2 reactions compared to methanesulfonate esters.

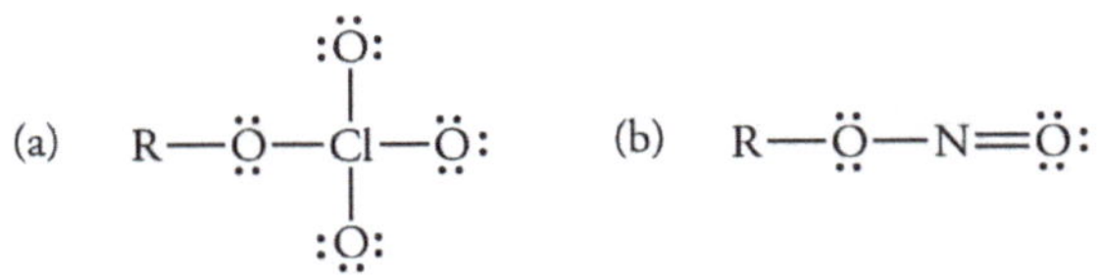

(a)   R—Ö—Cl—Ö: (b)   R—Ö—N=Ö:

Answer: The leaving groups are the perchlorate and nitrite ions, respectively. Perchloric acid is a strong acid and nitrous acid is a weak acid. Methanesulfonic acid is a strong acid. Thus, the perchlorate ion is comparable to the methanesulfonate ion as a leaving group. The nitrite ion is a much poorer leaving group.

15.9   The relative reactivities of alkyl p-nitrobenzoates and alkyl p-nitrobenzenesulfonates in S_N2 reactions relative to the reactivity of alkyl chlorides are 10^{-5} and 10^5, respectively. Explain the difference in the relative rates of reaction of the two esters.

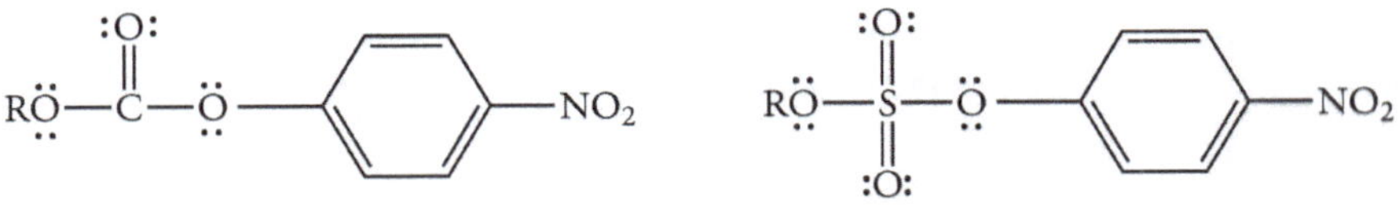

an alkyl p-nitrobenzoate   an alkyl p-nitrobenzenesulfonate

Answer: Benzenesulfonic acids are strong acids, and benzoic acids are weak acids. Thus, esters of benzenesulfonic acids are more reactive because the benzenesulfonate ion is a weaker base than the benzoate ion.

15.10   Predict whether p-bromobenzenesulfonate is a better or worse leaving group than p-toluenesulfonate.

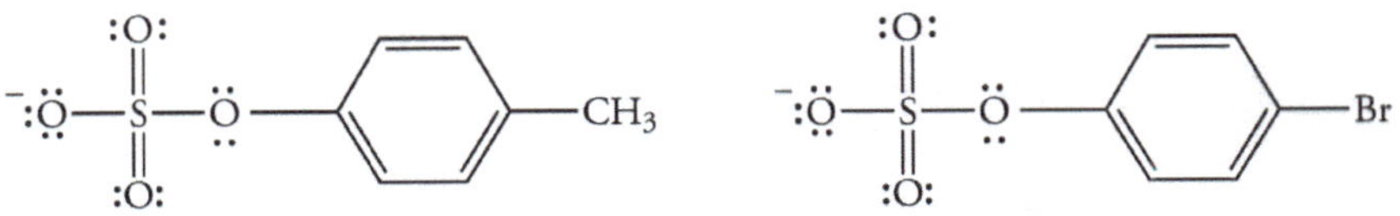

p-toluenesulfonate   p-bromobenzenesulfonate

Answer: Bromine is an electron withdrawing group and makes the p-bromobenzenesulfonate ion a weaker base than p-toluenesulfonate. Thus, the bromine-substituted ion is a better leaving group than the tosylate ion.

Reactions of Alcohols with Acid

15.11   Explain why (R)-2-butanol in aqueous acid gradually loses its optical activity.

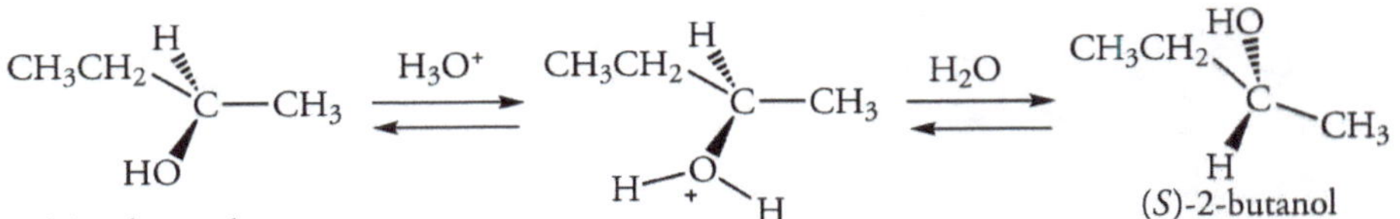

(R)-2-butanol   (S)-2-butanol

Answer: Protonation of the oxygen atom gives an oxonium ion, and water of the solvent acts as a nucleophile in an S_N2 reaction. The result is net inversion of configuration. Eventually, total racemization will occur.

15.12   Explain why l-phenyl-2-propen-1-ol rearranges to an isomer in the presence of a catalytic amount of H_2SO_4.

Answer: Protonation of the oxygen atom gives an oxonium ion which loses water from C-1 to give a resonance-stabilized allyl carbocation. Reaction of water at the original C-3 atom gives an alcohol in which the double bond is conjugated with the aromatic ring.

15.13   *cis*-2-Buten-1-ol isomerizes to form a mixture containing an isomeric alcohol when treated with dilute sulfuric acid. Write the structure of this alcohol.

Answer: Protonation of the oxygen atom gives an oxonium ion, and water can leave at C-1 giving a resonance-stabilized allyl carbocation. Reaction of water at the original C-3 atom gives 1-buten-3-ol.

15.14   When (*S*)-4-methyl-1,4-hexanediol is heated with acid, optically inactive 2-ethyl-2-methyltetrahydrofuran results. Write a mechanism for the reaction that accounts for the formation of the product and its lack of optical activity.

2-ethyl-2-methyltetrahydrofuran

Answer: Protonation of the tertiary alcohol at the C-4 atom and loss of water gives a tertiary carbocation. This carbocation is achiral, and subsequent attack of the oxygen atom of the hydroxyl group at C-1 can occur at either side of the plane to give the observed product, which is racemic.

Formation of Alkyl Halides

15.15 Draw the structure of the product of reaction for each of the following compounds with PBr₃.

Answers:

(a)

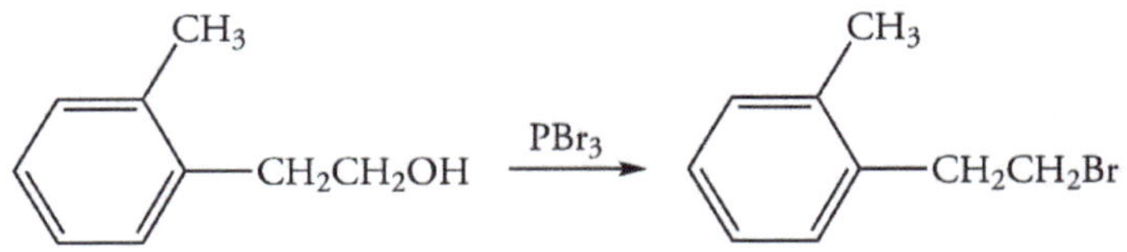

(b) 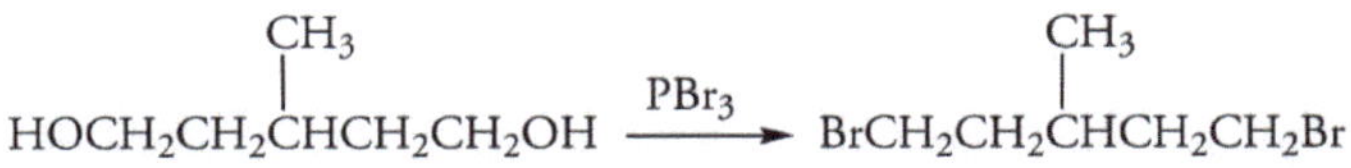 HOCH₂CH₂CHCH₂CH₂OH —PBr₃→ BrCH₂CH₂CHCH₂CH₂Br

(c)

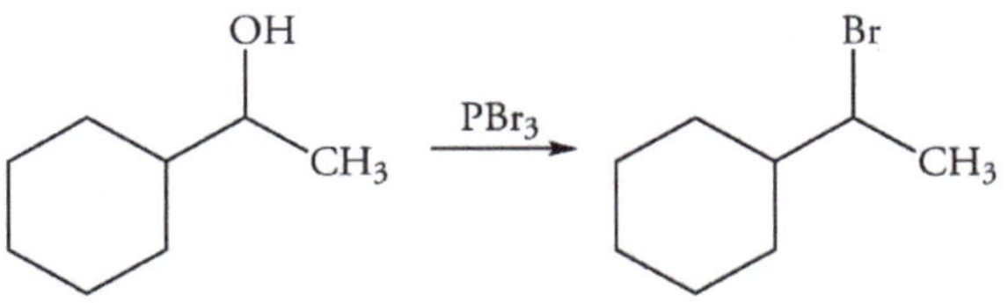

(d)

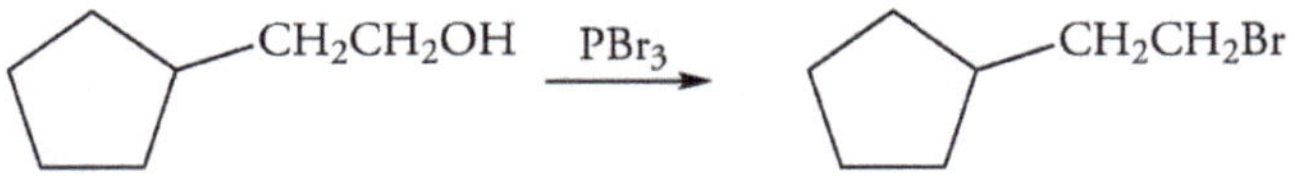

15.16 Draw the structure of the product of the reaction for each of the following compounds with SOCl₂ and pyridine.

Answers:

(a)

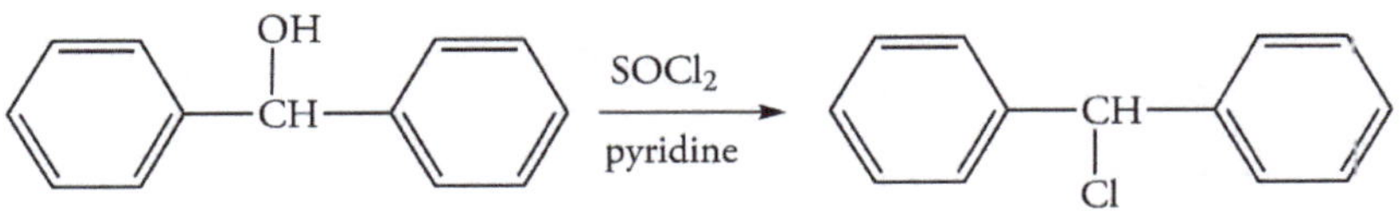

(b) 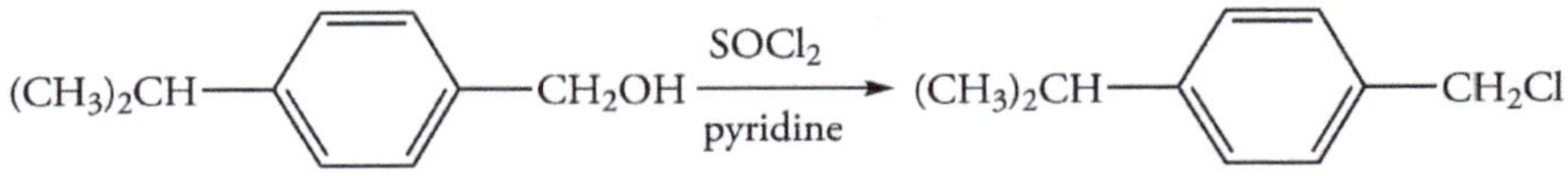 (CH₃)₂CH——CH₂OH —SOCl₂/pyridine→ (CH₃)₂CH——CH₂Cl

(c)

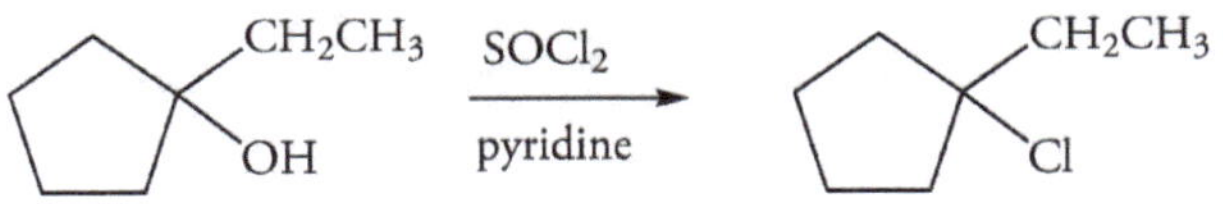

(d)

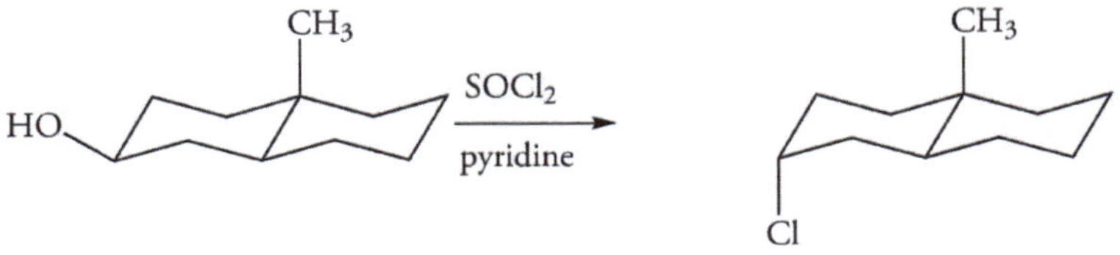

15.17 Both 2-methyl-2-buten-l-ol and 3-methyl-2-buten-1-ol are converted to chlorides using concentrated Hcl. Which compound reacts at the faster rate?

2-methyl-2-buten-1-ol

3-methyl-2-buten-1-ol

Answer: Both compounds give resonance-stabilized allyl carbocations. However, the carbocation derived from 3-methyl-2-buten-1-ol has its positive charge distributed between a primary and a tertiary center. This carbocation is more stable than the carbocation derived from 2-methyl-2-buten-1-ol, which has its positive charge distributed between a primary and a secondary center. Thus, 3-methyl-2-buten-1-ol reacts at a faster rate because the reaction has a lower-energy transition state that resembles the carbocation is formed.

15.18 3-Methyl-3-cyclopentenol reacts with aqueous HBr to yield a mixture of two isomeric bromo compounds. Draw the structures of the two products. Predict the major isomer, assuming the reaction is not reversible. How might the data be different if the products can equilibrate?

major   minor

Answer: The allylic carbocation has its positive charge distributed between a secondary center at the original C-1 atom and a tertiary center at C-3. Under conditions of kinetic control, the bromine atom attacks the more positive center and gives 3-bromo-3-methyl-cyclopentene. If the products can equilibrate, the compound with the more substituted double bond would be favored. That isomer is 1-methyl-3-bromo-cyclopentene.

15.19 The yields of alkyl bromides obtained by reaction of an alcohol with PBr_3 are reduced if some of the HBr formed escapes from the reaction. In what alternate product would the alkyl groups be found under these conditions?

Answer: If the HBr escapes, then the nucleophilic bromide ion required to displace the substituted phosphite ion as the leaving group is lost. Thus, the phosphite ester remains and will regenerate the original alcohol when the reaction mixture is treated with water.

15.20 The yields of alkyl bromides in the reaction of alcohols with PBr_3 are increased if HBr is bubbled into the reaction vessel after the PBr_3 and alcohol are mixed. Explain why.

Answer: The extra HBr provides an added source of nucleophilic bromide ion to displace the substituted phosphite ion as the leaving group.

15.21 How could *trans*-4-*tert*-butylcyclohexanol be converted into *trans*-4-chloro-l-*tert*-butylcyclohexane? How could *trans*-4-*tert*-butylcyclohexanol be converted into *cis*-4-chloro-l-*tert*-butylcyclohexane?

Answer: Use thionyl chloride in dioxane, which gives a substitution product with retention of configuration by an internal return mechanism, to prepare the *trans* isomer. Use thionyl chloride in pyridine to achieve inversion of configuration to prepare the *cis* isomer.

15.22 What is the product of the reaction of (*R*)-2-octanol with thionyl chloride in pyridine? What is the product in diethyl ether as solvent?

Answer: Thionyl chloride in pyridine gives inversion and yields the (*S*) isomer. In diethyl ether, as in dioxane, the (*R*) isomer results by retention of configuration via an internal return mechanism.

15.23 Draw the structure of the product of the following series of reactions. What product would result if the alcohol reacted with HCl?

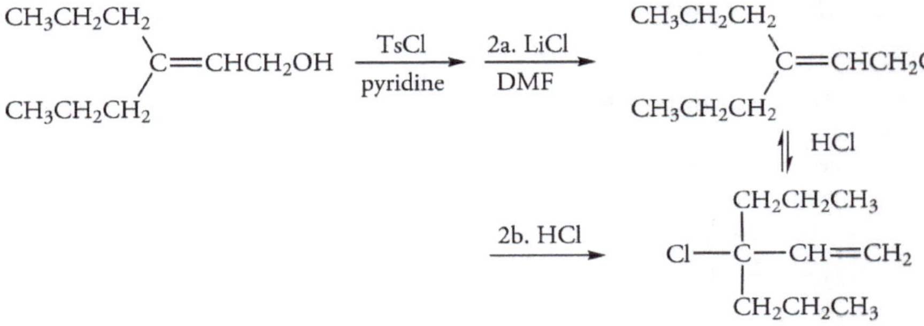

15.24 Draw the structure of the product of the following series of reactions. What product would result if the alcohol reacted with HBr?

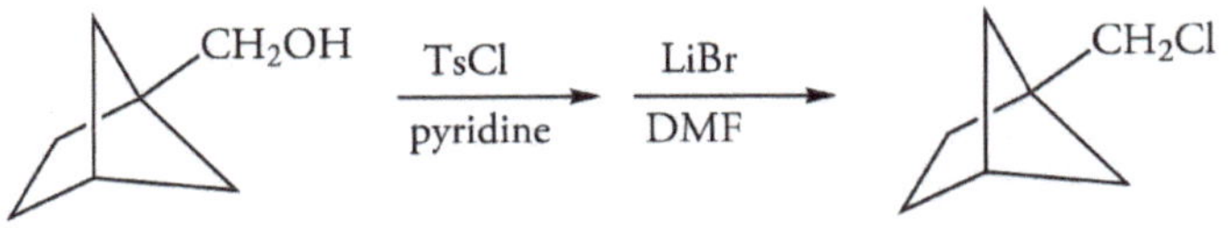

Answer: Formation of a methanesulfonate followed by reaction with the nucleophilic chloride ion in an aprotic solvent tends to give substitution at the original C-1 atom, as shown above. In HCl, a resonance-stabilized carbocation forms which can react with chloride ion at C-1 to give the product shown above, or a rearrangement can occur to give 1-bromobicyclo[3.1.1]heptane.

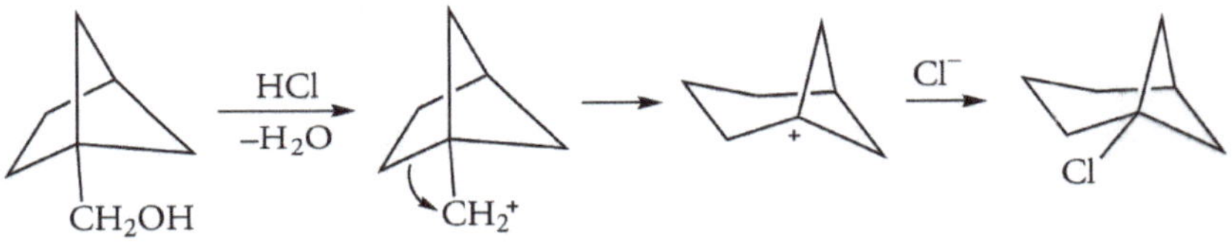

15.25 Sterically hindered alcohols react with phosphorus tribromide but tend to give large quantities of rearranged product. The product mixture obtained from 2,2-dimethyl-1-propanol (neopentyl alcohol) contains 63% 1-bromo-2,2-dimethylpropane, 26% 2-bromo-2-methylbutane, and 11% 2-bromo-3-methylbutane. Explain the origin of the products. Why are sterically hindered alcohols more prone to give rearranged products?

Answer: Sterically hindered alcohols are less likely to undergo S_N2 reactions. The competing S_N1 reaction gives a primary carbocation that can undergo a methide shift, which gives the 2-bromo-2-methylbutane product. Although the resulting carbocation is less stable, a hydride shift of the carbocation resulting from the methide shift gives a secondary carbocation that leads to the 2-bromo-3-methylbutane product.

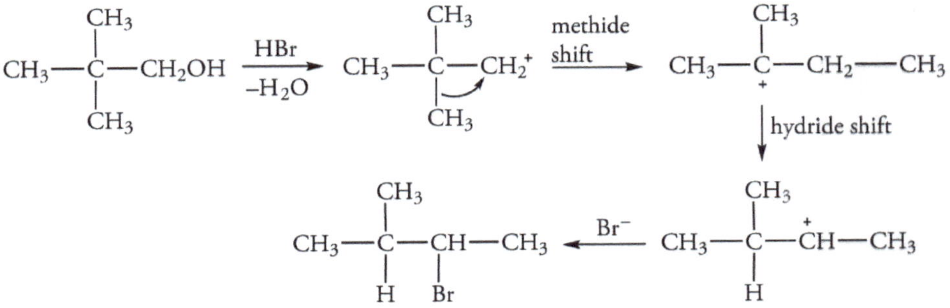

15.26 Both 2-chloropentane and 3-chloropentane are converted to a mixture of the two compounds in a concentrated HCl solution containing zinc chloride. The ratio of the 2-chloro to the 3-chloro compound is 2:1. Write a mechanism that explains how the zinc chloride accounts for the equilibration. Why does the observed ratio occur?

Answer: The zinc chloride acts as a Lewis acid that abstract a chloride ion to form a carbocation.

$$CH_3-\overset{\overset{\displaystyle Cl}{|}}{\underset{\underset{\displaystyle H}{|}}{C}}-\overset{\overset{\displaystyle H}{|}}{\underset{\underset{\displaystyle H}{|}}{C}}-\overset{\overset{\displaystyle H}{|}}{\underset{\underset{\displaystyle H}{|}}{C}}-CH_3 \quad\underset{\displaystyle ZnCl_2}{\rightleftharpoons}\quad CH_3-\overset{+}{\underset{\underset{\displaystyle H}{|}}{C}}-\overset{\overset{\displaystyle H}{|}}{\underset{\underset{\displaystyle H}{|}}{C}}-\overset{\overset{\displaystyle H}{|}}{\underset{\underset{\displaystyle H}{|}}{C}}-CH_3 \quad + ZnCl_3^-$$

Answer: Sterically hindered alcohols are less likely to undergo S_N2 reactions. The competing S_N1 reaction gives a primary carbocation that can undergo a methide shift, which gives the 2-bromo-2-methylbutane product. Although the resulting carbocation is less stable, a hydride shift of the carbocation resulting from the methide shift gives a secondary carbocation that leads to the 2-bromo-3-methylbutane product.

Answer: A hydride shift interconverts secondary carbocations with positive charge at C-2 and C-3. Since there are two equivalent carbocations (I and III) at C-2, and only one at C-3 (II), the ratio of 2-bromobutane to 3-bromobutane is 2:1.

Oxidation of Alcohols

15.27 Both l-octanol and 2-octanol react with aqueous basic potassium permanganate. The product of the reaction of 2-octanol is not soluble in aqueous base, but the product of reaction of l-octanol is soluble. What are the products? Explain the difference in solubility.

Answer: The product from 2-octanol is a ketone, which is insoluble in water because it is a nonpolar compound (even though there is a carbonyl group). The product from 1-octanol is a carboxylic acid, which reacts with base to form a carboxylate ion which is soluble in water.

15.28 Draw the structure of the product of each of the following reactions.

(a) $CH_3CH_2CH_2-C\equiv C-CH_2\overset{\overset{\displaystyle OH}{|}}{C}HCH_3 \xrightarrow[\text{reagent}]{\text{Jones}} CH_3CH_2CH_2-C\equiv C-CH_2\overset{\overset{\displaystyle O}{\|}}{C}CH_3$

(b) $CH_2\!\!=\!\!CH-CH\!\!=\!\!CH-CH_2CH_2OH$

$\downarrow PCC$

$CH_2\!\!=\!\!CH-CH\!\!=\!\!CH-CH_2\overset{\overset{\displaystyle O}{\|}}{C}-H$

(c) $\xrightarrow{PCC}$

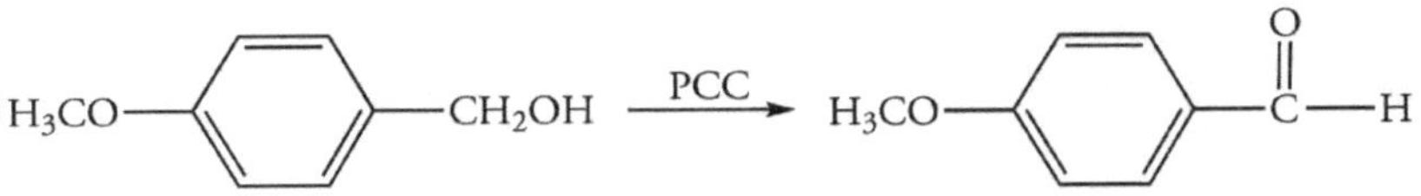

(d) $\xrightarrow{\text{Jones reagent}}$

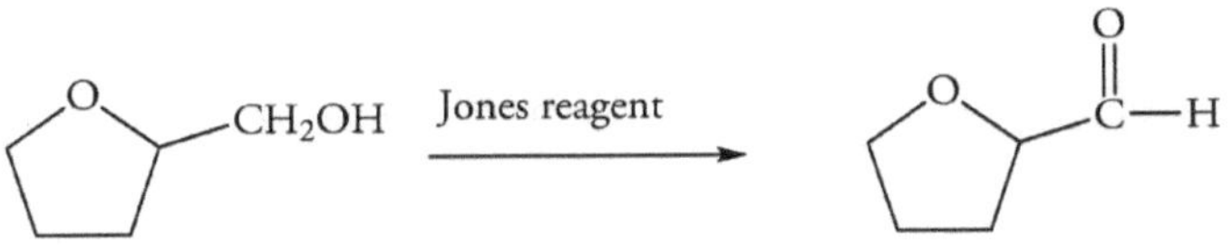

15.29 Write the product formed from the oxidation of each of the compounds in Exercise 15.15 using PCC.

(a) $\xrightarrow{PCC}$

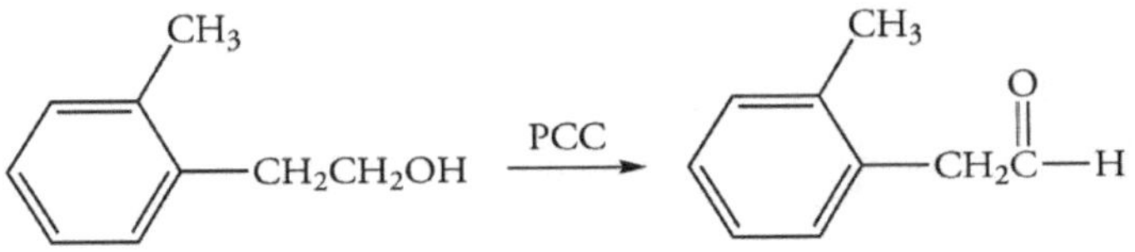

(b) $HOCH_2CH_2\overset{\overset{\displaystyle CH_3}{|}}{C}HCH_2CH_2OH \xrightarrow{PCC} H-\overset{\overset{\displaystyle O}{\|}}{C}CH_2\overset{\overset{\displaystyle CH_3}{|}}{C}HCH_2\overset{\overset{\displaystyle O}{\|}}{C}-H$

(c) $\xrightarrow{PCC}$

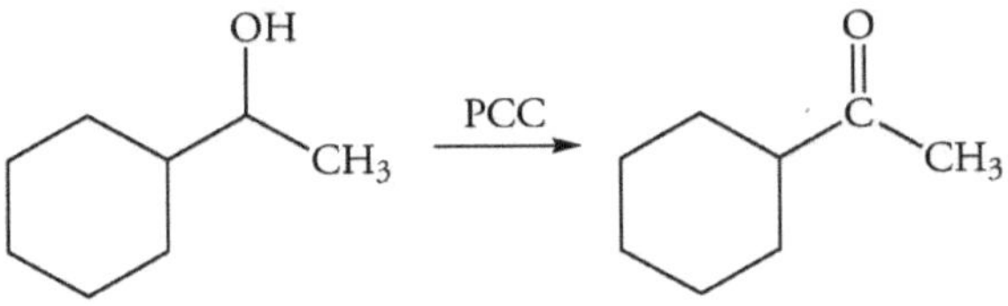

(d) $\xrightarrow{PCC}$

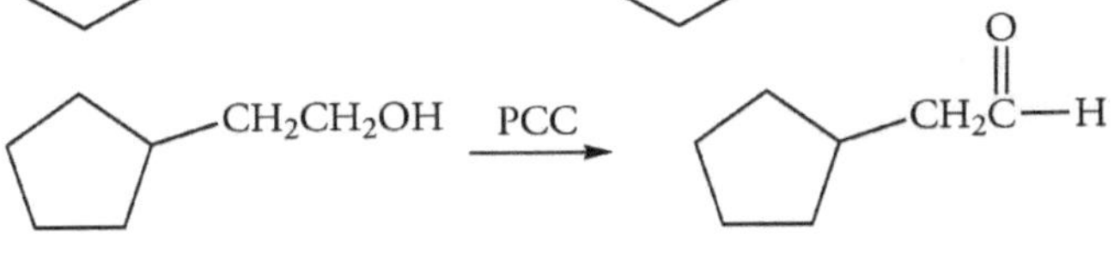

15.30 Write the product formed from the oxidation of each of the compounds in Exercise 15.16 using the Jones reagent.

(a)

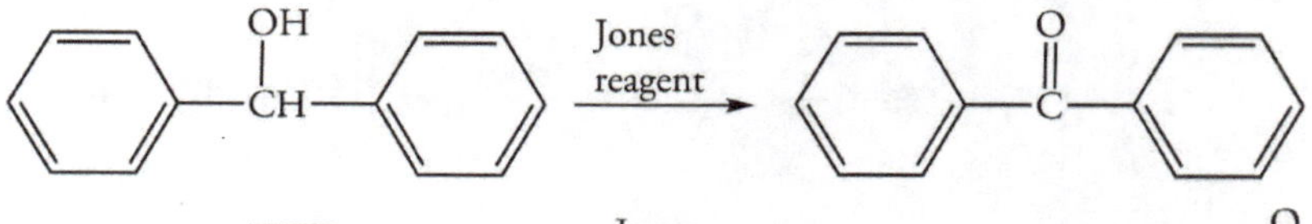

OH
|
CH   →(Jones reagent)→   (diphenyl ketone, C=O)

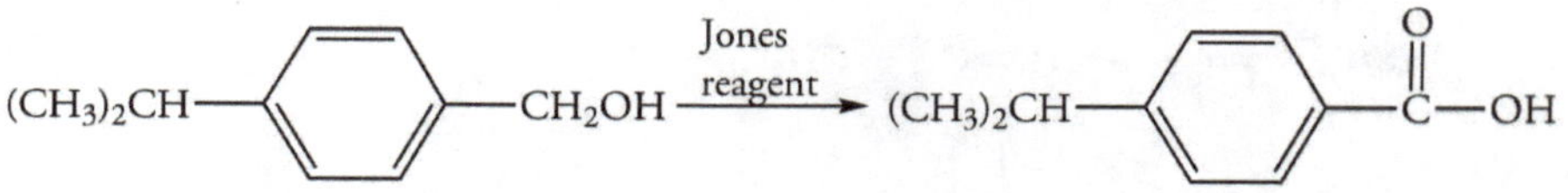

(b) (CH₃)₂CH—⟨benzene ring⟩—CH₂OH —(Jones reagent)→ (CH₃)₂CH—⟨benzene ring⟩—C(=O)—OH

(c)

CH₂CH₃
|
(cyclopentane)—OH   —(Jones reagent)→   tertiary alcohol, no reaction

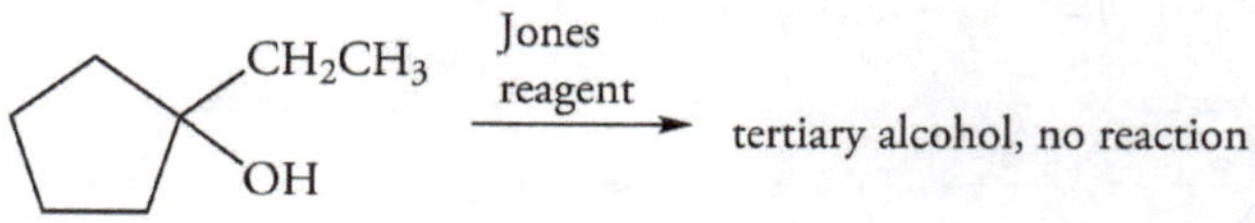

(d)

CH₃
HO—(decalin ring)   —(Jones reagent)→   CH₃—(decalin ring with =O ketone)

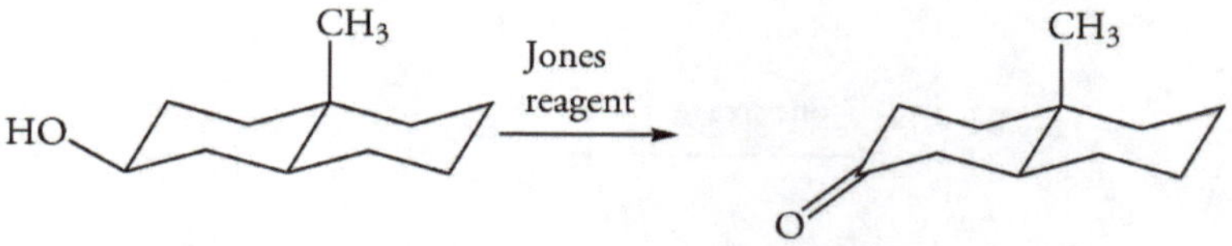

15.31 Write the product formed from the oxidation of the sex attractant of the Mediterranean fruit fly by PCC.

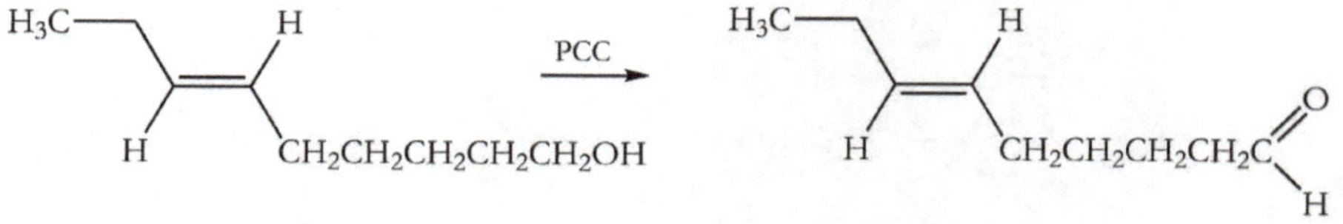

H₃C—   H
     \C=C/   —(PCC)→
H /   \CH₂CH₂CH₂CH₂CH₂OH

H₃C—   H
     \C=C/
H /   \CH₂CH₂CH₂CH₂C(=O)H

16.32 Write the product formed from the mosquito repellent by PCC.

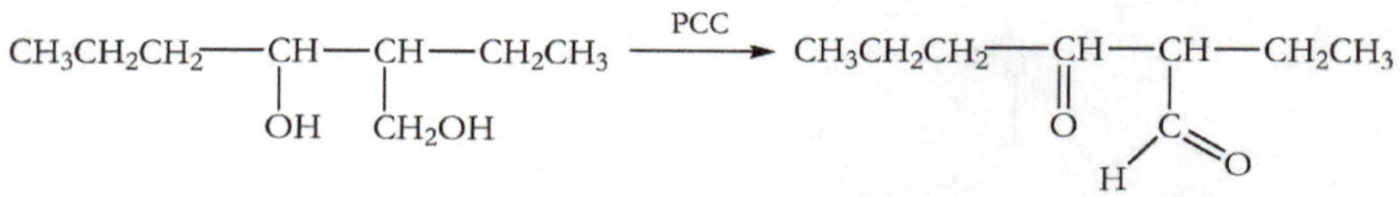

CH₃CH₂CH₂—CH—CH—CH₂CH₃   —(PCC)→   CH₃CH₂CH₂—CH—CH—CH₂CH₃
                | | ‖ |
                OH CH₂OH   O   C(=O)H

Answer: The slower rate for the compound with deuterium on C-2 indicates that the C—D bond is cleaved in the rate-determining step.

15.33 Consider the relative rates of oxidation of the following three compounds by chromium(VI). What do these data reveal about the rate-determining step of the reaction?

Compound	CH₃CH(OH)CH₃	CH₃CD(OH)CH₃	CD₃CH(OH)CH₃
Relative rate	1.0	0.16	1.0

15.34 The rate of oxidation of *endo*-bicyclo[2.2.1]heptan-2-ol is faster than the rate of oxidation of the *exo* isomer. What does this fact indicate about the rate-determining step for the reaction?

endo-bicyclo[2.2.1]hepan-2-ol   *exo*-bicyclo[2.2.1]hepan-2-ol

Answer: The rate-determining step is not the formation of the chromium ester because the more hindered *endo* alcohol reacts at the slower rate. The rate-determining step is cleavage of the C—H bond, which is *exo* in the *endo* alcohol, and is sterically more accessible than the endo C—H bond in the *exo* alcohol.

15.35 Write a mechanism of the oxidation of an alcohol by chromium(VI) that uses only an intramolecular process for the abstraction of the a hydrogen atom. Considering the size of the ring in the cyclic process, how likely is it that this process will occur?

Answer: The reaction occurs via a five-atom transition state in an intramolecular process, which is both strain free and highly probable.

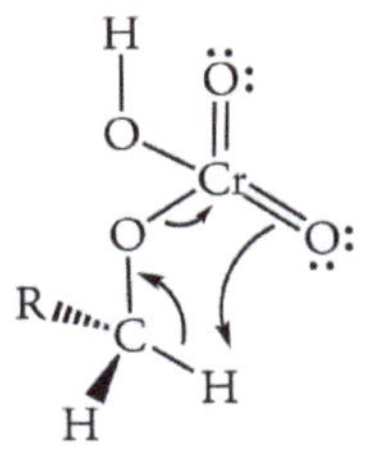

15.36 Which of the two sites within the following structure will be oxidized at the faster rate when only one equivalent of a chromium(VI) oxidizing agent is available?

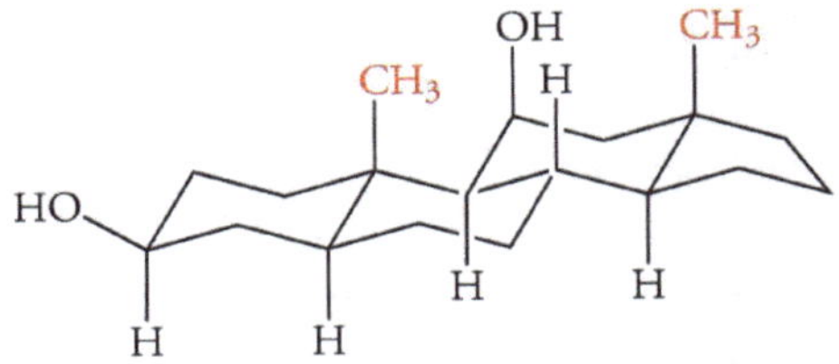

Answer: The rate-determining step is cleavage of the C—H bond, as seen in Exercise 15.33, above. That bond is equatorial for the alcohol of the C ring of the steroid. It is oxidized faster than the equatorial alcohol of the A ring which has an axial C—H bond and is much more sterically hindered.

Reactions of Vicinal Diols

15.37 Draw two possible structures of products formed by treating the following vicinal diol with sulfuric acid.

Answer: A pinacol-type rearrangement occurs. Protonation occurs at either of the two equivalent oxygen atoms and water readily leaves to give a tertiary carbocation. Migration of either a methyl or an ethyl group from the adjacent carbon atom gives a protonated carbonyl compound that then loses a proton to give a mixture of two products. The mechanism for formation of one of them, by a methide shift, is shown below.

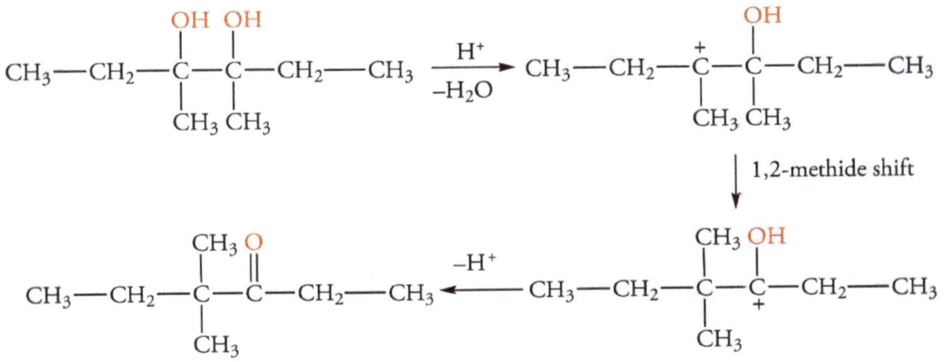

Answer: A pinacol-type rearrangement occurs. Protonation occurs at either of the two equivalent oxygen atoms and water readily leaves to give a tertiary carbocation. Migration of an ethyl group by a 1,2-ethide shift gives another carbocation that then loses a proton to give the product shown below.

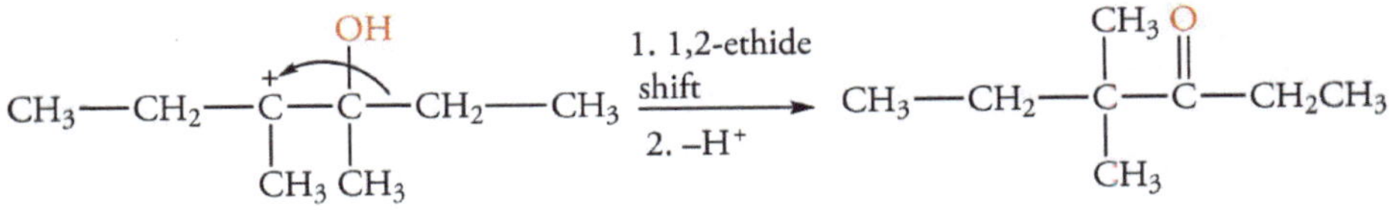

15.38 Only one product is formed by treating the following vicinal diol with sulfuric acid. Draw its structure. Why is it formed rather an isomeric product?

Answer: A pinacol-type rearrangement occurs. Protonation occurs at the hydroxyl group of the benzyl carbon atom bonded to the two phenyl groups, and water readily leaves to give a tertiary carbocation that is stabilized by the phenyl groups. Migration of a methylene group of the cyclopentane ring gives a hydroxy carbocation that then loses a proton.

15.39 Draw the structure of the product(s) of the reaction of each of the following compounds with periodic acid.

15.40 Which compound in the following pairs reacts fastest with periodic acid?

(a)

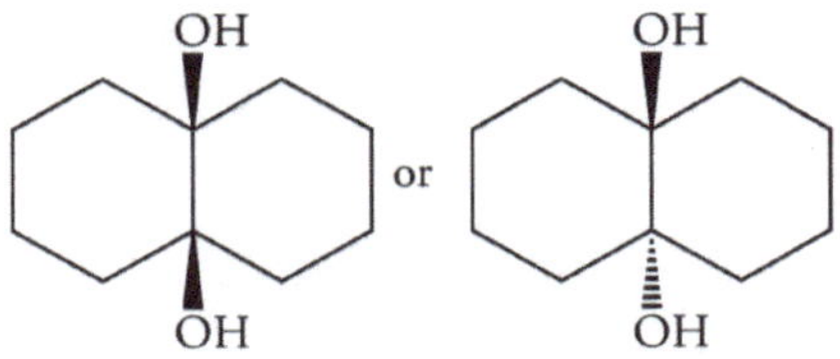

OH ... OH   or   OH ... OH

(b)

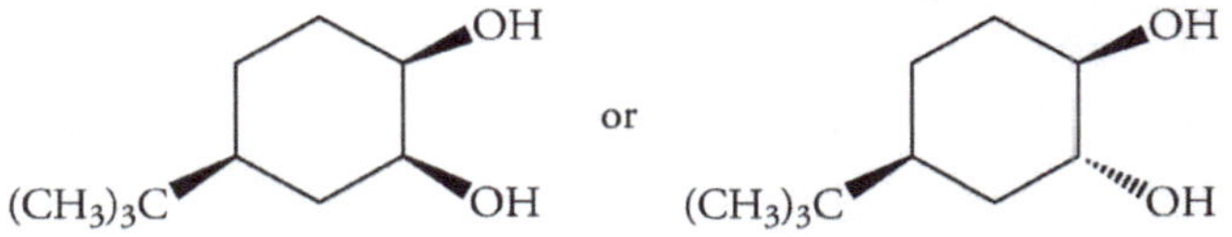

(CH$_3$)$_3$C ... OH, OH   or   (CH$_3$)$_3$C ... OH, OH

(c)

HO, H ... HO, H   or   HO, H ... HO, H

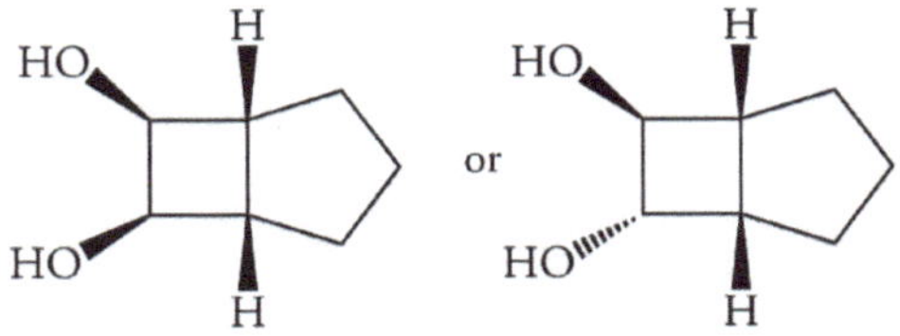

Answer: Compounds with two hydroxyl groups in a *syn* periplanar arrangement or a *cis* arrangement can form the cyclic iodate ester more easily and then react to cleave the carbon–carbon bond. In (a), (b), and (c), the first compound (on the left) reacts at the faster rate.

15.41 The reaction of oleic acid (C$_{18}$H$_{34}$O$_2$) with osmium tetraoxide followed by reaction with periodate yields the following two compounds. Draw the structure of oleic acid.

Answer: Osmium tetroxide reacts with a double bond to give a vicinal diol. Subsequent reaction with periodate cleaves the carbon–carbon bond that was originally a double bond. The compound has a double bond between C-9 and C-10 of an unsaturated carboxylic acid and could be either *cis* or *trans*. The natural product is *cis*.

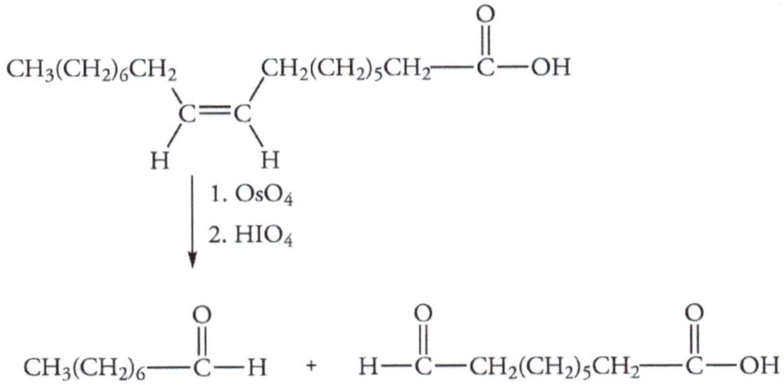

$$CH_3(CH_2)_6CH_2 \quad CH_2(CH_2)_5CH_2{-}\overset{\overset{O}{\|}}{C}{-}OH$$
$$C{=}C$$
$$H \qquad H$$

1. OsO$_4$
2. HIO$_4$

$$CH_3(CH_2)_6{-}\overset{\overset{O}{\|}}{C}{-}H \quad + \quad H{-}\overset{\overset{O}{\|}}{C}{-}CH_2(CH_2)_5CH_2{-}\overset{\overset{O}{\|}}{C}{-}OH$$

15.42 A hydrocarbon of molecular formula C$_9$H$_{14}$ is found in sandalwood oil. Reaction of the compound with osmium tetraoxide followed by reaction with period ate yields the following compound. Draw the structure of the hydrocarbon.

Answer: Osmium tetroxide reacts with a double bond to give a vicinal diol. Subsequent reaction with periodate cleaves the carbon–carbon bond that was originally a double bond. The product is a dicarbonyl compound, which indicates that the original double bond was contained in a ring that has been cleaved.

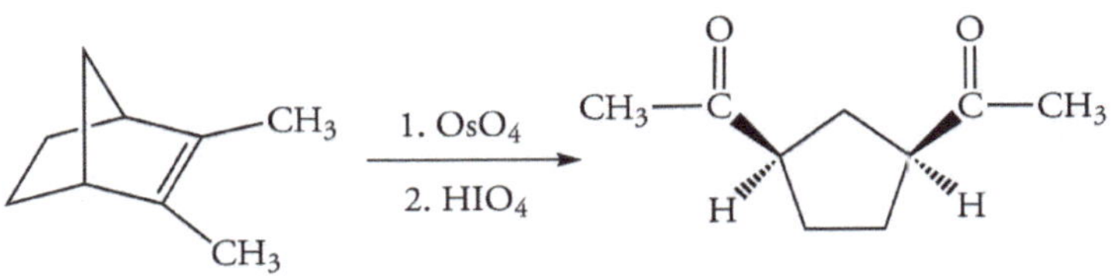

CH$_3$ ... CH$_3$ 1. OsO$_4$ 2. HIO$_4$ →   CH$_3${-}C(=O) ... C(=O){-}CH$_3$, H ... H

Synthesis of Alcohols from Alkyl Halides

15.43   Which compound of each of the following pairs will react with ethanoate ion at the faster rate?
   (a) 1-iodohexane or l-bromohexane
   (b) l-bromo-l-phenylethane or l-bromo-2-phenylethane
   (c) l-bromo-2,2-dimethylpropane or 1-bromopentane

Answer: (a) Iodide ion is a better leaving group than bromide ion. Thus, 1-iodohexane reacts faster than 1-bromohexane with the same nucleophile.
   (b) A secondary benzylic carbocation results from 1-bromo-l-phenylethane which reacts faster than l-bromo-2-phenylethane, which is a primary halogen compound.
   (c) Both compounds are primary but 1-bromo-2,2-dimethylpropane is sterically hindered and reacts via an S_N2 mechanism at a much lower rate.

15.44   Would DMF or ethanol be the better solvent for the displacement of halide ions from alkyl halides by ethanoate ion?

Answer: DMF would be a better solvent because it is aprotic and does not decrease the nucleophilicity of the ethanoate ion as does the ethanol, which forms hydrogen bonds to the nucleophile.

15.45   Attempted synthesis of bicyclo[2.2.2]octan-l-ol by reaction of ethanoate with l-bromobicyclo[2.2.2]octane fails. Why?

1-bromobicyclo[2.2.2]octane

Answer: Reaction by an S_N2 mechanism is impossible because the nucleophilic ethanoate ion cannot approach from the back side of the C—Br bond. Although the compound is a tertiary halide, the carbocation required for the S_N1 mechanism cannot form because it cannot be planar due to restriction of the bicyclic ring. As a result, the substitution reaction cannot occur.

15.46   Predict the stereochemistry of the alcohols obtained by the reaction of *cis*-1-bromo-4-*tert*-butylcyclohexane with ethanoate followed by hydrolysis under basic conditions.

Answer: Displacement of the axial bromo group by attack of the ethanoate occurs with inversion of configuration to give the *trans* ester which has an equatorial C—O bond. Hydrolysis of the ester occurs with nucleophilic attack at the carbonyl carbon atom and releases the alkoxy portion of the ester. The product is the *trans*-4-*tert*-butylcyclohexanol.

cis-1-bromo-4-*tert*-butylcyclohexane   *trans*-4-*tert*-butylcyclohexanol

16 ETHERS AND EPOXIDES

KEYS TO THE CHAPTER

The chemistry of ethers has substantially less variety than the chemistry of alcohols because several of the reactions of alcohols involve the O—H bond, namely dehydration and oxidation reactions are not possible for ethers. However, if comparable reactions are considered, such as substitution, then alcohols and ethers have similar reactivities. The reactions of epoxides are the result of ring strain, which leads to the formation of ring-opened products.

16.1 Structure of Ethers

Ethers contain an oxygen atom bonded to two alkyl groups, two aryl groups, or one of each. The geometry of ethers resembles that of alkanes, with the substitution of a methylene carbon atom by an sp^2-hybridized oxygen atom. Conformations of ethers resemble those of alkanes. The two nonbonding electron pairs of the ether oxygen are directed to the corners of a tetrahedron.

16.2 Nomenclature of Ethers

The common names of simple ethers are based on the names of the alkyl or aryl groups bonded to the oxygen atom. The name results from listing the alkyl (or aryl) groups in alphabetical order and appending the name ether.

The IUPAC name is based on the longest carbon chain bonded to the oxygen atom. The smaller group bonded to the oxygen atom is named as an alkoxy group and is regarded as a substituent on the longer chain.

The three-, five-, and six-membered cyclic ethers have common names. Three-membered ring compounds are called epoxides of the corresponding alkene from which they may be synthesized. The common names of five- and six-membered ring compounds are called tetrahydrofurans and tetrahydropyrans, respectively. In the IUPAC system, each ring size has a specific name. The names for cyclic ethers having three-, four-, five-, and six-membered rings are oxirane, oxetane, oxolane, and oxane, respectively. The oxygen atom in each ring is assigned the number 1, and the ring is numbered in the direction that gives the lowest numbers to substituents.

16.3 Physical Properties of Ethers

Ethers have two polar C—O bonds and have substantial dipole moments. They are more polar than alkanes, but less polar than alcohols. Ethers do not have an O—H bond and cannot act as hydrogen bond donors. They can, however, serve as hydrogen bond acceptors, which makes the low-molecular-weight ethers soluble in water. Because they cannot form intermolecular hydrogen bonds, ethers have boiling points that are substantially lower than those of alcohols of comparable molecular weight. The boiling points of ethers are very close to the boiling points of alkanes of similar molecular weight. Ethers are excellent solvents for both nonpolar and nonpolar solutes. They are aprotic, so they are used as solvents for reagents that react with acidic protons, as is the case for the Grignard reagent. Polyethers readily dissolve polar compounds and hydrogen bond donors.

16.4 Polyether Antibiotics

Polyether antibiotics act by selectively bind to cations, typically sodium or potassium. They disrupt the ion balance in bacterial cells, and therefore kill bacteria.

16.5 Ether synthesis by Alkoxymercuration–Demercuration of Alkenes

Alkoxymercuration occurs by the same mechanism as oxymercuration. The only difference is the alkyl group bonded to the oxygen atom of an alcohol compared to the hydrogen atom bonded to the oxygen atom of water. The regiospecificity of the reaction corresponds to net Markovnikov addition.

16.6 The Williamson Ether Synthesis

The Williamson ether synthesis is an S_N2 reaction in which an alkoxide ion is a nucleophile that displaces a halide ion from an alkyl halide to give an ether. The reaction occurs with inversion of configuration at chiral centers and can be limited by possible competing elimination reactions.

Intramolecular Williamson ether synthesis occurs at rates that depend on the number of atoms in the transition state. The rates are affected by the probability of the alkoxide approaching the carbon atom bearing the halide ion, as well as the strain of the resulting ring compound. The observed order of reactivity in terms of the ring size is 3 > 5 > 6 > 4.

16.7 Reactions of Ethers

Both ethers and alcohols can act as bases because they have two lone pairs of electrons on the oxygen atom. They are both very weak bases and can only be protonated to form the conjugate acid, an oxonium ion, by a strong acid. The formation of an oxonium ion is analogous to the reaction of water with a strong acid to give the hydronium ion. Oxonium ions are intermediates in many reactions catalyzed by strong acids in the reactions of both ethers and alcohols.

Ethers react with strong acids such as HBr and HI to give cleavage products. The reaction proceeds by a two-step process in which first the oxygen atom is protonated and then the halide ion attacks one of the alkyl groups to displace an alcohol by an S_N2 process. The alkyl group attacked by the halide ion is controlled by the reactivity order 1° > 2° > 3°. The product alcohol can react with additional HX to give a second mole of a haloalkane.

16.8 Ethers as Protecting Groups

Protecting groups are provided by reagents that easily form derivatives of the functional group to be "protected," but can also be easily removed when required. A protecting group is used to render a functional group unreactive toward specific reagents that are required to transform a second functional group in the molecule.

Alcohols are protected by preparing silyl ethers of the general formula $R'—O—SiR_3$. The silyl ether is obtained by reacting an alcohol $R'—OH$ with a chlorosilane $Cl—SiR_3$. The silyl ether is cleaved by fluoride ion to liberate the alcohol after other transformation are completed.

16.9 Synthesis of Epoxides

Epoxidation is the reaction of an alkene with certain peroxyacids such as MCPBA or MMPP. The stereochemistry of the groups of the alkene is retained in the epoxide. Halohydrins undergo an intramolecular Williamson ether synthesis. The reaction occurs inversion of configuration at the center where the halide ion is displaced. Halohydrins are formed by addition of halogen to a double bond in aqueous solution.

16.10 Reactions of Epoxides

The ring strain of the three-membered epoxide ring results in ring-opening reactions in which one of the C—O bonds breaks. The regiochemistry of the ring opening and the stereochemistry of the product depend on whether the reaction occurs under basic or acidic conditions.

Nucleophiles displace an alkoxide ion, a reaction not observed in acyclic compounds. As in the case of other S_N2 reactions, the order of reactivity for a substrate with a nucleophile is primary > secondary > tertiary. The resulting product has the nucleophile and the hydroxyl group on adjacent carbon atoms. The nucleophile is on the less substituted carbon atom. Grignard reagents react with epoxides to give an alcohol containing two additional carbon atoms between the alkyl group and the carbon atom bearing the hydroxyl group in the product. Thus, for the reaction of a Grignard reagent, $RMgBr$, with ethylene oxide, the product is RCH_2CH_2OH.

Under acidic conditions, the oxygen atom is protonated to give a cyclic intermediate that resembles the bromonium and mercurinium ions we have encountered in other mechanisms. Subsequently, the nucleophilic reagent attacks the more substituted carbon atom because it has the greater partial positive charge. Thus, the nucleophile is bonded to the more substituted carbon atom in the product.

The stereochemistry of the ring opening of an epoxide is easily predicted. Nucleophilic attack of a reagent such as methoxide ion occurs with inversion of configuration at the least substituted carbon atom. The stereochemistry of the other carbon atom of the original epoxide is unchanged because no bond to that center is broken in the reaction. In a substituted epoxide, the ring opening reaction under acidic conditions occurs by inversion of configuration at the more substituted center where the nucleophile attacks. The configuration at the least substituted center is unchanged.

16.11 Reactions of Epoxides

The sulfur analogs of ethers are sulfides. They are named using alkylthio groups as substituents to the parent chain. Cyclic sulfides have common names.

Sulfides are prepared by reaction of a thiolate, the conjugate base of a thiol, with an alkyl halide. Because the thiolate ion is less basic and is a better nucleophile than an alkoxide. The sulfur analog of a Williamson synthesis has fewer complications due to elimination reactions.
Sulfide are oxidized to sulfoxides and then to sulfones.

16.12 Spectroscopy of Compounds with C—O and C—S Bonds

Infrared spectroscopy is usually not used to confirm the presence of either C—O or C—S bonds because the stretching vibrations occur in a region complicated by other absorptions. The O—H stretching vibration of alcohols is easily seen as a strong broad absorption in the spectrum in the $3400–3600 \text{ cm}^{-1}$ region.

The chemical shift of hydrogen atoms bonded to the carbon atom bearing the oxygen atom of either alcohols or ethers occurs in the $3–4$ δ region. The chemical shift of hydrogen atoms bonded to the carbon atom bearing the sulfur atom of either thiols or sulfides is less deshielded and occurs in the 2.5 δ region. Both O—H and S—H hydrogen atoms have variable chemical shifts depending on concentration.

The a carbon atom of an alcohol or an ether has a chemical shift that reflects the deshielding of the electronegative oxygen atom. The deshielding due to a sulfur atom is smaller.

Summary of Reactions

1. Synthesis of Ethers by Addition of an Alcohol to an Alkene

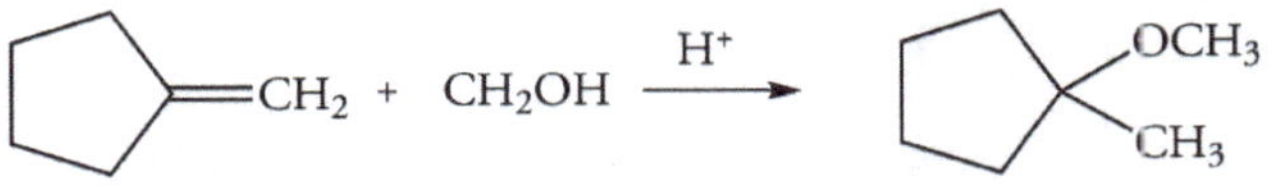
CH2
+
CH2OH
H+
OCH3
CH3

2. Synthesis of Ethers by Alkoxymercuration–Demercuration of Alkenes

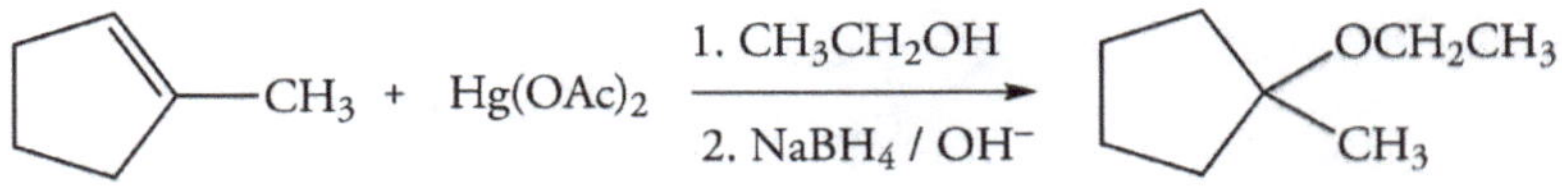
CH3
+
Hg(OAc)2
1. CH3CH2OH
2. NaBH4 / OH−
OCH2CH3
CH3

3. Williamson Ether Synthesis

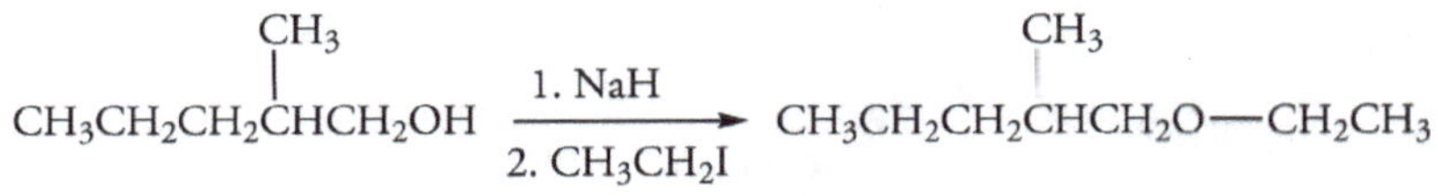
CH3
CH3CH2CH2CHCH2OH
1. NaH
2. CH3CH2I
CH3
CH3CH2CH2CHCH2O—CH2CH3

4. Cleavage of Ethers

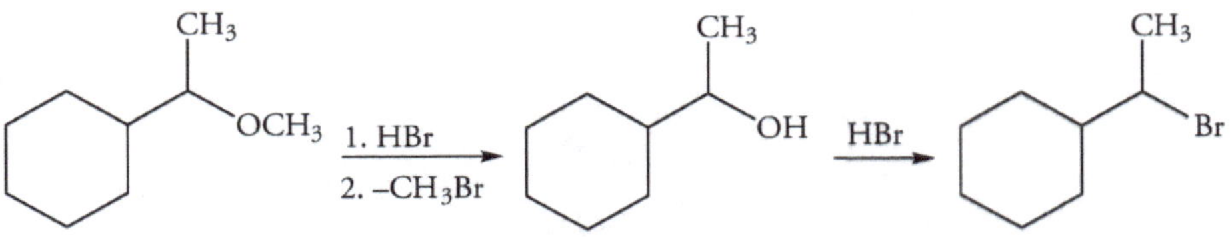
CH3
OCH3
1. HBr
2. –CH3Br
CH3
OH
HBr
CH3
Br

5. Silyl Ethers as Protecting Groups

Add protecting group

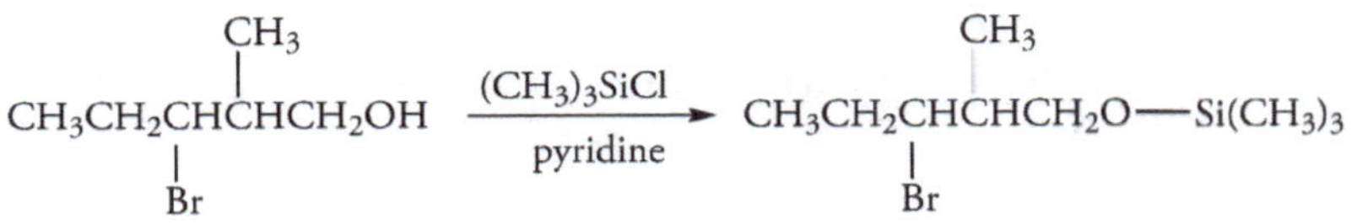
CH3
CH3CH2CHCHCH2OH
Br
(CH3)3SiCl
pyridine
CH3
CH3CH2CHCHCH2O—Si(CH3)3
Br

Remove protecting group

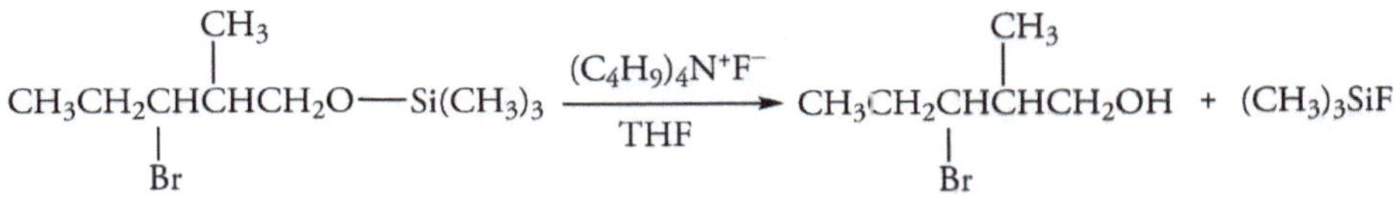
CH3
CH3CH2CHCHCH2O—Si(CH3)3
Br
(C4H9)4N+F−
THF
CH3
CH3CH2CHCHCH2OH
Br
+ (CH3)3SiF

6. Synthesis of Epoxides

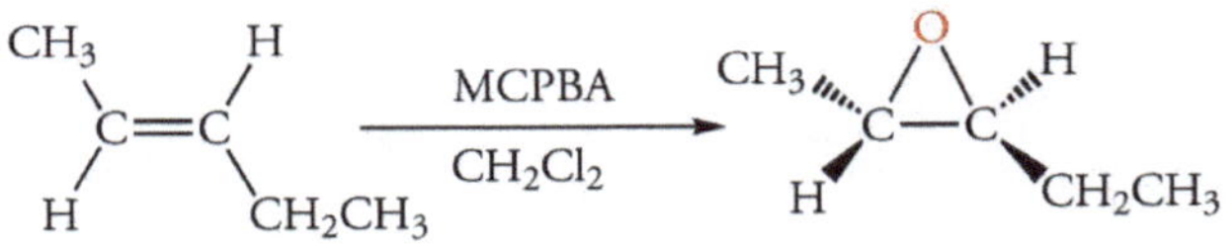
CH3
H
C=C
H
CH2CH3
MCPBA
CH2Cl2
O
CH3
H
C—C
H
CH2CH3

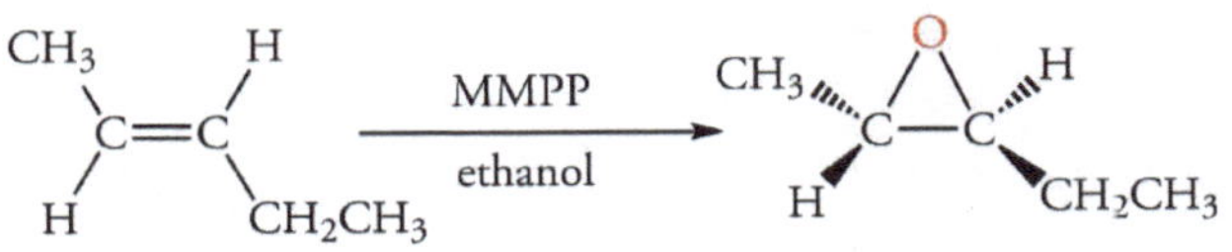
CH3
H
C=C
H
CH2CH3
MMPP
ethanol
O
CH3
H
C—C
H
CH2CH3

7. Ring Cleavage of Epoxides

Base catalyzed

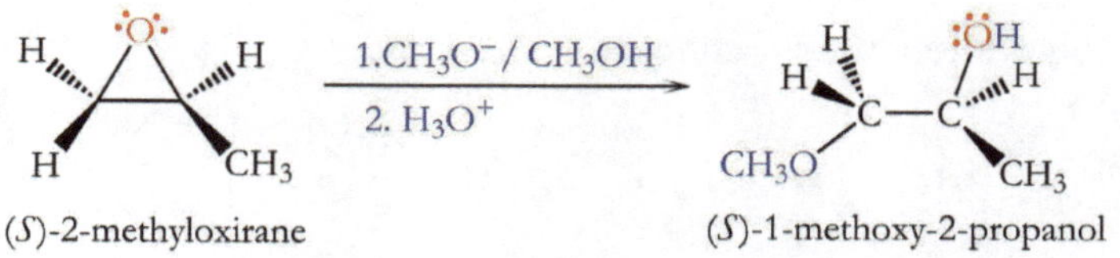

(S)-2-methyloxirane → 1.CH₃O⁻ / CH₃OH 2. H₃O⁺ → (S)-1-methoxy-2-propanol

Acid catalyzed

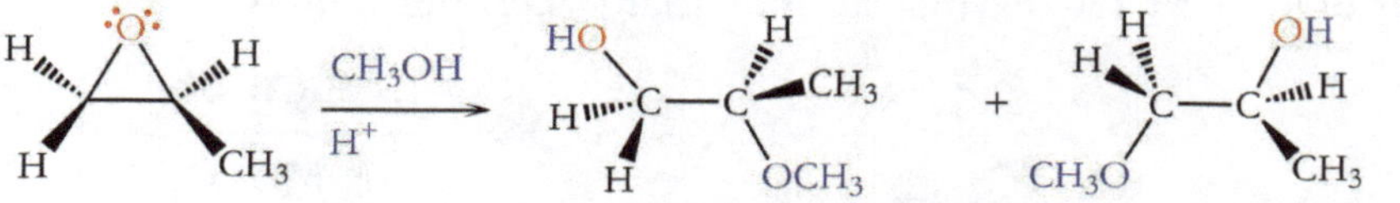

(R)-2-methoxy-1-propanol (major product) + (S)-1-methoxy-2-propanol (minor product)

8. Synthesis of Sulfides

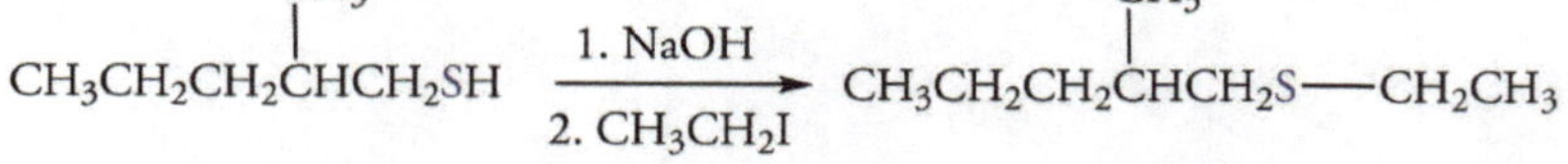

$$CH_3CH_2CH_2CHCH_2SH \xrightarrow[\text{2. } CH_3CH_2I]{\text{1. NaOH}} CH_3CH_2CH_2CHCH_2S-CH_2CH_3$$

(with CH₃ branch)

9. Oxidation of Sulfides

Sulfide to Sulfoxide

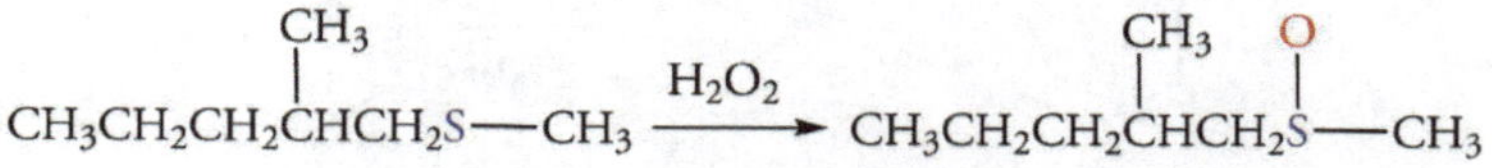

$$CH_3CH_2CH_2CHCH_2S-CH_3 \xrightarrow{H_2O_2} CH_3CH_2CH_2CHCH_2S-CH_3$$

(with CH₃ branch; product has =O on S)

Sulfoxide to Sulfone

$$CH_3CH_2CH_2CHCH_2S-CH_3 \xrightarrow{CH_3CO_3H} CH_3CH_2CH_2CHCH_2-S-CH_3$$

(with CH₃ branch; product has two =O on S)

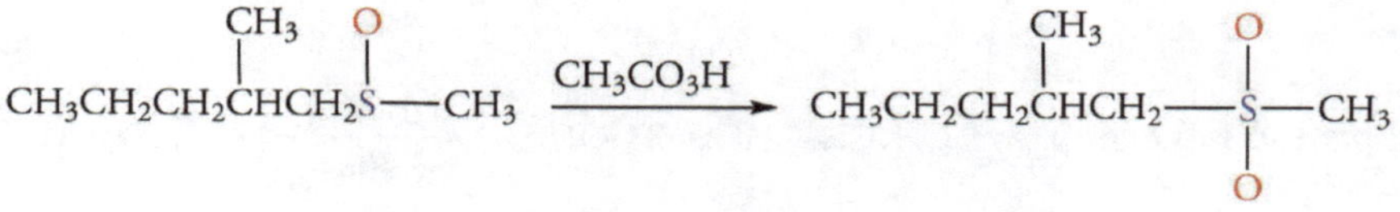

Ether Isomers

16.1   Draw the structures of the isomeric ethers with the following characteristics.
    (a)   molecular formula $C_4H_{10}O$
    (b)   a methyl ether with molecular formula $C_5H_{12}O$
    (c)   a saturated ether with the molecular formula C_3H_6O

Answers:

(a) $CH_3CH_2CH_2$—O—CH_3 (b) 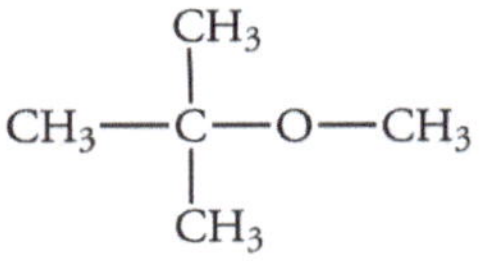 CH_3—C—O—CH_3 (with CH_3 above and CH_3 below the central C) (c) epoxide with CH_3

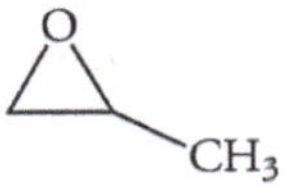

16.2   Draw the structures of the following ethers.
    (a)   oxane and two methyl oxolane
    (b)   2-methoxypropene
    (c)   *cis*-2,3-dimethyloxetane

Answers:

(a) 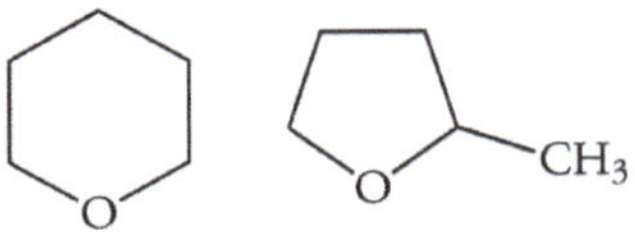 oxane ring, 2-methyl oxolane ring
(b) 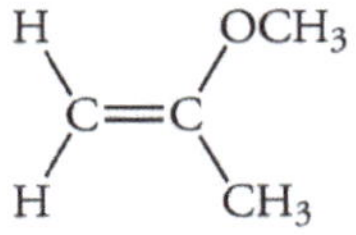 H,OCH_3 / C=C / H,CH_3
(c) oxetane ring with CH_3 and CH_3

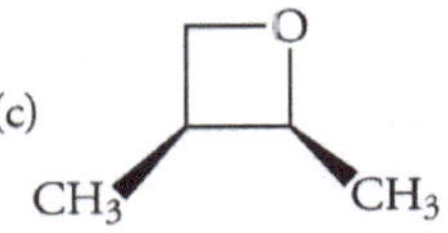

Ether Nomenclature

16.3   Give the common name of each of the following compounds.

Answers:
(a) dicyclopentyl ether
(b) phenyl propyl ether
(c) cyclopentyl propyl ether

(a) 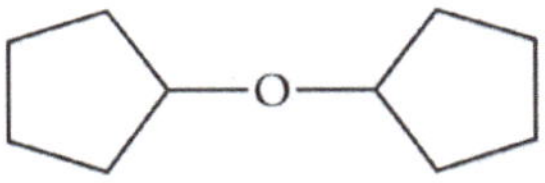 dicyclopentyl ether structure
(b) 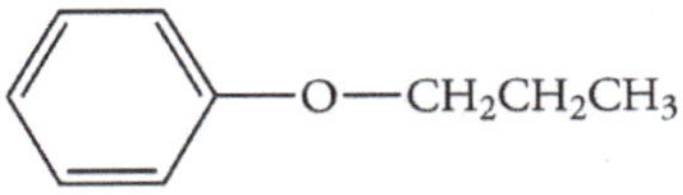 phenyl—O—$CH_2CH_2CH_3$
(c) 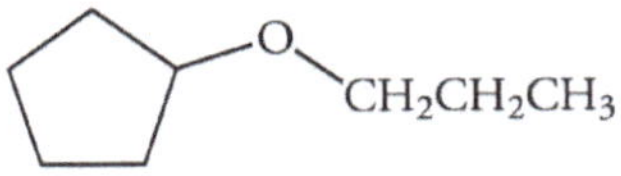 cyclopentyl—O—$CH_2CH_2CH_3$

16.4   Give the common name of each of the following compounds.

Answers:
(a) *tert*-butyl phenyl ether
(b) benzyl *tert*-butyl ether
(c) cyclohexyl propargyl ether

(a) 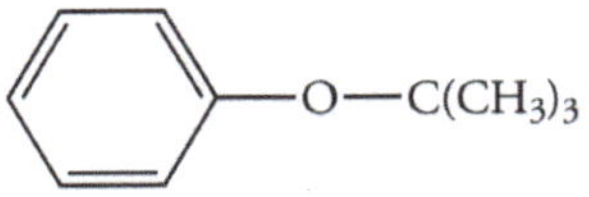 phenyl—O—$C(CH_3)_3$ (b) phenyl—CH_2—O—$C(CH_3)_3$
(c) cyclohexyl—CH_2—O—$CH_2C\equiv CH$

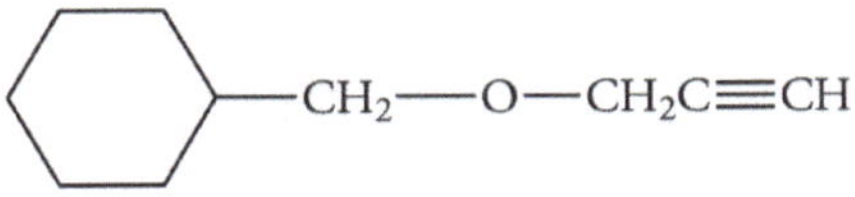

16.5   Assign the IUPAC name of each of the following compounds.

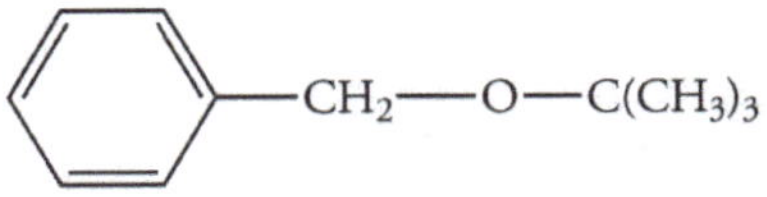

Answers:
(a) 2-methoxypentane
(b) 2-methoxy-4-methylpentane
(c) 2,4-diethoxyheptane

(a) 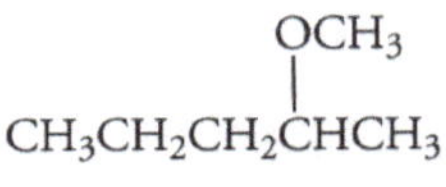 $CH_3CH_2CH_2\overset{OCH_3}{\underset{|}{C}}HCH_3$ (b) $CH_3\overset{CH_3}{\underset{|}{C}}HCH_2\overset{OCH_3}{\underset{|}{C}}HCH_3$ (c) $CH_3CH_2CH_2\overset{OCH_2CH_3}{\underset{|}{C}}HCH_2\underset{\underset{OCH_2CH_3}{|}}{C}HCH_3$

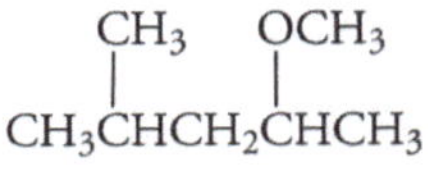

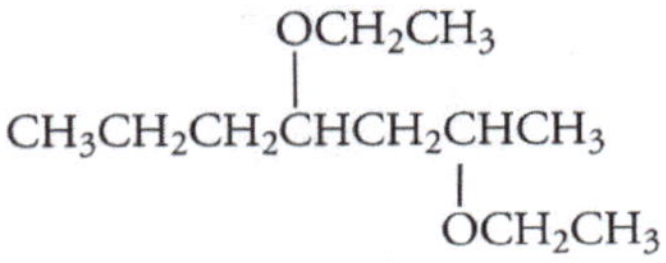

16.6   Assign the IUPAC name of each of the following compounds.

Answers:
(a) 1,1-dimethoxypentane
(b) 2,4-dimethoxypentane
(c) 3-ethoxyhexane

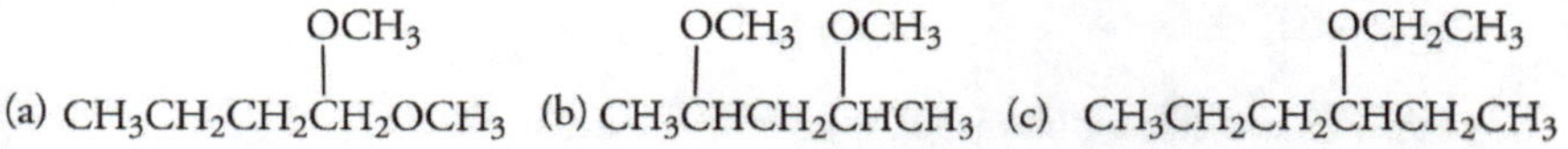

(a) $CH_3CH_2CH_2CH_2OCH_3$ with OCH_3 label (b) $CH_3CHCH_2CHCH_3$ with two OCH_3 labels (c) $CH_3CH_2CH_2CHCH_2CH_3$ with OCH_2CH_3 label

16.7   Draw the structure of each of the following general anesthetics.
        (a)   1,1,1,3,3,3-hexafluoroisopropyl methyl ether (isoindoklon)
        (b)   2-chloro-1,1,2-trifluoro-1-(difluoromethoxy)ethane (enflurane)

Answers:

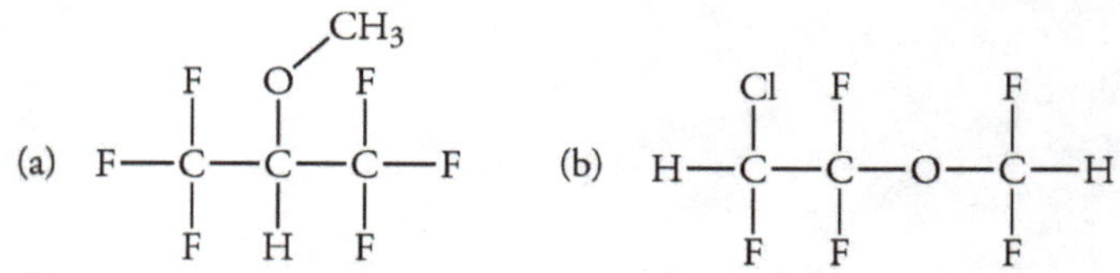

(a)   (b)

16.8   What is the common name of each of the following anesthetics?
        (a) $CH_2{=}CH{-}O{-}CH{=}CH_2$
        (b) $CF_3CHCl{-}O{-}CHF_2$

Answers:
(a) divinyl ether   (b) difluoromethyl 1-chloro-2,2,2-trifluoroethyl ether

16.9   Draw the structure of each of the following compounds.
        (a)   *trans*-4-methoxycyclohexanol
        (b)   3-ethoxy-1,1-dimethylcyclohexane
        (c)   12-crown-4

Answers:

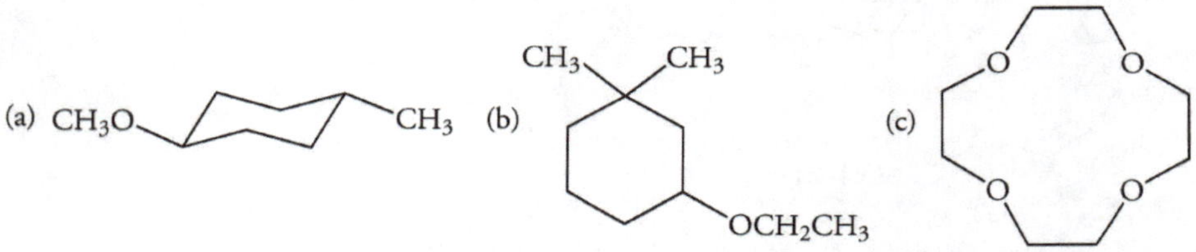

(a)   (b)   (c)

16.10   Draw the structure of each of the following compounds.
        (a)   3-methoxyoxolane
        (b)   *trans*-2-chloro-l-methoxycyclobutane
        (c)   *cis*-2-ethoxy-3-methyloxirane
        (d)   15-crown-5

Answers:

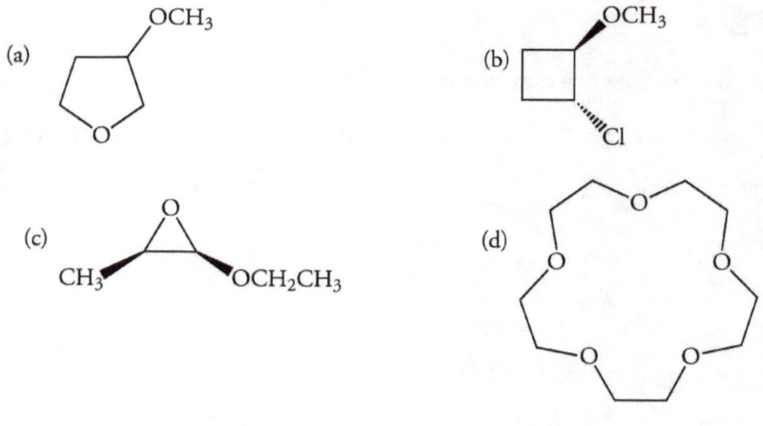

(a)   (b)

(c)   (d)

Properties of Ethers

16.11   Explain why 1,4-dioxane is more soluble in water (it is miscible) than diethyl ether.

Answer: Diethyl ether is somewhat soluble in ether. The added oxygen atom of 1,4-dioxane increases the extent of hydrogen bonding with water, leading to increased solubility. The ratio of carbon to oxygen atoms decreases from 4:1 in diethyl ether to 2:1 in 1,4-dioxane.

16.12   Explain why *p*-ethylphenol is more soluble in water than ethoxybenzene.

Answer: Ethoxybenzene is only a hydrogen bond acceptor, but *p*-ethylphenol is both a hydrogen bond acceptor and a hydrogen bond donor.

16.13   The boiling points of dipropyl ether and diisopropyl ether are 91 and 68 °C, respectively. Explain why the boiling points of these isomeric ethers differ.

Answer: Diisopropyl ether has a more compact conformation and has smaller London forces than the cylindrical conformation of dipropyl ether.

16.14   The boiling points of 1-ethoxypropane and 1,2-dimethoxyethane are 64 and 83 °C, respectively. Explain why.

Answer: Although they have similar molecular weights, 1,2-dimethoxyethane has an additional oxygen atom and is more polar.

16.15   Explain why dipropyl ether is soluble in concentrated sulfuric acid, whereas heptane is insoluble.

Answer: The oxygen atom of dipropyl ether can accept a proton from sulfuric acid. The conjugate base is an alkoxoxonium ion and is soluble in the polar medium. Heptane cannot be protonated.

16.16   Explain why aluminum trichloride dissolves in tetrahydropyran, releasing heat.

Answer: Aluminum trichloride is a Lewis base. It forms a coordinate covalent bond with lone pair electrons on the oxygen atom of tetrahydrofuran. The process is exothermic.

16.17   Explain why some potassium compounds dissolve in 18-crown-6, but the related rubidium compounds do not.

Answer: The atomic radius of rubidium is larger than the atomic radius of potassium. 18-crown-6 has a cavity the right size to solvate potassium, which means that it would be too small for the rubidium cation.

16.18   Some sodium compounds dissolve in 15-crown-5. Would the related lithium compounds be more or less likely to be soluble in 18-crown-6 or 12-crown-4?

Answer: The atomic radius of lithium is smaller than the atomic radius of sodium. 15-Crown-5 has a cavity the right size to solvate the sodium cation, which means that a smaller ring such as 12-crown-4 would be suitable for the lithium cation.

16.19   Draw the stable conformation of 1,4-dioxane.

16.20   Explain why the most stable chair conformation of 1,3-dioxan-5-ol has an axial hydroxyl group, but the most stable chair conformation of 5-methoxy-1,3-dioxane has an equatorial methoxy group.

Answer: The methoxy group should occupy the equatorial position of 1,3-dioxane much like it does in cyclohexane. The hydroxyl group occupies the axial position of 1,3-dioxane because it forms hydrogen bonds with the ring oxygen atoms.

16.21   Write a mechanism for the formation of 1,4-dioxane from ethylene glycol catalyzed by sulfuric acid.

16.22   Write a mechanism for the formation of 2,2-dimethyloxolane from 4-methyl-1,4-pentanediol and sulfuric acid. Which of the two oxygen atoms remains in the ether?

Answer: The oxygen atom bonded to C-1 remains in the ether.

16.23   Reaction of 1-hexene with mercuric acetate in methanol as solvent followed by reduction of the intermediate product with sodium borohydride yields 2-methoxyhexane. What is the structure of the intermediate product? How is it formed?

Answer: Electrophilic attack of $Hg(OAc)_2$ gives a mercurinium ion that then reacts with the nucleophilic methanol at the carbon atom with the larger positive charge, which is the more substituted carbon atom.

16.24   Reaction of 4-penten-l-ol with mercuric acetate followed by reduction with sodium borohydride yields 2-methyl-oxolane. Write the structure of the intermediate and explain why this ether forms.

Answer: Electrophilic attack of $Hg(OAc)_2$ gives a mercurinium ion that reacts with the nucleophilic oxygen atom of the alcohol at C-1. Attack occurs at the carbon atom with the larger positive charge, which is the more substituted carbon atom, C-4.

16.25   Reaction of 5-chloro-2-pentanol with sodium hydride yields 2-methyloxolane. Write the structure of the intermediate and explain why this ether forms.

Answer: An alkoxide ion, formed by reaction of the alcohol with the hydride ion, displaces chloride in an intramolecular S_N2 reaction.

16.26   Treatment of 3,4-dibromo-l-butanol with sodium hydroxide yields a cyclic ether. What is the structure of the ether? What alternate ether could form, and why is it not produced?

Answer: The alkoxide ion displaces a bromide ion from C-4, giving an oxolane. Displacement of a bromide ion from C-3 would give a more highly strained oxetane.

16.27   Which of the following compounds can be synthesized in good yield using the Williamson method? Explain why the method would fail for the remaining compounds.
(a) ethyl cyclopentyl ether   (b) 1-methyl-l-methoxycyclohexane   (c) *tert*-butyl cyclohexyl ether
(d) di-*sec*-butyl ether   (e) 2-methyl-3-phenoxyhexane

Answers:
(a) Reaction of the alkoxide of cyclopentanol with iodoethane gives a good yield because the S_N2 reaction occurs at a primary center.
(b) Although the alkoxide of 1-methylcyclohexanol is sterically hindered, the reaction of this nucleophile with iodomethane gives a good yield because the alkyl halide is unhindered, and it cannot undergo an elimination reaction.
(c) The reaction of the alkoxide of cyclohexanol with 2-bromo-2-methylpropane will give only elimination product. The reaction of *tert*-butoxide with bromocyclohexane may give some ether, but the major product will be cyclohexene, an elimination product.
(d) The reaction of the alkoxide of 2-butanol with 2-bromobutane gives largely elimination products.
(e) The reaction of the phenoxide with 3-bromo-2-methylhexane gives largely elimination products.

17 ORGANOMETALLIC CHEMISTRY OF TRANSITION METAL ELEMENTS AND INTRODUCTION TO RETROSYNTHESIS

KEYS TO THE CHAPTER

17.1 Transition Metal Complexes

Transition metal complexes consist of a transition metal bonded to molecules or ions called **ligands** in a **coordination complex**. In the formation of a coordination complex, the metal ion acts as a Lewis acid and the ligands act as Lewis bases. A coordination complex in solution is in equilibrium with its component metal atom (or ion) and its ligands.

The number of ligands that form coordinate covalent bonds in a transition metal complex is called the **coordination number** of the complex. The most common coordination numbers are 2, 4, and 6. The metal atom (or ion) and its bonded ligands constitute the **coordination sphere** of the complex.

A covalent bond in which one of the bonding groups contributes both bonding electrons to the metal is called a **coordinate covalent bond**.

Palladium metal, Pd(0), has a $[Kr]4d^{10}$ electron configuration. Coordination complexes in which Pd(0) has four ligands have tetrahedral geometry. Pd(0) is sp^3 hybridized in these complexes.

Pd(II) has a $[Kr]4d^8$ electron configuration. It has square planar geometry in which Pd(II) is dsp^2 hybridized. This geometry places the electron pairs in the coordinate covalent bonds, and therefore the ligands as far apart as possible.

Coordination complexes of palladium and other metals having six ligands usually have octahedral geometry. This geometry minimizes steric repulsion. The bonding in an octahedral complex is d^2sp^3.

17.2 Gilman Reagents

A **Gilman reagent** is an organometallic compound that has the general formula R_2CuLi, where the R group is derived from an alkyl halide. Gilman reagents undergo a wide variety of substitution reactions in which the R group of the reagent acts as a nucleophile that replaces a halogen in an alkyl, alkenyl, or aryl halide (Figure 17.2).

A Gilman reagent is prepared from an alkyl halide in a two-step process. First, lithium reacts with haloalkanes to give organolithium reagents. The organolithium compound then reacts with Cu(I) iodide to give a lithium dialkylcuprate.

Gilman reagents are nucleophiles that react by either an S_N2 mechanism or a more complex pathway of oxidative addition followed by reductive elimination to give products in which the alkyl group of the Gilman reagent couples with a substrate molecule.

17.3 Organopalladium-Catalyzed Cross-Coupling Reactions

Organopalladium complexes catalyze cross-coupling reactions between two aryl groups—Suzuki coupling; between an aryl group and an alkenyl group—the Heck reaction; and between an aryl group and an alkyne— the Sonogashira reaction.

17.4 The Suzuki Coupling Reaction

The Suzuki coupling reaction couples an aryl or vinyl halide to an aryl or vinyl boronic acid. The Suzki catalyst is $Pd(PPh_3)_4$.

Alkenyl and aryl boronic acids can be prepared by the reaction of a Grignard reagent with trimethoxyborane, $B(OCH_3)_3$ to give a dimethoxyaryl borane, $ArB(OCH_3)_2$. Hydrolysis of the borane gives the aryl boronic acid, $ArB(OH)_2$.

The vinyl halide adds to the Suzuki catalyst in an oxidative addition reaction in the first step of the catalytic cycle. Then, hydroxide displaces iodide. The base that is present in the reaction medium activates the boronic acid. The aryl group of the borate anion then adds to the catalyst. Cross-coupling of the aryl and vinyl groups occurs in a reductive elimination step. The palladium atom of the catalyst returns to its original oxidation state, $Pd(0)$, and the cycle continues.

17.5 The Heck Reaction

The Heck reaction couples aryl halides and alkenes. The product of the reaction is a conjugated aryl alkene.

17.6 The Sonogashira Reaction

The Sonogashira reaction couples the sp-hybridized carbon of a terminal an alkyne to a wide range of other substrates, including aryl and alkenyl halides. The reaction requires a palladium catalyst and a catalytic amount of CuI. It is carried out in an amine solvent.

17.7 The Wilkinson Catalyst: Homogeneous Catalytic Hydrogenation

Wilkinson's catalyst, $Ru[(PPh_3)_3Cl$, catalyzes homogeneous hydrogenation. The reaction is regiospecific. Alkenes are reduced in the following order.

Relative rates of hydrogenation by Wilkinson's catalyst

17.8 Noyori Asymmetric Reduction of Ketones

The Noyori asymmetric hydrogenation of ketones uses chiral ruthenium catalysts for the stereospecific hydrogenation of ketones. One of the most common ligands for the Noyori reaction is BINAP, a chiral catalyst. BINAP stereoisomers result from restricted rotation around single bonds. They are **atropisomers**. The chiral ruthenium catalysts carry out stereospecific **chiral synthesis** reactions.

17.9 The Grubbs Metathesis Reaction

The Grubbs reaction exchanges the groups attached to the double bond of alkenes. The two alkenes exchange partners to give two new products in which neither one is oxidized or reduced. This process is a metathesis reaction. Most of the time both reactants for the Grubbs are terminal alkenes. When two terminal alkenes react, they exchange methylene groups. Thus, one of the products is ethene, which escapes from the solution as a gas, converting a reversible process to one that is effectively irreversible.

Ethene escapes as a gas, pulling
the reaction to completion.

17.10 Introduction to Retrosynthesis

The design of a series of steps leading from simple, readily available starting materials to a more complex product that contains new functional groups and carbon–carbon bonds requires careful thought. One way to imagine a complex synthetic procedure is to go backward from the desired end-product to the initial reactants. This mode of thinking backward is called **retrosynthesis**. Consider the following synthetic scheme, viewed from right to left. The molecules leading to the **target molecule, T**, are called **precursor molecules**. In this scheme, there are four precursor molecules. Each double arrow pointing to the left is a **reverse step**.

Retrosynthetic Scheme I

Each precursor molecule is also a target molecule, but one with a simpler structure than the one preceding it. In terms of structural complexity, then, A < 4 < 3 < 2 < 1 < T. In each reverse step, we imagine how a bond can break to give two fragments; this is called **disconnection**.

The fragments of a hypothetical bond dissociation are called **synthons**. To have synthetic value, a synthon must correspond to an actual molecule called a **synthetic equivalent** that can be used in a synthetic reaction.

 **End of Chapter Exercises**

1. Preparation of Gilman Reagents

$$CH_3—Br + 2 Li \longrightarrow CH_3—Li + LiBr$$

methyl lithium

$$2 CH_3—Li + CuI \longrightarrow \left[CH_3—\overset{-}{Cu}—CH_3 \right] Li^+ + LiI$$

alkyl lithium   lithium dimethylcuprate
(a Gilman reagent)

2. Reactions of Gilman Reagents

3. Types of Cross-Coupling Reactions

Pd catalyst
Suzuki coupling

Pd catalyst
Heck reaction

Pd/Cu catalyst
Sonogashira reaction

4. Synthesis of Aryl Boronic Acids for the Suzuki Reaction

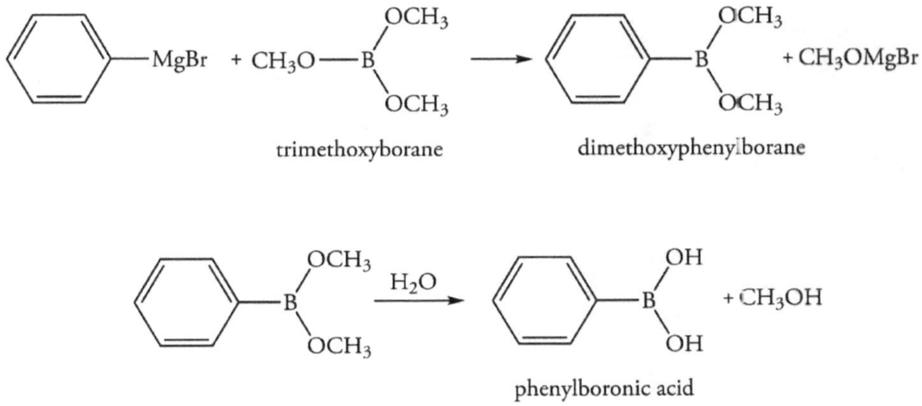

trimethoxyborane   dimethoxyphenylborane

phenylboronic acid

5. Suzuki Coupling Reactions

Suzuki coupling reactions

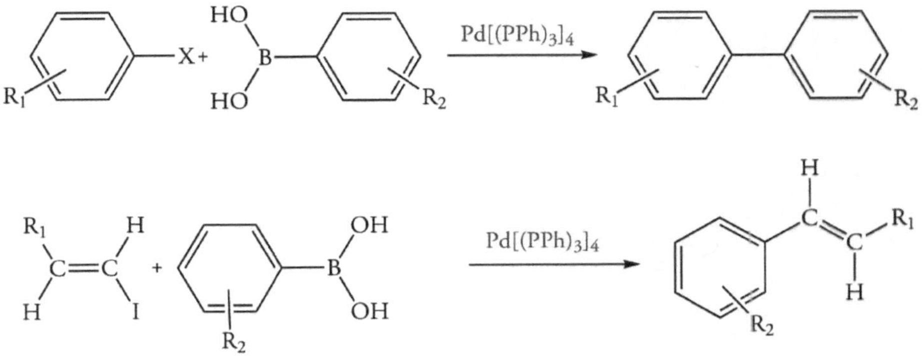

6. The Heck Reaction

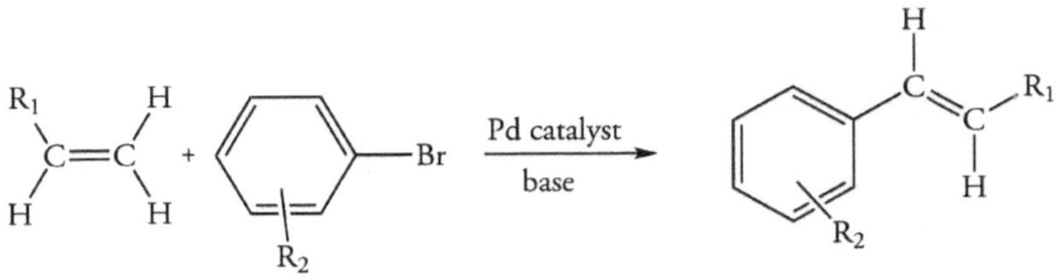

7. The Sonogashira Reaction

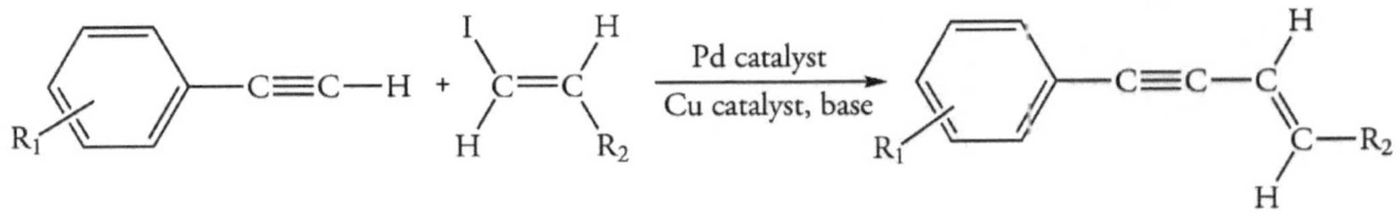

8. Wilkinson's Catalyst: Homogeneous Catalytic Hydrogenation

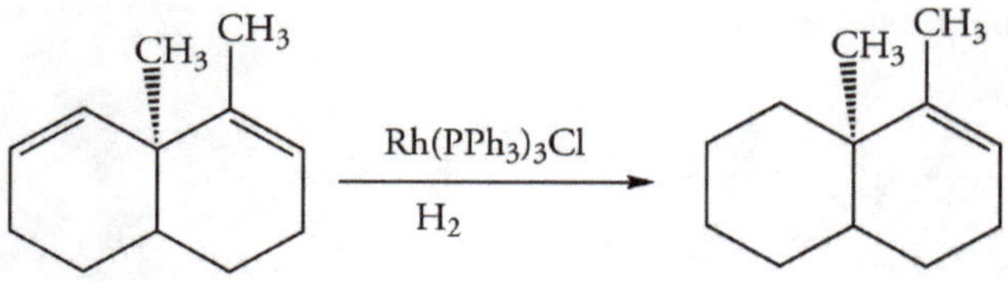

9. Noyori Asymmetric Reduction of Ketones

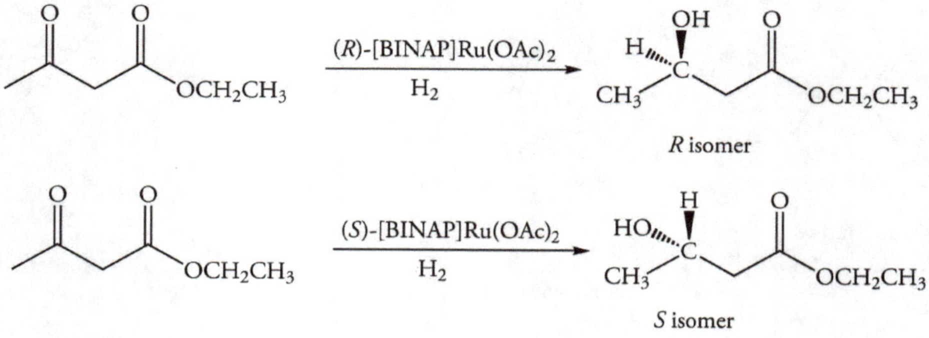

R isomer

S isomer

10. The Grubbs Metathesis Reaction

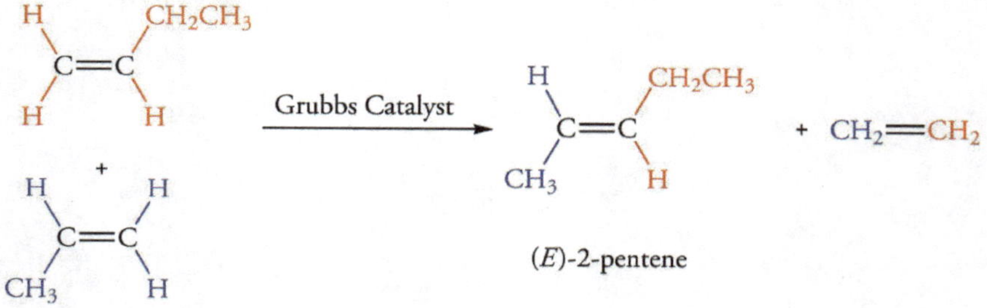

(*E*)-2-pentene

Gilman Reagent

17.1 What is the product of the following reaction?

Answer:

17.2 What is the product of the following reaction?

Answer:

17.3 What is the product of the following reaction?

Answer:

17.4 What is the product of the following reaction?

Answer:

(*S*)-2-bromobutane   (*R*)-2-phenylbutane

17.5 What is the product of the following reaction?

$(CH_3CH_2)_2CuLi$ → ?

Answer:

$(CH_3CH_2)_2CuLi$ →

Suzuki Coupling

17.6 What is the product of the following reaction?

+ $B(OH)_2$ $\xrightarrow[\text{NaOH}]{Pd(PPh_3)_4}$?

Answer:

+ $B(OH)_2$ $\xrightarrow[\text{NaOH}]{Pd(PPh_3)_4}$

17.7 What is the product of the following reaction?

+ $B(OH)_2$ $\xrightarrow[\text{NaOH}]{Pd(PPh_3)_4}$?

Answer:

+ $B(OH)_2$ $\xrightarrow[\text{NaOH}]{Pd(PPh_3)_4}$

17.8 What is the product of the following reaction?

NH_2 ... $B(OH)_2$ + Br $\xrightarrow[\text{NaOH}]{Pd(PPh_3)_4}$?

Answer: NH_2 ... $B(OH)_2$ + Br $\xrightarrow[\text{NaOH}]{Pd(PPh_3)_4}$

Heck Reaction

17.9 What is the product of the following reaction?

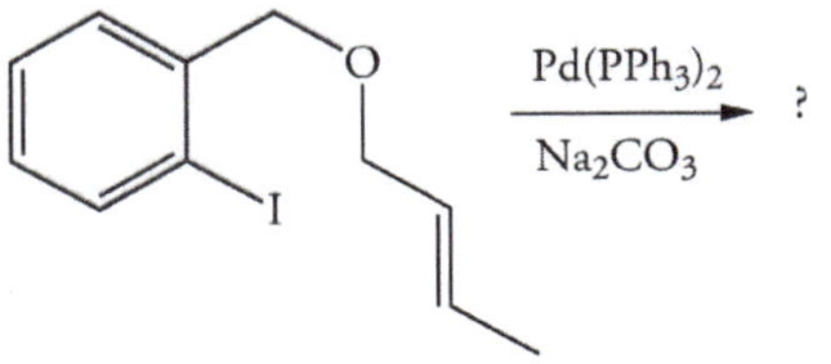

Answer:

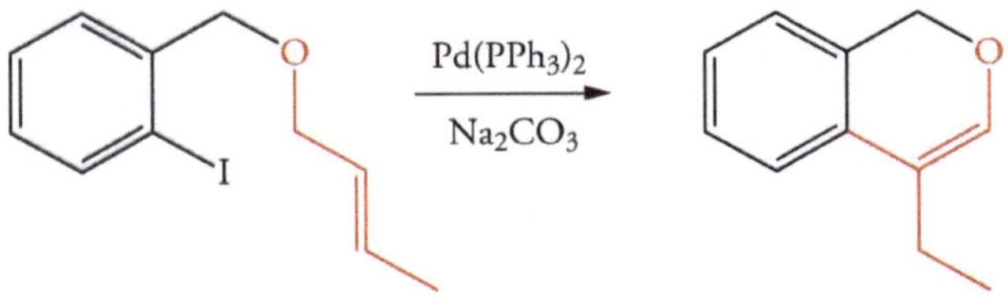

17.10 What is the product of the following reaction?

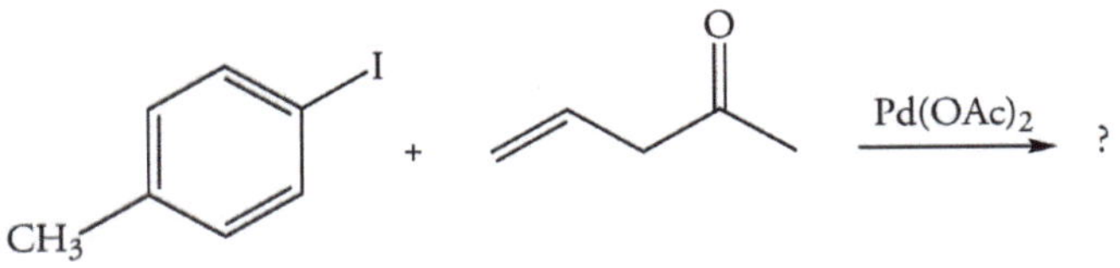

Answer:

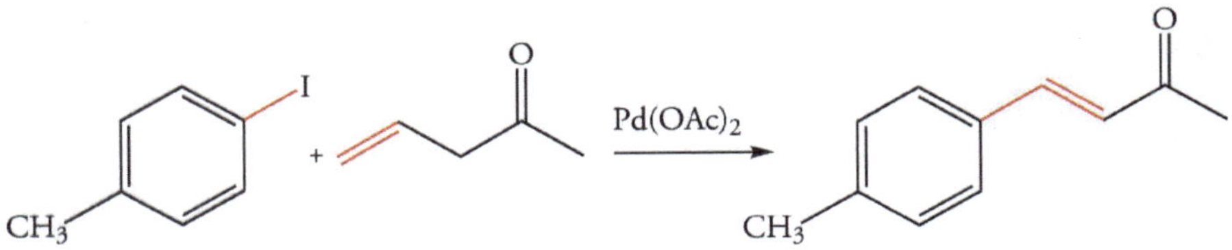

17.11 What is the product of the following reaction?

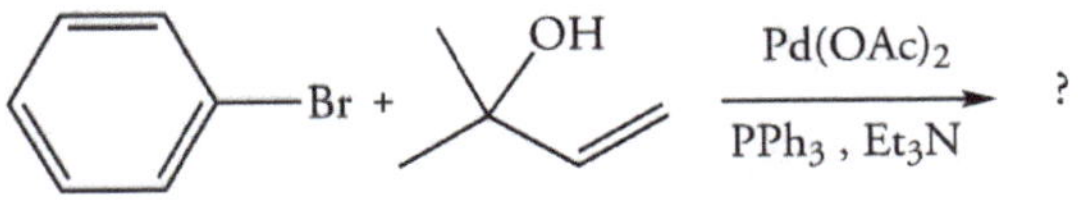

Answer:

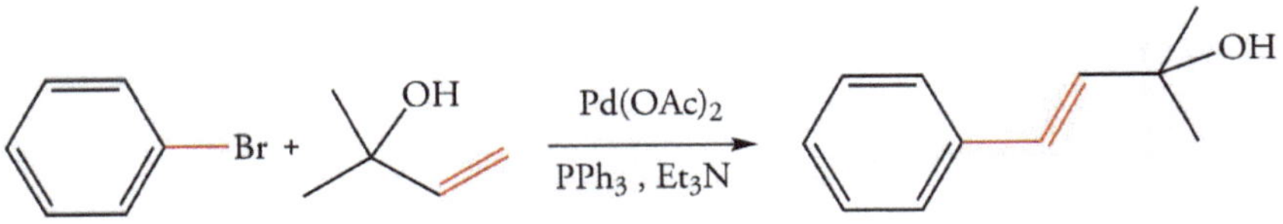

Sonogashira Coupling

17.12 What is the product of the following reaction?

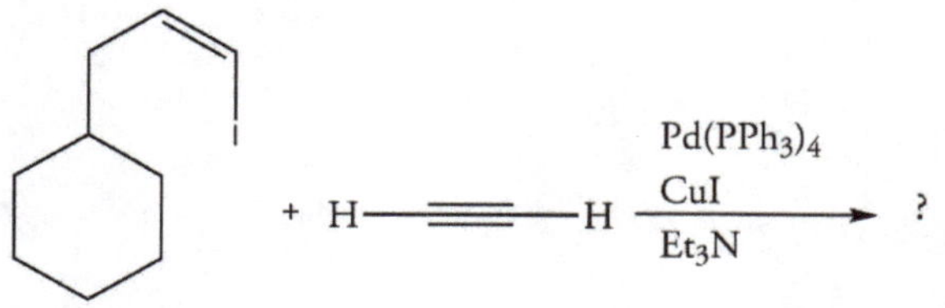

Answer:

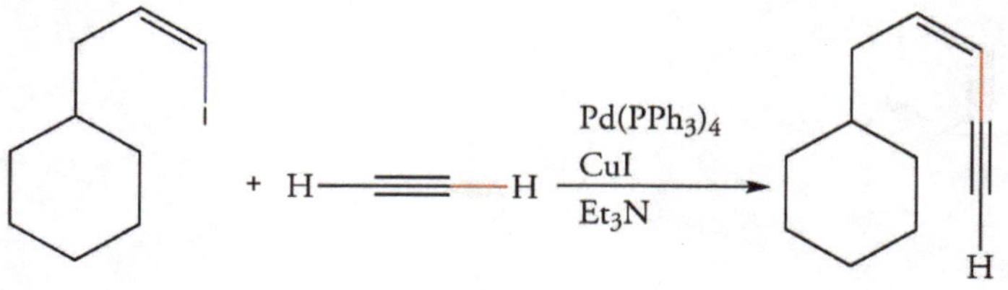

Wilkinson's Catalyst

17.13 What is the product of the following reaction?

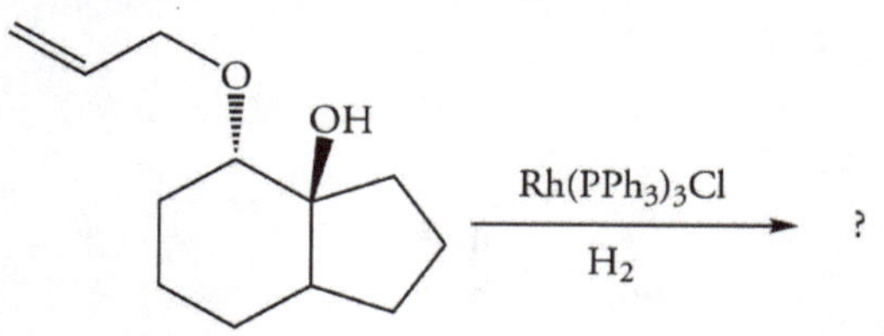

Answer:

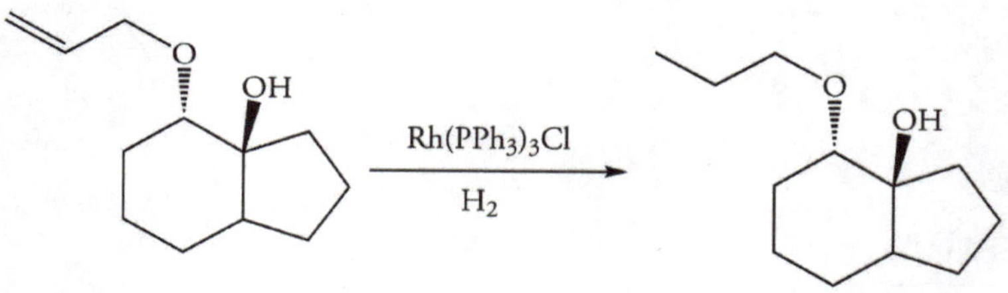

17.14 What is the product of the following reaction?

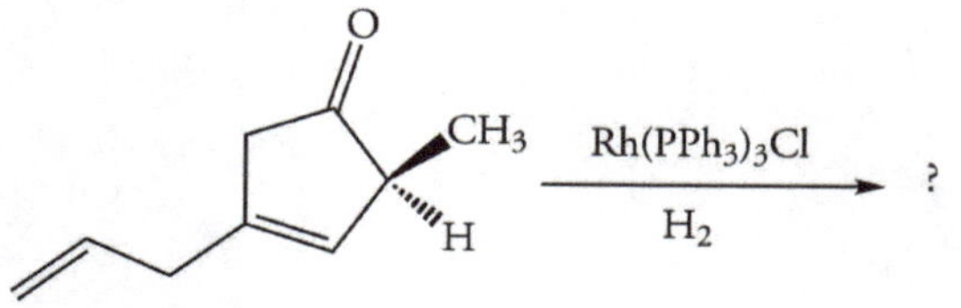

Answer:

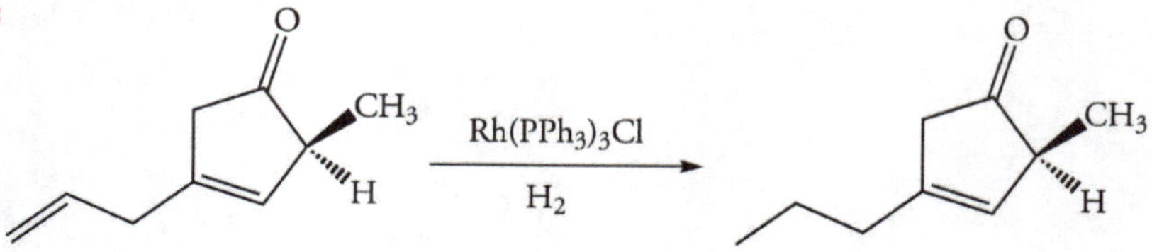

Grubbs Alkene Metathesis

17.15 What is the product of the following reaction?

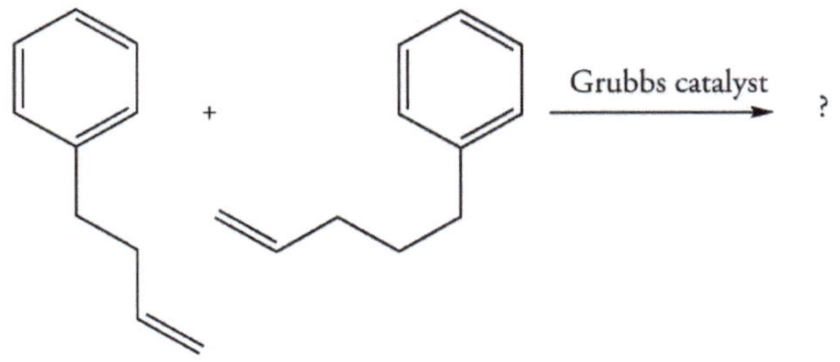

Answer:

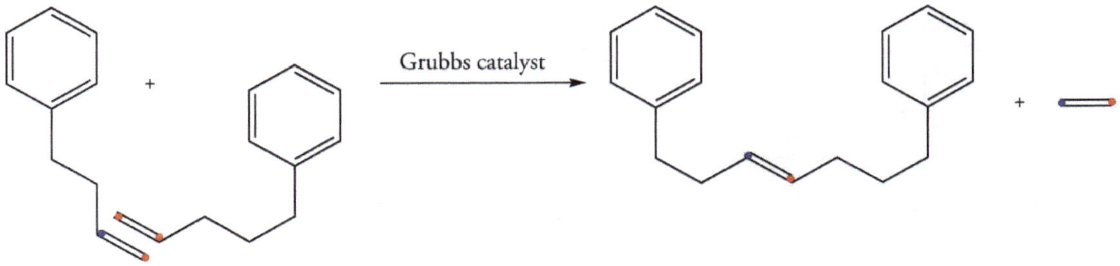

17.16 What is the product of the following reaction?

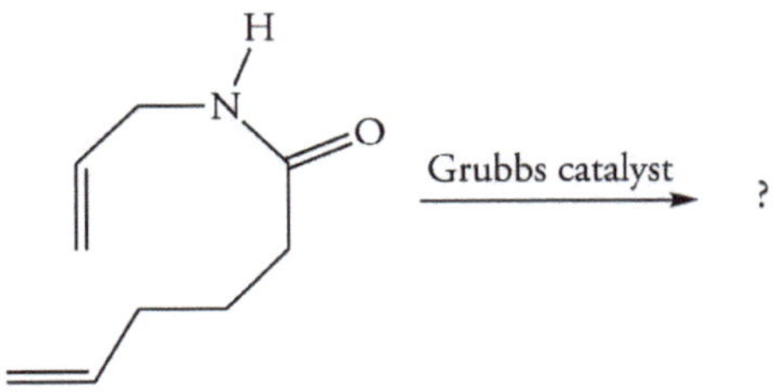

Answer:

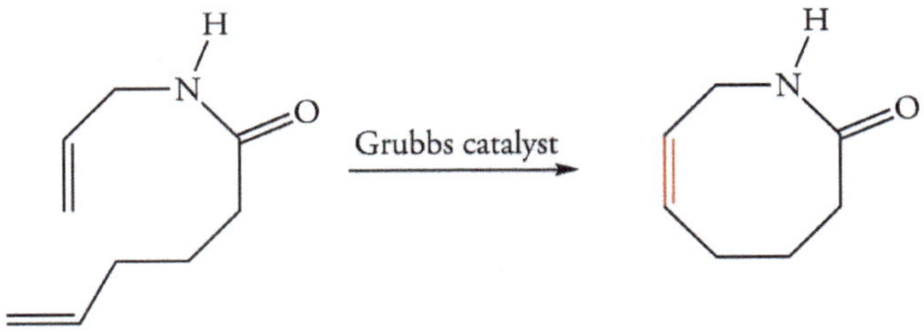

17.20 What is the product of the following reaction?

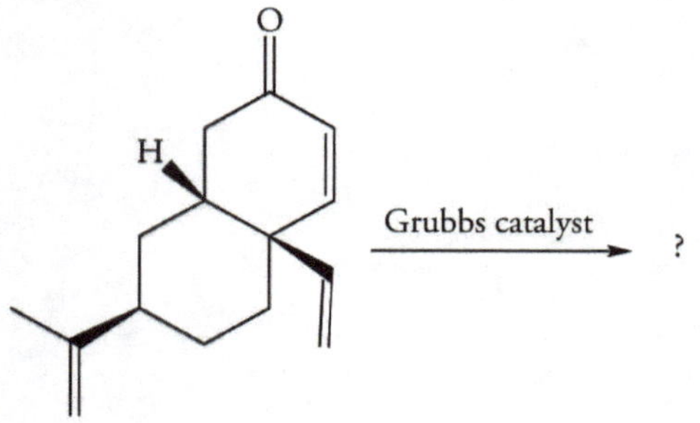

Grubbs catalyst

?

Answer:

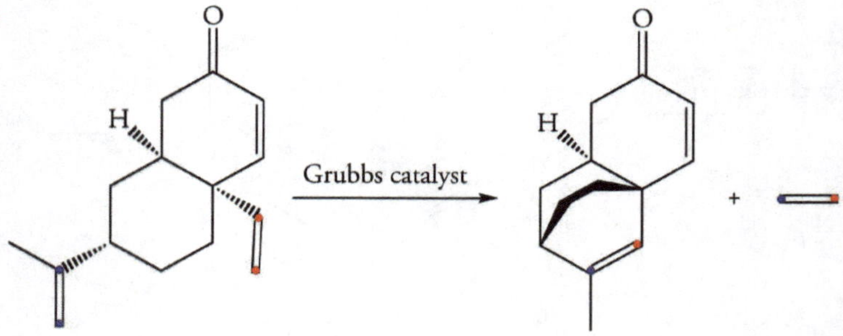

Grubbs catalyst

ALDEHYDES AND KETONES

The chemistry of carbonyl compounds will occupy us for most of the rest of this text. The key points of this chapter revolve around the structure of the carbonyl group and the influence of the structure of the carbonyl bond on the physical and chemical properties of aldehydes and ketones.

18.1 The Carbonyl Group

The carbonyl group is C=O with the carbon atom bonded to two other atoms. Carbonyl compounds with only hydrogen, alkyl, or aryl groups bonded to the carbonyl carbon atom are aldehydes or ketones. Aldehydes have one hydrogen atom and one alkyl or aryl group bonded to the carbonyl carbon atom. Ketones have only alkyl or aryl groups bonded to the carbonyl carbon atom.

The carbonyl carbon and oxygen are sp^2-hybridized. The reactivity of the carbonyl group is interpreted based on its π electrons and the two sets of nonbonded electrons. In addition, pay particular attention to the dipolar structure that is a contributing resonance structure for the carbonyl group. The structure of the carbonyl bond affects both the stability of carbonyl compounds and their reactivity.

18.2 Physical Properties of Aldehydes and Ketones

Carbonyl compounds are polar compounds, and as a result, they have higher boiling points than alkanes of similar molecular weight. The lower-molecular-weight carbonyl compounds are soluble in water because water can form hydrogen bonds to the carbonyl oxygen atom. Alkanes are insoluble in water.

Carbonyl compounds cannot form intermolecular hydrogen bonds like alcohols can, because they lack a hydrogen atom bonded to an electronegative atom such as oxygen. Thus, they have lower boiling points than alcohols of similar molecular weight.

18.3 Redox Reactions of Carbonyl Compounds

Aldehydes are easily oxidized by a variety of oxidizing agents, such as Tollens reagent, Benedict's solution, and Fehling's solution. A reaction is detected by the formation of metallic silver with Tollens reagent or a red precipitate of Cu_2O with Benedict's or Fehling's solution. Ketones are not oxidized by these reagents. The product of oxidation is a carboxylate ion which yields the carboxylic acid when neutralized. Recall that stronger oxidizing agents such as the Jones reagent (Chapter 15) oxidize primary alcohols to carboxylic acids. Thus, aldehydes are also oxidized by this reagent.

Aldehydes and ketones are reduced to primary and secondary alcohols, respectively. Hydrogen gas with a platinum catalyst may be used as a reducing agent, but high pressures are required, and any carbon–carbon double bond would be reduced first. Lithium aluminum hydride and sodium borohydride reduce carbonyl groups without affecting carbon–carbon double bonds. Aldehydes yield primary alcohols; ketones yield secondary alcohols.

The carbonyl group in both aldehydes and ketones can be reduced to a methylene group in a Clemmensen reduction with Zn/Hg and HCl, or with NH_2NH_2 and KOH in a Wolff–Kishner reduction.

18.4 Synthesis of Carbonyl Compounds

An effective synthesis of *any* type of functional group depends on the type of substrates available, and on the specificity of several possible reagents that could be used. We have already described four general methods to prepare carbonyl compounds that have already been presented.

1. Oxidation of alcohols
2. Friedel–Crafts acylation of aromatic compounds
3. Ozonolysis of alkenes
4. Oxidative cleavage of diols
5. Hydroboration–oxidation of alkynes

Primary alcohols are oxidized by PCC to give aldehydes. Secondary alcohols are oxidized by the same reagent to give ketones. The Jones reagent (CrO_3 in acetone/H_2SO_4) further oxidizes primary alcohols beyond the aldehyde state to give carboxylic acids. The Jones reagent reacts with secondary alcohols to give ketones, which are not further oxidized.

Friedel–Crafts acylation gives a carbonyl compound with the carbonyl group directly attached to an aromatic ring. The reaction is limited to aromatic compounds lacking strongly deactivating groups. Two variations are an intramolecular cyclization reaction of carboxylic acids using HF, and the Gatterman–Koch reaction using CO and HCl, which behave together like formyl chloride to give an aldehyde.

Ozonolysis of an alkene gives two carbonyl fragments. If the fragments are identical or if one of the two fragments can be isolated, then the method may be useful. For example, the ozonolysis of a methylenecycloalkane give formaldehyde and a cycloalkanone. This method depends on the availability of the alkene, which in turn must usually be prepared by an elimination reaction of another substrate.

The oxidative cleavage of diols also gives two carbonyl fragments. As a synthesis of carbonyl compounds, this reaction is limited because usually a synthetic method focuses on formation of a single product. Furthermore, the diol must be prepared by oxidation of an alkene.

Indirect hydration of alkynes can be controlled to add one mole of a borane. The resulting enol has a hydroxyl group bonded to an unsaturated carbon atom. The enol rapidly rearranges to the isomeric ketone. Because an alkyne can react twice with any reagent, the addition reaction of a borane is controlled by using disiamylborane, which has two sterically hindering 1,2-dimethylpropyl groups. Only one mole of the borane compound adds to the double bond at the least hindered position.

18.5 A Preview of Carbonyl Synthesis

Each of the synthetic methods given in this section depends on the chemistry of functional groups we will encounter in future chapters.

Acid chlorides can be prepared by the reaction of carboxylic acids with thionyl chloride. They are reduced to aldehydes by either of two methods. The Rosenmund reduction uses hydrogen gas and a modified palladium catalyst. Lithium tri(*tert*-butoxy)aluminum hydride also converts acid chlorides to aldehydes. The aldehyde is not further reduced by this hydride reagent as it would be by LiAlH$_4$.
Esters are reduced to aldehydes using diisobutylaluminum hydride (DIBAL). The aldehyde itself is not formed in the reaction but is formed in the aqueous workup reaction to give a hemiacetal which decomposes to the aldehyde. We will discuss these latter reactions in Chapter 18.

Carboxylic acids and acid chlorides react with certain organometallic compounds to give ketones. Addition of an organolithium compound to a carboxylic acid gives a salt of a geminal diol that gives a diol upon hydrolysis. This rapidly eliminates water to give a ketone. Reaction of an acid chloride with a Gilman reagent gives a ketone directly. The product does not react with the Gilman reagent.

Nitriles are reduced by modified hydride reagents to give imine anion intermediates. Upon aqueous workup, the imine hydrolyzes to give the more stable carbonyl compound, which is an aldehyde. Grignard reagents react with nitriles to give a salt of an imine. It is converted to an imine and then to a ketone in the aqueous workup.

18.6 Spectroscopy of Aldehydes and Ketones

The infrared spectra of aldehydes and ketones have strong carbonyl group absorptions in the neighborhood of 1700 cm^{-1}. Aldehydes absorb at slightly higher wavenumber than ketones. The aldehyde C—H bond also has a characteristic absorption at 2710 cm^{-1}. The position of the carbonyl absorption occurs at lower wavenumber with conjugation due to a greater contribution of the dipolar resonance form because it decreases the double bond character of the carbonyl group. The position of the carbonyl absorption of cycloalkanones depends on ring size.

The proton NMR of aldehydes has a characteristic absorption near 10 δ due to the aldehyde C—H bond. The a hydrogen atoms of both aldehydes and ketones have NMR absorptions in the 2.0–2.5 δ region

The α carbon atoms of both aldehydes and ketones have C-13 NMR absorptions in the 30–50 δ region. The carbonyl carbon atom is easily identified by its absorption in the 190–220 δ region.

Summary of Reactions

1. Oxidation of Aldehydes

$$\text{2-naphthaldehyde} \xrightarrow{\text{Ag(NH}_3)_2{}^+ \text{ (aq)}} \text{2-naphthoic acid} + \text{Ag(s)}$$

$$\text{cyclohexanecarbaldehyde} \xrightarrow{\text{Cu}_2{}^+ / \text{HO}^-} \text{cyclohexanecarboxylic acid} + \text{Cu}_2\text{O}$$

2. Reduction of Aldehydes and Ketones to Alcohols

$$\xrightarrow[\text{ethanol}]{\text{NaBH}_4}$$

$$\xrightarrow[\text{2. H}_3\text{O+}]{\text{1. LiAlH}_4}$$

3. Reduction of Aldehydes and Ketones to Methylene Groups

$$\xrightarrow{\text{Zn(Hg) / HCl}}$$

$$\xrightarrow{\text{H}_2\text{NNH}_2 / \text{KOH}}$$

4. Synthesis of Aldehydes and Ketones by Oxidation of Alcohols

(a) CH_3O—⟨ ⟩—$\text{CH}_2\text{OH} \xrightarrow{\text{PCC}} \text{CH}_3\text{O}$—⟨ ⟩—CHO

(b) CH_2—CH_2—$\overset{\overset{\displaystyle \text{OH}}{|}}{\underset{\underset{\displaystyle \text{H}}{|}}{\text{C}}}$—$\text{CH}_3 \xrightarrow{\text{Jones reagent}} \text{CH}_2$—$\text{CH}_2$—$\overset{\overset{\displaystyle \text{O}}{\|}}{\text{C}}$—$\text{CH}_3$

5. Synthesis of Aryl Ketones by Friedel–Crafts Acylation

6. Synthesis of Carbonyl Compounds by Ozonolysis of Alkenes

7. Synthesis of Carbonyl Compounds by Oxidative Cleavage of Vicinal Diols

2-methyl-1-phenyl-1,2-butanediol   benzaldehyde   2-butanone

cis-1,2-cyclohexanediol   6-oxohexanal

8. Synthesis of Carbonyl Compounds by Hydration of Alkynes

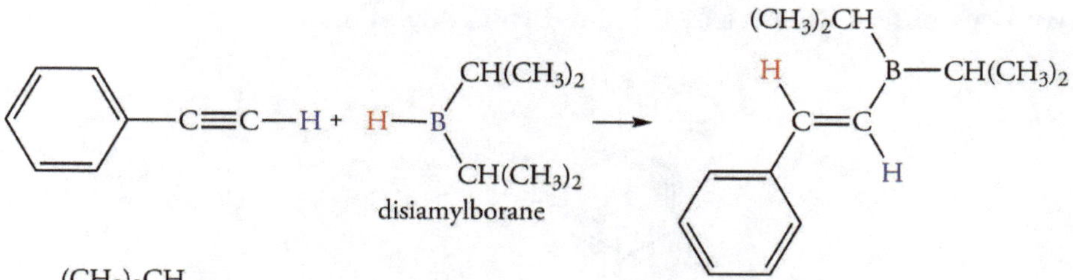

$C\equiv C-H$ + $H-B$ disiamylborane $\rightarrow$ (with $CH(CH_3)_2$ groups)

disiamylborane

$\xrightarrow[\text{NaOH}]{H_2O_2}$

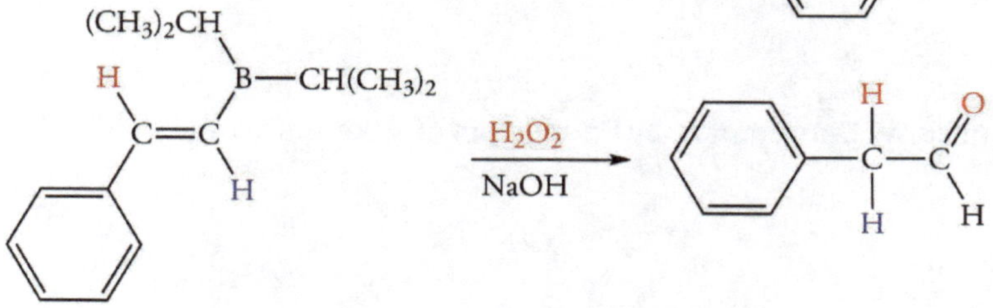

9. Synthesis of Carbonyl Compounds by Reduction of Acid Derivatives

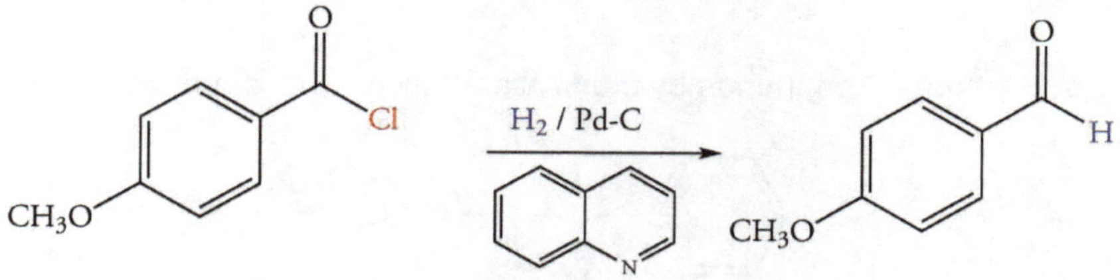

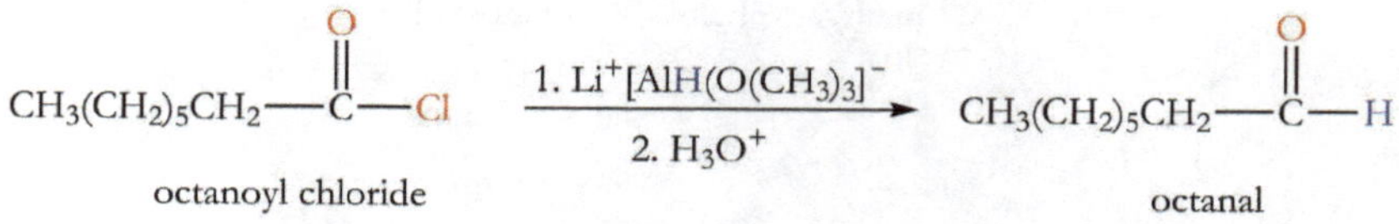

$\xrightarrow{H_2 \ / \ Pd\text{-}C}$ (quinoline)

CH_3O- (acid chloride) $\rightarrow$ CH_3O- (aldehyde)

$CH_3(CH_2)_5CH_2-\overset{O}{\overset{\|}{C}}-Cl$ $\xrightarrow[\text{2. } H_3O^+]{\text{1. } Li^+[AlH(O(CH_3)_3)]^-}$ $CH_3(CH_2)_5CH_2-\overset{O}{\overset{\|}{C}}-H$

octanoyl chloride octanal

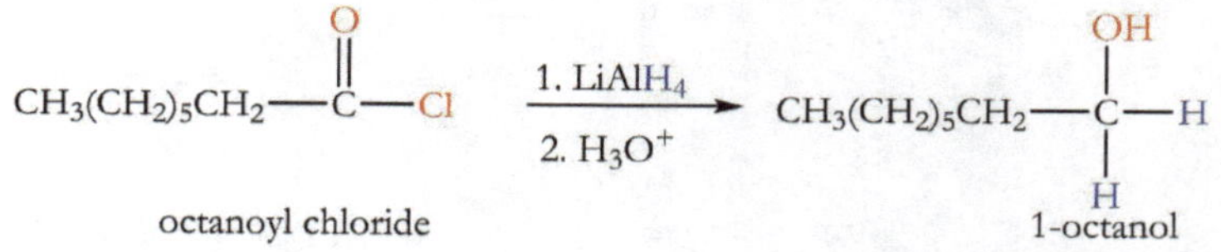

$CH_3(CH_2)_5CH_2-\overset{O}{\overset{\|}{C}}-Cl$ $\xrightarrow[\text{2. } H_3O^+]{\text{1. } LiAlH_4}$ $CH_3(CH_2)_5CH_2-\overset{OH}{\underset{H}{\overset{|}{C}}}-H$

octanoyl chloride 1-octanol

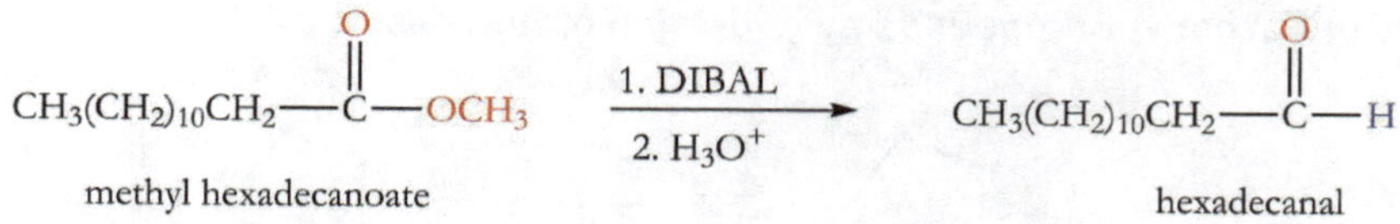

$CH_3(CH_2)_{10}CH_2-\overset{O}{\overset{\|}{C}}-OCH_3$ $\xrightarrow[\text{2. } H_3O^+]{\text{1. DIBAL}}$ $CH_3(CH_2)_{10}CH_2-\overset{O}{\overset{\|}{C}}-H$

methyl hexadecanoate hexadecanal

10. Reactions of Organometallic Reagents with Acid Derivatives

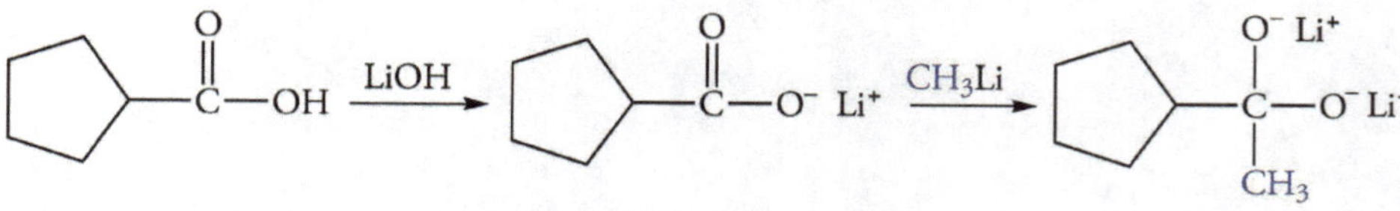

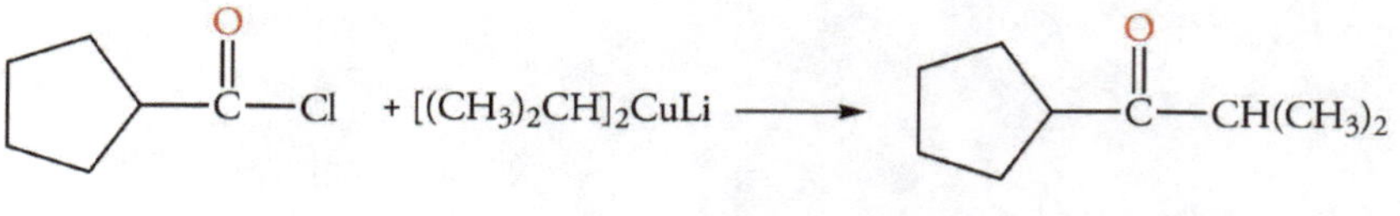

(cyclopentyl)$-\overset{O}{\overset{\|}{C}}-OH$ $\xrightarrow{LiOH}$ (cyclopentyl)$-\overset{O}{\overset{\|}{C}}-O^- Li^+$ $\xrightarrow{CH_3Li}$ (cyclopentyl)$-\overset{O^- Li^+}{\underset{CH_3}{\overset{|}{C}}}-O^- Li^+$

(cyclopentyl)$-\overset{O}{\overset{\|}{C}}-Cl$ + $[(CH_3)_2CH]_2CuLi$ $\longrightarrow$ (cyclopentyl)$-\overset{O}{\overset{\|}{C}}-CH(CH_3)_2$

11. Formation of Carbonyl Compounds from Nitriles

$$\text{C}_6\text{H}_5-\text{CH}_2-\text{C}\equiv\text{N} \xrightarrow[\text{2. H}_3\text{O}^+]{\text{1. LiAlH(OCH}_2\text{CH}_3)_3} \text{C}_6\text{H}_5-\text{CH}_2-\overset{\displaystyle O}{\text{C}}\text{H}$$

$$\text{C}_6\text{H}_5-\text{CH}_2-\text{C}\equiv\text{N} \xrightarrow[\text{2. H}_3\text{O}^+]{\text{1. CH}_3\text{MgBr}} \text{C}_6\text{H}_5-\text{CH}_2-\overset{\displaystyle O}{\text{C}}\text{CH}_3$$

End of Chapter Exercises

Nomenclature of Aldehydes and Ketones

18.1 Write the structure for each of the following compounds.

(a) 2-methylbutanal   (b) 3-ethylpentanal   (c) 2-bromopentanal

(d) 3,4-dimethyloctanal   (e) 1-bromocyclobutanecarbaldehyde

Answers:

(a)
$$\begin{array}{c} CH_3 \\ | \\ CH_3CH_2CHCHO \end{array}$$
   (b)
$$\begin{array}{c} CH_2CH_3 \\ | \\ CH_3CH_2CHCH_2CHO \end{array}$$
   (c)
$$\begin{array}{c} Br \\ | \\ CH_3CH_2CH_2CHCHO \end{array}$$

(d)
$$\begin{array}{c} CH_3 \\ | \\ CH_3CH_2CH_2CH_2CHCHCH_2CHOCHO \\ | \\ CH_3 \end{array}$$
   (e) [cyclobutane ring with Br and CHO]

18.2 Write the structure of each of the following compounds.

(a) 3-bromo-2-pentanone   (b) 2,4-dimethyl-3-pentanone   (c) 4-methyl-2-pentanone

(d) 3,4-dimethyl-2-pentanone   (e) 2-methyl-1,3-cyclohexanedione

Answers:

(a)
$$\begin{array}{c} Br\ \ O \\ |\ \ || \\ CH_3CH_2CHCCH_3 \end{array}$$
   (b)
$$\begin{array}{c} CH_3\ CH_3 \\ |\ \ \ | \\ CH_3CHCCHCH_3 \\ || \\ O \end{array}$$
   (c)
$$\begin{array}{c} CH_3\ \ O \\ |\ \ \ || \\ CH_3CHCH_2CCH_3 \end{array}$$

(d)
$$\begin{array}{c} CH_3\ \ O \\ |\ \ \ || \\ CH_3CHCHCCH_3 \\ | \\ CH_3 \end{array}$$
   (e) [cyclohexanedione ring with CH_3]

18.3 Give the IUPAC name of each of the following compounds.

(a) $CH_3CH_2CH_2CHO$ (b)
$$\begin{array}{c} CH_3 \\ | \\ CH_3C-CH_2CHO \\ | \\ CH_3 \end{array}$$
   (c)
$$\begin{array}{c} CH_3 \\ | \\ CH_3CHCHO \end{array}$$
   (d)
$$\begin{array}{c} CH_3 \\ | \\ CH_3CH_2CHCHCHO \\ | \\ CH_2H_3 \end{array}$$

Answers:

(a) butanal   (b) 3,3-dimethylbutanal   (c) 2-methylpropanal   (d) 2-ethyl-3-methylpentanal

18.4 Give the IUPAC name of each of the following compounds.

(a)
$$\begin{array}{c} O \\ || \\ CH_3CH_2CCH_2CH_3 \end{array}$$
   (b)
$$\begin{array}{c} O\ \ \ CH_3 \\ ||\ \ \ \ | \\ CH_3C-C-CH_3 \\ | \\ CH_3 \end{array}$$
   (c)
$$\begin{array}{c} CH_3 \\ | \\ CH_3CHCCH_2CH_3 \\ || \\ O \end{array}$$
   (d)
$$\begin{array}{c} CH_3\ \ O \\ |\ \ \ || \\ CH_3CHCH_2CCH_3 \end{array}$$

Answers:

(a) 3-pentanone   (b) 3,3-dimethyl-2-butanone   (c) 2-methyl-3-pentanone   (d) 5-methyl-3-hexanone

18.5   Give the IUPAC name of each of the following compounds.

(a) 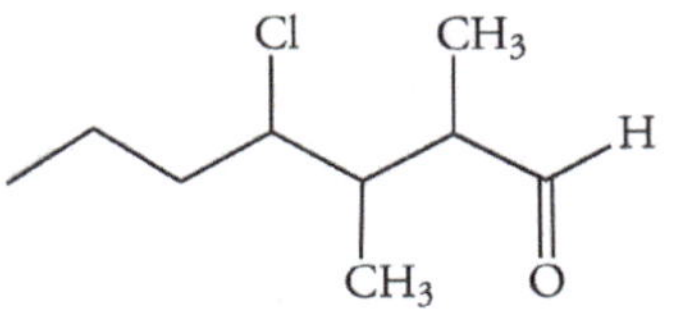   (b) 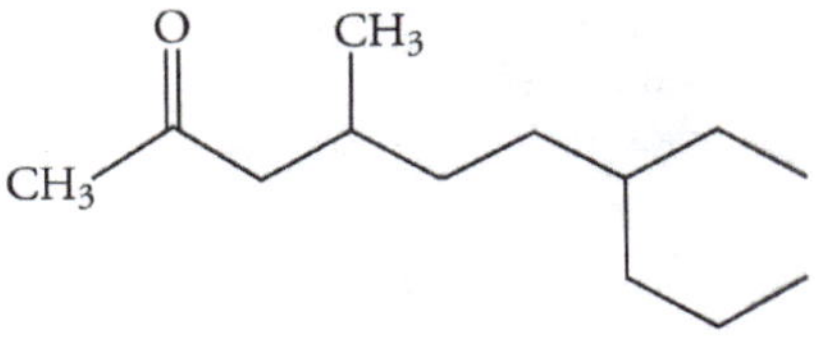   (c) 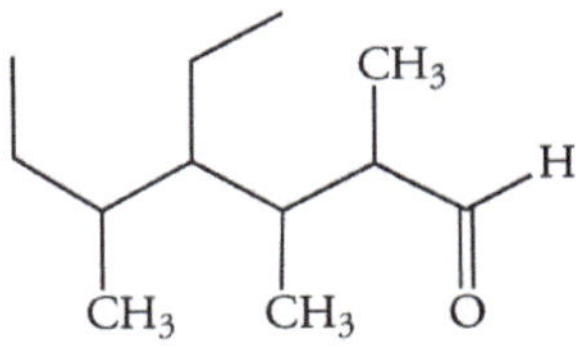   (d)

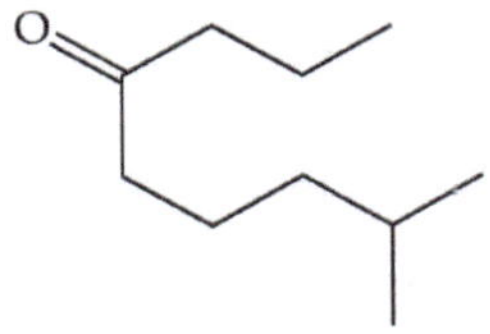

Answers:
(a) 4-chloro-2,3-dimethylheptanal   (b) 6-ethyl-3-methyl-2-nonanone   (c) 4-ethyl-2,3,5-trimethylheptanal   (d) 8-methyl-4-nonanone

18.6   Give the IUPAC name of each of the following compounds.

(a) 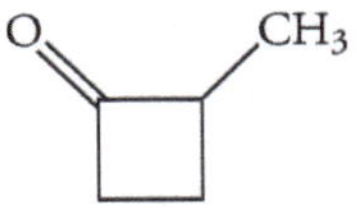   (b) 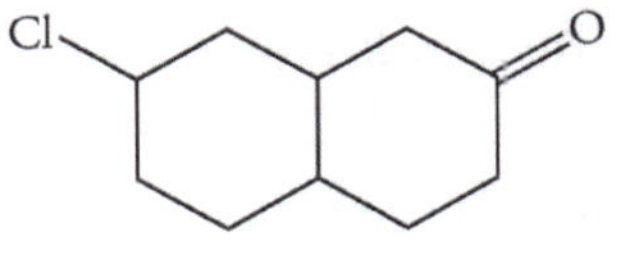   (c) 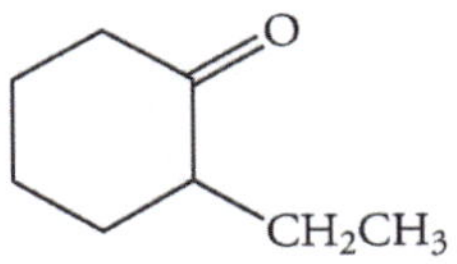   (d)

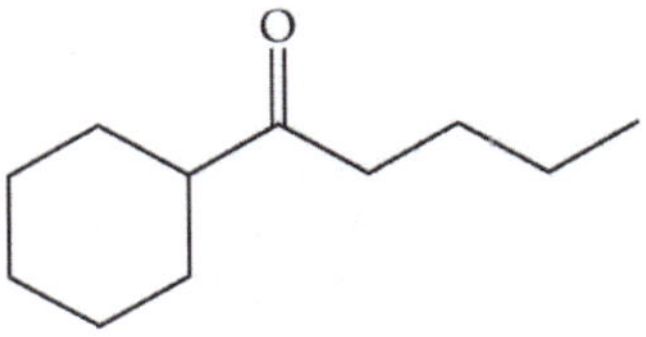

Answers:
(a) 2-methylcyclobutanone (b) 5-chlorocyclodecanone (c) 2-ethylcyclohexanone   (d) 1-cyclohexyl-1-pentanone

18.7   Many aldehydes and ketones are better known by their common names. Draw the structural formula of each of the following carbonyl compounds. Their common names are given within parentheses.
    (a)   2,2-dimethylpropanal (pivaldehyde)
    (b)   2-hydroxy-1,2-diphenyl-1-ethanone (benzoin)
    (c)   2-propenal (acrolein)
    (d)   4-methyl-3-penten-2-one (mesityl oxide)
    (e)   5,5-dimethyl-1,3-cyclohexanedione (dimedone)

(a) 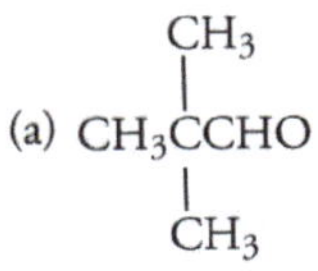 $CH_3\overset{\displaystyle CH_3}{\underset{\displaystyle CH_3}{C}}CHO$ (b) 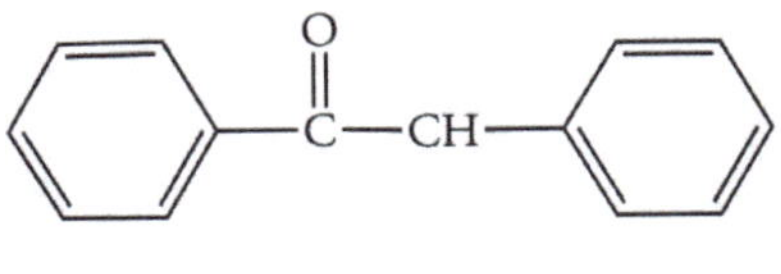   (c) $CH_2{=}CHCHO$

(d) 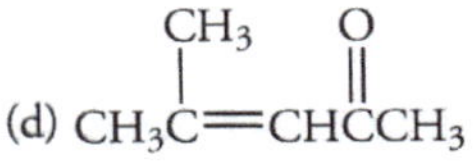 $CH_3\overset{\displaystyle CH_3}{C}{=}CH\overset{\displaystyle O}{C}CH_3$ (e)

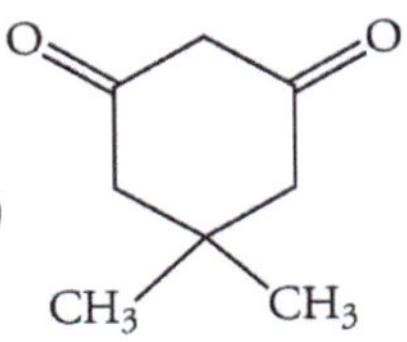

18.8 Draw the structural formula of each of the following carbonyl compounds. The common name of each compound is given within parentheses.

(a) 3,3-dimethyl-2-butanone (pinacolone) (b) 4-hydroxy-4-methyl-2-pentanone (diacetone alcohol)
(c) (E)-2-butenal (crotonaldehyde) (d) 1,3-dipheny1-2-buten-l-one (dypnone)
(e) 2,3-butanedione (biacetyl)

Answers:

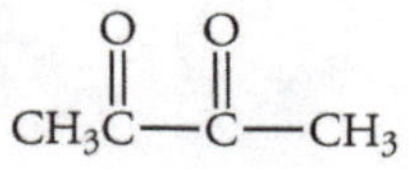

Properties of Aldehydes and Ketones

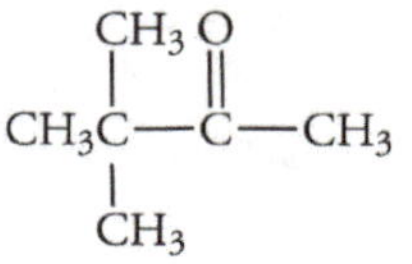

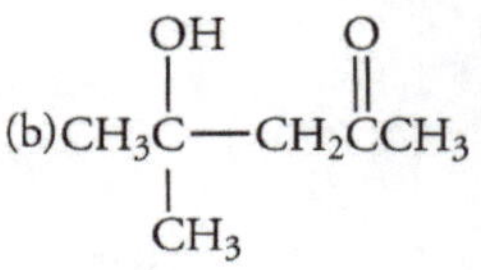

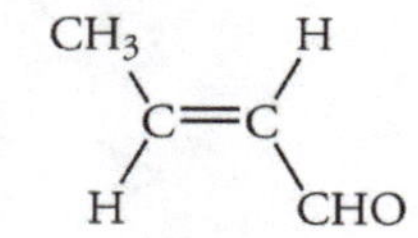

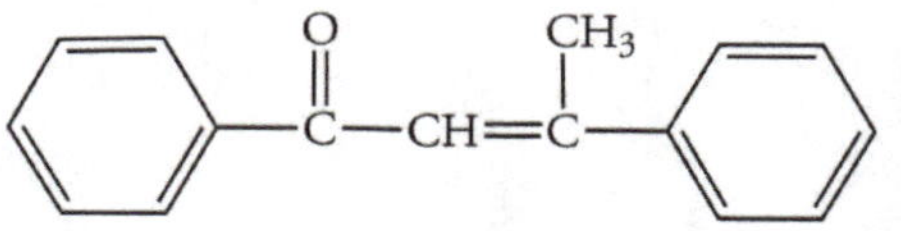

18.9 The H—C—H bond angle of formaldehyde is 116.5°. The H—C—C bond angle of acetaldehyde is 118.2°. Explain this difference.

Answer: Although sterically small, the methyl and hydrogen atoms bonded to the carbonyl carbon atom occupy more space than two hydrogen atoms of formaldehyde and the bond angle widens to accommodate the methyl group.

18.10 The C=C bond length in alkenes and the C=O bond length in aldehydes are 134 and 123 pm, respectively. Explain this difference.

Answer: The atomic radius of oxygen is smaller than the atomic radius of carbon, and the bond length between two atoms reflects their respective atomic radii.

18.11 The dipole moments of acetone and isopropyl alcohol are 2.7 and 1.7 D, respectively. Explain this difference.

Answer: The oxygen atom of the "two" carbon–oxygen bonds of the carbonyl group pulls electron density away from the carbonyl carbon atom more than the single oxygen atom of the carbon–oxygen bond of the alcohol.

18.12 The dipole moments of propanal and propenal are 2.52 and 3.12 D, respectively. Consider the resonance forms of these compounds and explain the difference in their dipole moments.

Answer: An additional dipolar resonance form of propenal gives added stabilization of the positive charge and increases the polarity of the molecule.

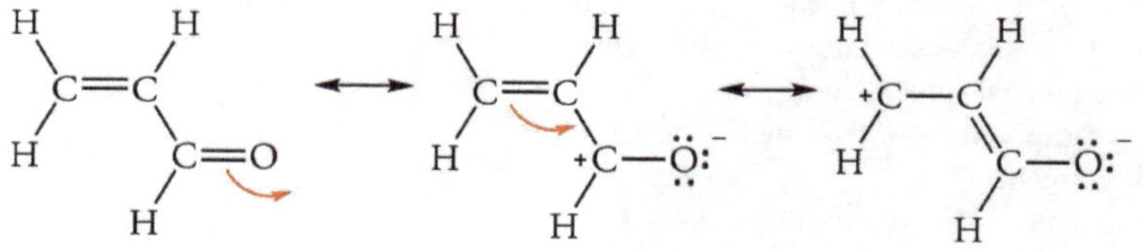

18.13 The boiling points of butanal and 2-methylpropanal are 75 and 61 °C, respectively. Explain this difference.

Answer: 2-Methylpropanal is a more compact molecule that has a more spherical shape than the cylindrically shaped butanal, so its London forces are smaller.

18.14 The boiling points of 2-heptanone, 3-heptanone, and 4-heptanone are 151, 147, and 144 °C, respectively. What is responsible for this trend?

Answer: As the polar carbonyl group moves to the interior of the molecule, the compound more closely resembles an alkane structure and the intermolecular attractive forces are smaller.

18.15 The boiling points of 2-hydroxy- and 3-hydroxybenzaldehydes are 197 and 240 °C, respectively. Suggest a reason for this difference.

Answer: As the polar carbonyl group moves to the interior of the molecule, the compound more closely resembles an alkane structure and the intermolecular attractive forces are smaller.

18.16 The boiling points of 2-hydroxy- and 3-hydroxyacetophenones are 218 and 296 °C, respectively. Suggest a reason for this difference.

Answer: 2-Hydroxyacetophenone can form intramolecular hydrogen bonds, which decrease the number of intermolecular hydrogen bonds that would cause a higher boiling point like that of 3-hydroxyacetophenone.

18.17 The solubilities of butanal and 1-butanol in water are 7 and 9 g/100 mL, respectively. Explain this difference.

Answer: Butanal is a hydrogen bond acceptor of water molecules but 1-butanol is both a hydrogen bond acceptor and a hydrogen bond donor, so it is somewhat more soluble.

18.18 The solubilities of butanal and 2-methylpropanal in water are 7 and 11 g/100 mL, respectively. Explain this difference.

Answer: 2-Methylpropanal is a more compact molecule that has a more spherical shape than the cylindrically shaped butanal, so it interferes less with the hydrogen bonding arrays of water.

Oxidation and Reduction of Carbonyl Compounds

18.19 What is observed when an aldehyde reacts with Benedict's solution? What is observed when an aldehyde reacts with Tollens reagent?

Answer: A red precipitate of Cu_2O forms when an aldehyde reacts with Benedict's solution. A precipitate of silver usually seen as a silver mirror forms when an aldehyde reacts with Tollens reagent.

18.20 What class of compounds results from the reduction of ketones with sodium borohydride? What class of compounds results from the reduction of aldehydes by lithium aluminum hydride?

Answer: Ketones yield secondary alcohols when reduced using sodium borohydride. Aldehydes yield primary alcohols when reduced using lithium aluminum hydride.

18.21 Draw the structure of the product of each of the following reactions.

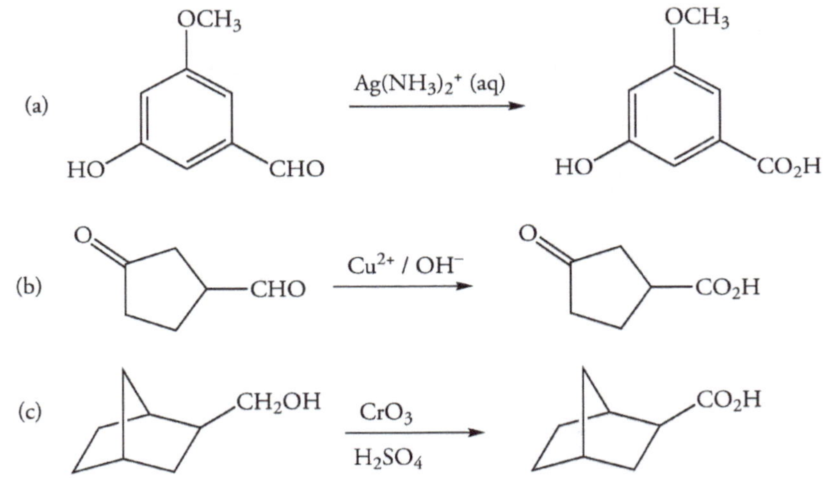

Answers:
(a) Oxidation of an aldehyde by Tollenss reagent yields a carboxylate ion in solution. Upon neutralization, the isolated product is the carboxylic acid.
(b) Oxidation of an aldehyde by Benedict's solution or Fehling's solution yields a carboxylate ion in solution. Upon neutralization, the isolated product is the carboxylic acid.
(c) Oxidation of a primary alcohol by CrO_3 in sulfuric acid yields a carboxylic acid.

18.22 Determine the best choice of reactants to synthesize each of the following ethers using the Williamson method.

(a) $CH_3-\overset{\overset{O}{\parallel}}{C}$—〈benzene〉—$\overset{\overset{O}{\parallel}}{C}$—H $\xrightarrow{Cu^{2+}\ /\ OH^-}$ $CH_3-\overset{\overset{O}{\parallel}}{C}$—〈benzene〉—$CO_2H$

(b) reaction with $Ag(NH_3)_2^+$ (aq)

(c) CH_2OH $\xrightarrow{PCC}$ CHO

Answers:

(a) Oxidation of an aldehyde by Benedict's solution or Fehling's solution yields a carboxylate ion in solution. Upon neutralization, the isolated product is the carboxylic acid.

(b) Oxidation of an aldehyde by Tollenss reagent yields a carboxylate ion in solution. Upon neutralization, the isolated product is the carboxylic acid.

(c) Oxidation of a primary alcohol by PCC acid yields an aldehyde

18.23 What is the product when each of the following reacts with lithium aluminum hydride?

(a) $\xrightarrow[\text{2. }H_3O^+]{\text{1. LiAlH}_4\ /\text{ether}}$

(b) $\xrightarrow[\text{2. }H_3O^+]{\text{1. LiAlH}_4\ /\text{ether}}$ CH_2OH

(c) $-CH_2-\overset{\overset{O}{\parallel}}{C}-$ $\xrightarrow[\text{2. }H_3O^+]{\text{1. LiAlH}_4\ /\text{ether}}$ $-CH_2CH-C_6H_5$ with OH

Answer: Lithium aluminum hydride reduces aldehydes to primary alcohols and ketones to secondary alcohols. Carbon–carbon double bonds and aromatic rings are unaffected.

18.24   What is the product when each of the following reacts with sodium borohydride?

Answer: Sodium borohydride reduces aldehydes to primary alcohols and ketones to secondary alcohols. Carbon–carbon double bonds and aromatic rings are unaffected.

18.25   Explain why the reduction of carvone by lithium aluminum hydride yields two products.

Answer: The hydride reagent can attack from either face of the carbonyl group, so both *cis* and *trans* isomers may form.

18.26   Explain why the reduction of the following compound by sodium borohydride yields two products.

Answer: The hydride reagent can attack from either face of the carbonyl group, so both *cis* and *trans* isomers may form.

18.27 What is the product of each of the following reactions?

(a)

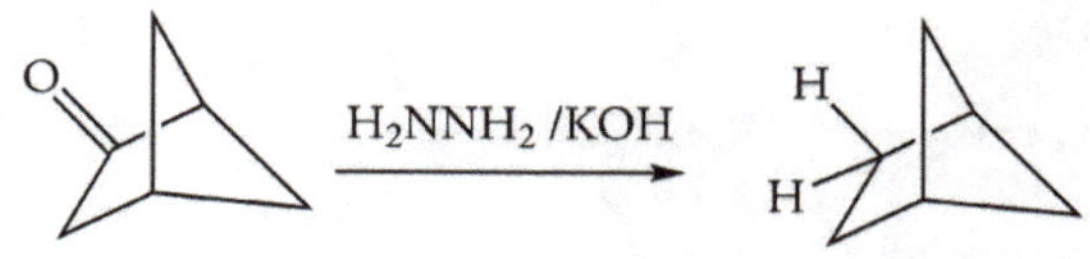

H_2NNH_2 /KOH

(b) 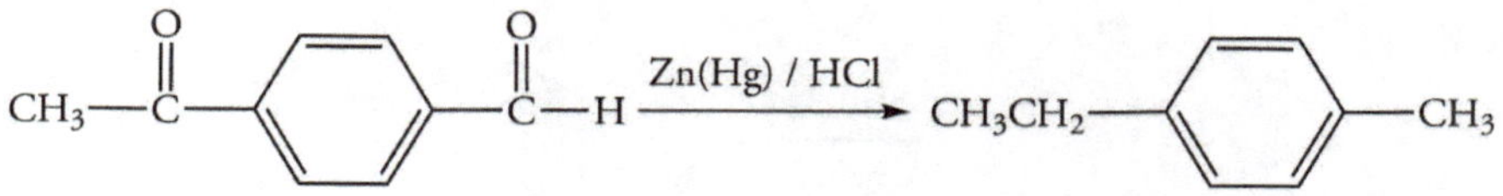 $CH_3-\overset{\overset{\displaystyle O}{\|}}{C}-$ $\overset{\overset{\displaystyle O}{\|}}{C}-H$ $\xrightarrow{\text{Zn(Hg) / HCl}}$ CH_3CH_2- $-CH_3$

(c)

$\xrightarrow[\text{Pd}]{H_2 \text{ (1 atm)}}$

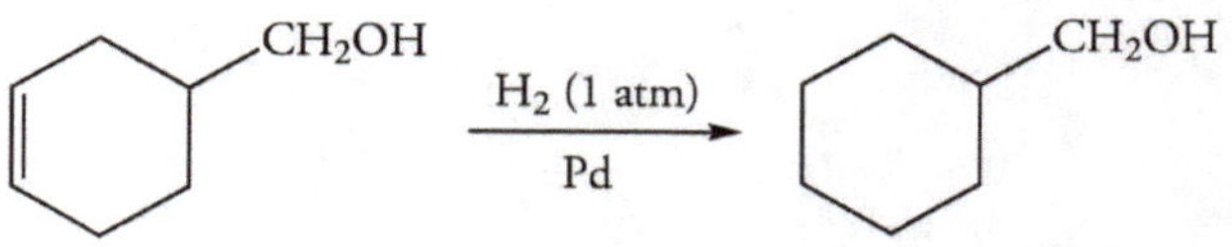

Answers:
(a)   The Wolff–Kishner reduction converts the ketone into a methylene group.
(b)   The Clemmensen reduction converts the ketone into a methylene group and the aldehyde group into a methyl group.
(c)   At atmospheric pressure the carbon–carbon double bond is reduced.

18.28 What is the product of each of the following reactions?

(a)

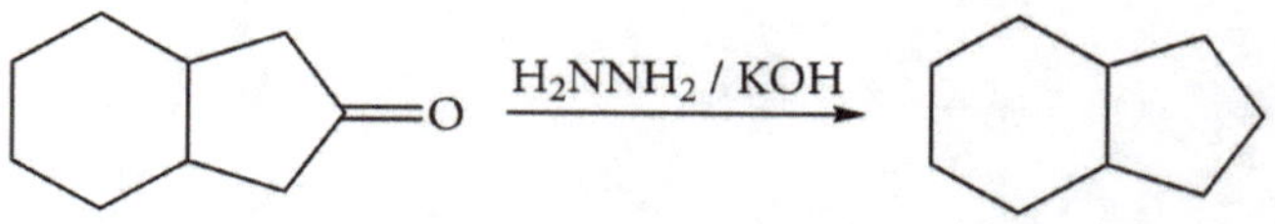

$=O$ $\xrightarrow{H_2NNH_2 \text{ / KOH}}$

(b)

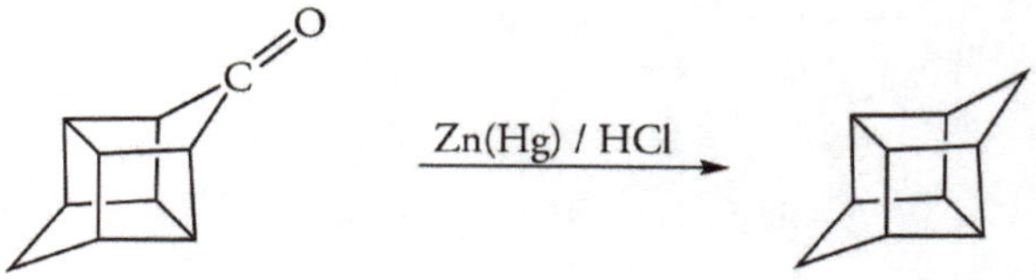

$\xrightarrow{\text{Zn(Hg) / HCl}}$

(c)

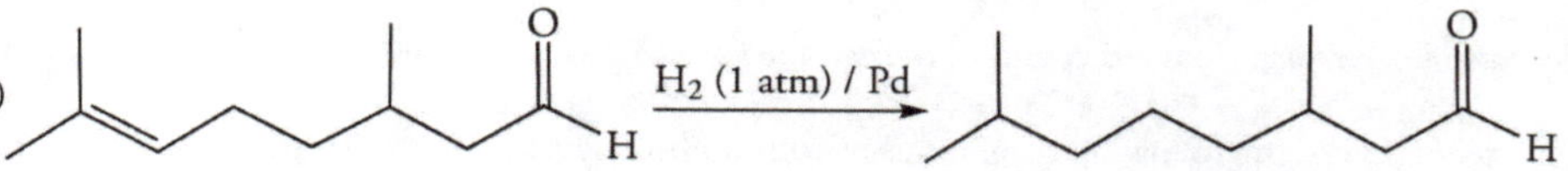

$\xrightarrow{H_2 \text{ (1 atm) / Pd}}$

Answers:
(a)   The Wolff–Kishner reduction converts the ketone into a methylene group.
(b)   The Clemmensen reduction converts the ketone into a methylene group.
(c)   At atmospheric pressure, the carbon–carbon double bond is reduced and the carbonyl group is unaffected.

19 ALDEHYDES AND KETONES: NUCLEOPHILIC ADDITION REACTIONS

KEYS TO THE CHAPTER   The chemistry of carbonyl compounds depends on the bonding characteristics of the carbonyl group. A dipolar resonance form contributes to the carbonyl group. Much of the chemistry of carbonyl compounds involves addition of a nucleophile to the partially positive carbon atom of the carbonyl group. Aldehydes and ketones undergo addition reactions with unsymmetrical reagents such as H—Nu. The nucleophilic part of the reagent adds to the carbonyl carbon, and the electrophilic part adds to the oxygen atom. These products are obtained in both acid- and base-catalyzed reactions.

19.1 Synthesis of Cyanohydrins

Hydrogen cyanide adds to an aldehyde or ketone to give a compound called a **cyanohydrin**. The equilibrium constant for the addition of HCN is much more favorable for aldehydes than for ketones. Thus, as we noted above, because ketones are more stable than aldehydes, the addition reactions of ketones are less favorable (have smaller equilibrium constants) than addition reactions of aldehydes.

19.2 Hydration of Carbonyl Compounds

The equilibrium constant for the addition of water to a carbonyl compound is greater than one for only a few compounds such as formaldehyde. Although hydration is not a useful synthetic reaction, the equilibrium constant for hydration provides a basis for evaluating the effect of structure on carbonyl addition reactions in general.

The equilibrium constants for hydration formation is greater for aldehydes than for ketones because the carbonyl carbon of an aldehyde has a greater partial positive charge than the carbonyl carbon of a ketone. Nucleophiles are attracted to a carbonyl carbon atom because of its partial positive charge. The partial positive charge is larger in an aldehyde, which has only one alkyl group bonded to the carbonyl carbon atom, than in a ketone, which has two alkyl groups. There is also a steric effect. The carbonyl carbon atom has a larger bond angle than the tetrahedral product. As a result of increased steric hindrance, the addition product is less stable for ketones than for aldehydes.

Groups bonded to the α carbon atom can affect the electron density of the carbonyl carbon atom by either resonance or inductive effects. Resonance stabilization of the carbonyl group decreases the equilibrium constant for addition reactions. Inductive electron withdrawal by electronegative groups such as halogen atoms destabilizes the carbonyl group and increases the equilibrium constant for addition reactions.

19.3 Mechanisms of Carbonyl Compounds Addition Reactions

Addition reactions to carbonyl groups can occur by either of two mechanisms depending on whether the reaction conditions are acidic or basic. In both cases, a nucleophile adds to the carbonyl carbon atom, and an electrophile adds to the oxygen carbon. The difference between the mechanisms for addition under acidic or basic conditions is in the sequence of two possible reactions. Under acidic conditions, a proton adds to the carbonyl oxygen atom to give a conjugate acid that has a greater positive charge on the carbonyl carbon atom, which then reacts with a nucleophile. Under basic conditions, the nucleophile attacks the carbonyl carbon atom to give an alkoxide ion, which subsequently reacts with the electrophile—usually a proton.

19.4 Kinetic Effects in Addition Reactions

The same structural features that control the equilibrium constant for the addition reaction of carbonyl compounds also affect the rate of the reaction. In the case of the rate of a reaction, the stabilization of the partial positive charge on the carbonyl carbon diminishes its attraction to a nucleophile. Thus, aldehydes react faster than ketones. The steric features of the alkyl groups in aldehydes, and to a greater degree in ketones, affect the reaction by decreasing the rate of reaction with increasing size of the groups. The groups bonded to the carbonyl carbon sterically hinder the approach of the nucleophile in the same way observed in S_N2 reactions.

19.5 Addition of Alcohols to Carbonyl Compounds

Alcohols add to aldehydes to give acetals and to ketones to give ketals. The equilibrium constant is less than one. However, if an intramolecular acetal or ketal can form, then the equilibrium constant is greater than one. There is very little difference in the bond energies of the reactants and products in either case. However, there is an entropy change for the intermolecular reaction since two moles of reactants give one mole of products. For the intramolecular reaction, the entropy change is nearly zero because there is no difference in the number of moles of reactant and product.

Hemiacetals result from the reaction of an aldehyde, RCHO, with an alcohol, R'OH. In a hemiacetal a hydroxyl group, an —OR' group from the alcohol, an —R group from the aldehyde, and a hydrogen atom all bonded to the same carbon atom. Hemiketals result from the reaction of a ketone, R_2CO, with an alcohol, R'OH. In a hemiketal, the original carbonyl carbon is bonded to an —OH group an —OR' group from the alcohol, and two —R groups from the ketone. The —R groups may be alkyl or aryl groups in either hemiacetals or hemiketals.

19.6 Formation of Acetals and Ketals

Hemiacetals can be converted into acetals and hemiketals to ketals. An acetal results from the reaction of a hemiacetal with an alcohol in acidic solution. An acetal has two —OR' groups from the alcohol, an —R group from the original aldehyde, and a hydrogen atom bonded to the original carbonyl carbon. A ketal results from the reaction of a hemiketal with an alcohol in acidic solution. A ketal has two —OR' groups from the alcohol and two —R groups from the ketone bonded to the original carbonyl carbon. The —R groups may be alkyl or aryl groups

The formation of acetals from aldehydes and alcohols is driven to completion by removing the water that forms during the reaction.

The mechanism of conversion of a hemiacetal into an acetal (or a hemiketal into a ketal) proceeds by protonation of the alcohol to convert it into a better leaving group. The resulting oxocarbocation is resonance stabilized. The oxocarbocation subsequently reacts with the nucleophilic oxygen atom of the alcohol, and the last step is proton transfer from the oxonium ion. The reaction requires acidic conditions. As a result, acetals (or ketals) can regenerate the carbonyl compound under acidic conditions in the presence of water. However, both acetals and ketals are stable in basic solutions.

19.7 Acetals and Ketals as Protecting Groups

A protecting group must easily form a derivative of a functional group and easily removed to regenerative it when required. A protecting group makes a functional group unreactive toward specific reagents that are required to transform a second functional group in the molecule. Acetals and ketals are ideal protecting groups because they are easily formed in acidic solution and easily removed when the compound is again exposed to acid.

Cyclic acetals and ketals derived from ethylene glycol are used as protecting groups for carbonyl compounds. In dicarbonyl compounds, it is possible to selectively form a derivative of one carbonyl group if the structural features between the two are sufficiently different.

Alcohols can also be protected by their incorporation into an acetal or ketal. Both 1,2-diols and 1,3-diols react with acetone to give a ketal. Alcohols react under acid-catalyzed conditions with dihydropyran to give a THP derivative which is easily hydrolyzed.

19.8 Thioacetals and Thioketals

The sulfur analogs of acetals and ketals are synthesized using the dithiols 1,2-ethanedithiol or 1,3-ethanedithiol. Thioacetals and thioketals are stable under acidic and basic conditions. Unprotecting the carbonyl group requires mercury(II) chloride. Thioacetals and thioketals can be reduced to methylene compounds using Raney nickel. Thus, the sequence of thioacetal formation followed by reduction is another method to reduce carbonyl compounds to hydrocarbons.

19.9 Addition of Nitrogen Compounds

Amines and other nitrogen derivatives that are sufficiently nucleophilic attack the carbonyl carbon atom to give a tetrahedral addition product, called a hemiaminal, but it is unstable with respect to dehydration. As a result, an imine forms. The overall reaction is an addition–elimination reaction In every case, the net result of adding a nitrogen compound, $R-NH_2$, to an to an aldehyde or ketone gives a product in which the original carbonyl carbon is bonded to the nitrogen with a double bond.

19.10 The Wittig Reaction

Alkenes can be synthesized from carbonyl compounds by reaction with a phosphorus ylide in the Wittig reaction. It is produced by the reaction of a primary or secondary alkyl halide with triphenylphosphine to obtain a phosphonium ion. This compound is then treated with a strong base such as butyl lithium, which removes a proton from the alkyl group. The negatively charged carbon atom of the ylide reacts as a nucleophile, and attacks the carbonyl carbon atom to give an intermediate that triphenylphosphine oxide, leaving the desired alkene.

Summary of Reactions

1. Formation of Cyanohydrins

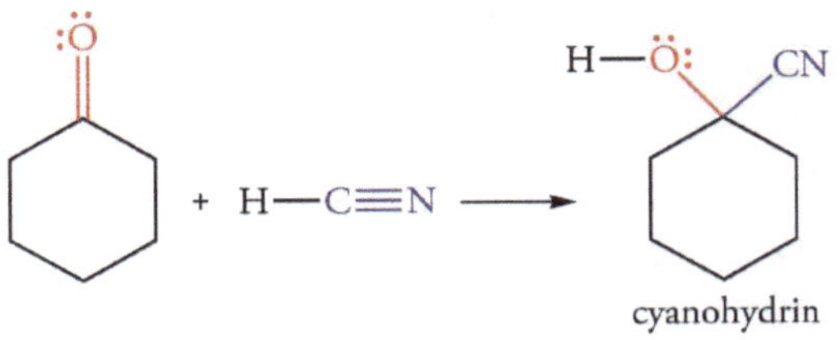

2. Formation of Acetals and Ketals

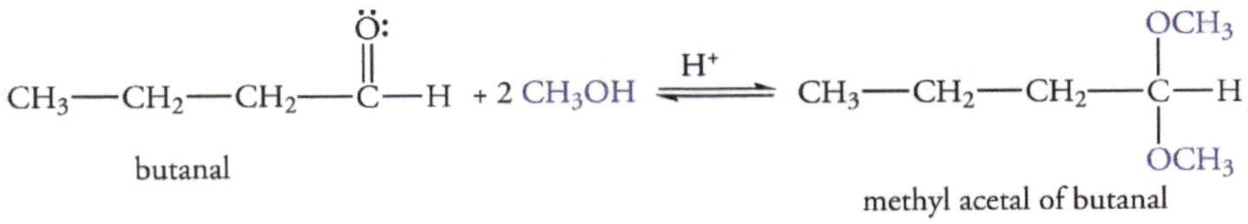

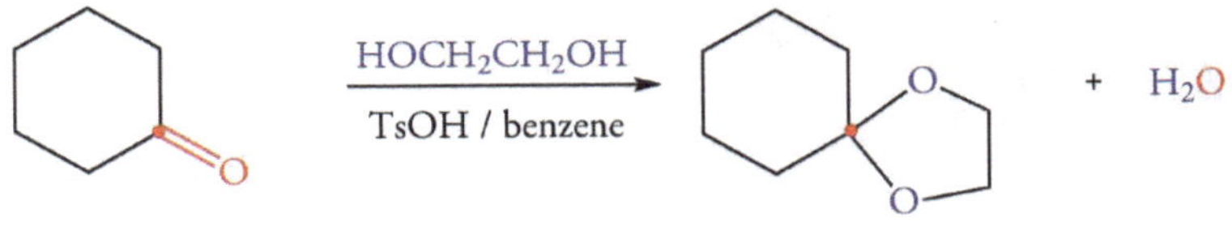

butanal

methyl acetal of butanal

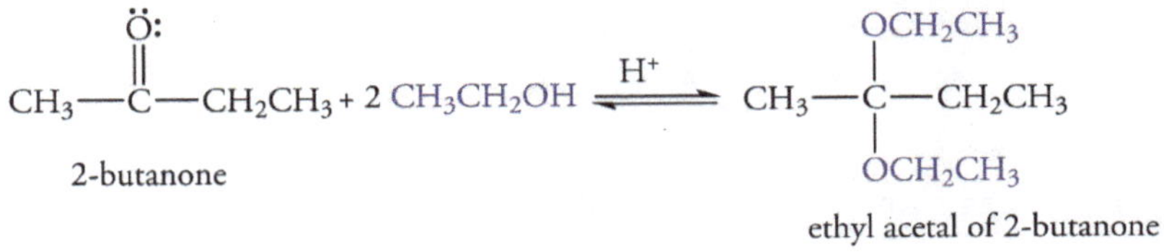

2-butanone

ethyl acetal of 2-butanone

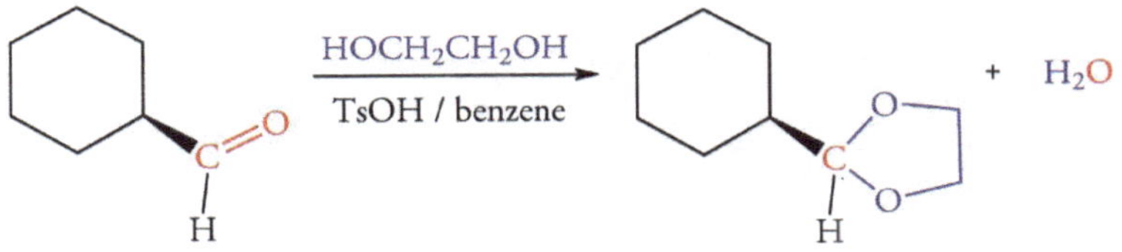

3. Protection of Alcohols by Acetal Formation

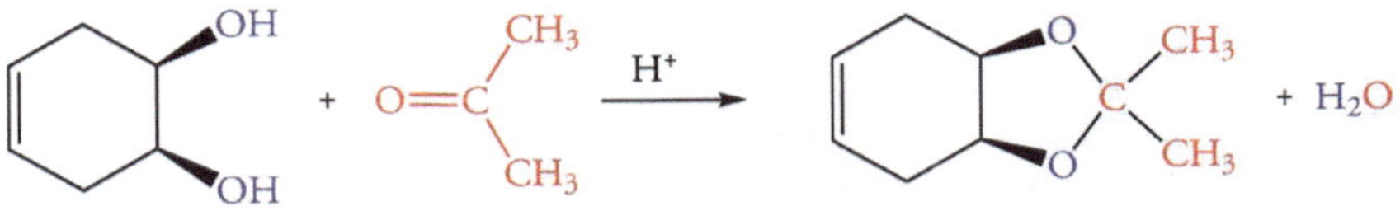

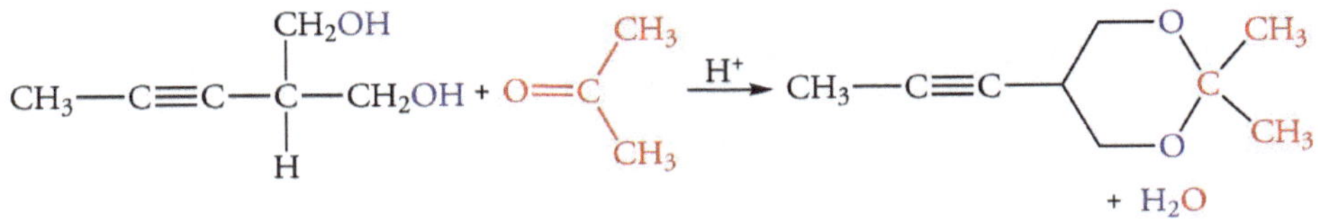

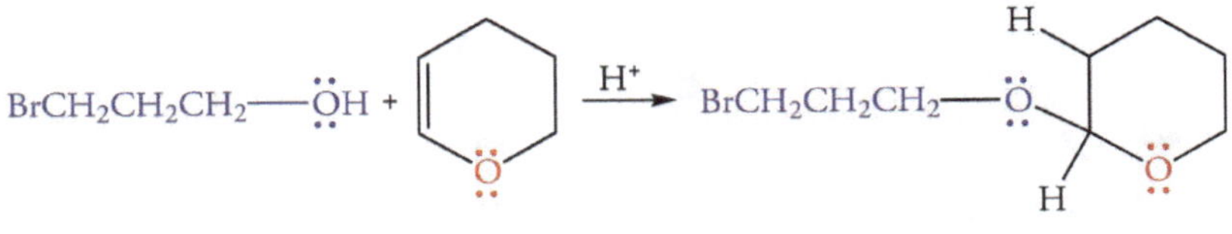

4. Thioacetals and Thioketals as Protecting Groups

5. Reduction of Thioacetals and Thioketals to Methylene Groups

6. Addition of Nitrogen Compounds to Carbonyl Compounds

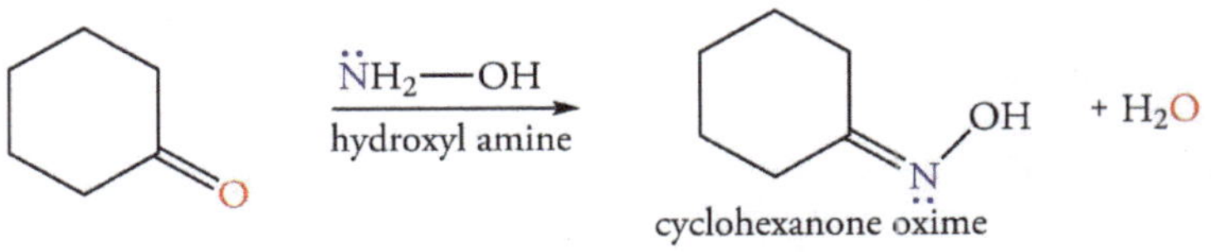

cyclohexanone oxime

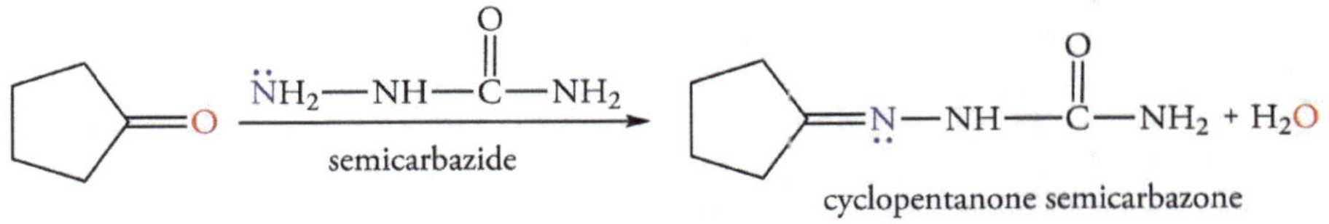

cyclopentanone semicarbazone

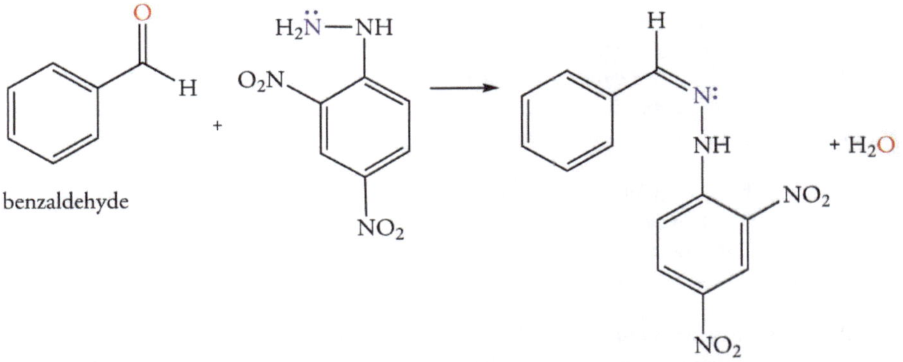

benzaldehyde

benzaldehyde 2,4-DNP

7. Synthesis of Alkenes With the Wittig Reaction

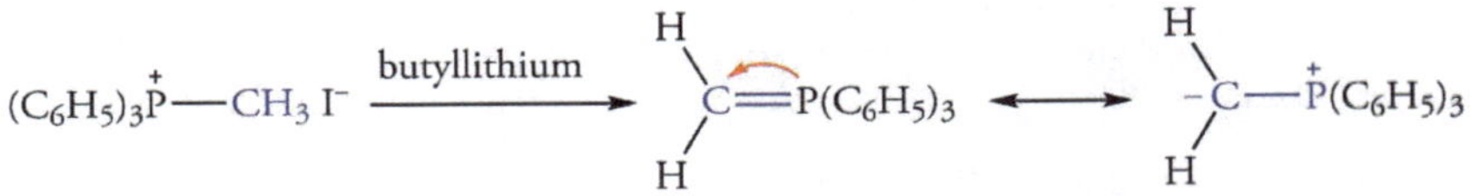

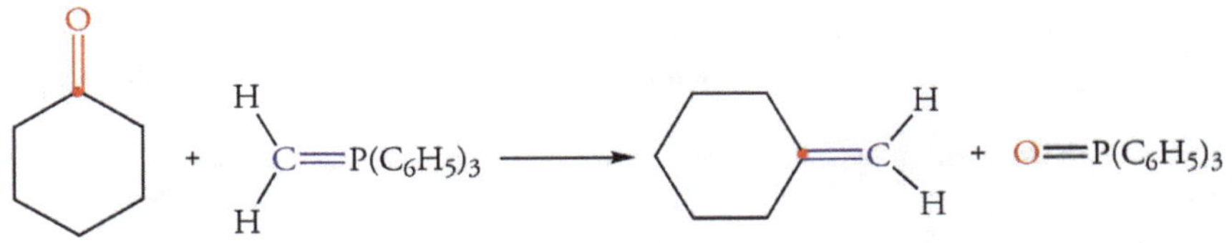

 **End of Chapter Exercises**

Reactivity of Carbonyl Compounds

19.1   Which member of each of the following pairs of compounds reacts faster with sodium borohydride?
   (a)   cyclopropanone or cyclopentanone
   (b)   acetophenone or benzaldehyde
   (c)   acetone or 3,3-dimethyl-2-butanone

Answers:

   (a) Cyclopropanone is the more reactive because of its greater ring strain. Thus, cyclopropanone reacts faster with sodium borohydride.
   (b) Both compounds have resonance-stabilized carbonyl groups. However, acetophenone also has a methyl group bonded to the carbonyl carbon atom and is a ketone. Thus, acetophenone has a more stabilized carbonyl group and is less reactive. The carbonyl group of acetophenone is also more sterically hindered than the carbonyl group of benzaldehyde. Thus, benzaldehyde reacts faster with sodium borohydride.
   (c) Both compounds are ketones, and the inductive effects of the alkyl groups are similar in both. However, one of the alkyl groups of 3,3-dimethyl-2-butanone is a *tert*-butyl group. Therefore, the carbonyl group of this compound is more sterically hindered than the carbonyl group of acetone. As a result, acetone reacts faster with sodium borohydride.

19.2   Which member of each of the following pairs of compounds reacts faster with sodium borohydride?
   (a)   benzaldehyde or acetaldehyde
   (b)   cyclopentanone or cyclohexanone
   (c)   *p*-trifluoromethylbenzaldehyde or benzaldehyde

Answers:

   (a) Benzaldehyde has a resonance-stabilized carbonyl group and is less reactive than an aldehyde such as acetaldehyde that does not have such stabilization. The carbonyl group of benzaldehyde is also more sterically hindered than the carbonyl group of acetaldehyde. As a result, acetaldehyde reacts faster with sodium borohydride.
   (b) Reduction of cyclopentanone gives an alcohol with additional steric hindrance. In the case of cyclohexanone, the product does not have additional steric hindrance because the C—H and C—OH bonds are each staggered with respect to bonds on adjacent carbon atoms. Thus, cyclohexanone reacts faster with sodium borohydride.
   (c) The trifluoromethyl group is inductively electron withdrawing, and it destabilizes the dipolar resonance form of the carbonyl group. As a result, *p*-trifluoromethylbenzaldehyde is more reactive than benzaldehyde.

Equilibrium Constants of Hydration Reactions

19.3   The equilibrium constants for the formation of hydrates of acetaldehyde and chloroacetaldehyde are 1 and 37, respectively. Explain whether you expect the equilibrium constant for formation of the hydrate of trichloroacetaldehyde to be greater or less than 37.

Answer:

   Each chlorine atom withdraws electron density from the carbonyl carbon by an inductive effect, and therefore increases the equilibrium constant for the reaction. The cumulative effect is huge: if each additional chlorine atom increase the equilibrium constant by a factor of 37, the estimated equilibrium constant is $(37)^3$, which is approximately 5×10^4.

19.4   The equilibrium constant for formation of a hydrate of acetone is 1.4×10^{-3}. Explain whether you expect the equilibrium constant for formation of the hydrate of 1,3-dichloroacetone to be greater or less than 1.4×10^{-3}.

Answer:

   As explained in exercise 19.3, each additional chlorine atom increases the equilibrium constant for the reaction. Therefore, the equilibrium constant for formation of the hydrate of 1,3-dichloroacetone is greater than 1.4×10^{-3}.

19.5   Explain why the methoxy group of *p*-methoxybenzaldehyde decreases the equilibrium constant for hydration relative to benzaldehyde, whereas the methoxy group of *m*-methoxybenzaldehyde increases the equilibrium constant.

Answer:

   The *p*-methoxy group stabilizes the carbonyl group by a resonance interaction in which the methoxy group releases electrons through the benzene ring to the carbonyl group. The *m*-methoxy group destabilizes the carbonyl group somewhat by an inductive effect in which the methoxy group withdraws electron density. No donation of electrons by resonance is possible for two groups located in *meta* positions.

19.6   The equilibrium constants for the formation of hydrates of acetone, acetophenone, and benzophenone are 1.4×10^{-3}, 6.6×10^{-6}, and 1.7×10^{-7}, respectively. Explain why the second phenyl group of benzophenone has a much smaller effect on the equilibrium constant than the phenyl group of acetophenone compared to acetone.

Answer:
One phenyl group can stabilize the carbonyl group because it can exist in a coplanar conformation with the carbonyl group, allowing the π electrons to overlap. Two phenyl groups cannot both be coplanar with the carbonyl group because there is a steric repulsion between the *ortho* hydrogen atoms of the two rings. The *ortho* hydrogen atoms of the two rings would have to occupy the same space if rings were coplanar. Rotation of the ring about the bond to the carbonyl carbon atom relieves this repulsion but eliminates the resonance interaction with the carbonyl group. The resonance interaction between the carbonyl group and the first phenyl group stabilizes acetophenone compared to acetaldehyde, but the second phenyl group has only a small effect.

One phenyl group rotates out of the plane, so the
π electrons in the two benzene rings do not overlap.

19.7   Explain why the equilibrium constant for hydration of cyclopropanone is significantly larger than for hydration of cyclopentanone.
Answer:
Cyclopropanone has a larger equilibrium constant because it has more ring strain cyclopentanone. The hydrate of cyclopropanone is less strained, so the reaction is favored.

19.8   Considering the role of torsional strain in determining cycloalkane stability, predict the order of the equilibrium constants for hydration of cyclopentanone and cyclohexanone.
Answer:
Hydration of cyclopentanone gives a hydrate with additional torsional strain between the C—OH bonds and the adjacent C—H bonds. The hydrate of cyclohexanone does not have added torsional strain because the C—OH bonds are staggered with respect to adjacent C—H bonds. Thus, the equilibrium constant for formation of the hydrate of cyclohexanone is larger.

Formation of Cyanohydrins

19.9   Explain why hydrogen cyanide reacts with 2-propanone to give a good yield of an addition product, but 2,2,4,4-tetramethyl-3-pentanone gives a poor yield in the same reaction.
Answer:
The carbonyl carbon atom of 2,2,4,4-tetramethyl-3-pentanone has two *tert*-butyl groups bonded to it at 120° angles. Formation of the cyanohydrin decreases the bond angle between the two *tert*-butyl groups to 109°, resulting in increased steric crowding. Thus, the reaction is less favored than the reaction with 2-propanone.

19.10   Explain why the equilibrium constant for formation of a cyanohydrin of cyclohexanone is much larger than the equilibrium constant for formation of a cyanohydrin of cyclopentanone.
Answer:
Adding HCN to cyclopentanone gives a cyanohydrin with additional torsional strain between the C—OH and C—CN bonds and the adjacent C—H bonds. The cyanohydrin of cyclohexanone does not have added torsional strain because the C—OH and C—CN bonds are each staggered with respect to adjacent C—H bonds. Thus, cyclohexanone has the larger equilibrium constant.

19.11   Explain why the equilibrium constant for formation of a cyanohydrin of butanone is much larger than the equilibrium constant for the formation of a cyanohydrin of 3,3-dimethylbutanone.
Answer:
The carbonyl carbon atom of 3,3-dimethyl-2-butanone has a *tert*-butyl and a methyl group bonded to it at 120° angles. Formation of the cyanohydrin decreases the bond angle between the two groups to 109°, resulting in increased steric strain. The alkyl groups in 2-butanone are methyl and ethyl, so the steric repulsion is far less. Thus, the reaction with 3,3-dimethylbutanone is less favored than the reaction with butanone.

19.12 Is the equilibrium constant for formation of a cyanohydrin of *p*-methoxylbenzaldehyde larger or smaller than the equilibrium constant for benzaldehyde?

Answer:
The *p*-methoxy group stabilizes the carbonyl group by a resonance interaction in which the methoxy group releases electrons through the benzene ring to the carbonyl group. Thus, its conversion to a cyanohydrin, in which this resonance stabilization is lost, is unfavorable. Therefore, the equilibrium constant for cyanohydrin formation is smaller for *p*-methoxylbenzaldehyde than for benzaldehyde.

19.13 Is the equilibrium constant for formation of a cyanohydrin of *p*-methylbenzaldehyde larger or smaller than the equilibrium constant for benzaldehyde?

Answer:
The *p*-methyl group donates electron density through the aromatic ring to the carbonyl group; hence, it stabilizes the carbonyl group. Thus, the conversion of *p*-methylbenzaldehyde to a cyanohydrin, in which this stabilization is lost, is less favorable than for benzaldehyde. The equilibrium constant for formation of the cyanohydrin of *p*-methylbenzaldehyde is smaller.

19.14 Is the equilibrium constant for formation of a cyanohydrin of *p*-ethoxybenzaldehyde larger or smaller than the equilibrium constant for *p*-dimethylaminobenzaldehyde?

Answer:
Nitrogen is a better donor of electrons by resonance than oxygen. Therefore, the dimethylamino group stabilizes the carbonyl group by a resonance interaction more effectively than the *p*-ethoxy group. Thus, the conversion of *p*-dimethylaminobenzaldehyde to a cyanohydrin, in which this stabilization is lost, is less favorable than for *p*-ethoxybenzaldehyde.

19.15 Explain why the equilibrium constant for formation of a cyanohydrin of 3,3,5-trimethylcyclohexanone is smaller than the equilibrium constant for cyclohexanone.

Answer:
Conversion into a cyanohydrin must place either a hydroxyl or a cyano group in an axial position, so there will be a 1,3-diaxial interaction with the axial methyl group at C-3. This steric repulsion between groups decreases the stability of the product and therefore decreases the equilibrium constant compared to that of cyclohexanone.

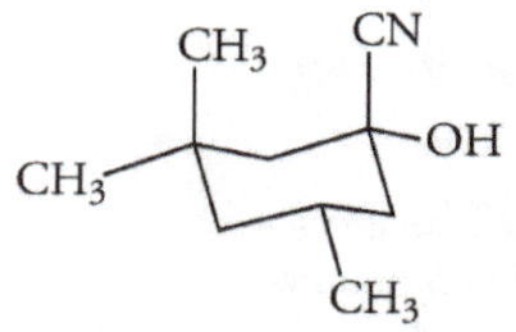

19.16 Two cyanohydrins of 4-*tert*-butyl-butylcyclohexanone exist. Which is the kinetic product? Which is the thermodynamic product?

Answer:
Attack of the cyanide occurs fastest in the equatorial direction to form the cyanohydrin with the hydroxyl group in the axial position. This is the kinetic product. However, the hydroxyl group has a larger conformational preference for the equatorial position than the cyano group. As a consequence, the reversible formation of the cyanohydrin under equilibrium conditions favors formation of the compound with the cyano group axial and the hydroxyl group equatorial.

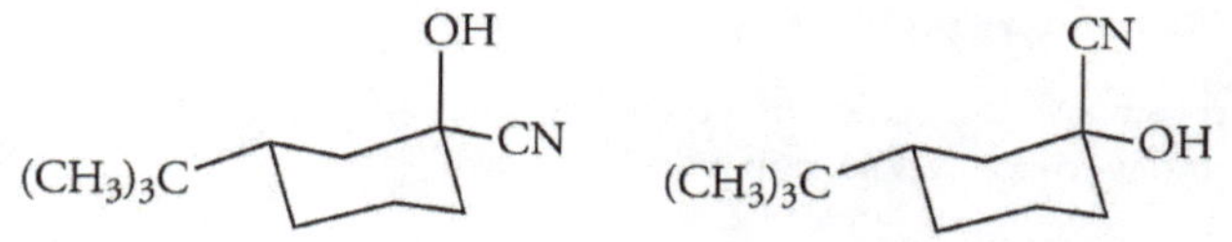

19.17 When benzaldehyde is heated with acetone cyanohydrin and a catalytic amount of base, the following reaction occurs. Explain whether you expect the equilibrium constant to be greater or less than 1.0. Write a mechanism for the reaction.

Answer:
There is no direct reaction between the two reactants. Each carbonyl compound can exist in equilibrium with its cyanohydrin and HCN. Thus, the HCN is simply distributed between the two carbonyl compounds. The mechanism is the standard one for addition of HCN to a carbonyl compound. The equilibrium constants for formation of the cyanohydrin for benzaldehyde and propanone are 210 and 30, respectively. Thus, the reaction as written has $K = 7$.

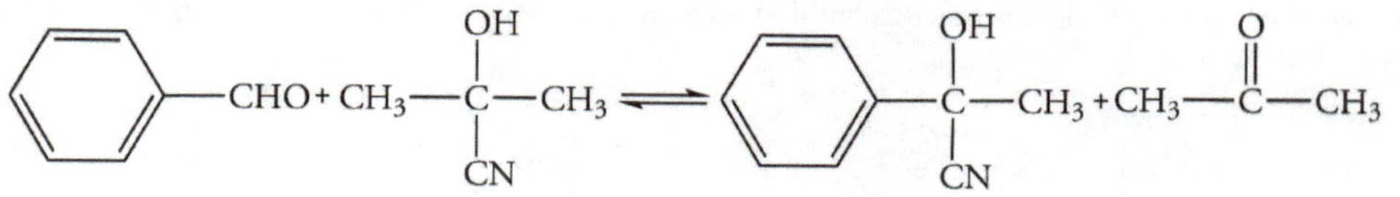

19.18 Write the structure of the product for the reaction of one equivalent of HCN with each of the following compounds.

Answers:

(a) Both carbonyl groups are equivalent, and only one reacts with one equivalent of HCN.

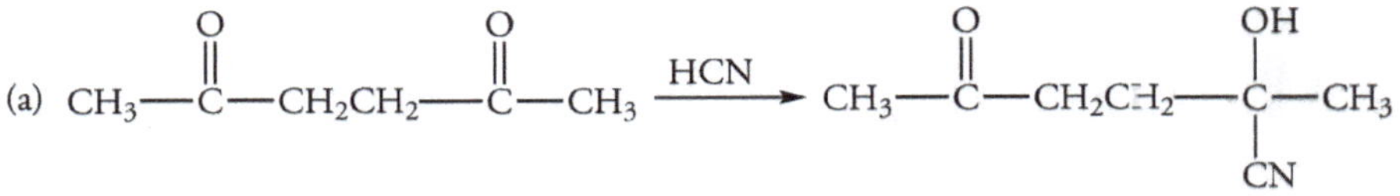

$$\text{(a)} \quad CH_3-\overset{O}{\underset{\|}{C}}-CH_2CH_2-\overset{O}{\underset{\|}{C}}-CH_3 \xrightarrow{\text{HCN}} CH_3-\overset{O}{\underset{\|}{C}}-CH_2CH_2-\overset{OH}{\underset{\underset{CN}{|}}{\overset{|}{C}}}-CH_3$$

(b) The equilibrium constant for the saturated aldehyde side chain is larger than for benzaldehyde, so reaction occurs at that site.

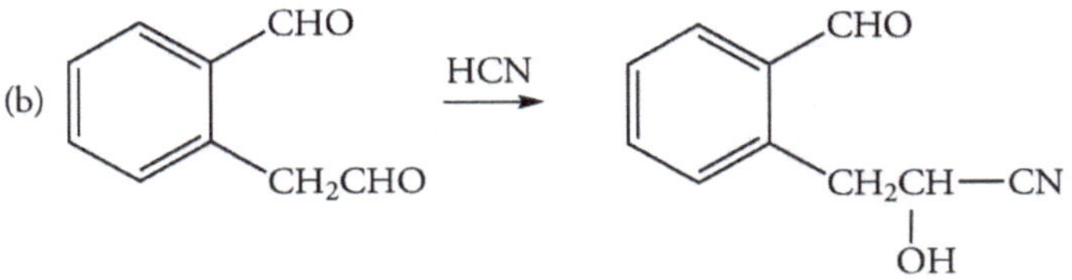

(c) The ketone group that is para to the methoxy group is resonance stabilized, so the equilibrium constant for its reaction with HCN is smaller than for the keto group that is *meta* to the methoxy group. The methoxy group is inductively electron withdrawing, and it destabilizes the dipolar resonance form of the carbonyl group *meta* to it, so it is more reactive.

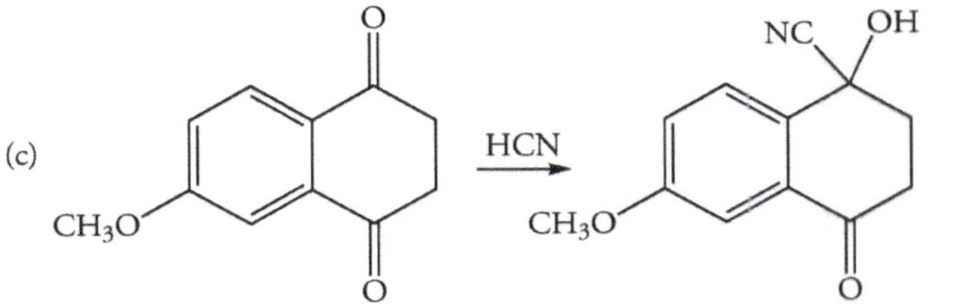

(d) The cyclic ketone is sterically hindered by the two methyl groups at the adjacent carbon atom. Thus, the cyanohydrin formed occurs at the methyl ketone of the chain bonded to the ring.

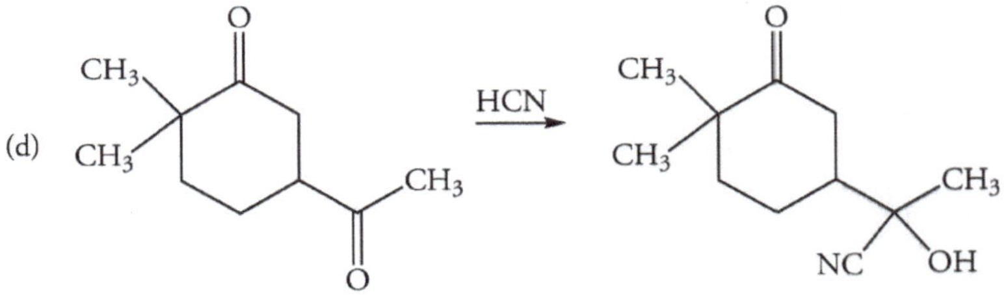

Addition of Alcohols to Carbonyl Compounds

19.19 Identify each of the following as a hemiacetal, hemiketal, acetal, or ketal.

Answers:
(a) acetal
(b) ketal
(c) acetal
(d) hemiacetal

(a) $CH_3CH_2CH(OCH_3)_2$

(b) $CH_3CH_2C(OCH_3)_2CH_3$

(c) $CH_3\overset{OCH_2CH_3}{\underset{|}{C}HOCH_2CH_3}$

(d) $CH_3CH_2\overset{OCH_3}{\underset{|}{C}HOH}$

19.20   Identify each of the following as a hemiacetal, hemiketal, acetal, or ketal.

Answers:
(a) ketal
(b) diacetal
(c) hemiketal
(d) ketal

(a) 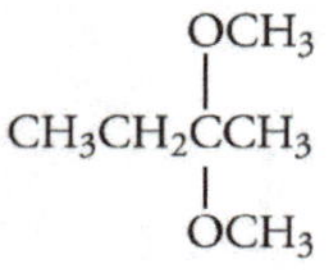 $CH_3CH_2\overset{\displaystyle OCH_3}{\underset{\displaystyle OCH_3}{C}}CH_3$

(b) $(CH_3O)_2CHCH_2CH(OCH_3)_2$

(c)

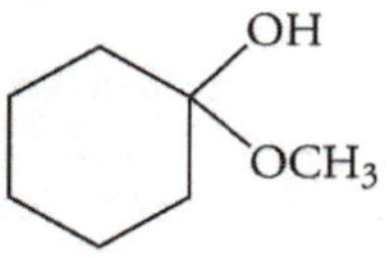

(d)

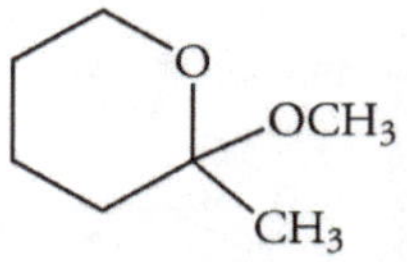

19.21   Identify each of the following as a hemiacetal, hemiketal, acetal, or ketal.

Answers:
(a) acetal
(b) hemiacetal
(c) hemiketal
(d) ketal

(a)

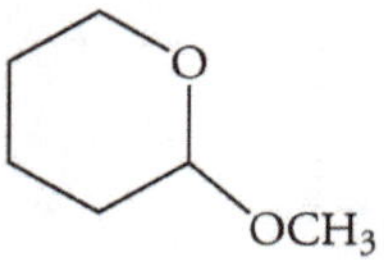

(b)

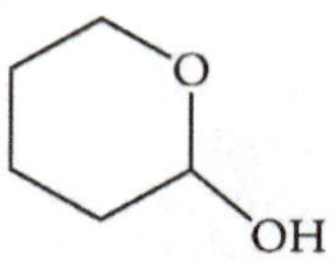

(c)

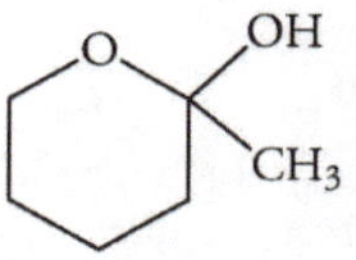

(d)

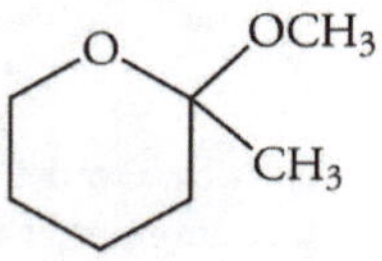

19.22   Identify each of the following as a hemiacetal, hemiketal, acetal, or ketal.

Answers:
(a) hemiacetal
(b) ketal
(c) acetal
(d) acetal

(a)

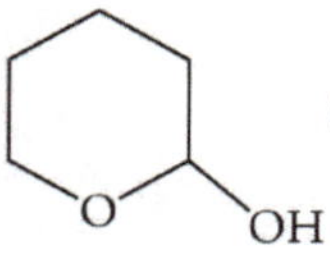

(b)

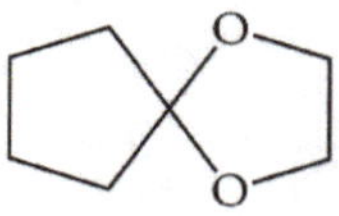

(c)

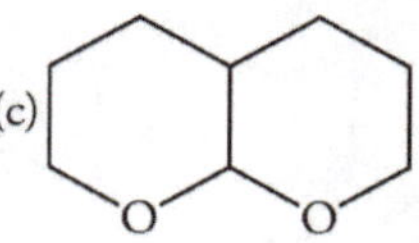

(d)

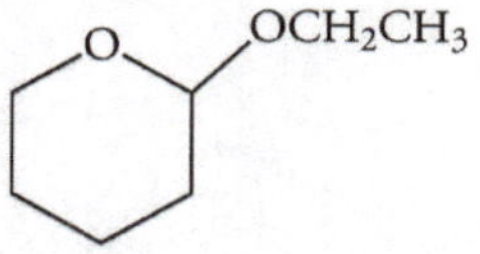

19.23   Identify the functional groups in talaromycin A, a substance found in the fungus that grows in poultry litter.

Answers:
(a) primary alcohol
(b) secondary alcohol
(c) ketal

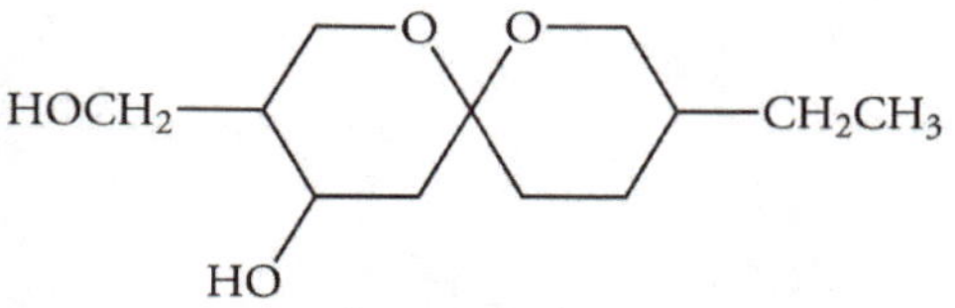

talaromycin A

19.24   Identify the functional groups in daunosamine, a component of Adriamycin, used in cancer chemotherapy.

Answers:
(a) secondary alcohol
(b) amino group
(c) hemiacetal

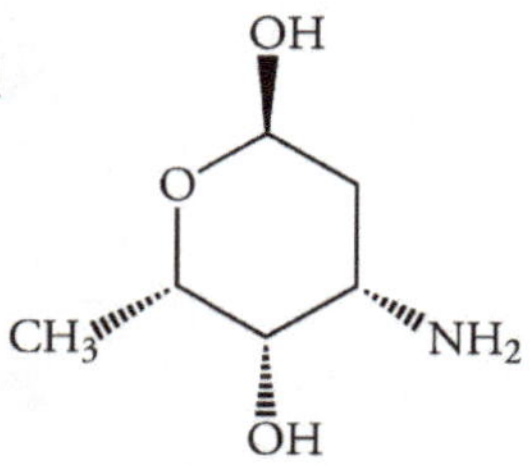

daunosamine

19.25   Is the equilibrium constant for the following reaction greater than or less than 1.0?

$$CF_3CHO + CH_3CH(OCH_3)_2 \rightleftharpoons CH_3CHO + CF_3CH(OCH_3)_2$$

Answer: The equilibrium constants for the formation constants for acetals parallel those for the formation of hydrates. The fluoro groups favors formation of the acetal of trifluoroethanal compared to ethanal. Thus, the equilibrium constant is greater than one.

19.26 Which compound should exist to the larger extent as a hemiacetal, 4-hydroxybutanal or 5-hydroxypentanal?

Answer:

Formation of a five-membered hemiacetal ring of 4-hydroxybutanal is less favored than formation of a six-membered hemiacetal ring of 5-hydroxypentanal because there is torsional strain in the five-membered ring. The six-membered ring has no torsional strain and is favored.

19.27 Benzaldehyde reacts with 1,2-propanediol to give two isomeric acetals. Draw their structures.

Answer:

The benzene ring and the methyl group can be *cis* or *trans* in the cyclic acetal.

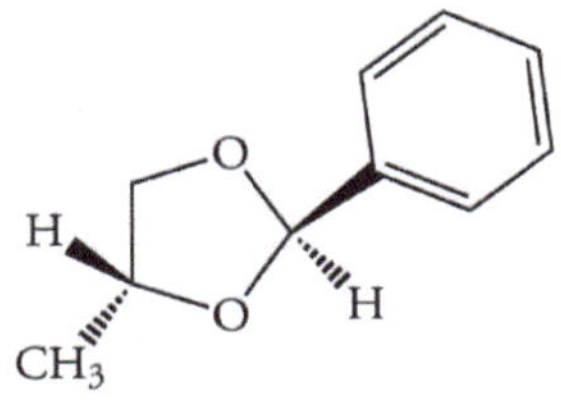

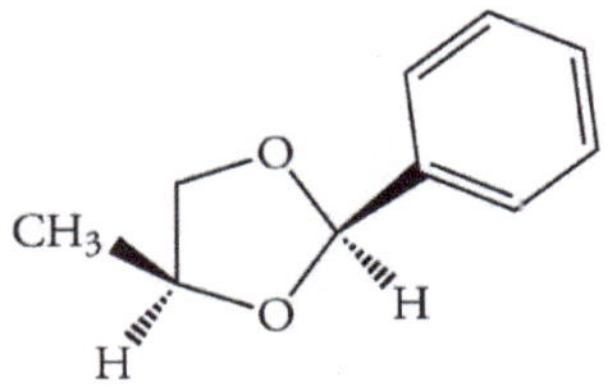

19.28 Acetone reacts with 1,2,3-propanetriol to give two isomeric ketals. Draw their structures.

Answer: Acetone can form a ketal with a six-membered ring by reacting with the C-1 and C-3 hydroxyl groups, or a ketal with a five-membered ring by reacting with the C-1 and C-2 hydroxyl groups.

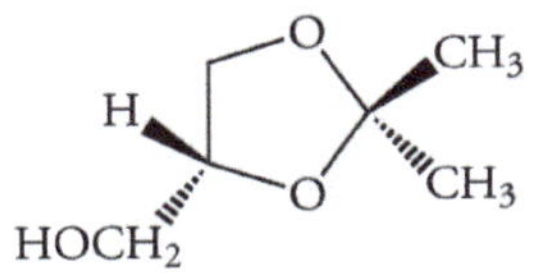

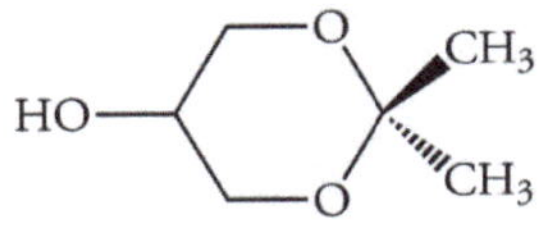

19.29 2-Oxopropanal reacts with excess methanol in an acid-catalyzed reaction to give a compound with molecular formula $C_5H_{10}O_3$. Draw its structure.

Answer:

An acetal forms. The ketone is less reactive, so ketal formation is not favored.

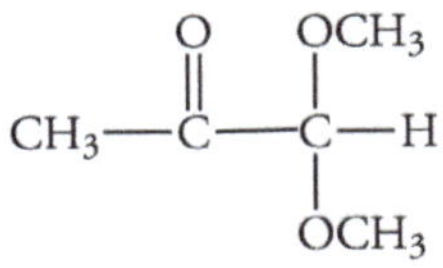

19.30 2-Oxopropanal reacts with excess ethylene glycol in an acid-catalyzed reaction to give a compound with molecular formula $C_7H_{12}O_4$. Draw its structure. What is the difference between the product of this reaction and the product in Exercise 19.29?

Answer:

Both an acetal forms at C-1 and a ketal forms at C-2 because the equilibrium constants for formation of cyclic derivatives are more favorable than those for reaction with two moles of an alcohol to form acyclic derivatives.

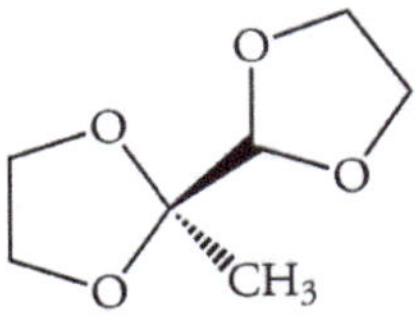

20 CARBOXYLIC ACIDS

KEYS TO THE CHAPTER

20.1 The Carboxyl and Acyl Groups

The reactivity of the carboxyl group depends on the interplay of its π electrons, its nonbonded electron pairs, and the hydroxyl group. The carboxyl group is resonance stabilized, with two contributing dipolar resonance structures. The lone-pair electrons of the hydroxyl oxygen atom are donated to the electron-deficient carbon atom in one dipolar resonance structure. As a result, the electron density of the carbonyl carbon atom in a carboxylic acid is larger than for aldehydes and ketones.

Carboxylic acids are the "parents" of acid derivatives. The acyl group, RCO, is bonded to an electronegative atom in carboxylic acids and their derivatives. The acyl group of an ester is bonded to the oxygen atom of an alkoxy group. The acyl group of an amide is bonded to a nitrogen atom. Cyclic esters and amides are lactones and lactams, respectively. The acyl group is bonded to a chlorine atom, a carboxyl group, and a thiolate group in acid chlorides, acid anhydrides, and thioesters, respectively.

20.2 Nomenclature of Carboxylic Acids

The common names of the unbranched carboxylic acids do not provide information about the number of carbon atoms contained in the chain, so the common names simply have to be learned. Carbon atoms in the parent chain are named with Greek lower case letters (α, β, γ, etc.) starting with the carbon atom directly bonded to the carboxyl carbon atom. Branches on the chain are indicated using these Greek letters to identify the location of substituents or alkyl groups.

IUPAC names of carboxylic acids are based on the alkane names, for which the suffice *-oic* acid replaces the final *-e* in the name of the alkane. The carbon chain is numbered starting with the carboxyl carbon atom, but the number 1 is not included in the name, since C-1 is automatically reserved for the carboxyl group. The carboxyl group takes priority over halogen atoms, alkyl groups, double or triple bonds, and other functional groups containing a carbonyl group. Other wise, the names of carboxylic acids follow the rules for the other functional groups we have considered.

The names of salts of carboxylic acids begin with the name of the metal ion followed by the name of the acid, modified by changing the suffice *-oic acid* to *-oate*. For example, the sodium salt of propanoic acid is sodium propanoate.

20.3 Physical Properties of Carboxylic Acids

Carboxylic acid molecules form intermolecular hydrogen bonds. Carboxylic acids exist as dimers, and consequently, they have higher boiling points than other hydrocarbons of similar molecular weight. The carboxyl group forms hydrogen bonds with water molecules, and low-molecular-weight carboxylic acids are soluble in water.

20.4 Acidity of Carboxylic Acids

Carboxylic acids are more acidic than alcohols. Loss of a proton from the carboxyl group leaves the carboxylate anion, which is stabilized by delocalization of two electron pairs over the carbon atom and two oxygen atoms. The formation of stable conjugate bases enhances the acidity of carboxylic acids.

Electronegative groups attached to the α-carbon atom pull electron density away from the carboxyl group. This inductive effect increases the acidity of carboxylic acids. Increased acidity increases the acid ionization constant, K_a, and decreases the pK_a.

Both the methoxy group and the nitro group of substituted ethanoic acids increase their acidity by an inductive effect. The oxygen atom of a methoxy group is more electronegative than carbon, hence it withdraws electron density carboxyl group. The nitrogen atom of the nitro group, which has a formal positive charge, is also electron withdrawing.

The acidity of a substituted benzoic acid is affected by the change in the electron density of the aromatic ring. The substituents do not interact by resonance with the carboxylate group, nor does the carboxylate group interact by resonance with the aromatic ring.

20.5 Carboxylate Ions

In water, unsubstituted carboxylic acids are present largely as the nonionized form. However, if the pH of a solution is greater than the pK_a of the acid, then the carboxylic acid exists as its conjugate base. At pH 7, physiological pH, carboxylic acids exist predominantly as carboxylate ions. The solubility of carboxylate anions in water provides an easy way to separate them form other, water-insoluble organic compounds.

20.6 Synthesis of Carboxylic Acids

Carboxylic acids are prepared from substrates with the proper hydrocarbon skeleton by oxidation of either an alcohol or an aldehyde. Because special methods are required to prepare aldehydes, the more common substrate for preparation of a carboxylic acid is the structurally related alcohol. The Jones reagent is used as the oxidizing agent

Oxidation of an alkylbenzene by $KMnO_4$ produces a benzoic acid. The haloform reaction oxidizes a methyl ketone by removing the methyl group. The reagent is a halogen in basic solution. Preparation of carboxylic acids with one more carbon atom than the substrate is accomplished by two alternate procedures, both of which commence with an alkyl halide. Conversion of an alkyl halide into a Grignard reagent followed by reaction with carbon dioxide gives a carboxylic acid. A haloalkane, or an aryl halide, can be used. The second method is based on nucleophilic substitution of a halide ion by a cyanide ion. The resulting nitrile is hydrolyzed to form a carboxylic acid. The limitation on the reaction is the first step, which is effective for primary haloalkanes and to a limited extent for secondary haloalkanes. The competing reaction is dehydrohalogenation of the haloalkane.

20.7 Reduction of Carboxylic Acids

It is more difficult to reduce carboxylic acids than to reduce aldehydes or ketones. The acidic carboxyl proton reacts with the metal hydride reagents commonly used for reduction of carbonyl groups. Only lithium aluminum hydride is sufficiently reactive to reduce a carboxylic acid to a primary alcohol. Sodium borohydride does not reduce carboxylic acids.

Diborane reduces carboxylic acids to give primary alcohols. Diborane does not readily reduce other unsaturated functional groups such as nitrile or nitro groups. Diborane cannot be used if the substrate contains a double bond because hydroboration would occur.

20.8 Decarboxylation Reactions

Loss of a carboxyl group and replacement by an atom such as hydrogen is decarboxylation. Carboxylic acids containing a β keto group undergo decarboxylation in a cyclic mechanism. Both β keto acids and malonic acids undergo this reaction.

The Hunsdiecker reaction replaces a carboxyl group by a halogen. The reagent is bromine, which reacts with the silver salt of the carboxylic acid. Mercury(II) oxide and bromine also can be used with the carboxylic acid. An unsaturated carboxylic acid cannot be used because bromine adds to double bonds.

20.9 Reactions of Carboxylic Acids and Derivatives—A Preview

The salts of carboxylic acids are weak nucleophiles but can be used to displace halide ion from a haloalkane to give esters. The most common reaction of carboxylic acids and their derivatives is attack of nucleophiles at the carbonyl carbon atom. However, in contrast to aldehydes and ketones, the tetrahedral intermediate is unstable, and it ejects a leaving group. This overall process is nucleophilic acyl substitution.

Although nucleophilic acyl substitution has the same stoichiometry as nucleophilic substitution at a saturated carbon atom, the mechanisms are very different. Nucleophilic acyl substitution occurs in two steps via a tetrahedral intermediate.

20.10 Conversion of Carboxylic Acids Into Acyl Halides

Carboxylic acids react with thionyl chloride to give acyl chlorides. The mechanism resembles the reaction of alcohols with thionyl chloride. The hydroxyl group acts as a nucleophile to displace a chloride ion from thionyl chloride. However, the subsequent displacement of the sulfur-containing moiety occurs by a nucleophilic substitution in the case of alcohols, and nucleophilic acyl substitution in the case of acids.

20.11 Ester Synthesis

The synthesis of esters can be carried out in several ways.
1. Alkylation of a carboxylate ion using a haloalkane, which resembles the Williamson ether synthesis. It occurs by a S_N2 mechanism and therefore is limited to primary haloalkanes.
2. Methylation of a carboxylic acid using diazomethane. The reaction is largely limited to this simplest diazo compound but will occur with other diazo compounds.
3. Reaction of acyl chlorides with alcohols, which occurs by nucleophilic acyl substitution. The reaction is general for all alcohols and any acyl chloride.
4. The Fischer esterification, which uses a carboxylic acid and an alcohol, is an acid-catalyzed equilibrium reaction that is driven to product by removal of the by-product, water.

20.12 Mechanism of Esterification

The Fischer esterification reaction occurs by nucleophilic attack of the alcohol on the carbonyl carbon atom of the carboxylic acid to give a tetrahedral intermediate. The acid catalyst protonates the carbonyl oxygen atom to increase the partial positive charge of the carbonyl carbon atom. Isotope studies show that the oxygen atom of the alcohol is the bridging atom of the ether. The configuration of a chiral alcohol is retained in this esterification reaction because the bond to the oxygen atom of the alcohol is not cleaved.

20.13 Brief Synthetic Review

The reactions we discussed in this chapter and in Chapter 19 can be combined in many ways in organic synthesis. Our brief review reminds us that there is almost always more than one way to convert one functional group into another. Sometimes a single regioselective reaction will act on one functional group and leave the other one unaltered. Other times, potentially reactive groups have to be protected.

20.14 Spectroscopy of Carboxylic Acids

Carboxylic acids are characterized by the strong absorption due to the carbonyl group in the infrared spectra of these compounds. The absorption occurs in the same region as the carbonyl groups of aldehydes and ketones, but the absorption for carboxylic acids occurs at slightly higher wavenumber, and tends to be somewhat broadened. The O—H bond of carboxylic acids absorbs in the same region as that for alcohols. However, the absorption is very much broader for carboxylic acids, and it overlaps the C—H absorptions.

The proton NMR spectra of carboxylic acids show a characteristic absorption in the 9–12 δ region for the strongly deshielded carboxyl proton. The α hydrogen atoms of carboxylic acids occur in the 2.0–2.5 δ region, which is the same region for the α hydrogen atoms of aldehydes and ketones.

The α carbon atoms of carboxylic acids have C-13 NMR absorptions in the 20 d region, which is at slightly higher field than for aldehydes and ketones. The carbonyl carbon atom is easily identified by its absorption in the 200 δ region.

Summary of Reactions

1. Synthesis of Carboxylic Acids by Oxidative Methods

$$\text{cyclopentyl-CH}_2\text{OH} \xrightarrow[\text{Jones Reagent}]{\text{CrO}_3 \, / \, \text{H}_2\text{SO}_4} \text{cyclopentyl-CO}_2\text{H}$$

$$\text{—CHO} \xrightarrow{\text{Ag(NH}_3)_2{}^+} \text{—CO}_2\text{H}$$

$$\xrightarrow{\text{KMnO}_4}$$

$$\xrightarrow{\text{Br}_2 \, / \, \text{OH}^-} \quad + \text{CHBr}_3$$

2. Synthesis of Carboxylic Acids from Haloalkanes

$$\xrightarrow[\substack{\text{2. CO}_2 \\ \text{3. H}_3\text{O}^+}]{\text{1. Mg / ether}}$$

$$\text{CH}_3\text{CH}_2\overset{\overset{\text{CH}_3}{|}}{\text{CH}}\text{CH}_2\text{—Br} \xrightarrow[\text{2. H}_3\text{O}^+]{\text{1. NaCN / DMF}} \text{CH}_3\text{CH}_2\overset{\overset{\text{CH}_3}{|}}{\text{CH}}\text{CH}_2\text{—C—OH}$$

3. Reduction of Carboxylic Acids

$$\xrightarrow[\text{2. H}_3\text{O}^+]{\text{1. LiAlH}_4 \text{ /ether}}$$

$$\xrightarrow[\text{diglyme}]{\text{B}_2\text{H}_6}$$

306

4. Decarboxylation of Carboxylic Acids

$$\text{heat} \longrightarrow \quad + CO_2$$

$$\xrightarrow[\text{1. } Br_2]{\text{1. } Ag_2O} \quad + CO_2 + AgBr$$

5. Synthesis of Acyl Halides

$$\xrightarrow{SOCl_2}$$

6. Synthesis of Esters

$$CH_3CH_2\!-\!Cl \ + \ {}^-O\!-\!\overset{\overset{\displaystyle O}{\|}}{C}\!-\!CH_2\overset{\overset{\displaystyle CH_3}{|}}{C}HCH_3 \longrightarrow CH_3CH_2\!-\!O\!-\!\overset{\overset{\displaystyle O}{\|}}{C}\!-\!CH_2\overset{\overset{\displaystyle CH_3}{|}}{C}HCH_3$$

$$+ \ CH_2N_2 \longrightarrow \qquad + N_2$$

$$CH_3\overset{\overset{\displaystyle CH_3}{|}}{C}HCH_2\!-\!OH \ + Cl\!-\!\overset{\overset{\displaystyle O}{\|}}{C}\!-\!CH_2CH_3 \longrightarrow CH_3\overset{\overset{\displaystyle CH_3}{|}}{C}HCH_2\!-\!O\!-\!\overset{\overset{\displaystyle O}{\|}}{C}\!-\!CH_2CH_3$$

$$CH_3\overset{\overset{\displaystyle CH_3}{|}}{C}HCH_2\!-\!OH \ + HO\!-\!\overset{\overset{\displaystyle O}{\|}}{C}\!-\!CH_2CH_3 \xrightarrow{H^+} CH_3\overset{\overset{\displaystyle CH_3}{|}}{C}HCH_2\!-\!O\!-\!\overset{\overset{\displaystyle O}{\|}}{C}\!-\!CH_2CH_3$$

End of Chapter Exercises

Nomenclature

20.1 Give the common name for each of the following acids.
   (a) $CH_3CH_2CO_2H$ (b) HCO_2H (c) $CH_3(CH_2)_4CO_2H$
   (d) $CH_3(CH_2)_{10}CO_2H$ (e) $CH_3(CH_2)_{16}CO_2H$ (f) $CH_3(CH_2)_{14}CO_2H$

Answers:
   (a) propionic acid (b) formic acid   (c) caproic acid   (d) lauric acid   (e) stearic acid   (f) palmitic acid

20.2 Give the common name for each of the following acids.

Answers:
(a) β-chlorobutyric acid
(b) α-bromopropionic acid
(c) β-bromo-γ-methylvaleric acid

(a) $CH_3CHCH_2CH_2CO_2H$ (with Cl on the second carbon)

(b) $BrCHCO_2H$ (with CH_3 substituent)

(d) α-chloro-γ,γ-dimethylvaleric acid

(c) $CH_3CHCHCH_2CO_2H$ (with CH_3 and Br substituents)

(d) $CH_3CCH_2CHCO_2H$ (with CH_3, Cl, and CH_3 substituents)

20.3 Give the IUPAC name for each of the following acids.

Answers:
(a)   *trans*-2-hydroxycyclohexanecarboxylic acid
(b)   (*E*)-3-chloro-3-phenyl-2-propenoic acid
(c)   5-hexynoic acid
(d)   cyclodecanecarboxylic acid

(a) [structure: cyclohexane with OH and CO_2H groups]

(b) [structure: phenyl with Cl, H, and CO_2H on a double bond]

(c) $HC\equiv C-CH_2CH_2CH_2CO_2H$ (d) [structure: cyclodecane with CO_2H group]

20.4 Give the IUPAC name for each of the following acids.

Answers:
(a)   3-methoxybenzoic acid
(b)   2-(3-methoxyphenyl)propanoic acid
(c)   3-ethyl-2-pentenoic acid
(d)   cyclopentylethanoic acid

(a) [structure: benzene ring with CH_3O and CO_2H groups]

(b) [structure: benzene ring with CH_3O and CO_2H/CH_3 group]

(c) [structure: CH_3CH_2 and CH_3CH_2 on $C=C$ with CO_2H and H]

(d) [structure: cyclopentane with CH_2CO_2H group]

20.5 The IUPAC name of ibuprofen, the analgesic in Motrin, Advil, and Nuprin, is 2-(4-isobutylphenyl)propanoic acid. Draw its structure.

[structure: $CH_3-C(CH_3)(H)-CH_2-$ benzene ring $-C(CH_3)(H)-CO_2H$]

Molecular Formulas

20.6   What is the general molecular formula for each of the following classes of compounds?
    (a) saturated acyclic carboxylic acid
    (b) saturated acyclic dicarboxylic acid
    (c) saturated monocyclic carboxylic acid
    (d) monounsaturated acyclic carboxylic acid

Answers:
(a) $C_nH_{2n}O_2$
(b) $C_nH_{2n-2}O_4$
(c) $C_nH_{2n-2}O_2$
(d) $C_nH_{2n-2}O_2$

20.7   Draw the structure of two isomers having the following characteristics
    (a) dicarboxylic acids with molecular formula $C_4H_4O_4$
    (b) carboxylic acids with molecular formula $C_4H_8O_2$
    (c) saturated carboxylic acids with molecular formula $C_5H_8O_2$

Answers:

(a) $HO_2CCH_2CH_2CO_2H$ HO_2CCHCO_2H with CH_3 substituent

(b) $CH_3CH_2CH_2CO_2H$ CH_3CHCO_2H with CH_3 substituent

(c)

20.8   10-Undecenoic acid is the anti fungal agent contained in Desenex and Cruex. Write the structure.

Answer: $CH_2{=}CH(CH_2)_8CO_2H$

Properties of Acids

20.9   Explain why 1-butanol is less soluble in water than butanoic acid.
Answer:
    Since butanoic acid has one more oxygen atom than butanol, it has more unshared pairs of electrons that act as hydrogen bond acceptors with water molecules. Therefore, it is more soluble than 1-butanol in water.

20.10   Explain why adipic acid is much more soluble in water than hexanoic acid.

Answer: Adipic acid has two carboxylic acid groups that act as hydrogen bond acceptors with water molecules, as well as another hydrogen atom bonded to oxygen that is a hydrogen bond donor to a water molecule. Therefore, it is much more soluble in water than hexanoic acid.

20.11   Explain why the boiling point of decanoic acid is higher than that of nonanoic acid.

Answer:Decanoic acid has one more methylene group than nonanoic acid, so it has stronger van der Waals attractive forces than nonanoic acid and a higher boiling point.

20.12   Explain why the boiling point of 2,2-dimethylpropanoic acid (164 °C) is lower than that of pentanoic acid (186 °C).

Answer: The shape of 2,2-dimethylpropanoic acid is approximately spherical, so it has a smaller surface area than the cylindrically shaped pentanoic acid. Thus, the van der Waals forces are smaller for 2,2-dimethylpropanoic acid, and its boiling point is lower.

20.13 Explain why the boiling point of 4-methoxybenzoic acid (278 °C) is higher than that of 2-methoxybenzoic acid (200 °C).

Answer: 4-Methoxybenzoic acid has a high boiling point because it can form intermolecular hydrogen bonds between carboxylic acid groups, forming a dimer.

20.14 Explain why the boiling point of *trans*-2-butenoic acid (185 °C) is higher than that of *cis*-2-butenoic acid (169 °C).

The *trans* isomer has a larger dipole moment than the *cis* isomer due to the contribution of both the methyl and carboxyl groups.

Acidity of Carboxylic Acids

20.15 The K_a of methoxyacetic acid is 2.7×10^{-4}. Explain why this value differs from the K_a of acetic acid (1.8×10^{-5}).

Answer: The methoxy group of methoxyacetic acid inductively attracts electron density, which stabilizes the conjugate base. Therefore, it is more acidic (has a larger K_a) than acetic acid.

20.16 The K_a values of benzoic acid and *p*-nitrobenzoic acid are 6.3×10^{-5} and 3.8×10^{-4}, respectively. Explain why these values differ.

Answer: The higher K_a for *p*-nitrobenzoic acid means than the nitro group stabilizes its carboxylate ion compared to benzoate ion. The nitro group inductively attracts electron density, but there is a long distance between it and the carboxylate ion. The stabilization results from electron withdrawal from the aromatic ring by resonance. As a consequence of lower electron density at C-1 in the resonance-stabilized structure, the electrons of the carboxylate ion are inductively pulled toward the aromatic ring.

20.17 Estimate the pK_a values of the two carboxyl groups in 3-chlorohexanedioic acid.

Answer: The location of the chlorine atom with respect to the C-1 carboxyl corresponds to that in 3-chlorobutanoic acid, so its pK_a must be close to 4.06. The location of the chlorine atom with respect to the carboxylic acid of the C-6 carboxyl group corresponds to that in 4-chlorobutanoic acid, so its pK_a must be close to 4.52.

20.18 The pK_a of 3-cyanobutanoic acid is 4.44. Using the pK_a values of chlorine-substituted butanoic acids as a guide, estimate the pK_a of 2-cyanobutanoic acid.

Answer: The pK_a increases from 4.06 to 2.84 for 3-chlorobutanoic acid and 2-chlorobutanoic acid, respectively. A similar increase would be expected for the cyano compounds. The estimated pK_a is 3.22.

20.19 The pK_a for the first dissociation of dicarboxylic acids levels off at approximately 4.85. The pK_a of long-chain carboxylic acids levels off at approximately 4.55. What relationship exists between these two numbers? What structural features are responsible for this difference?

Answer: They differ by 0.3 units on the log scale, which corresponds to a factor of 2 in K_a. The long chain dicarboxylic acids are twice as acidic as the long chain carboxylic acids because there are two hydrogen atoms per molecule in the dicarboxylic acids. The difference reflects a statistical factor, not an influence of structure.

20.20 The difference between the pK_a values for dissociation of the first and second protons of the long-chain dicarboxylic acids is about 1 unit. The difference between the pK_a values for both oxalic and malonic acids is about 3 units. Explain these data, focusing on the pK_a for the second ionization step.

Answer:
The ionization constant for the transfer of a second proton from an acid to water is expected to be smaller than for the first proton because it is more difficult to remove a proton from a negatively charged ion than from a neutral molecule. In the case of the oxalate ion and the malonate ion, the carboxylate group and carboxylic acid group are close enough to form a hydrogen bond. As a consequence, more energy is required to remove the proton, and the acid is weaker.

20.21 The methoxy group is an effective donor of electrons and as a consequence is an activating group in electrophilic aromatic substitution. Explain why the pK_a of methoxyacetic acid (3.5) is less than that of acetic acid (4.7).

Answer:
The methoxy group cannot donate electrons by resonance in methoxyacetic acid as it does in an aromatic ring because there is an intervening methylene group in the acetic acid. The smaller pK_a of methoxyacetic acid indicates that the methoxy group stabilizes the carboxylate ion, but this stabilization is the result of inductive electron withdrawal by the oxygen atom.

20.22 The pK_a values of cyanoacetic acid and nitroacetic acid are 2.45 and 1.65, respectively. What do these data indicate about the substituent properties of —CN and —NO$_2$?

Answer:
The groups cannot affect the acidity of the acetic acids by resonance because there is an intervening methylene group. Therefore, the difference indicates that the nitro group is inductively more electron withdrawing than the cyano group. The nitro group has a formal positive charge on nitrogen. The carbon atom of the cyano group is partially positive as a result of the electronegativity difference between carbon and nitrogen atom. However, it does not have a formal charge.

20.23 The substituent effects of the hydroxyl and methoxy groups are quite similar, as evidenced by the pK_a values of p-hydroxy- and p-methoxybenzoic acids, which are 4.48 and 4.47, respectively. However, the pK_a values of o-hydroxy- and o-methoxybenzoic acids are 2.97 and 4.09, respectively. Explain why the values for the *ortho* isomers are so different.

Answer:
The enhanced acidity of the o-hydroxy compound compared to the p-methoxy compound indicates that its carboxylate group is stabilized. The hydrogen atom of the hydroxyl group can form a hydrogen bond with the oxygen atom of the carboxylate group. This type of interaction does not exist in the p-methoxy compound.

20.24 The pK_a values of *para*-substituted benzoic acids for the —PCl$_2$ and —Si(CH$_3$)$_3$ groups are 3.6 and 4.3, respectively. Based on these data, determine whether these groups are activating or deactivating in electrophilic aromatic substitution.

Answer:
The pK_a value of the phosphorus compound is smaller than that of benzoic acid. Thus, the phosphorus group withdraws electron density from the aromatic ring and stabilizes the carboxylate ion. The phosphorus group should be a deactivator in aromatic substitution reactions. The pK_a value of the silicon compound is slightly larger than that of benzoic acid. Thus, it donates electron density to the aromatic ring and destabilizes the carboxylate ion. The silicon group is a weak activator in aromatic substitution reactions.

20.25 p-Methoxybenzoic acid is a weaker acid than benzoic acid, but p-(methoxymethyl)benzoic acid is a stronger acid than p-methylbenzoic acid. Why does the methoxy group have opposite effects in these two cases?

Answer:
The smaller K_a for p-methoxybenzoic acid means that the methoxy group destabilizes its carboxylate ion. The methoxy group inductively attracts electron density and could stabilize the ion, but there is a long distance between it and the carboxylate group. The destabilization results from electron donation to the aromatic ring by resonance. As a consequence of higher electron density at C-1 in the resonance-stabilized structure, the carboxylate ion is destabilized. In the case of the methoxymethyl group, there is no resonance interaction between the oxygen atom and the aromatic ring. A methyl group is inductively electron donating, but the added methoxy group makes the methoxymethyl group somewhat inductively electron withdrawing. The net effect is a decrease in electron density at C-1, thus stabilizing the carboxylate ion.

20.26 The van del Waals radii of fluorine and hydrogen atoms arc similar. The pK_a values of o-, m-, and p-fluoro benzoic acids are 4.1, 3.9, and 3.3, respectively. The pK_a value of benzoic acid is 4.2. Explain the order of the pK_a values of the fluorobenzoic acids. Estimate the contribution of fluorine as an electron donor in terms of resonance.

Answer:

The pK_a values of o-, m-, and p-fluorobenzoic acid are 3.3, 3.9, and 4.1, respectively. All of the fluorine compounds are stronger acids than benzoic acid, suggesting that in each case the fluorine atom withdraws electron density from the ring and stabilizes the carboxylate ion. The effect is largest for the *ortho* compound and decreases with distance between the fluorine atom and the carboxylate ion, which is typical of an inductive effect. If the fluorine atom effectively donated electrons by resonance, the acidity of the *para* substituted compound would be greater than that of benzoic acid.

20.27 Compare the pK_a values of biphenyl-3-carboxylic acid (4.14) and biphenyl-4-carboxylic acid (4.21) to that of benzoic acid (4.20) and explain the different effect of the phenyl group on the pK_a values.

Answer:

The two phenyl rings do not interact by resonance because the *ortho* hydrogen atoms do not allow the two rings to be coplanar. As a result, the observed effect is that of an sp²-hybridized carbon atom of the phenyl group, which is inductively electron withdrawing. That effect stabilizes the carboxylate ion. The effect, although small, is greater in the case of the 3-carboxylic acid because the phenyl ring is closer to the carboxyl group.

20.28 Which is the stronger acid in each of the following pairs of aromatic carboxylic acids? What accounts for the difference in acid strength?

Answers:

(a) The fluoro compound is more acidic because the fluorine atoms inductively withdraw electron density from the aromatic ring and stabilize the carboxylate ion.
(b) Oxygen is more electronegative than nitrogen. As a result of inductive electron withdrawal, the carboxylate ion of the oxygen compound is more stable than the carboxylate ion of the nitrogen compound. The furan derivative is more acidic.
(c) The inductive electron withdrawal by oxygen decreases with distance from the carboxylate ion. Thus, the first compound is more acidic.

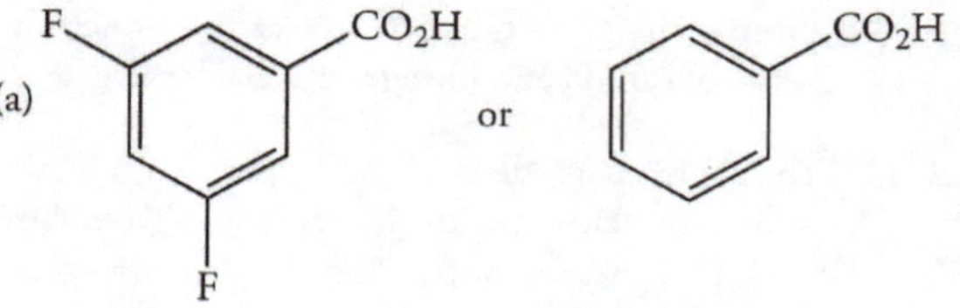

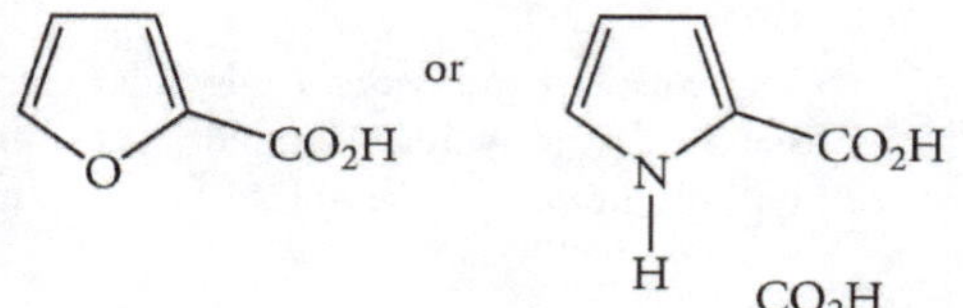

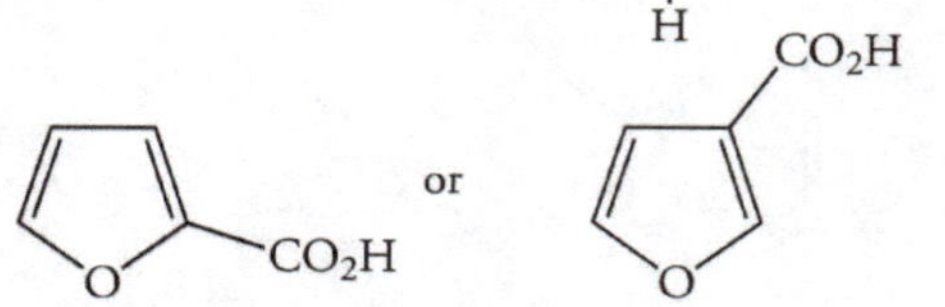

20.29 The pK_a of benzoic acid is 4.2. The pK_a of probenecid is 3.4. Explain why probenecid is the stronger acid. (Probenecid is a drug that is used to treating gout and hyperuricemia.)

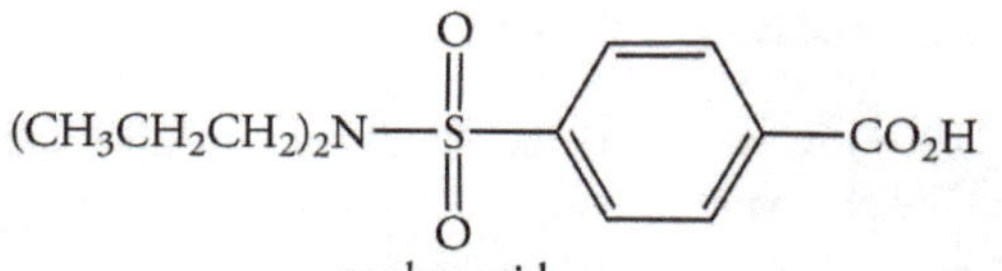

Answer:

The nitrogen atom and the two oxygen atoms bonded to the sulfur atom make this group strongly electron withdrawing. As a result, the carboxylate ion is stabilized by a decrease in electron density at C-1.

20.30 Predict the pK_a of indomethacin, an anti-inflammatory agent.

Answer:
There are no strongly electron withdrawing groups near the carboxyl group. However, there is an sp²-hybridized carbon atom bonded to the spa methylene group, and it can withdraw electron density from that group, which is bonded to the carboxyl group. The effect on the acidity should be small, but the compound should be a stronger acid than acetic acid. Its pK_a is less than 4.72.

indomethacin

Carboxylate Anions

20.31 The pK_a of penicillin G is 2.8. Is it more soluble in stomach acid (pH ~2) or in blood (pH =7.4)?

Answer:
Since the pH of the stomach is less than the pK_a of the acid, it will be predominantly protonated at pH 2. However, in blood it will be predominantly a carboxylate ion. So penicillin G is more soluble in blood than in the stomach.

penicllin G

20.32 Sodium benzoate is used as a preservative in foods, but only if the pH is greater than 5. In what form is the compound present at pH 7?

Answer:
Carboxylic acids exist to a larger degree as carboxylate salts in a more basic medium and hence are more soluble. Thus, sodium benzoate is more soluble in foods whose pH is greater than 5.

20.33 Explain why benzoic acid with an ¹⁸O isotopic label in the hydroxyl oxygen atom can be prepared, but that it cannot be used in mechanistic studies in aqueous solutions.

Answer:
As the carboxylic acid, the two oxygen atoms are nonequivalent. However, in aqueous solution the carboxylate ion forms, and the two oxygen atoms become equivalent.

Synthesis of Carboxylic Acids

20.34 Outline the steps required to prepare cyclohexanecarboxylic acid from each of the following reactants.
(a) bromocyclohexane (b) cyclohexanol (c) cyclohexene (d) vinylcyclohexane (e) cyclohexylmethanol

Answers:
(a)   Prepare the Grignard reagent using magnesium and ether and then add it to carbon dioxide, followed acidification in workup.
(b)   Convert the alcohol into bromocyclohexane using PBr$_3$ and then proceed as in part (a).
(c)   Add HBr to the double bond to form bromocyclohexane and then proceed as in part (a).
(d)   Use ozone under oxidative workup conditions.
(e)   Oxidize the alcohol to the carboxylic acid using the Jones reagent

20.35 Outline the steps required to prepare hexanoic acid from each of the following reactants.
(a) 1-chloropentane   (b) 1-hexanol   (c) hexanal   (d) 1-hexene   (e) 1-heptene

Answers:
(a)   Prepare the Grignard reagent using magnesium and ether and then add it to carbon dioxide, followed by acidification in workup.
(b)   Oxidize the alcohol to the carboxylic acid using the Jones reagent.
(c)   Oxidize the aldehyde to the carboxylic acid using the Jones reagent.
(d)   Prepare 1-hexanol using B$_2$H$_6$ followed by treatment with basic hydrogen peroxide (the hydroboration–oxidation procedure). Then oxidize the alcohol using the Jones reagent.
(e)   Use ozone under oxidative workup conditions.

20.36   Outline the steps required to convert methylenecyclohexane to each of the following compounds.
(a) cyclohexanecarboxylic acid   (b) cyclohexylacetic acid   (c) 1-methylcyclohexanecarboxylic acid

Answers:

(a) Prepare cyclohexylmethanol using B_2H_6 followed by treatment with basic hydrogen peroxide (the hydroboration–oxidation procedure). Then oxidize the alcohol to the carboxylic acid using the Jones reagent.

(b) Add HBr in the presence of peroxide to give a primary halogen compound (anti-Markovnikov addition). Prepare the Grignard reagent using magnesium and ether and then add it to carbon dioxide, followed by acidification in workup. Alternatively, displace bromide ion by cyanide ion, followed by hydrolysis of the nitrile.

(c) Add HBr to give a tertiary halogen compound. Prepare the Grignard reagent using magnesium and ether and then add it to carbon dioxide, followed by acidification in workup.

20.37   Outline the steps required to convert p-ethylanisole into each of the following compounds.
(a) p-methoxybenzoic acid (b) 2-(p-methoxyphenyl)propanoic acid   (c) 3-(p-methoxyphenyl)butanoic acid

Answers:

(a) Oxidize the ethyl group to a carboxylic acid using a strong oxidizing agent such as potassium permanganate.

(b) Use NBS to prepare the secondary bromo compound of the ethyl side chain. Then, prepare the Grignard reagent using magnesium and ether. Third, add it to carbon dioxide, followed by acidification in workup.

(c) Use NBS to prepare the secondary bromo compound of the ethyl side chain. Second, prepare the Grignard reagent using magnesium and ether; third, add it to ethylene oxide to give 3-(p-methoxyphenyl)butanol. Oxidize the alcohol using the Jones reagent.

20.38   Fatty acids from natural sources are long-chain unbranched carboxylic acids that contain an even number of carbon atoms. Outline steps to convert the readily available dodecanoic acid (lauric acid) into the rare tridecanoic acid.

Answer:

First, reduce the carboxylic acid to dodecanol. Second, use PBr_3 to give 1-bromododecane. Third, prepare the Grignard reagent using magnesium and ether. Fourth, add the Grignard reagent to carbon dioxide, followed by acidification in workup.

20.39   Pivalic acid, $(CH_3)_3CCO_2H$, can be prepared from *tert*-butyl chloride. What method should be used?

Answer:

First, prepare the Grignard reagent using magnesium and ether. Second, add the Grignard reagent to carbon dioxide, followed by acidification in workup. The alternate method of displacing bromide ion by cyanide ion to give a nitrile cannot be accomplished with a tertiary halide.

20.40 Draw the structure of the product of each of the following reactions.

Answers:

(a)

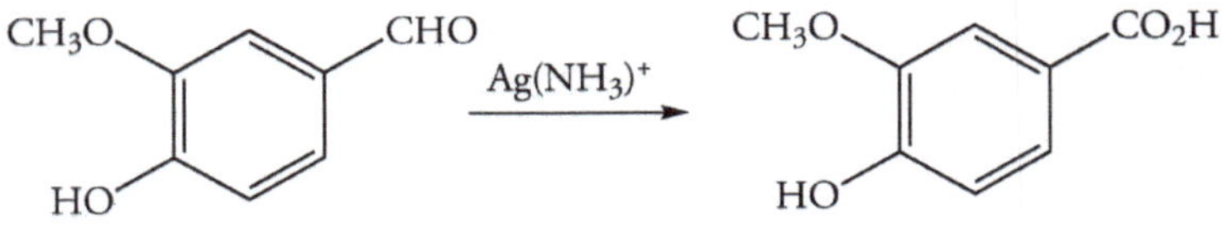

CH3O
CHO
HO
Ag(NH3)+
CH3O
CO2H
HO

(b)

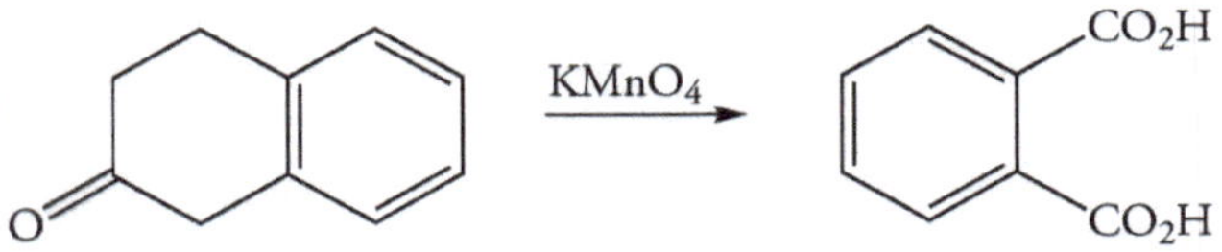

O
KMnO4
CO2H
CO2H

(c)

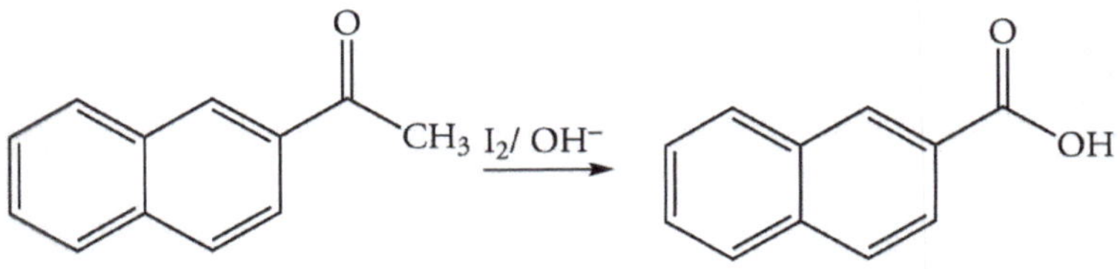

O
CH3
I2/ OH−
O
OH

(d)

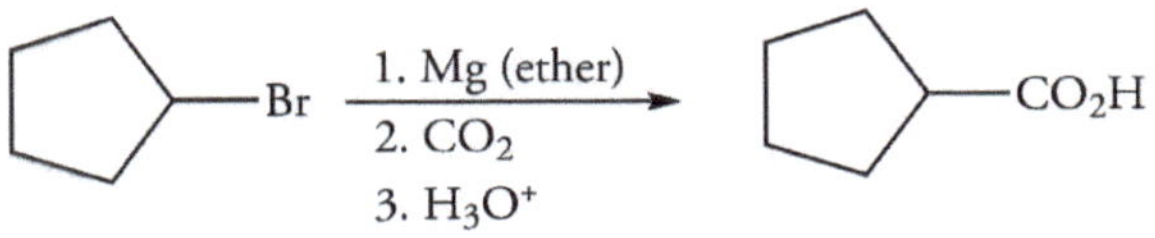

Br
1. Mg (ether)
2. CO2
3. H3O+
CO2H

(e)

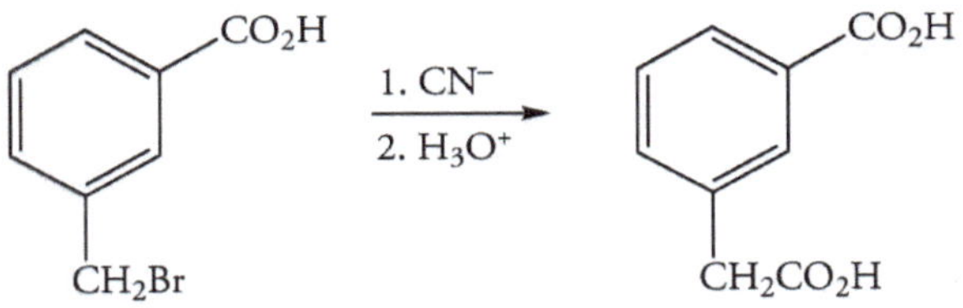

CO2H
CH2Br
1. CN−
2. H3O+
CO2H
CH2CO2H

20.41 Draw the structure of the product of each of the following reactions.

Answers:

(a)  furan-CH₂OH $\xrightarrow[\text{H}_2\text{SO}_4]{\text{CrO}_3}$ furan-CO₂H

(b) 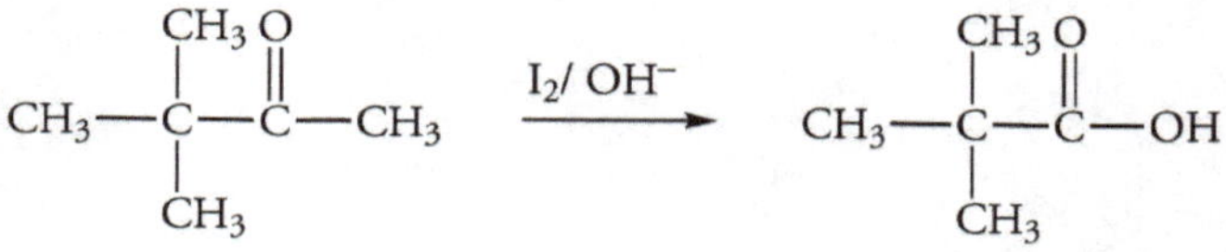 $CH_3-\overset{\overset{\displaystyle CH_3}{|}}{\underset{\underset{\displaystyle CH_3}{|}}{C}}-\overset{\overset{\displaystyle O}{||}}{C}-CH_3$ $\xrightarrow{\text{I}_2/\text{OH}^-}$ $CH_3-\overset{\overset{\displaystyle CH_3}{|}}{\underset{\underset{\displaystyle CH_3}{|}}{C}}-\overset{\overset{\displaystyle O}{||}}{C}-OH$

(c) [cyclohexadiene ring with CHO group and CH₃ group] $\xrightarrow{\text{KMnO}_4}$ [cyclohexadiene ring with CO₂H and CO₂H groups]

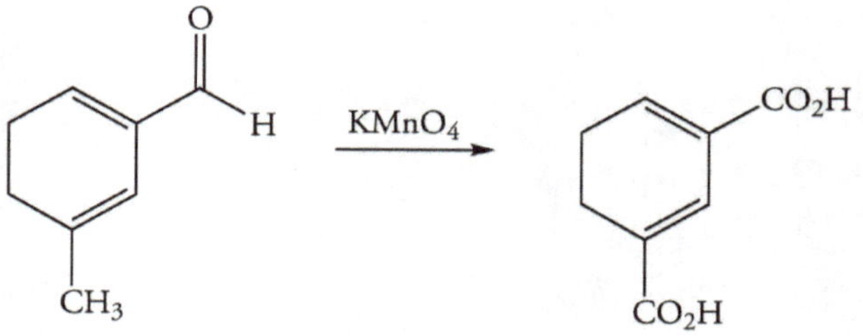

(d) 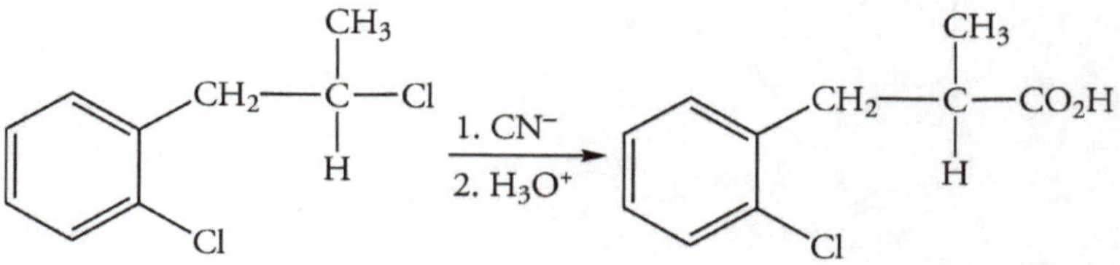 [o-chlorophenyl-CH₂-C(CH₃)(H)-Cl] $\xrightarrow[\text{2. H}_3\text{O}^+]{\text{1. CN}^-}$ [o-chlorophenyl-CH₂-C(CH₃)(H)-CO₂H]

(e) 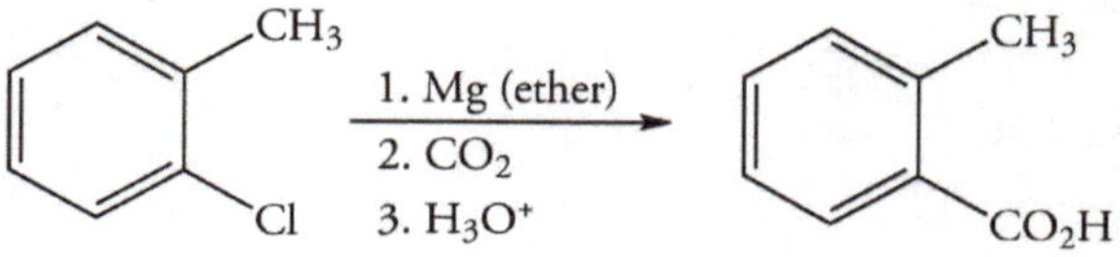 [o-chlorotoluene, CH₃ and Cl] $\xrightarrow[\substack{\text{2. CO}_2 \\ \text{3. H}_3\text{O}^+}]{\text{1. Mg (ether)}}$ [o-methylbenzoic acid, CH₃ and CO₂H]

Reduction of Carboxylic Acids

20.42 Metal hydride reductions occur by nucleophilic attack at the carbonyl carbon atom of acyl derivatives. Reduction of carboxylic acids with hydride reagents occurs slowly, but reduction by diborane occurs rapidly. Based on the structure of BH_3, the active reagent in diborane reductions suggests the structure of the first intermediate formed in the reaction.

Answer:

Boron is electron deficient in BH_3 and can form a coordinate covalent bond to the oxygen atom of the carbonyl group. As a result, the hydrogen atom can be transferred to the carbonyl carbon atom. The cyclic intermediate would resemble that of the reaction of BH_3 with an alkene in the hydroboration reaction.

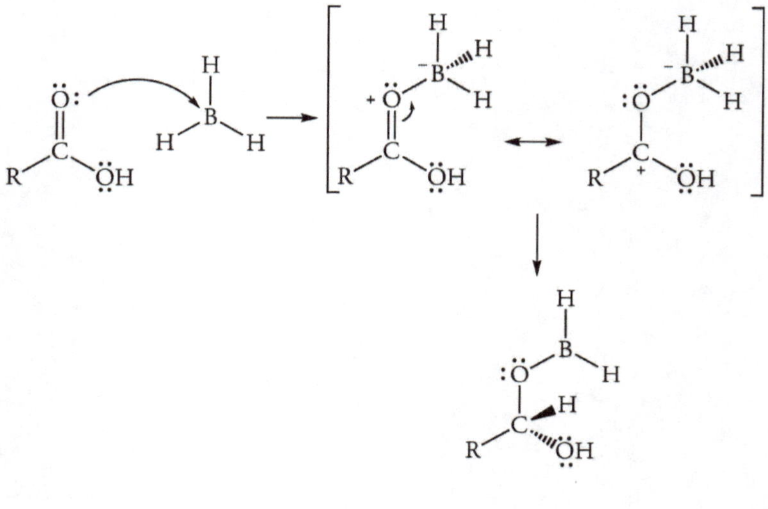

316

20.43 Diborane slowly reduces nitriles to amines but rapidly reduces aldehydes and ketones. Using the structure of BH_3, and the mechanism your wrote in Exercise 20.42, explain why nitriles react more slowly than aldehydes and ketones.

Answer:
The unshared pair of electrons of the nitrile can coordinate to the boron atom, but the linear geometry of the nitrile group does not allow close approach of the hydrogen atom to the carbon atom.

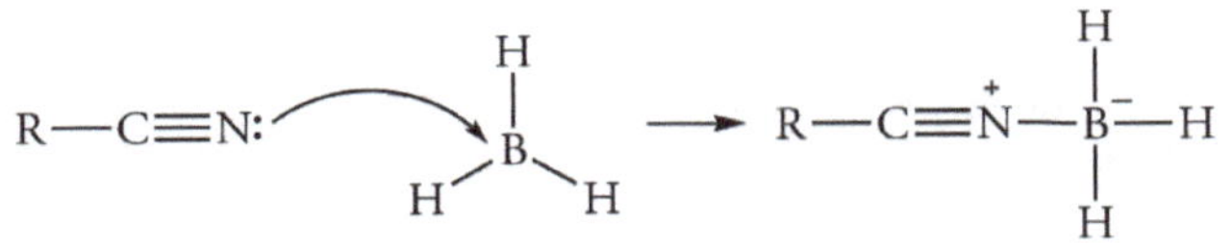

20.44 Lithium borohydride is a more active reducing agent than sodium borohydride, but less active than lithium aluminum hydride. Lithium borohydride reduces the ester group of the following compound selectively. Explain this selectivity.

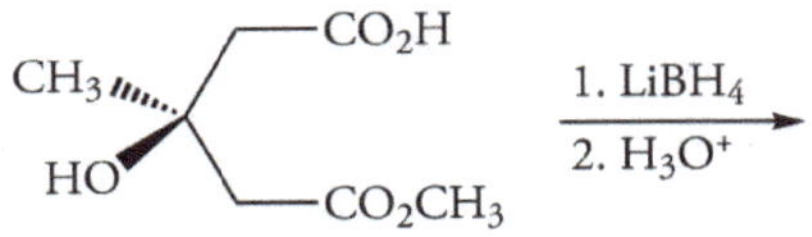

Answer: Aluminum is more electropositive than boron. As a result, the electron pair in an Al—H bond is more available to the hydrogen atom, which can depart as a hydride ion in a reaction. Lithium borohydride is more active than sodium borohydride because the coordination of the smaller lithium ion is stronger than that for the larger sodium ion. Coordination of the metal ion polarizes the carbonyl bond and increases the partial positive charge on the carbonyl carbon atom. As a result, the center is more reactive toward hydride ion derived from the borohydride ion.

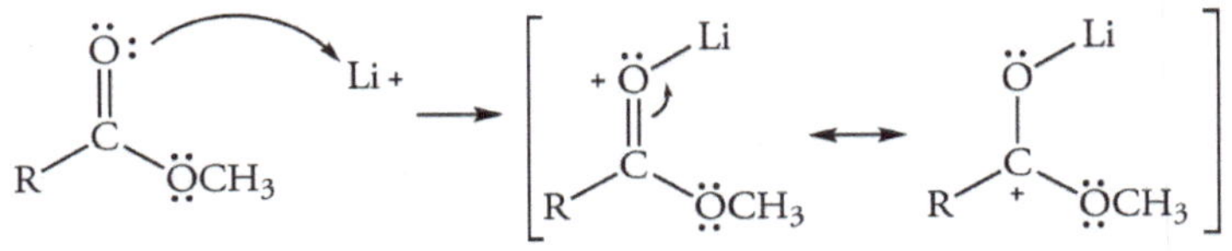

20.45 Draw the structure of the product of the following reaction. What relationship exists between this compound and the product of the reaction of Exercise 20.44?

Answer: BH_3 reduces the carboxylic acid group, but the ester group is unchanged. This compound is the methyl ester of the enantiomer of the product in Exercise 20.44.

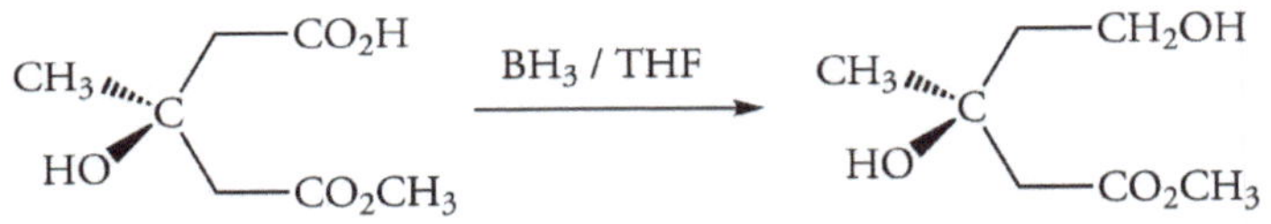

Decarboxylation

20.46 Could the Hunsdiecker reaction be used to decarboxylate an unsaturated carboxylic acid?

Answer: No, because the bromine would react with the unsaturated carbon–carbon π bond and give an addition product.

20.47 Which carboxylic acid should decarboxylate the more easily in a Hunsdiecker reaction, benzoic acid or cyclohexanecarboxylic acid?

Answer: Cyclohexanecarboxylic acid decarboxylates more readily because the carboxyl carbon atom is bonded to an sp^3-hybridized carbon atom, which is a weaker bond than the bond to the sp^2-hybridized carbon atom of benzoic acid.

Multistep Synthesis

22.48 Write the structure of the final product of each of the following sequences of reactions.

Answers:

(a) Oxidation gives a carboxylic acid, which is converted into an acid chloride by $SOCl_2$. Subsequent reaction with methanol gives a methyl ester.

(b) Reduction by lithium aluminum hydride gives a primary alcohol, which is then converted to a chloro compound by $SOCl_2$. Formation of a Grignard reagent followed by addition to carbon dioxide gives a carboxylic acid with one more carbon atom than the starting material.

(c) Oxidation by PCC gives an aldehyde. Subsequent reaction with methanol gives an acetal.

(a)

CH_2CH_2OH $\xrightarrow[H_2SO_4]{CrO_3}$ $\xrightarrow{PCl_5}$ $\xrightarrow[H^+]{CH_3OH}$ $CH_2CH_2O-\overset{\overset{\displaystyle O}{\|}}{C}-CH_3$

(b)

$-CHO$ $\xrightarrow[2.\ H^+]{1.\ LiAlH_4}$ $\xrightarrow{SOCl_2}$ $\xrightarrow[\substack{2.\ CO_2 \\ 3.\ H_3O^+}]{1.\ Mg\ /\ ether}$ $-CH_2CO_2H$

(c)

$-CH_2OH$ $\xrightarrow{PCC}$ $\xrightarrow[H^+]{CH_3OH}$ $\overset{\displaystyle O-CH_3}{\underset{\displaystyle O-CH_3}{C}}$

22.49 Write the structure of the final product of each of the following sequences of reactions.

Answers:

(a) Oxidation gives a carboxylic acid. Subsequent reaction with methanol in a Fischer esterification reaction gives a methyl ester.

(b) Reduction by lithium aluminum hydride give a secondary alcohol, which is then converted to an acetate ester by Fischer esterification.

(b) Reduction by the hydride reagent gives an aldehyde. Subsequent reaction with ethylene glycol gives a cyclic acetal.

(a)

CH_3 ... CHO $\xrightarrow[H_2SO_4]{CrO_3}$ $\xrightarrow[H^+]{CH_3OH}$ CH_3 ... $\overset{\overset{\displaystyle OCH_3}{\displaystyle C}}{\underset{\displaystyle O}{}}$

(b)

$-\overset{\overset{\displaystyle O}{\|}}{C}-CH_3$ $\xrightarrow[2.\ H_3O^+]{1.\ LiAlH_4}$ $\xrightarrow[H_2SO_4]{CH_3CO_2H}$ $-\underset{\displaystyle H}{\overset{\displaystyle O-\overset{\overset{\displaystyle O}{\|}}{C}-CH_3}{C}}-CH_3$

(c)

$-\overset{\overset{\displaystyle O}{\|}}{C}-Cl$ $\xrightarrow[2.\ HOCH_2CH_2OH]{1.\ LiAlH(t\text{-}OBu)_3}$ (cyclic acetal)

21 CARBOXYLIC ACID DERIVATIVES

KEYS TO THE CHAPTER

21.1 Nomenclature of Carboxylic Acid Derivatives

The names of the acid derivatives with the exception of nitriles resemble those of the structurally related carboxylic acid. The acid halides are named using *-oyl halide* in place of *-oic acid* of carboxylic acids. Compounds with an acid halide bonded to a ring carbon atom are named by appending *carbonyl halide* to the name of the alkane. The acid anhydrides are named using *-oic anhydride* in place of *-oic acid* of carboxylic acids.

Esters are named by the name of the alkyl group bonded to the bridging oxygen atom followed by the name of the acid, with *-oate* replacing *-oic acid*. Compounds with the acid portion bonded to a ring carbon atom are named by appending *carboxylate* to the name of the alkane. Lactones are named by adding *lactone* as a separate word to the name of the hydroxy acid.

Amides are named using *amide* in place of *-oic acid* of carboxylic acids. Compounds with an amide bonded to a ring carbon atom are named by appending *carboxamide* to the name of the alkane. The prefix *N-* is used to identify alkyl or aryl groups bonded to the nitrogen atom of the amide. Lactams are named by adding *lactam* as a separate word to the name of the amino acid. Nitrites are named by adding *nitrile* to the name of the related alkane. The name includes the carbon of the cyano group.

21.2 Physical Properties of Acyl Derivatives

Acid halides and acid anhydrides are polar compounds and have physical properties that resemble those of structurally similar carbonyl derivatives of similar molecular weight.

Esters are polar molecules, but they cannot form intermolecular hydrogen bonds like carboxylic acids do. Hence, esters have lower solubility in water and have lower boiling points than carboxylic acids.

The lower-molecular-weight amides are soluble in water as a result of hydrogen bonding to water molecules. The melting points of primary and secondary amides are higher than those for alkanes of similar molecular weight because of intermolecular hydrogen bonding. Tertiary amides have lower melting points than isomeric primary and secondary amides because they do not have an N—H bond to form intermolecular hydrogen bonds.

Nitriles are very polar acid derivatives due to the large bond moment of the triple bond between carbon and nitrogen. Lower-molecular-weight nitriles are soluble in water.

21.3 Basicity of Acyl Derivatives

Protonation of the carbonyl oxygen atom is a reaction common to all acid-catalyzed reactions of acid derivatives. The lone-pair electrons of oxygen are in sp^2-hybridized orbitals, and the oxygen atom is less basic than the oxygen atom of alcohols or ethers, whose lone-pair electrons are in sp^3-hybridized orbitals.

The basicity of acid derivatives is related to the ability of the electronegative atom bonded to the carbonyl carbon atom to stabilize the conjugate acid by resonance donation of electron density from that atom. Amides are the most basic because nitrogen is an effective donor of electrons. Esters are less basic than amides because oxygen is more electronegative and is a less effective electron donor. Acid chlorides are very much less basic because chlorine is not effective at stabilizing a positive charge.

Nitriles are very weak bases because the lone-pair electrons of the nitrogen atom are in an sp-hybridized orbital. Also, there is no alternative stabilized resonance form for the conjugate acid.

21.4 Nucleophilic Acyl Substitution

Nucleophilic acyl substitution (acyl transfer reaction) occurs by a two-step mechanism. First, attack of the carbonyl carbon atom of an acyl derivative by a nucleophile yields a tetrahedral intermediate. The tetrahedral intermediate can then eject a leaving group. The net result is a substitution reaction. The first step is rate determining. Thus, the rate of the reaction depends on the effect of the hydrocarbon group and the attached electronegative atom on the stability of the acid derivative and the transition state, not on the leaving group characteristics of the group displaced by the nucleophile.

The acid-catalyzed reaction of acid derivatives occurs by protonation of the carbonyl oxygen atom, which increases the electrophilicity of the carbonyl carbon atom. Attack of a nucleophile gives a tetrahedral intermediate that subsequent ejects the leaving group. Loss of a proton from the carbonyl oxygen atom completes the reaction.

The base-catalyzed reaction of acid derivatives occurs by abstraction of a proton from the nucleophile H—Nu, which prepares the nucleophile for attack of the carbonyl oxygen atom. The attack of the nucleophile gives a tetrahedral intermediate that has a negative charge on the oxygen atom. Return of an electron pair from that oxygen atom to reform the carbon–oxygen double bond results in ejection of the leaving group.

The different effects of the electronegative atoms on the reactivity of acid derivatives are explained using resonance and inductive effects on the stability of the reactant. The transition state resembles the tetrahedral intermediate, which cannot be stabilized by resonance. The stability of the transition states of all acyl derivatives for a reaction with the same nucleophile is approximately the same. Thus, the stability of the reactant controls the rate of its reaction. The most stable acyl derivatives are the least reactive. Amides are the most stable, and the least reactive, because nitrogen is an effective donor of electrons to the carbonyl group. Anhydrides and esters are somewhat less stable, because oxygen is more electronegative than nitrogen and is a less effective donor of electrons. Anhydrides are less stable because the donation of electrons to one carbonyl group is in competition with the donation of electrons to the second carbonyl group. Thus, in comparison to esters, where the oxygen atom need only stabilize one carbonyl group, anhydrides are more reactive than esters. Acid chlorides are very much less stable because chlorine is not effective at stabilizing positive charge by donation of electron density by resonance.

21.5 Hydrolysis of Acyl Derivatives

Acyl halides and acid anhydrides react readily with water to give carboxylic acids. Esters react with water in an equilibrium reaction to give an alcohol and a carboxylic acid. Amides are stable to water under neutral conditions.

The degree of completion of the reaction of an ester with water is increased by use of an equivalent amount of hydroxide ion. The saponification reaction is spontaneous because the product is a carboxylate ion that is a weaker base than hydroxide ion.

Amides are difficult to hydrolyze, and an equivalent amount of acid or base must be used. In the case of acid, the products are the carboxylic acid and the conjugate acid of the amine. In the case of base, the products are the amine and the conjugate base of the carboxylic acid.

Nitriles hydrolyze with difficulty using either concentrated acid or base. In the case of acid, the product is the ammonium ion and the carboxylic acid. In the case of base, the product is the conjugate base of the carboxylic acid and ammonia.

21.6 Reaction of Acyl Derivatives with Alcohols

The reactivity of acyl derivatives with alcohols parallels their reactivity with water. The reaction of either acid chlorides or acid anhydrides with alcohols is an excellent way to prepare esters.

Esters react with alcohols to interchange alkoxy groups in a *transesterification* reaction. The reaction is one to be avoided. Reactions of esters in an alcohol solvent are selected so that the alcohol and the ester contain the same alkoxy group.

21.7 Reaction of Acyl Derivatives with Amines

Amines (or ammonia) are better nucleophiles than alcohols (or water), so the reactions of amines with acyl derivatives are faster than the corresponding reaction with alcohols. Acid chlorides react with ammonia, primary amines, and secondary amines to produce primary, secondary, and tertiary amides. A second mole of the nitrogen compound is required to neutralize the HCl formed. Usually, pyridine is added to conserve the amine and allow the use of only one mole of the amine.

Acid anhydrides react with ammonia, primary amines, and secondary amines to produce primary, secondary, and tertiary amides. The by-product of the reaction is one mole of the carboxylic acid related to the acid anhydride.

Esters react similarly with ammonia, primary amines, and secondary amines to give an alcohol as the by-product.

21.8 Reduction of Acyl Derivatives

The salts of carboxylic acids are weak nucleophiles but can be used to displace halide ion from a haloalkane to give esters. The most diverse reaction of carboxylic acids and their derivatives is attack of nucleophiles at the carbonyl carbon atom. However, in contrast to aldehydes and ketones, the tetrahedral intermediate is unstable, and it ejects a leaving group, This overall process is nucleophilic acyl substitution, or an acyl transfer reaction.

Esters are reduced only by lithium aluminum hydride to give a primary alcohol related to the acid portion of the ester, with the alcohol of the original ester as a by-product. The process occurs by nucleophilic attack of hydride at the carbonyl carbon atom to give a tetrahedral intermediate that subsequently ejects an alkoxide ion. An aldehyde results, which is then rapidly reduced to the primary alcohol. Diisobutylaluminum hydride (DIBAL) reacts with esters to give the aldehyde.

Amides are reduced by lithium aluminum hydride to give amines containing the groups originally bonded to the nitrogen atom and an alkyl group derived from the acid minus its oxygen atom.

Nitriles are reduced by a variety of reagents. Raney nickel and hydrogen at high pressures give primary amines. Lithium aluminum hydride also reduces nitriles to amines.

21.9 Reaction of Acyl Derivatives with Organometallic Reagents

The carbanion derived from an organometallic compound is a nucleophile that can attack the carbonyl carbon atom of acyl derivatives. Ejection of a leaving group from the tetrahedral intermediate gives the final product, which may or may not react a second time with the organometallic compound.

Acid chlorides react with Grignard reagents in a manner similar to that for esters to give an intermediate aldehyde that subsequently reacts further to give an alcohol. However, this reaction is not used because esters are better starting materials. Reaction of acid chlorides with the Gilman reagent gives ketones that do not react further.

Esters react with Grignard reagents to add an alkyl (or aryl) group to the carbonyl carbon atom and eject an alkoxide ion. The resulting aldehyde reacts further with the reagent to give a tertiary alcohol containing two equivalents of the alkyl (or aryl) group of the Grignard reagent.

21.10 Spectroscopy of Acid Derivatives

The infrared spectra of nitriles contain an absorption for the carbon–nitrogen triple bond in the 2200–2250 cm^{-1} region. The absorption is more intense than that for carbon–carbon triple bond in the 2100–2200 cm^{-1} region.

Acyl derivatives are characterized by a strong absorption due to the carbonyl group in the infrared spectra of these compounds. The absorption occurs in the same region as for the carbonyl groups of aldehydes and ketones. However, the position of the absorption is strongly affected by the electronegative atom and its contribution to the stability of the dipolar resonance form by a combination of inductive and resonance affects. Esters have absorptions at 1735 cm^{-1}, but that position is affected by conjugation of double bonds with the acyl group. The absorptions of lactones show the same changes with ring size as for cycloalkanones.

Acid chlorides absorb at 1800 cm^{-1} because the chlorine atom cannot stabilize the dipolar resonance form by donation of electrons. In fact, the inductive electron withdrawal of electrons destabilizes that resonance form and increases the double bond character of the carbonyl group.

Amides very effectively donate electrons to the carbonyl carbon atom by resonance. Hence, the double bond character of the carbonyl group is reduced, and the resulting absorption of the carbonyl group occurs in the 1650–1655 cm^{-1} region.

The proton NMR of the a hydrogen atoms of acyl derivatives occurs in the 2 δ region, which is the same region for the α hydrogen atoms of aldehydes and ketones. The chemical shift of the hydrogen atoms on the carbon atom bonded to the oxygen atom of the alkoxy part of esters are at somewhat lower field (4 δ) than alcohols. The chemical shift of the hydrogen atoms bonded to the alkyl carbon atom bearing the nitrogen atom in amides occur in the 2.6–3.0 δ region.

The absorption of a carbon atoms of acid derivatives have ^{13}C NMR absorptions in the 20 δ region. The carbonyl carbon atom is easily identified by its absorption in the 165–180 δ region. The absorptions of the carbon atom bonded to the oxygen atom of the alkoxy part of an ester are in the 60 δ region.

Summary of Reactions

1. Hydrolysis of Acid Derivatives

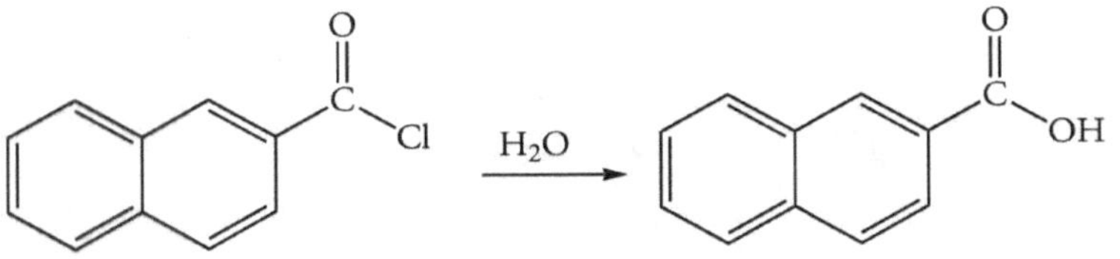

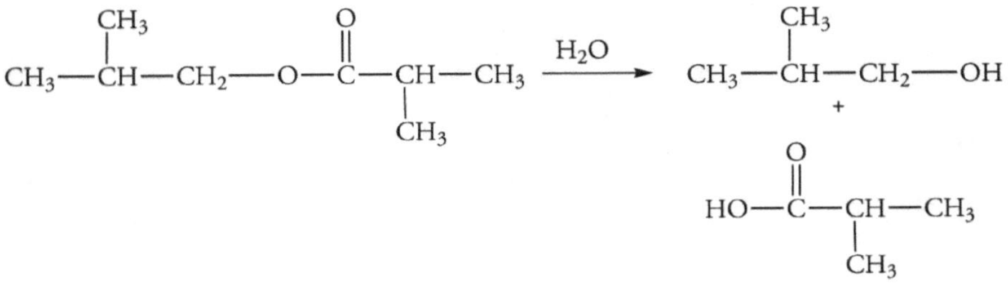

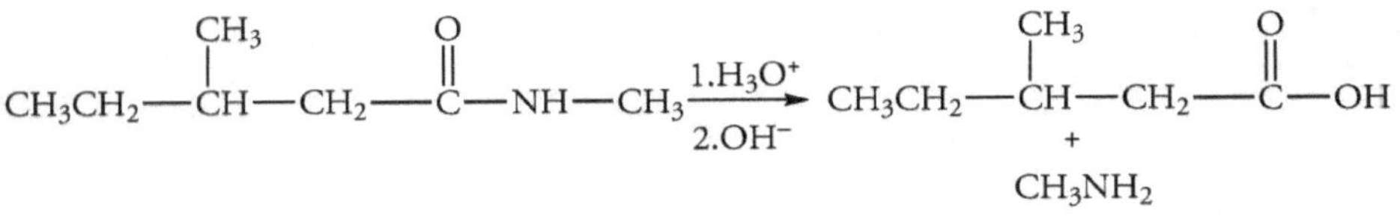

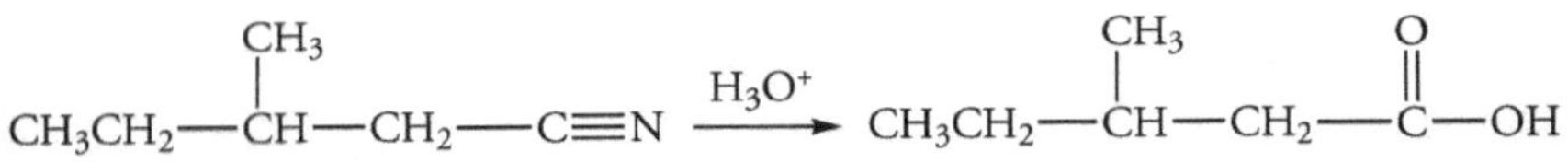

naphthalene-$C(=O)Cl$ $\xrightarrow{H_2O}$ naphthalene-$C(=O)OH$

$$CH_3-\underset{CH_3}{CH}-CH_2-O-\underset{O}{\overset{\|}{C}}-\underset{CH_3}{\overset{|}{CH}}-CH_3 \xrightarrow{H_2O} CH_3-\underset{CH_3}{\overset{|}{CH}}-CH_2-OH$$

$+$

$$HO-\underset{O}{\overset{\|}{C}}-\underset{CH_3}{\overset{|}{CH}}-CH_3$$

$$CH_3CH_2-\underset{CH_3}{\overset{|}{CH}}-CH_2-\underset{O}{\overset{\|}{C}}-NH-CH_3 \xrightarrow[2.OH^-]{1.H_3O^+} CH_3CH_2-\underset{CH_3}{\overset{|}{CH}}-CH_2-\underset{O}{\overset{\|}{C}}-OH$$

$+$

$$CH_3NH_2$$

$$CH_3CH_2-\underset{CH_3}{\overset{|}{CH}}-CH_2-C\equiv N \xrightarrow{H_3O^+} CH_3CH_2-\underset{CH_3}{\overset{|}{CH}}-CH_2-\underset{O}{\overset{\|}{C}}-OH$$

2. Reactions of Acid Derivatives with Alcohols and Phenols

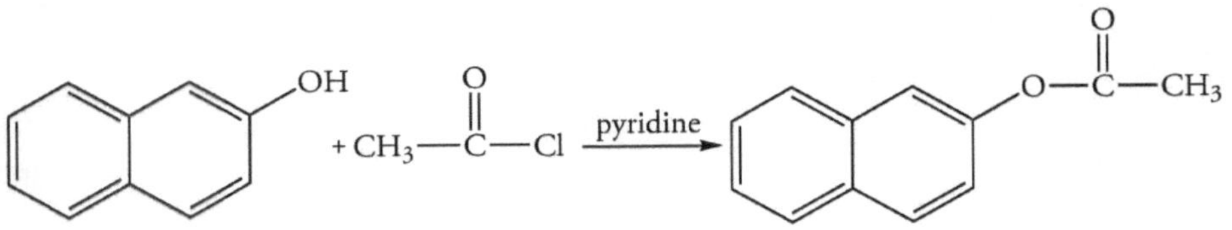

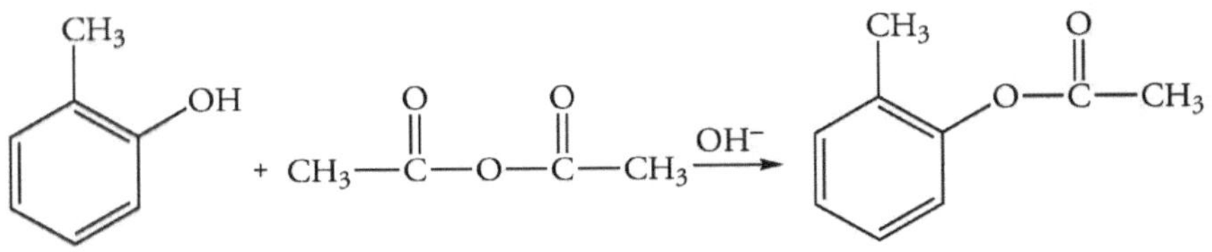

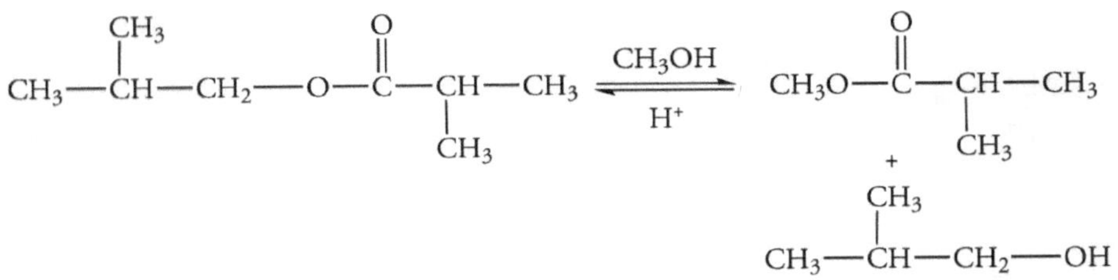

naphthalen-2-ol (OH) $+ CH_3-\underset{O}{\overset{\|}{C}}-Cl \xrightarrow{pyridine}$ naphthalen-2-yl $O-\underset{O}{\overset{\|}{C}}-CH_3$

2-methylphenol (CH_3, OH) $+ CH_3-\underset{O}{\overset{\|}{C}}-O-\underset{O}{\overset{\|}{C}}-CH_3 \xrightarrow{OH^-}$ 2-methylphenyl $O-\underset{O}{\overset{\|}{C}}-CH_3$

$$CH_3-\underset{CH_3}{\overset{|}{CH}}-CH_2-O-\underset{O}{\overset{\|}{C}}-\underset{CH_3}{\overset{|}{CH}}-CH_3 \underset{H^+}{\overset{CH_3OH}{\rightleftharpoons}} CH_3O-\underset{O}{\overset{\|}{C}}-\underset{CH_3}{\overset{|}{CH}}-CH_3$$

$+$

$$CH_3-\underset{CH_3}{\overset{|}{CH}}-CH_2-OH$$

3. Reactions of Acid Derivatives with Amines

$$CH_3CH_2\overset{\underset{\displaystyle |}{CH_3}}{C}HCH_2-\overset{\underset{\displaystyle ||}{O}}{C}-Cl \xrightarrow[\text{pyridine}]{CH_3CH_2NH_2} CH_3CH_2\overset{\underset{\displaystyle |}{CH_3}}{C}HCH_2-\overset{\underset{\displaystyle ||}{O}}{C}-NHCH_2CH_3$$

(2-methoxyphenyl)amine, OCH_3 substituent, NH_2 group $\quad + \quad CH_3-\overset{O}{C}-O-\overset{O}{C}-CH_3 \quad \longrightarrow \quad$ (2-methoxyphenyl) $NH-\overset{O}{C}-CH_3$

$$CH_3CH_2\overset{\underset{\displaystyle |}{CH_3}}{C}HCH_2-\overset{\underset{\displaystyle ||}{O}}{C}-OH + CH_3NH_2 \longrightarrow CH_3CH_2\overset{\underset{\displaystyle |}{CH_3}}{C}HCH_2-\overset{\underset{\displaystyle ||}{O}}{C}-NHCH_3$$

4. Reduction of Acid Chlorides

$$CH_3CH_2\overset{\underset{\displaystyle |}{CH_3}}{C}HCH_2-\overset{\underset{\displaystyle ||}{O}}{C}-Cl \xrightarrow[\text{2. }H_3O^+]{\text{1. LiAlH}_4} CH_3CH_2\overset{\underset{\displaystyle |}{CH_3}}{C}HCH_2-\overset{H}{\underset{H}{C}}-OH$$

$$CH_3CH_2\overset{\underset{\displaystyle |}{CH_3}}{C}HCH_2-\overset{\underset{\displaystyle ||}{O}}{C}-Cl \xrightarrow[\text{2. }H_3O^+]{\text{1. LiH(OC(CH}_3))_3} CH_3CH_2\overset{\underset{\displaystyle |}{CH_3}}{C}HCH_2-\overset{\underset{\displaystyle ||}{O}}{C}-H$$

5. Reduction of Esters

benzoic acid methyl ester (C(=O)OCH$_3$) $\xrightarrow[\text{2. }H_3O^+]{\text{1. LiAlH}_4}$ benzyl alcohol (CH_2OH) $\quad + \quad CH_3OH$

benzoic acid methyl ester (C(=O)OCH$_3$) $\xrightarrow[\text{2. }H_3O^+]{\text{1. DIBAL}}$ benzaldehyde (C(=O)H) $\quad + \quad CH_3OH$

6. Reduction of Amides

naphthalene-2-carboxamide ($-\overset{O}{C}-NHCH_3$) $\xrightarrow[\text{2. }H_3O^+]{\text{1. LiAlH}_4}$ 2-naphthyl $-\overset{H\quad H}{C}-NHCH_3$

 **End of Chapter Exercises**

Nomenclature

21.1 Give the IUPAC name for each of the following carboxylic acid derivatives.

Answers:
(a) ethyl phenylethanoate
(b) 3-cyclohexylbutanenitrile
(c) N,N-diethylcyclobutanecarboxamide
(d) 2-bromoethyl 3-bromobenzoate
(e) 3,4-dimethoxybenzoyl chloride

(a) $\text{C}_6\text{H}_5-\text{CH}_2-\overset{\overset{\displaystyle O}{\|}}{\text{C}}-\text{O}-\text{CH}_2\text{CH}_3$

(b) cyclohexyl$-\overset{\overset{\displaystyle CH_3}{|}}{\text{CH}}\text{CH}_2\text{C}\equiv\text{N}$

(c) cyclobutyl$-\overset{\overset{\displaystyle O}{\|}}{\text{C}}-\text{N(CH}_2\text{CH}_3)_2$

(d) $3\text{-Br-C}_6\text{H}_4-\overset{\overset{\displaystyle O}{\|}}{\text{C}}-\text{O}-\text{CH}_2\text{CH}_2\text{Br}$

(e) CH_3O, $\text{CH}_3\text{O}-\text{C}_6\text{H}_3-\overset{\overset{\displaystyle O}{\|}}{\text{C}}-\text{Cl}$

21.2 Give the IUPAC name for each of the following carboxylic acid derivatives.

Answers:
(a) cyclohexyl benzoate
(b) 4-methylbenzonitrile
(c) N-cyclohexyl-2-fluoroethanamide
(d) 3,4-dimethylbenzoyl chloride
(e) 2-methylbutyl 5-cyclopentylpentanoate

(a) $\text{C}_6\text{H}_5-\overset{\overset{\displaystyle O}{\|}}{\text{C}}-\text{O}-\text{cyclohexyl}$

(b) $\text{CH}_3-\text{C}_6\text{H}_4-\text{C}\equiv\text{N}$

(c) cyclohexyl$-\text{NH}-\overset{\overset{\displaystyle O}{\|}}{\text{C}}-\text{CH}_2\text{F}$

(d) $\text{CH}_3, \text{CH}_3-\text{C}_6\text{H}_3-\overset{\overset{\displaystyle O}{\|}}{\text{C}}-\text{Cl}$

(e) cyclopentyl$-\text{CH}_2(\text{CH}_2)_3-\overset{\overset{\displaystyle O}{\|}}{\text{C}}-\text{O}-\text{CH}_2\overset{\overset{\displaystyle CH_3}{|}}{\text{CH}}\text{CH}_2\text{CH}_3$

21.3 Write the structure of each of the following compounds.
(a) phenyl octanoate   (b) butanoic anhydride   (c) N-ethyl-4,4-dimethylcyclohexanecarboxamide
(d) 2-bromo-3-methylbutanoyl chloride   (e) trans-4-methylcyclohexanecarbonitrile

Answers:

(a) $\text{C}_6\text{H}_5-\overset{\overset{\displaystyle O}{\|}}{\text{C}}-\text{CH}_2(\text{CH}_2)_6\text{CH}_3$

(b) $\text{CH}_3(\text{CH}_2)_2\text{CH}_2-\overset{\overset{\displaystyle O}{\|}}{\text{C}}-\text{O}-\overset{\overset{\displaystyle O}{\|}}{\text{C}}-\text{CH}_2\text{CH}_2\text{CH}_3$

(c) $\text{(CH}_3)_2$-cyclohexyl$-\overset{\overset{\displaystyle O}{\|}}{\text{C}}-\text{NH}-\text{CH}_2\text{CH}_3$

(d) $\text{CH}_3-\overset{\overset{\displaystyle CH_3}{|}}{\text{CH}}-\overset{\overset{\displaystyle }{|}\\\underset{\underset{\displaystyle Br}{|}}{}}{\text{CH}}-\overset{\overset{\displaystyle O}{\|}}{\text{C}}-\text{Cl}$

(e) CH_3-cyclohexyl$-\text{C}\equiv\text{N}$

21.4   Write the structure of each of the following compounds.
      (a) 2-chloropropyl 3-bromobutanoate   (b) 4-methoxyphthalic anhydride (c) *N,N*-dimethyl-3-cyclopropylpentanamide
      (d) cyclobutanecarbonyl bromide   (e) (*R*)-2-methylbutanenitrile

Answers:

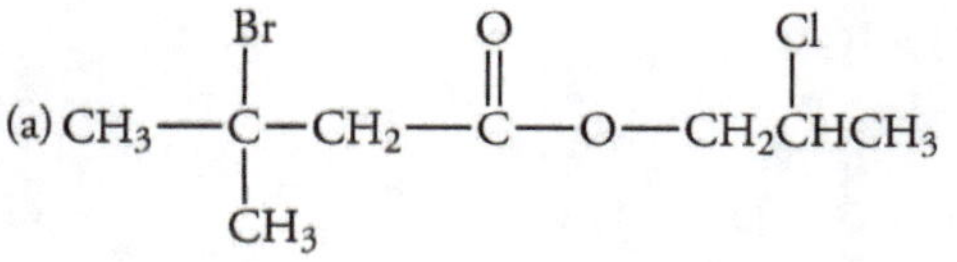

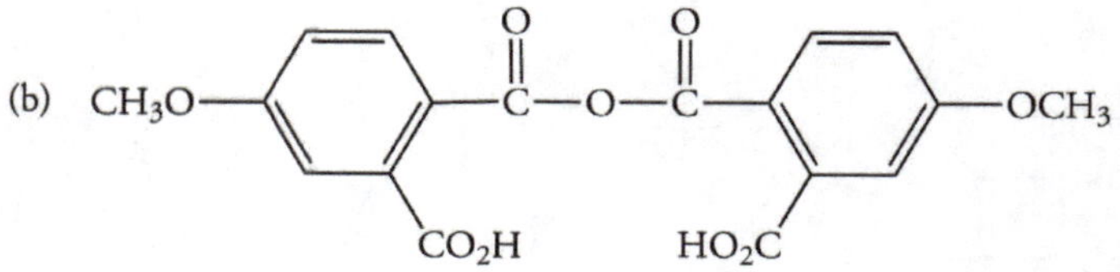

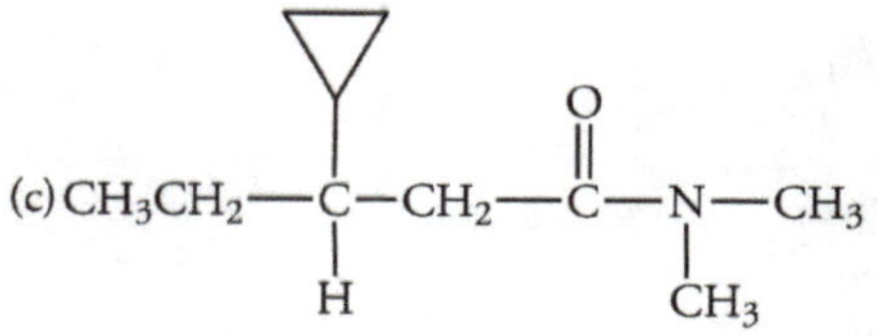

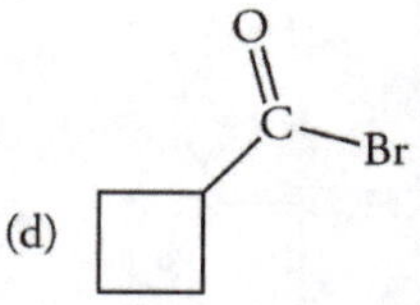

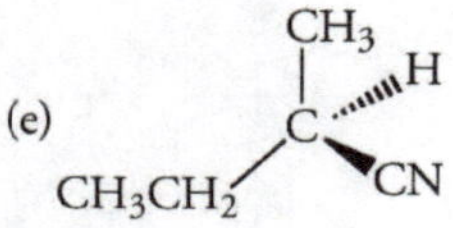

21.5   The common name of the vasodilator cyclandelate is 3,5,5-trimethylcyclohexyl mandelate. Give the structure and name of the acid contained in the ester.

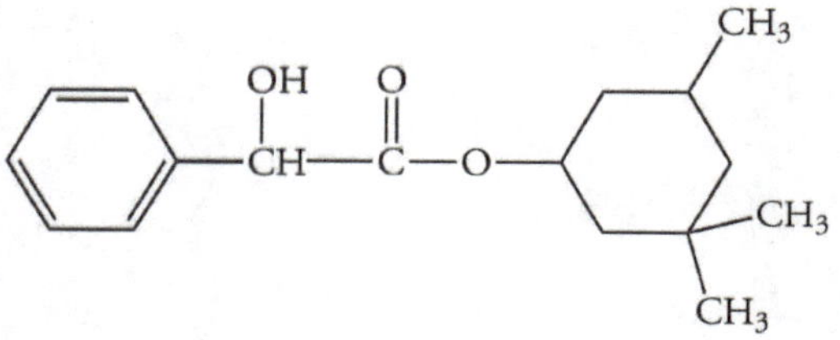

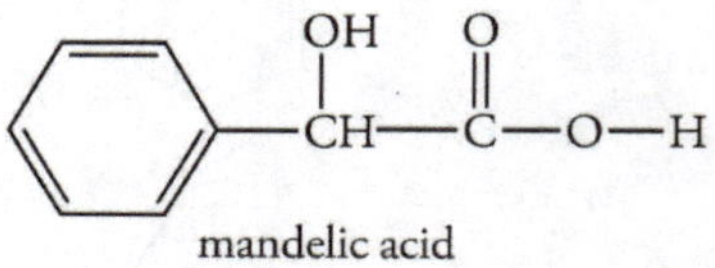

mandelic acid

21.6   Hydrolysis in the body is required for diloxanide furanoate to be effective against intestinal amebiasis. What is the acid component of the drug? Considering the name of the drug, name the acid.

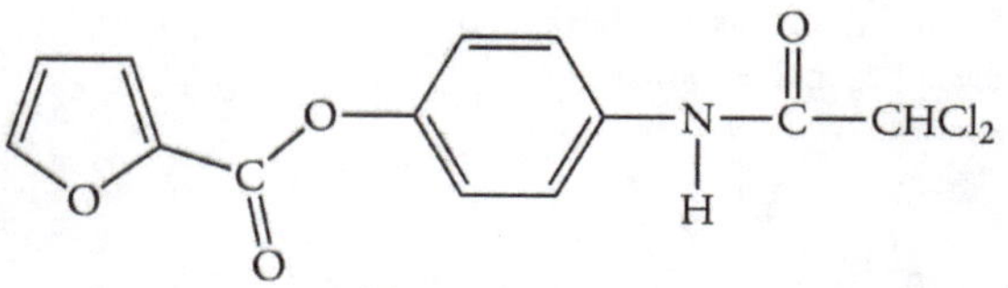

diloxanide furanoate 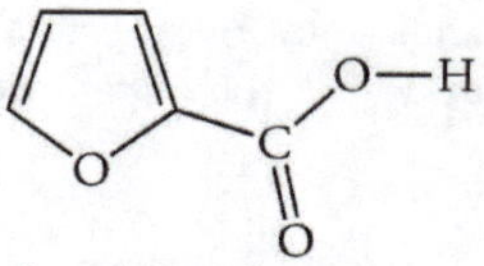   furanoic acid

Answer: The carboxylic acid is derived from the part of the compound that contains the carboxylate group. The name of the acid is furanoic acid, which is obtained by replacing *-oate* with *-ic acid*.

Cyclic Acyl Derivatives

21.7 (a) Identify the oxygen-containing functional group in the following structure, a pheromone of the female Japanese beetle. (b) What is the configuration around the alkene moiety?

Answers: (a) The functional group is an ester contained within a ring and is therefore a lactone. (b) The configuration about the double bond is (Z).

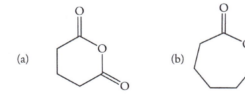

$CH_2)CH_2)_6CH_3$

21.8 Identify the nitrogen-containing functional group within the four-membered ring of cephalosporin C, antibiotic.

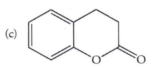

cephalosporin C

Answer: The functional group within the four-membered ring is an amide and is therefore a lactam.

21.9 Name each of the following lactones.

Answers:
(a) 5-hydroxypentanoic acid lactone
(b) 9-hydroxynonanoic acid lactone
(c) 4-hydroxypentanoic acid lactone

(a)

(b)

(c)

21.10 Draw the structure of each of the following lactams.
(a) 3-aminopropanoic acid lactam   (b) 4-aminopentanoic acid lactam   (c) 5-aminopentanoic acid lactam

Answers:

(a)

(b)

(c)

21.11 Which of the following compounds are lactones?

(a)

(b)

(c)

Answer: Compound (a) has two carbonyl groups joined by a common oxygen atom and is a cyclic anhydride. Only (b) and (c) are lactones.

21.12   Which of the following compounds are lactams?

(a)   (b)   (c)

Answer: In compound (b), a methylene group separates the nitrogen atom and the carbonyl group. It is not a lactam. Only (a) and (c) are lactams.

Properties of Acid Derivatives

21.13   The boiling points of methyl pentanoate and butyl ethanoate are 126 and 125 °C, respectively. Explain the similarity of these boiling points.

Answer: The compounds are isomers and have similar polarities. They also have similar molecular shapes. Thus, the dipole–dipole as well as London forces are similar, and as a result, they have similar boiling points.

21.14   The boiling points of methyl pentanoate and methyl 2,2-dimethylpropanoate are 126 and 102 °C, respectively. Explain why these values differ.

Answer: The compounds are isomeric methyl esters and thus have similar polarities. However, they have different molecular shapes. The acid portion of methyl 2,2-dimethyipropanoate has a nearly spherical shape, but the acid moiety of methyl pentanoate has a cylindrical shape. Thus, the van der Waals forces of methyl 2,2-dimethylpropanoate are smaller, and the boiling point is lower.

21.15   The boiling points of acetonitrile and 1-propyne are 81.5 and −23 °C, respectively. Account for this difference in boiling point between two compounds with similar molecular weights.

Answer: The molecular weights differ only by 1 amu, and both compounds have similar cylindrical shapes. However, the nitrile has a polar carbon–nitrogen triple bond and as a result has larger dipole–dipole forces than propyne, which has a much less polar carbon–carbon triple bond.

21.16   The boiling points of acetamide and acetic acid are 221 and 118 °C, respectively. Account for this difference in boiling point between two compounds with similar molecular weights.

Answer: Both compounds can form intermolecular hydrogen bonds between a hydrogen atom bonded to an electronegative atom on one molecule and the carbonyl group on a neighboring molecule. However, the N—H group is a better hydrogen bond donor, and it forms stronger hydrogen bonds, so the amide has a higher boiling point.

21.17   Explain why protonation of *N,N*-dimethylformamide occurs at the oxygen atom rather than the nitrogen atom.

Answer: The amide functional group is resonance stabilized. As a result, the oxygen atom has increased electron density and the nitrogen atom has decreased electron density.

Protonation at the electron pair of nitrogen would give a conjugate acid in which there is no resonance stabilization of the carbonyl group. Resonance stabilization of the carbonyl group is still possible in the conjugate acid obtained by protonating oxygen.

no resonance stabilization   resonance stabilization

21.18   The rotational barrier around the nitrogen-carbonyl carbon bond of *N,N*-dimethylformamide is approximately 87 kJ mole^{-1}. Why is this energy barrier substantially higher than values for other single bonds?

Answer: The carbon–nitrogen bond has some double bond character as a result of resonance stabilization of the carbonyl group by donation of electrons from the nitrogen atom, as shown in the answer to Exercise 21.17.

Cyclic Acyl Derivatives

21.19   Indicate whether each of the following reactions will occur.

(a) 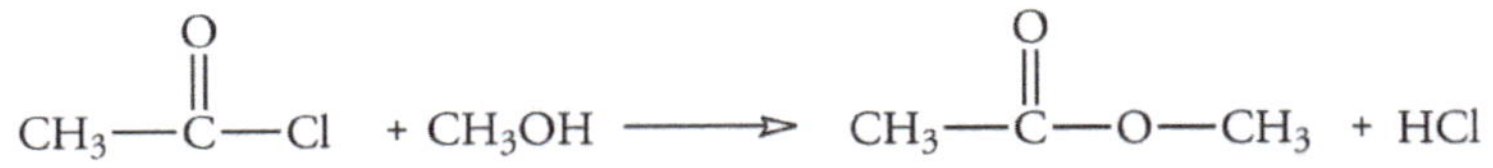 CH_3—$\overset{\displaystyle O}{\overset{\|}{C}}$—Cl + CH_3OH ⟶ CH_3—$\overset{\displaystyle O}{\overset{\|}{C}}$—O—$CH_3$ + HCl

(b) CH_3—$\overset{\displaystyle O}{\overset{\|}{C}}$—$NH_2$ + CH_3OH ⟶ CH_3—$\overset{\displaystyle O}{\overset{\|}{C}}$—O—$CH_3$ + NH_3

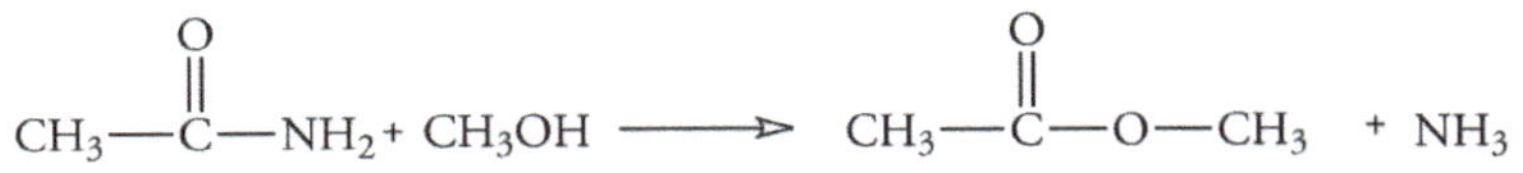

(c) CH_3—$\overset{\displaystyle O}{\overset{\|}{C}}$—O—$CH_3$ + CH_3NH_2 ⟶ CH_3—$\overset{\displaystyle O}{\overset{\|}{C}}$—$\overset{\displaystyle H}{\overset{\|}{N}}$—$CH_3$ + CH_3OH

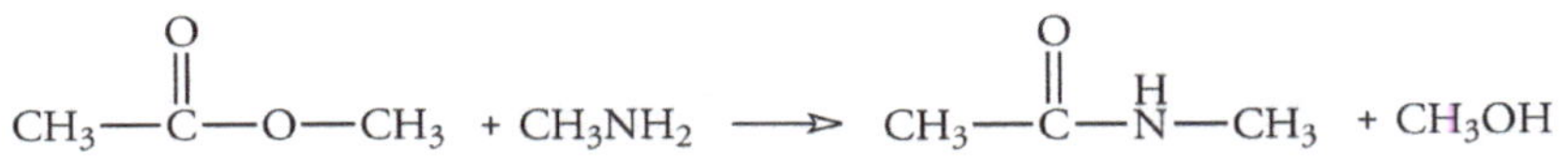

Answers:
(a) Esters are more stable than acid chlorides because the carbonyl group is more stabilized by donation of electrons by resonance from oxygen than from chlorine. The reaction occurs to give the more stable ester.
(b) Amides are more stable than esters because the carbonyl group is more stabilized by donation of electrons by resonance from nitrogen than from oxygen. The reaction will not occur.
(c) Amides are more stable than esters because the carbonyl group is more stabilized by donation of electrons by resonance from nitrogen than from oxygen. The reaction will occur.

21.20   Indicate whether each of the following reactions will occur.

(a) 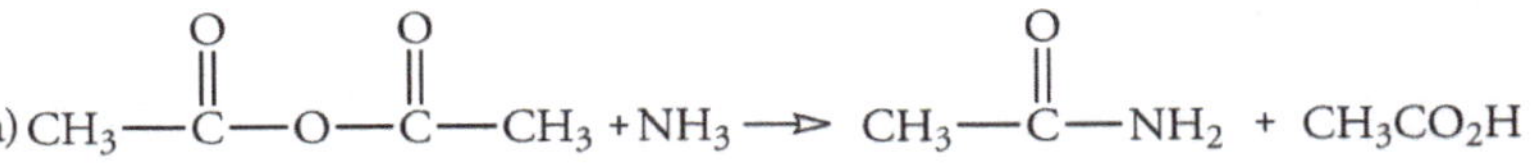 CH_3—$\overset{\displaystyle O}{\overset{\|}{C}}$—O—$\overset{\displaystyle O}{\overset{\|}{C}}$—$CH_3$ + NH_3 ⟶ CH_3—$\overset{\displaystyle O}{\overset{\|}{C}}$—$NH_2$ + CH_3CO_2H

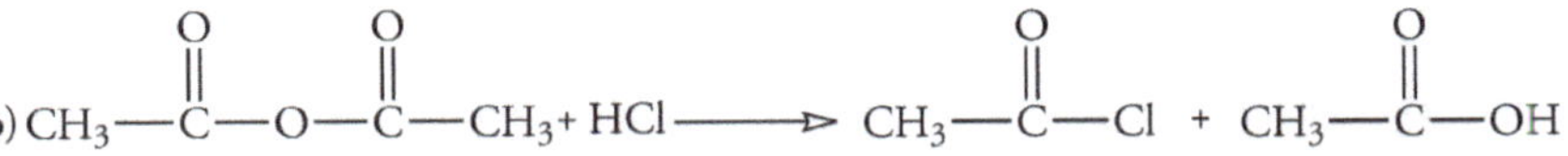

(b) CH_3—$\overset{\displaystyle O}{\overset{\|}{C}}$—O—$\overset{\displaystyle O}{\overset{\|}{C}}$—$CH_3$ + HCl ⟶ CH_3—$\overset{\displaystyle O}{\overset{\|}{C}}$—Cl + CH_3—$\overset{\displaystyle O}{\overset{\|}{C}}$—OH

(c) CH_3—$\overset{\displaystyle O}{\overset{\|}{C}}$—O—$\overset{\displaystyle O}{\overset{\|}{C}}$—$CH_3$ + CH_3OH ⟶ CH_3—$\overset{\displaystyle O}{\overset{\|}{C}}$—O—$CH_3$ + CH_3CO_2H

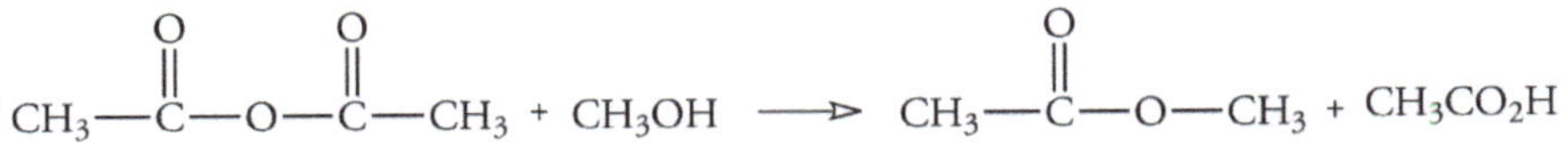

Answers:
(a) Amides are more stable than acid anhydrides because the carbonyl group is more stabilized by donation of electrons by resonance from nitrogen than from oxygen. The reaction will occur.
(b) Acid anhydrides are more stable than acid chlorides because the carbonyl group is more stabilized by donation of electrons by resonance from oxygen than from chlorine. The reaction will not occur.
(c) Esters are more stable than acid anhydrides because the carbonyl group is more stabilized by donation of electrons by resonance from oxygen bonded to an alkyl group than from oxygen bonded to another carbonyl group. The reaction will occur.

21.21   Considering the stability of the reactant, explain why thioesters react more readily than esters in acyl substitution reactions.

Answer: Esters are more stable than thioesters because the carbonyl group is more stabilized by donation of electrons by resonance from oxygen than from sulfur. The electrons of sulfur are in the third energy level, and overlap of 3p and 2p orbitals in thioesters is less effective than overlap of 2p and 2p orbitals in esters.

21.22 Considering the stability of the reactant, explain why thioesters are less reactive than acid chlorides in acyl substitution reactions.

Answer: Both sulfur and chlorine are third row elements and neither contributes electrons very effectively by resonance. However, to the extent that they do, sulfur is a better donor of electrons by resonance because it is less electronegative than chlorine. Thus, the resonance effect makes the thioester more stable than the acid chloride. There is also a stronger inductive electron withdrawal of electrons by the more electronegative chlorine, which destabilizes the carbonyl group leading to increased reactivity of the acid chloride.

21.23 Explain the order of reactivity in a saponification reactions of each of the following pairs of compounds.

(a) $CH_3-\overset{\overset{\displaystyle O}{\|}}{C}-O-CH_3$ > $CH_3-\overset{\overset{\displaystyle O}{\|}}{C}-O-C(CH_3)_3$

(b) $CH_3-\overset{\overset{\displaystyle O}{\|}}{C}-O-$⬡ > $CH_3-\overset{\overset{\displaystyle O}{\|}}{C}-O-$⬡

(c) $(CH_3)_3C$⬡CO_2CH_3 > $(CH_3)_3C$⬡CO_2CH_3 (H)

(d) ⬡ > ⬡

Answers:
(a) The *tert*-butyl group of the alcohol portion of the ester on the right sterically hinders approach of a nucleophile to the carbonyl group. Thus, the compound on the left is the more reactive.
(b) The resonance interaction of the nonbonded electrons of the oxygen atom with the aromatic ring of the phenolic portion of the ester decreases the electron density at oxygen. Thus, there is a decreased availability of electrons to the carbonyl group, the carbonyl group is not as stable for the compound on the left and it is more reactive
(c) The axial carbomethoxy group of the compound on the right is sterically hindered and approach of a nucleophile is also sterically hindered. Thus, the compound on the left is the more reactive.
(d) The carbonyl group of the six-membered lactone can be approached in an equatorial direction. The five-membered lactone has its carbonyl group in the plane of the ring, so approach of a nucleophile must occur perpendicular to that plane and is sterically hindered.

21.24 Explain the order of reactivity in saponification reactions of each of the following pairs of compounds.

(a) $CH_3-\overset{\overset{\displaystyle O}{\|}}{C}-O-CH_2CH_3$ > $(CH_3)_3C-\overset{\overset{\displaystyle O}{\|}}{C}-O-CH_3$

(b) $CH_3-\overset{\overset{\displaystyle O}{\|}}{C}-O-$⬡$-NO_2$ > $CH_3-\overset{\overset{\displaystyle O}{\|}}{C}-O-$⬡

(c) $CF_3-\overset{\overset{\displaystyle O}{\|}}{C}-O-$⬡$-OCH_3$ > $CH_3-\overset{\overset{\displaystyle O}{\|}}{C}-O-$⬡$-OCH_3$

(d) ⬡ > ⬡

(a) The *tert*-butyl group bonded to the carbonyl carbon atom of the compound on the right sterically hinders approach of a nucleophile. Thus, the compound on the left is the more reactive
(b) The resonance interaction of the nonbonded electrons of the oxygen atom is decreased by the resonance electron withdrawal of those electrons by the *p*-nitro group. As a result, the carbonyl group is not as stable for the compound on the left and it is more reactive
(c) The trifluoromethyl group bonded to the carbonyl carbon atom inductively withdraws electron density. As a result, the carbonyl group is not as stable in the compound on the left, and it is more reactive toward nucleophiles.
(d) Ring strain of the compound is increased by the incorporation of an sp²-hybridized carbon atom. In the transition state, that carbon atom becomes sp³-hybridized, and the ring strain is decreased. Thus, the activation energy is smaller for the compound on the left.

21.25   Explain the position of the following equilibrium.

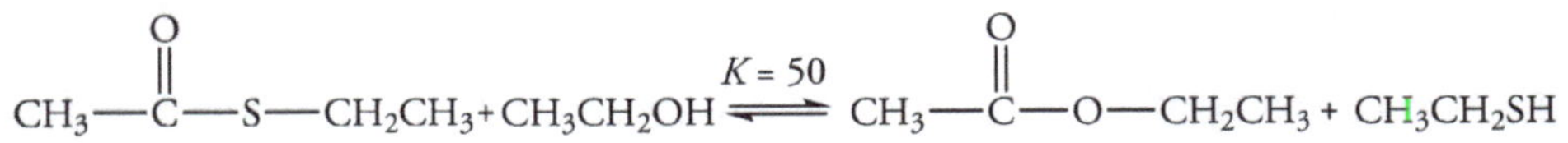

$$CH_3-\overset{\overset{O}{\|}}{C}-S-CH_2CH_3 + CH_3CH_2OH \underset{}{\overset{K\,=\,50}{\rightleftharpoons}} CH_3-\overset{\overset{O}{\|}}{C}-O-CH_2CH_3 + CH_3CH_2SH$$

Answer:
Esters are more stable than thioesters because the carbonyl group is more stabilized by donation of electrons by resonance from oxygen than from sulfur. The electrons of sulfur are in the third energy level, and overlap of 3p and 2p orbitals in thioesters is less effective than overlap of 2p and 2p orbitals in esters. Therefore, the equilibrium favors the more resonance stabilized ester.

21.26   Which equilibrium constant for the following reactions is larger, K_I or K_{II}?

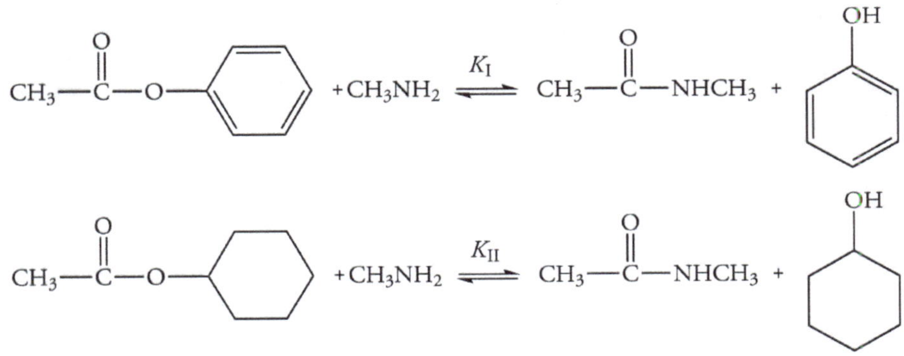

Answer:
In both reactions, a more stable amide is formed from an ester. The resonance interaction of the nonbonded electrons of the oxygen atom with the aromatic ring of the phenolic moiety of the aromatic ester decreases the electron density at oxygen. Thus, there is a decreased availability of electrons to the carbonyl group. Therefore, the carbonyl group is not as stable for the phenyl ester. Thus, the equilibrium constant is larger for the reaction of the phenyl ester than for the cyclohexyl ester.

21.27   Explain why the tautomeric equilibrium between an imidic acid and an amide lies on the side of the amide.

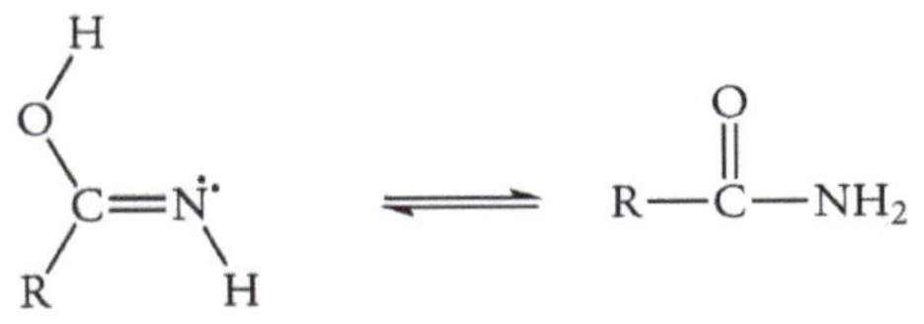

Answer:
The carbon–oxygen double bond is more polar and more stable than the carbon–nitrogen double bond. Also, the nitrogen atom of the amide stabilizes the carbonyl group by resonance donation of electrons. Because the oxygen atom of the imidic acid is more electronegative than a nitrogen atom, it is less effective in donation of electrons by resonance. Therefore, the amide is more stable than the imidic acid.

21.28 Explain why esters react with hydroxylamine (NH_2OH) to give hydroxamic acids rather than *O*-acyl hydroxyl amines.

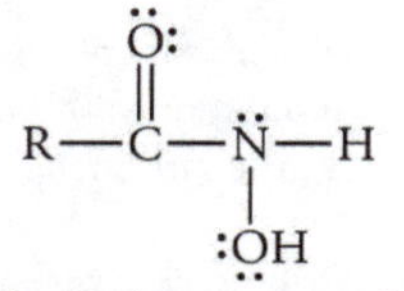

(an hydroxamic acid) 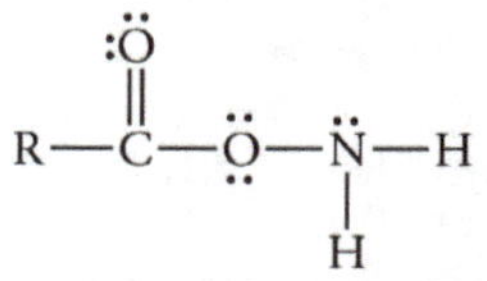   (an *O*-acyl hydroxyl amine)

Answer:

The carbonyl group of the hydroxamic acid is stabilized by donation of electrons of nitrogen by resonance. Because the oxygen atom of the *O*-acyl hydroxylamine is more electronegative than a nitrogen atom, it is less effective in donation of electrons by resonance.

21.29 One equivalent of methylamine reacts with *S*-ethyl-*O*-methylthiocarbonate as shown by the following equation. What other product is possible? Explain the observed selectivity of the reaction.

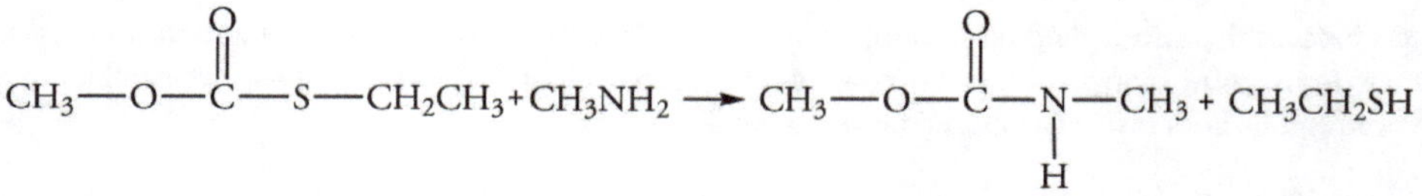

Answer:

The other possible product results from displacement of the ethoxy group as ethanol. The activation energy of the second step of the two step nucleophilic acyl substitution reaction is affected by the identity of the leaving group. Thiols are stronger acids than alcohols. Hence, the thiolate ion is a weaker base than the alkoxide ion. Weaker bases are generally better leaving groups, so thiolate is a better leaving group than methoxide ion.

21.30 Methanol reacts with glutaric anhydride to give a good yield of a monomethyl ester. Explain why the diiester does not form.

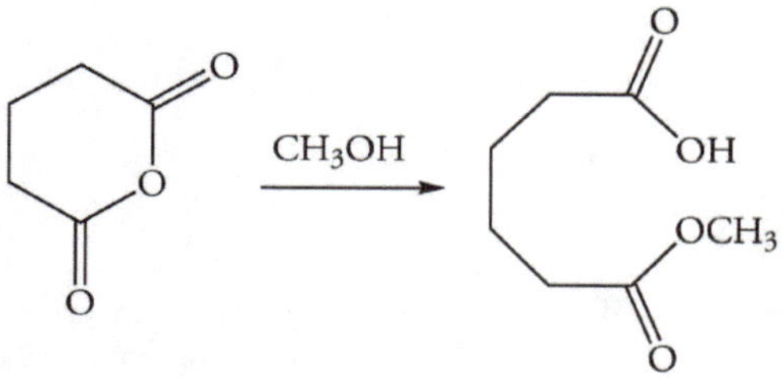

Answer:

The other possible product results from displacement of the ethoxy group as ethanol. The activation energy of the second step of the two step nucleophilic acyl substitution reaction is affected by the identity of the leaving group. Thiols are stronger acids than alcohols. Hence, the thiolate ion is a weaker base than the alkoxide ion. Weaker bases are generally better leaving groups, so thiolate is a better leaving group than methoxide ion.

21.31 Explain why alcohols react with the following mixed anhydride to give good yield of acetate esters.

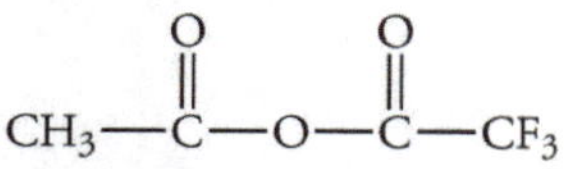

Answer:

The trifluoroacetate moiety of the compound withdraws electrons from the carbonyl carbon atom on the right to a greater extent than the acetate moiety of the compound withdraws electrons from the carbonyl carbon atom on the left. Thus, the nucleophile attacks the carbonyl group on the left to release the trifluoroacetate group.

21.32 Ethanol reacts with the following mixed anhydride to give two esters in a 36:64 ratio. Which of the two possible esters forms in the larger amount?

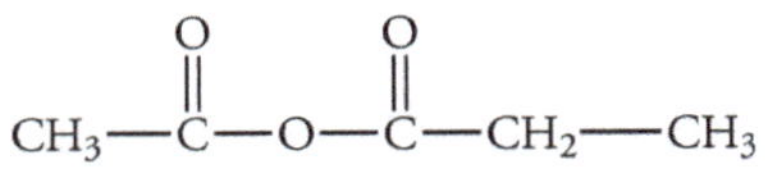

Answer:
A methyl group is sterically smaller than an ethyl group. Therefore, attack of a nucleophile is more favorable at the carbonyl group on the left to release a propanoate ion. The major ester is ethyl ethanoate; the minor ester is ethyl propanoate.

21.33 *p*-Hydroxyaniline reacts with acetic anhydride to give *N*-(4-hydroxyphenyl)acetamide. Explain why the reaction is selective, and acetylation does not occur at oxygen.

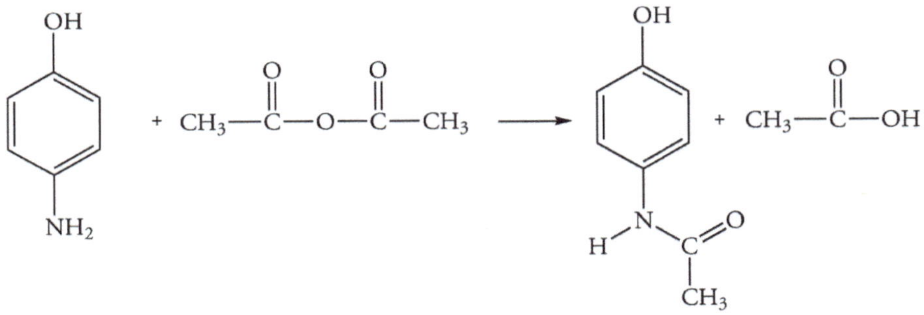

Answer:
The nitrogen atom of an amino group is more basic than the oxygen atom of a hydroxyl group. For atoms in the same period, the more basic atom is the better nucleophile.

21.34 Explain why the following bicyclic lactam hydrolyzes at a significantly faster rate than 5-aminopentanoic acid lactam.

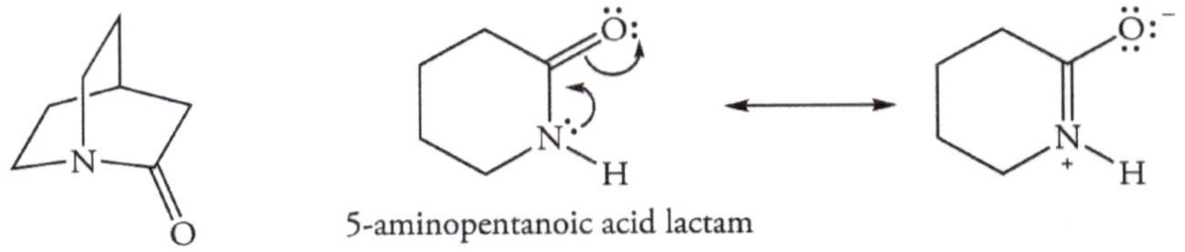

5-aminopentanoic acid lactam

Answer:
The amide cannot be resonance stabilized by donation of electrons from nitrogen to the carbonyl group. Such a resonance contributor would have a carbon–nitrogen double bond at the bridgehead of the bicyclic structure, which would be highly strained. Resonance stabilization is possible for 5-aminopentanoic acid lactam, so it is more stable, and it hydrolyzes more slowly. The product of the hydrolysis reaction of 5-aminopentanoic acid lactam is shown below.

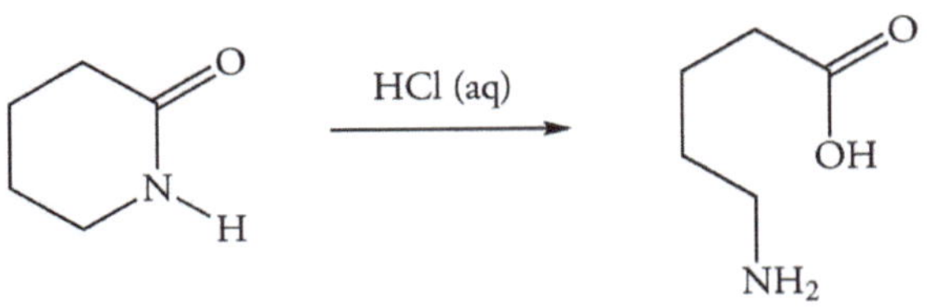

21.35 Mevalonic acid readily forms a lactone when heated. Draw two possible structures for the lactone. Which of the two is formed?

Answer:
The lactone is formed from the primary alcohol to give a six-membered ring. Reaction with the tertiary alcohol would give a more strained, four-membered lactone.

21.36 Explain why the rate of acid-catalyzed esterification of 2,2-dimethylcyclohexanecarboxylic acid is slower than that of cyclohexanecarboxylic acid.

2,3-dimethylcyclohexanecarboxylic acid

Answer:
The approach of the nucleophile to the carboxyl carbon atom is sterically hindered by the quaternary center at C-2.

Reactions of Acyl Derivatives

21.37 Draw the structures of the products of hydrolysis of each of the following esters.

Answers:

21.38 Draw the structures of the hydrolysis products of each of the following esters.

Answers:

(a) 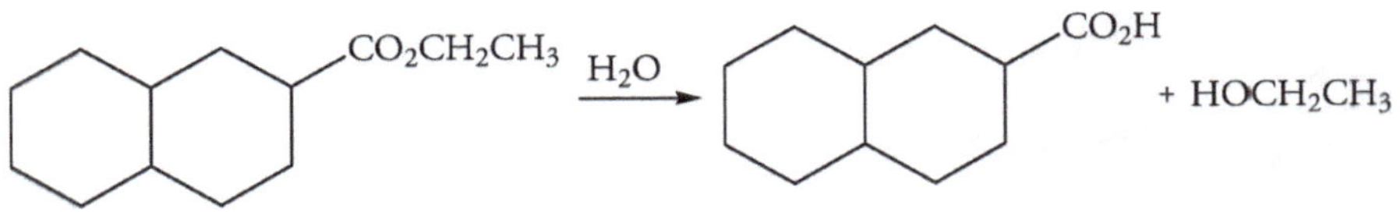
CO2CH2CH3
H2O
CO2H
+ HOCH2CH3

(b) 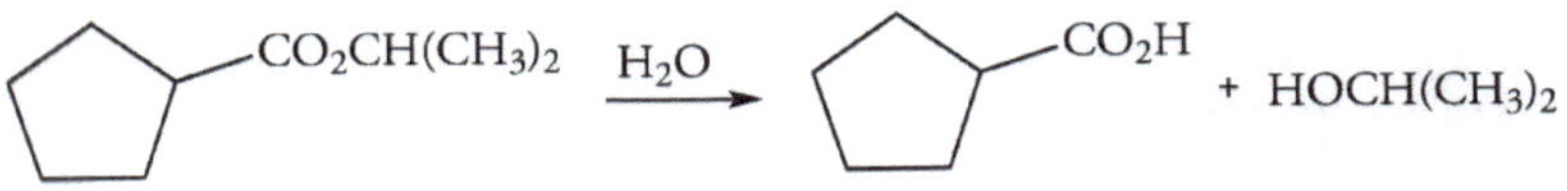
CO2CH(CH3)2
H2O
CO2H
+ HOCH(CH3)2

(c) 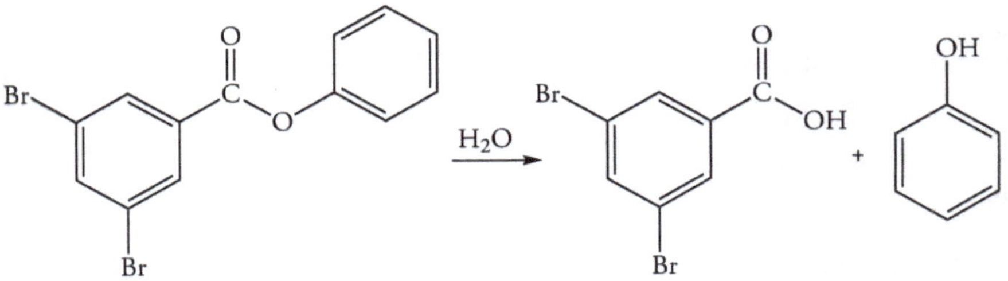
Br
O
C
O
Br
H2O
Br
O
C
OH
Br
+
OH

(d) 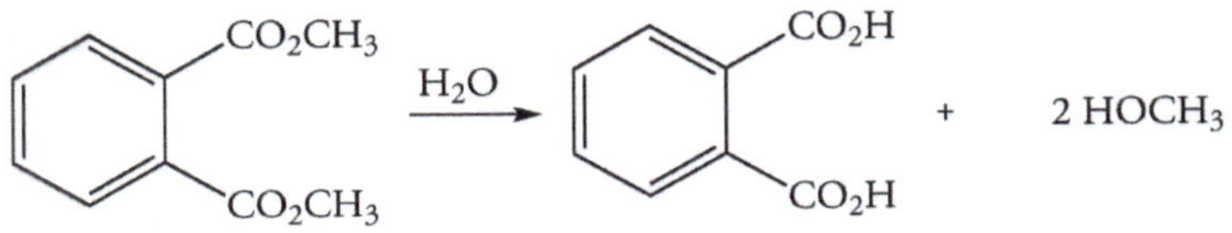
CO2CH3
CO2CH3
H2O
CO2H
CO2H
+
2 HOCH3

(e) 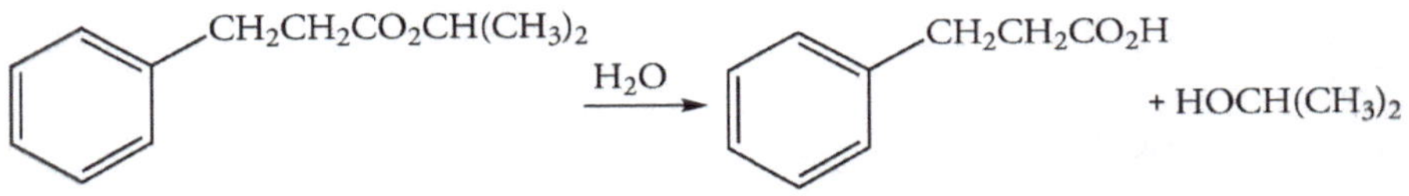
CH2CH2CO2CH(CH3)2
H2O
CH2CH2CO2H
+ HOCH(CH3)2

(f) 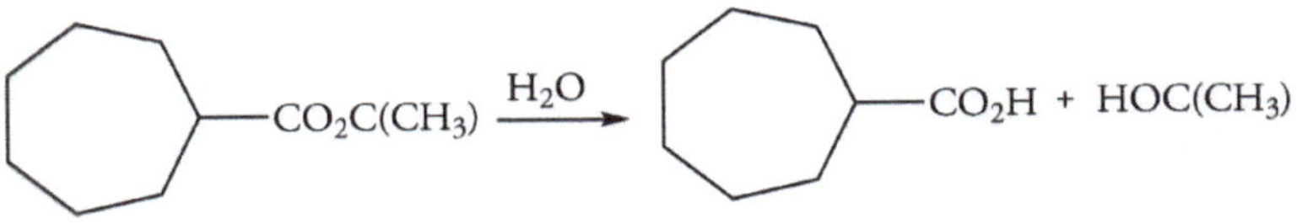
CO2C(CH3)
H2O
CO2H + HOC(CH3)

21.39 Hydrolysis of ambrettolide, contained in hibiscus, yields (E)-16-hydroxy-7-hexadecenoic acid. Draw the structure of ambrettolide.

Answer:

Since there is only one product, ambrettolide must be a lactone.

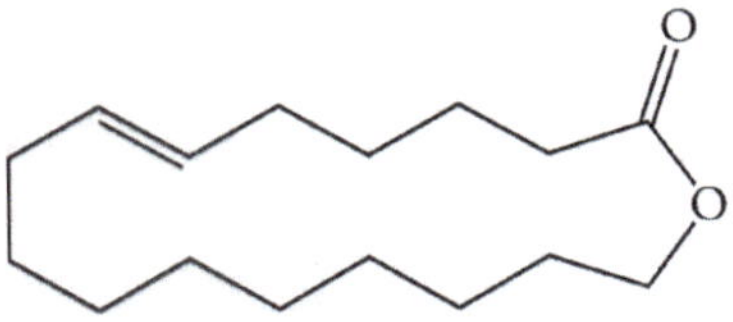
O
O

ambrettolide

21.40 Hydrolysis of beeswax gives a mixture containing unbranched acids with 26 and 28 carbon atoms and unbranched alcohols with 30 and 32 carbons atoms. Draw the structures of all possible components of beeswax

Answers:

$$CH_3(CH_2)_{24}CH_2-\overset{\overset{\textstyle O}{\|}}{C}-O-CH_2(CH_2)_{29}CH_3$$

$$CH_3(CH_2)_{26}CH_2-\overset{\overset{\textstyle O}{\|}}{C}-O-CH_2(CH_2)_{29}CH_3$$

$$CH_3(CH_2)_{24}CH_2-\overset{\overset{\textstyle O}{\|}}{C}-O-CH_2(CH_2)_{31}CH_3$$

$$CH_3(CH_2)_{26}CH_2-\overset{\overset{\textstyle O}{\|}}{C}-O-CH_2(CH_2)_{31}CH_3$$

21.41 Draw the structures of the hydrolysis products of each of the following compounds.

Answers:

(a)

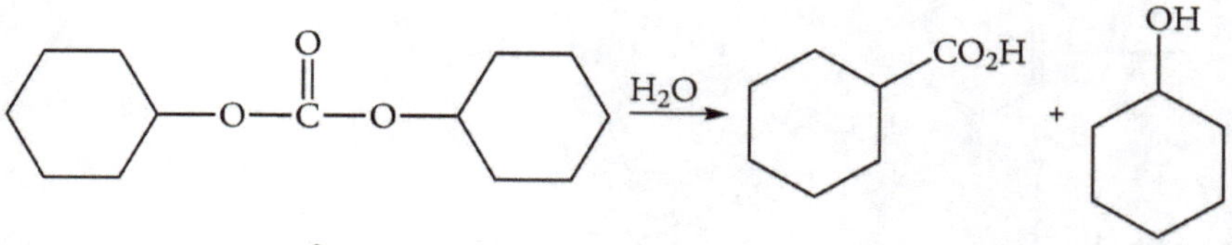

(b)

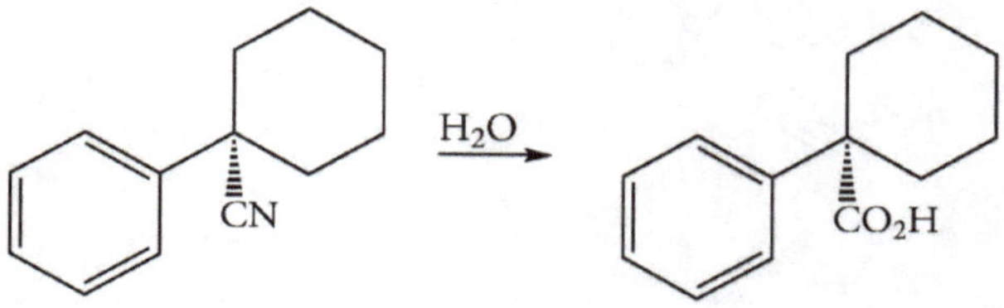

(c)

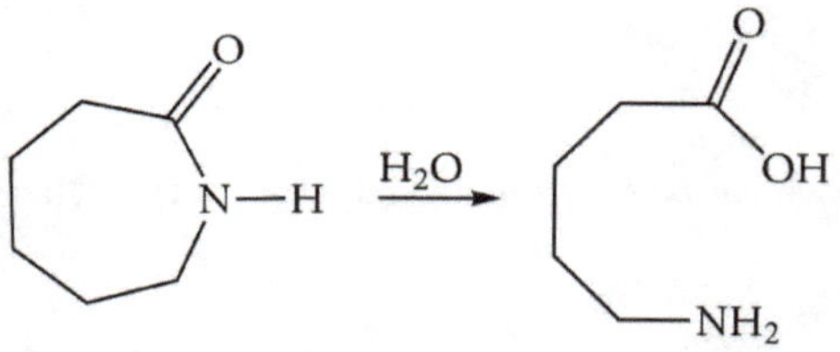

(d)

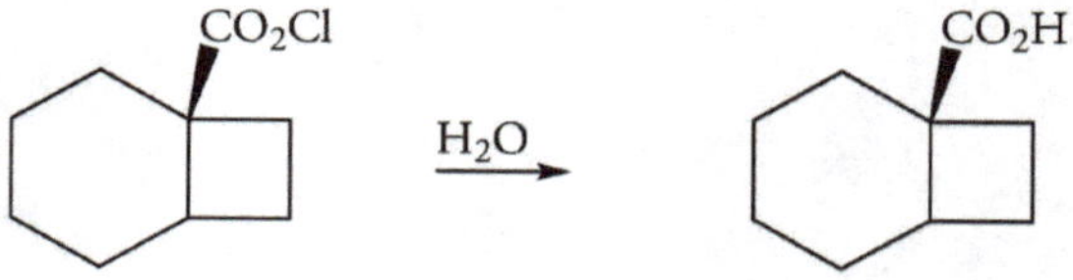

21.42   Draw the structures of the acid-catalyzed hydrolysis products of each of the following compounds.

Answers:

(a)

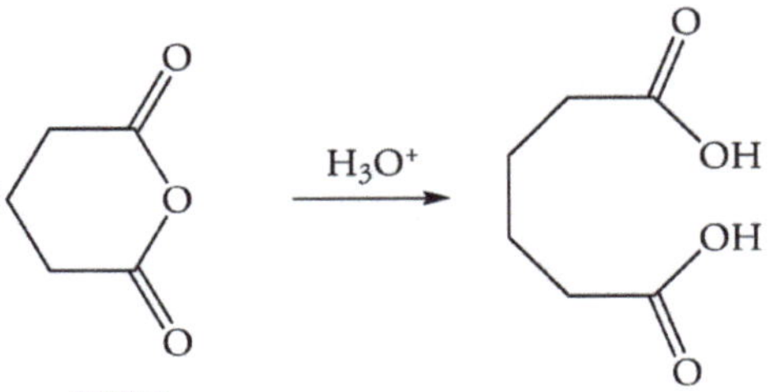

(b)

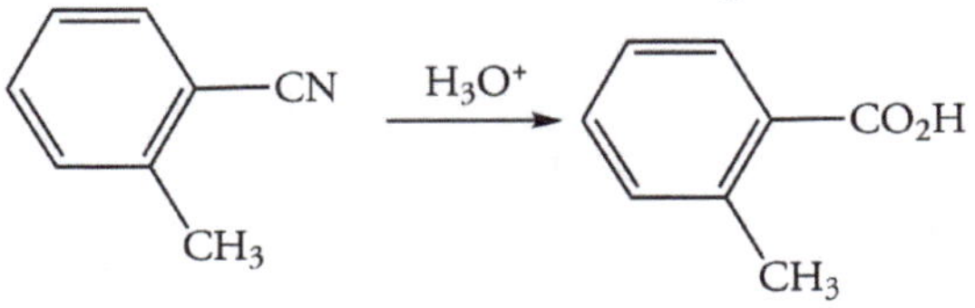

(c)

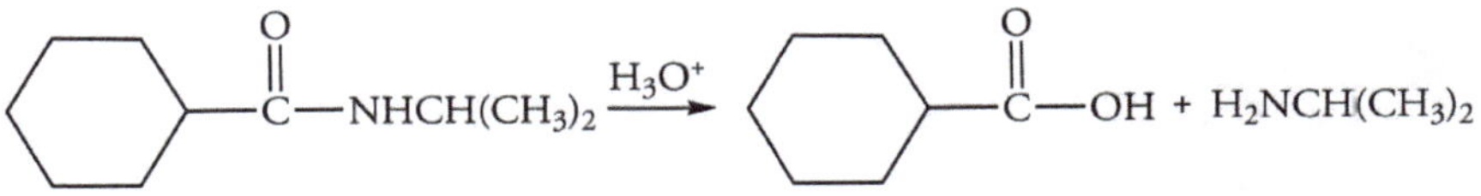

(d)

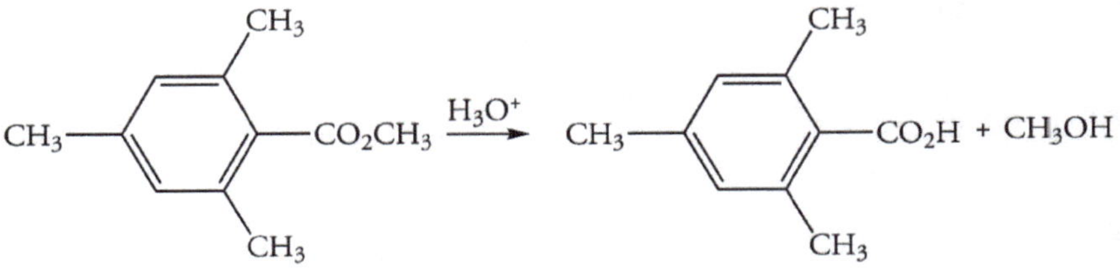

21.43   Draw the structure of the product of each of the following reactions.

(a) 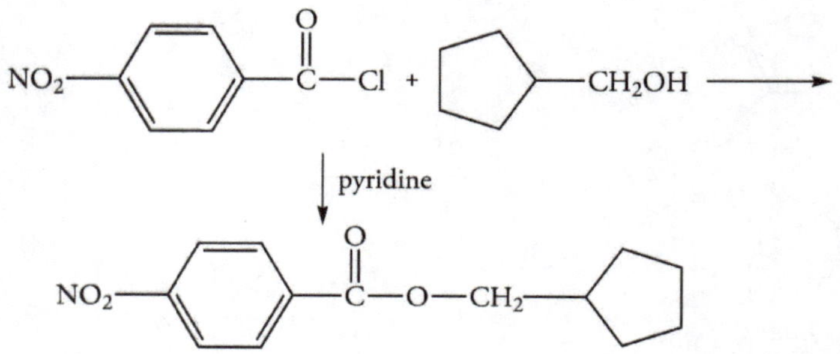 NO_2— (benzene ring) —C(=O)—Cl + (cyclopentane)—CH_2OH →

pyridine

NO_2— (benzene ring) —C(=O)—O—CH_2—(cyclopentane)

(b)   (chromanone/lactone ring) + CH_3CH_2OH $\xrightarrow{H^+}$ (benzene ring with $CH_2CH_2C(=O)OCH_2CH_3$ and OH)

(c)   (cyclohexane)—C(=O)—O—CH_2—(cyclohexane)

CH_3CH_2OH $\Big\downarrow$ H^+

(cyclohexane)—C(=O)—O—CH_2CH_3 + HO—CH_2—(cyclohexane)

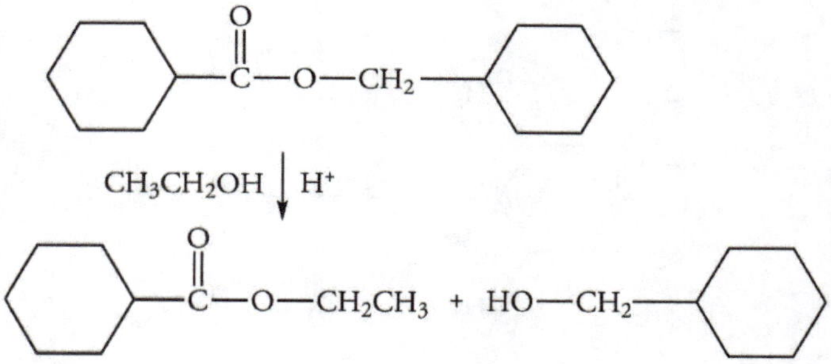

(d) 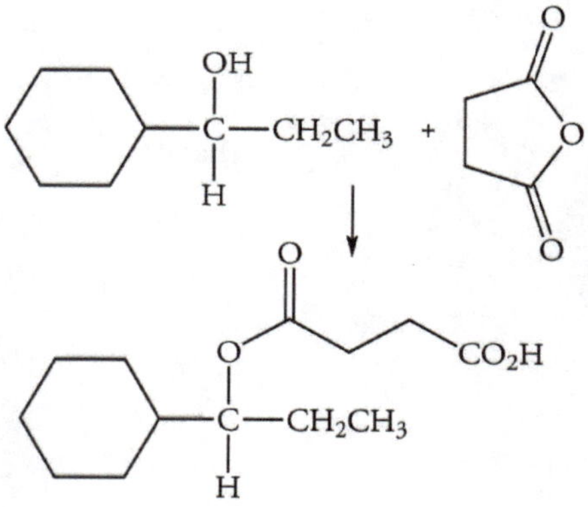   (cyclohexane)—C(OH)(H)—CH_2CH_3 + (succinic anhydride)

↓

(cyclohexane)—C(H)(—CH_2CH_3)—O—C(=O)—CH_2CH_2—CO_2H

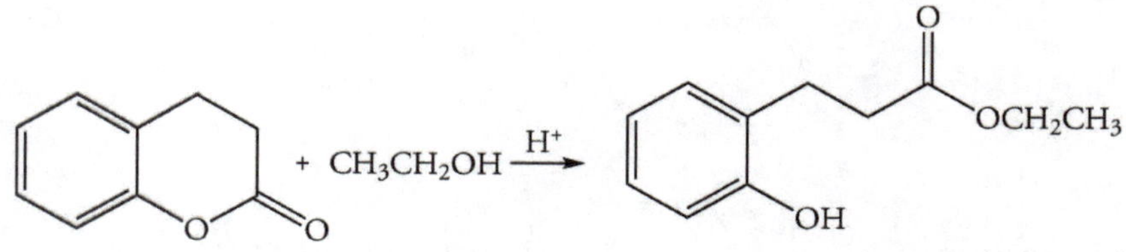

338

21.44 Draw the structure of the product of each of the following reactions.

Answers:

(a)

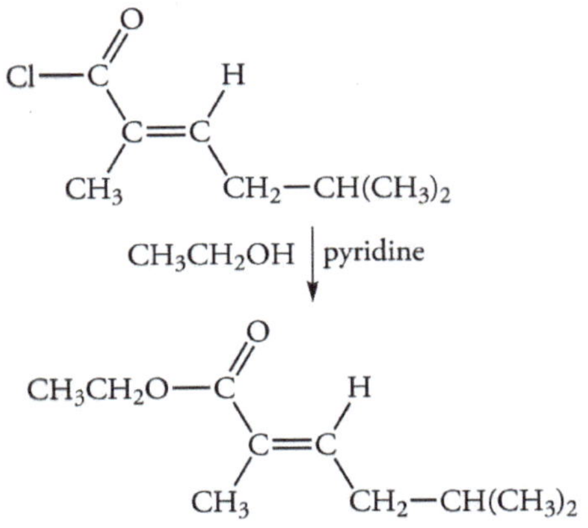

Cl—C(=O)—C(CH$_3$)=C(H)—CH$_2$—CH(CH$_3$)$_2$

CH$_3$CH$_2$OH │ pyridine

CH$_3$CH$_2$O—C(=O)—C(CH$_3$)=C(H)—CH$_2$—CH(CH$_3$)$_2$

(b)

[lactone ring structure with O and C=O] + CH$_3$CH$_2$OH $\xrightarrow{H^+}$ [open chain structure with OH and CO$_2$CH$_3$]

(c)

[benzene]—C(=O)—O—CH$_2$CH$_2$—[benzene]

CH$_3$CH$_2$OH │ H$^+$

[benzene]—C(=O)—OH + HO—CH$_2$CH$_2$—[benzene]

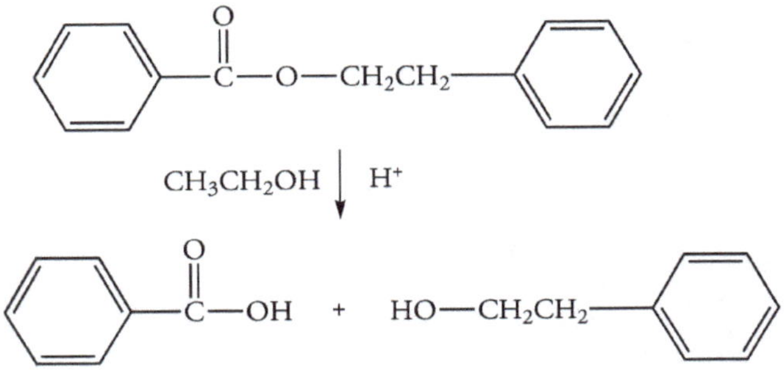

(d)

[decalin structure with CH$_3$ and HO''''] + (CH$_3$CO$_2$)$_2$O $\longrightarrow$ [decalin structure with CH$_3$ and CH$_3$—C(=O)—O'''']

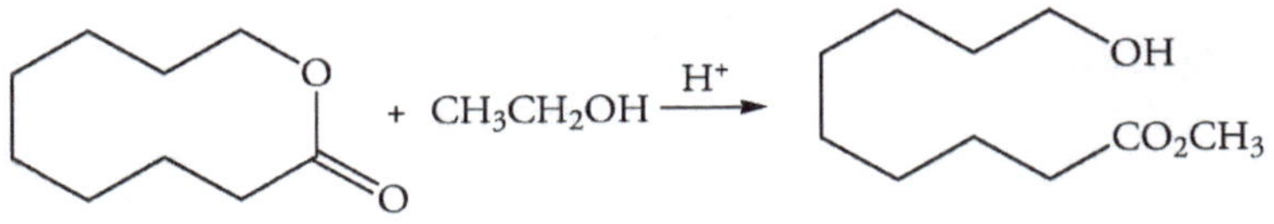

339

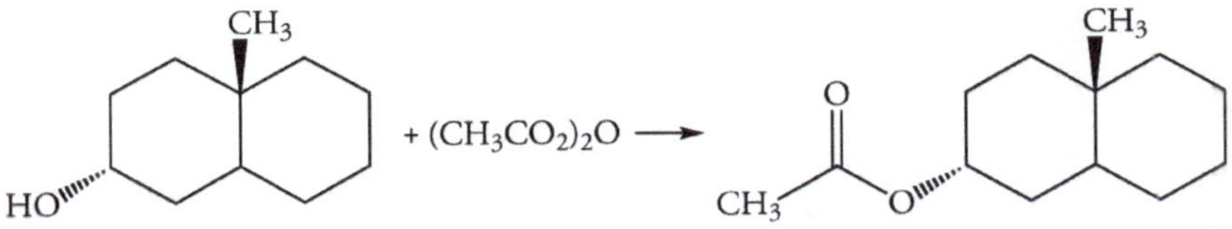

21.45   Draw the structure of the product of each of the following reactions.

Answers:

(a)

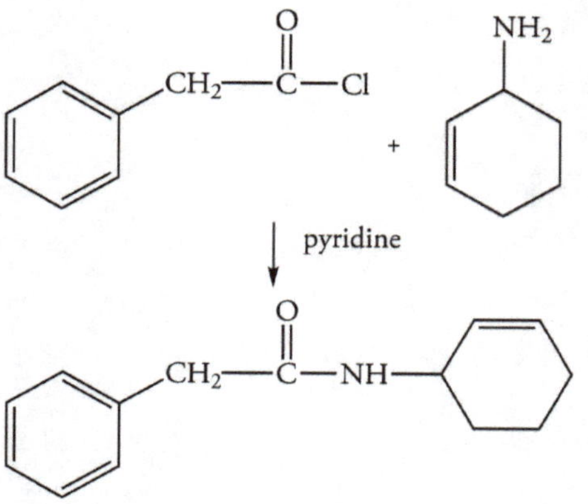

pyridine

(b) + CH₃CH₂NH₂ →

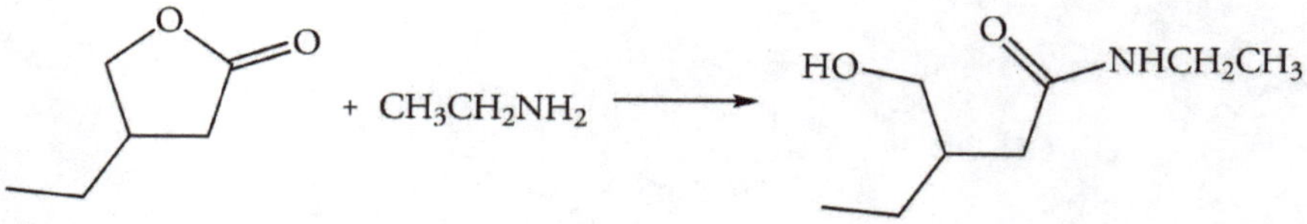

(c) +

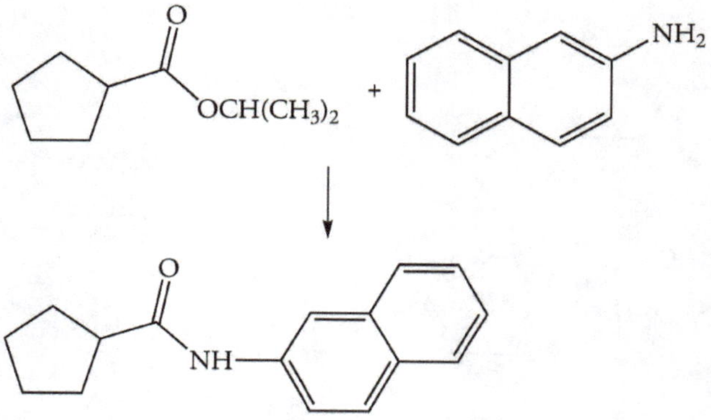

(d) + H—N →

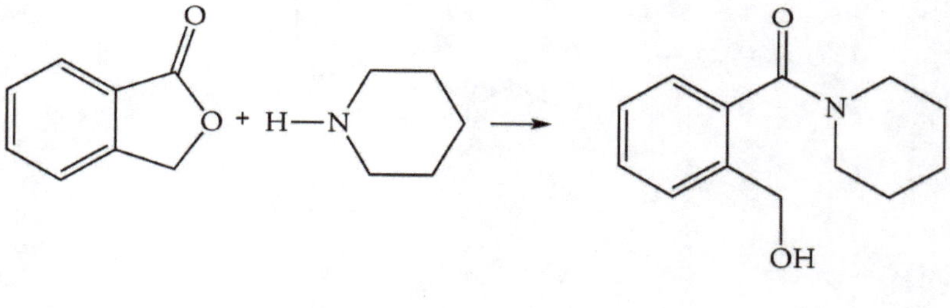

22

CONDENSATION REACTIONS OF CARBONYL COMPOUNDS

Keys to the Chapter

22.1 The α Carbon Atom of Carbonyl Compounds

The carbon atom bonded to the carbonyl carbon atom is known as the α carbon atom. It is a reactive site because its hydrogen atom is acidic. Extraction of the α hydrogen atom results in formation of a carbanion that serves as a nucleophile. The pK_a of the α hydrogen atom is approximately 18, which means that the K_a is approximately 30 powers of 10 larger than the K_a for hydrocarbons. The increased acidity is the result of resonance stabilization of the enolate ion. One of the two resonance forms of the enolate ion has the negative charge on the α carbon atom; the other resonance form has the negative charge on the oxygen atom.

Enolates are formed by reaction of a carbonyl compound with a base. The concentration of enolate formed depends on the K_b of the base, which in turn is related to the K_a of the conjugate acid of the base. Sodium hydroxide is a weaker base than the enolate ion, and it is not sufficiently basic to give a high concentration of enolate. Sodium amide is a much stronger base, and it quantitatively converts carbonyl compounds to their enolates.

Enolates can react as nucleophiles that attack an electrophilic center at the oxygen atom or the carbon atom of a second enolate. Although the electron density of the enolate is highest at the oxygen atom, the most common reaction site of enolates is at the carbon atom. This selectivity is related to the bonds formed in the transition state. Reaction at the oxygen atom forms an enol product that contains a carbon–carbon double bond. Reaction at the carbon atom forms a keto product that contains a carbon–oxygen double bond. The greater stability of the carbonyl group favors formation of the keto product.

22.2 Basicity of Acyl Derivatives

Aldehydes and ketones both exist in two isomeric forms known as keto and enol tautomers. They differ in the location of a hydrogen atom and the type of double bond. The keto form is more stable than the enol form. Tautomerization is a net process by which protons are transferred from one site to another by a series of steps in which the solvent is an intermediary. In acidic solution, the steps are protonation of the carbonyl oxygen atom by an acid to give a conjugate acid, followed by deprotonation of the α carbon atom by the conjugate base of the acid. The acid and base are the hydronium ion and water. In basic solution, the steps are deprotonation of the α carbon atom to give an enolate ion, followed by protonation of the oxygen atom by the conjugate acid of the base. The base and acid are the hydroxide ion and water.

The stability of an enol is reflected in its concentration in equilibrium with the keto form. Ketones have a smaller concentration of enol than aldehydes. This fact reflects the greater stability of ketones compared to aldehydes as a result of electron donation of alkyl groups to the carbonyl carbon atom. The stability of isomeric enols from a ketone reflects the stability due to the degree of substitution of the double bond. Conjugation of the double bond of the enol increases its stability.

22.3 Consequences of Enolization

The hydrogen atom of the α carbon atom of a carbonyl compound is called an **enolizable hydrogen atom**. If the α carbon atom is chiral, the formation of the isomeric enol results in loss of optical activity because a racemic mixture is formed when the keto form is regenerated. If the keto–enol equilibrium occurs in a protic solvent containing deuterium in place of hydrogen, then the α hydrogen atoms are exchanged by deuterium.

22.4 α-Halogenation Reactions of Aldehydes and Ketones

The α hydrogen atoms of carbonyl compounds can be replaced by halogen atoms. The regioselectivity and the number of hydrogen atoms substituted depend on whether acidic or basic conditions are used. Under acidic conditions, one hydrogen atom is substituted without the complication of multiple substitution. Under basic conditions, multiple substitution occurs.

Acid-catalyzed halogenation occurs by halogenation of the enol, whose formation is the rate determining step. The double bond of the enol is attacked by the halogen in a reaction similar to the electrophilic attack of a simple alkene. However, the resulting intermediate is not the bromonium ion, but an oxocarbocation. Loss of a proton from the hydroxyl group gives the halogenated product. Multiple substitution is disfavored because the halogen atom makes the carbonyl oxygen atom less basic and decreases the rate of formation of the enol. The halogen also destabilizes the conjugate acid because it tends to withdraw electron density from the oxocarbocation. In ketones with two nonequivalent α carbon atoms, the more substituted carbon atom is halogenated. This regioselectivity reflects the greater stability of the enol with the more substituted carbon–carbon double bond.

Halogenation under basic conditions occurs by nucleophilic attack of the enolate at the halogen molecule. Because the halogen atom is inductively electron withdrawing, the α hydrogen atom of the halogenated ketone is more acidic. Therefore, not only does multiple substitution occur, but it continues at the α carbon atom originally substituted in preference to a second α carbon atom. Continued halogenation of a methyl ketone forms a trihalomethyl derivative that is cleaved into a carboxylate and a haloform as the result of nucleophilic attack of the carbonyl carbon atom.

22.5 Alkylation of Enolate Ions

Enolates, formed by the abstraction of the α hydrogen atom by a strong base, are nucleophiles. Lithium diisopropylamide (LDA) or sodium hydride are required as bases. The site of proton abstraction is related to the acidity of the two possible α hydrogen atoms, which is in the order primary > secondary > tertiary. Reaction of the enolate with an alkyl halide forms a alklylated ketones. Multiple alkylation can occur as the result of proton exchange between the original enolate and the alkylated ketone, followed by alkylation of that enolate ion.

22.6 The Aldol Condensation of Aldehydes

The aldol condensation is the reaction of two moles of an aldehyde to form a β-hydroxyaldehyde, or aldol, in the presence of a base. The product is formed by addition of the enolate, formed by abstraction of an a hydrogen atom of one aldehyde by hydroxide ion, to the carbonyl carbon atom of the second aldehyde. Protonation of the alkoxide by exchange of a proton from water gives the aldol. The first step is an addition reaction to form a tetrahedral product. Subsequent dehydration in the reaction mixture often occurs to give an α,β-unsaturated aldehyde. The combination of the two steps constitutes a condensation reaction.

Under the basic conditions of the aldol condensation reaction, a dehydrated product forms. As a result of this step, the formation of an α,β-unsaturated aldehyde is favorable.

22.7 Mixed Aldol Condensation

A mixed aldol condensation is the formation of an aldol incorporating two different aldehydes. The reaction gives a mixture of four possible products if both aldehydes have α hydrogen atoms. If only one has α hydrogen atoms, then two products can result. If the carbonyl group of the aldehyde without α hydrogen atoms is more reactive toward nucleophiles, then one product results.

22.8 Intramolecular Aldol Condensation

Intramolecular aldol condensations are more favorable than intermolecular aldol condensations. Cyclization occurs if the α carbon atom and the second carbonyl carbon atom can bond to form a five- or six-membered ring. If two or more reactions can yield these rings, it is necessary to consider which process is favored. The various possible enolates exist in low concentration under equilibrium conditions. Thus, the enolate that is the better nucleophile attacks the more reactive carbonyl carbon atom and dominates the product formed. In general, for example, intramolecular aldol condensations where the enolate attacks the carbonyl carbon atom of an aldehyde are favored over addition to the carbonyl carbon atom of a ketone.

22.9 Conjugation in α,β Unsaturated Aldehydes and Ketones

The carbon–carbon double bond in an α,β-unsaturated aldehyde or ketone affects the stability of the carbonyl group, and hence its reactivity. The positive charge on the carbonyl carbon atom in the dipolar resonance form can be stabilized by donation of π electrons from the carbon–carbon double bond. Thus, there is some partial positive charge at the β carbon atom. This contributing resonance structure decreases the reactivity of the carbonyl carbon atom toward nucleophiles and offers an alternate site for reactivity at the β carbon atom.

22.10 Conjugate Addition Reactions

Addition of compounds represented by H—Nu to the carbon–oxygen double bond of an *O*-unsaturated aldehyde or ketone gives a 1,2-addition product. Addition of the nucleophilic portion of the reagent at the β carbon atom and a hydrogen atom at the carbonyl oxygen atom is a 1,4-conjugate addition.

Strong nucleophiles such as the hydride ion of metal hydrides, and the carbanion of a Grignard reagent, react to give 1,2-addition products with α,β-unsaturated aldehydes and ketones. Weak electrophiles such as cyanide ion, amines, alcohols, and thiols give 1,4-addition products. The Gilman reagent also gives 1,4-addition products.

22.11 The Michael Reaction and Robinson Annulation

The reaction of an enolate with α,β-unsaturated aldehydes and ketones is the **Michael reaction**. The enolate is called a Michael donor, and the α,β-unsaturated carbonyl compound is called a Michael acceptor. The carbonyl group of the Michael donor remains in the condensation product and may undergo an intramolecular aldol condensation reaction with the α carbon atom of the original Michael acceptor. This process, which forms a ring, is termed the **Robinson annulation**.

22.12 α Hydrogen Atoms of Acid Derivatives

The acidity of α hydrogen atoms of acid derivatives is affected by the amount of positive charge on the carbonyl carbon atom, which in turn is affected by the stabilization of that charge by the electronegative atom bonded to the carbonyl carbon atom. The acidity of α hydrogen atoms of esters (pK_a = 25) is less than that of aldehydes and ketones (pK_a = 20) because the oxygen atom of the alkoxy group supplies electrons by resonance to the partially positive carbonyl carbon atom. The resulting delocalization places some positive charge on oxygen and decreases the pK_a. As a result, the acidity of α hydrogen atoms of esters is less than that of aldehydes and ketones.

Enolates of esters can be prepared in low concentration at equilibrium by using the alkoxide ion in an alcohol corresponding to the alkoxy group contained in the ester. The equilibrium constant for the reaction is approximately 10^{-9} because the pK_a values of the ester and the alcohol differ by 10^9. High concentrations of the ester enolate are prepared by using LDA, which is a poor nucleophile and a strong base. The pK_a of diisopropyl amine is 40. Thus, the equilibrium constant for the reaction is approximately 10^{19}.

Dicarbonyl compounds such as ethyl acetoacetate and dimethyl malonate are significantly stronger acids (the pK_a values are 11 and 13, respectively). The increased acidity results from delocalization of negative charge in the conjugate base by the additional carbonyl group.

22.13 Reactions at the α Carbon Atom of Acid Derivatives

A hydrogen atom bonded to the α carbon atom of an ester is an **enolizable hydrogen atom**. If the α carbon atom is chiral, the formation of the ester enolate results in loss of optical activity because a racemic mixture is formed when the ester is regenerated. If the equilibrium occurs in a protic solvent containing deuterium in place of hydrogen, then the α hydrogen atoms are exchanged by deuterium.

Ester enolates, formed by the abstraction of the α hydrogen atom by a strong base, are nucleophiles. Lithium diisopropylamide (LDA) is required as the base. Reaction of the ester enolate with an alkyl halide forms α alkylated esters. Only primary haloalkanes can be used.

The α hydrogen atoms of carboxylic acids can be replaced by a single halogen atom using bromine and a small amount of phosphorus in the **Hell–Volhard–Zelinsky reaction**. The reaction occurs by bromination of a small amount of the acyl bromide. Under the reaction conditions, the bromoacyl bromide is converted into the bromocarboxylic acid. If one equivalent of PBr_3 is used with the Br_2, an a bromoacyl bromide is formed. Reaction of this compound with an alcohol gives an α-bromo ester.

22.14 The Claisen Condensation

The reaction of two moles of an ester in the presence of the alkoxide base corresponding to the alkoxyl group of the ester produces a β-keto ester. The ester must have two α hydrogen atoms, and one equivalent of base is required. The Claisen reaction occurs in four steps.

1. Abstraction the α hydrogen atom by the alkoxide ion.
2. Attack of the carbonyl carbon atom of the ester by an ester enolate.
3. Ejection of an alkoxide ion from the conjugate base of a hemiketal.
4. Abstraction of an a hydrogen atom of the keto ester by the alkoxide ion.

The addition of dilute acid at the end of the reaction protonates the conjugate base of the keto ester. Proton exchange between the β-keto ester and the alkoxide provides the driving force for the reaction.

The **Dieckmann condensation** is an intramolecular variation of the Claisen condensation. The ester must have two α hydrogen atoms and one equivalent of base is required. The intramolecular reaction is favorable because one mole of the diester gives one mole of keto ester and one mole of alcohol.

A mixed Claisen condensation is the formation of a keto ester incorporating two different esters. The reaction gives a mixture of four possible products if both esters have α hydrogen atoms. If only one has α hydrogen atoms, then two products can result. If the carbonyl group of the ester without α hydrogen atoms is more reactive toward nucleophiles, then one product results.

Nonenolizable esters react with the enolate of ketones to give β-diketones. The enolate of the ketone attacks the carbonyl carbon atom of the ester in a reaction similar to that of the Claisen condensation.

22.15 Aldol-Type Condensations of Acid Derivatives

Aldol-type condensations occur between a carbonyl compound and the enolate of an ester. The product is either a β-hydroxy ester or its related α,β-unsaturated ester. The **Knoevenagel condensation** occurs between a malonate ester and an aldehyde or ketone. This aldol-type reaction gives an α,β-unsaturated ester. The **Reformatskii reaction** occurs between a zinc enolate of an ester and an aldehyde or ketone to produce a β-hydroxy ester.

22.16 β-Dicarbonyl Compounds In Synthesis

The alkylation of acetoacetate or malonate esters is a useful synthetic process that is synthetically equivalent to the direct alkylation of a ketone or an ester. The acidity of both compounds is higher than that of ketones and esters and allows the abstraction of the a proton by an alkoxide ion to quantitatively form the conjugate base. Alkylation of the conjugate base occurs by nucleophilic attack on a haloalkane. Hydrolysis of the alkylated acetoacetate leads to decarboxylation of the keto acid and formation of a ketone. Hydrolysis of the alkylated malonate ester leads to decarboxylation of the diacid, and formation of an acid.

22.17 Michael Condensation of Acid Derivatives

The reaction of an enolate with α,β-unsaturated aldehydes and ketones is termed the **Michael reaction**. The enolate is called a Michael donor, and the α,β-unsaturated carbonyl compound, such as 3-buten-2-one, is called a Michael acceptor. 1,3-Dicarbonyl derivatives, such as dimethyl malonate, easily form enolates that act as Michael donors. Subsequent hydrolysis of the addition product and decarboxylation yields 1,5-dicarbonyl compounds.

Summary of Reactions

1. Exchange of α Hydrogen Atoms

$$\xrightarrow[\text{CH}_3\text{OD}]{\text{CH}_3\text{O}^-}$$

2. Isomerization of Carbonyl Compounds

$$\underset{\text{H}^+}{\rightleftharpoons}$$

3. α Halogenation of Carbonyl Compounds

$$\xrightarrow[\text{CH}_3\text{CO}_2\text{H}]{\text{Br}_2}$$

$$\text{CH}_3\text{-}\text{-}\overset{\text{O}}{\underset{}{\text{C}}}\text{-CH}_2\text{Br} + \text{HBr}$$

$$+ \text{Br}_2 \xrightarrow[\text{THF}]{\text{OH}^-}$$

$$\xrightarrow[\text{2. H}_3\text{O}^+]{\text{1. Cl}_2 \text{ / OH}^-}$$

$$\overset{\text{O}}{\underset{}{\text{C}}}\text{-OH} + \text{HCl}_3$$

$$\xrightarrow[\text{2. CH}_3\text{I}]{\text{1. LDA}}$$

$$\text{CH}_3\text{CH}_2\text{CH}_2\text{Br} +$$

$$\xrightarrow{\text{OH}^-}$$

$$\text{CH}_3\text{CH}_2\text{CH}_2\text{-}$$

4. Aldol Condensation

5. Conjugate Addition Reactions of Carbonyl Compounds

6. Michael Condensation of Carbonyl Compounds

7. Exchange of α Hydrogen Atoms of Esters

$$C_6H_5-\underset{\underset{CH_3}{|}}{\overset{\overset{H}{|}}{C}}-\overset{\overset{O}{\|}}{C}-OCH_2CH_3 \quad \xrightarrow[CH_3CH_2OD]{CH_3CH_2O^-} \quad C_6H_5-\underset{\underset{CH_3}{|}}{\overset{\overset{D}{|}}{C}}-\overset{\overset{O}{\|}}{C}-OCH_2CH_3$$

8. α Alkylation of Esters

$$C_6H_5-\underset{\underset{H}{|}}{\overset{\overset{H}{|}}{C}}-\overset{\overset{O}{\|}}{C}-OCH_2CH_3 \quad \xrightarrow[2.\ CH_3I]{1.\ LDA\ /\ THF} \quad C_6H_5-\underset{\underset{CH_3}{|}}{\overset{\overset{H}{|}}{C}}-\overset{\overset{O}{\|}}{C}-OCH_2CH_3$$

9. α Halogenation of Carboxylic Acids (Hell–Volhard–Zelinsky Reaction)

$$CH_3CH_2CH_2-\underset{\underset{CH_3}{|}}{\overset{\overset{H}{|}}{C}}-\overset{\overset{O}{\|}}{C}-OH \quad \xrightarrow[P\ (trace)]{Br_2} \quad CH_3CH_2CH_2-\underset{\underset{CH_3}{|}}{\overset{\overset{Br}{|}}{C}}-\overset{\overset{O}{\|}}{C}-OH$$

$$CH_3CH_2CH_2-\underset{\underset{CH_3}{|}}{\overset{\overset{H}{|}}{C}}-\overset{\overset{O}{\|}}{C}-OH \quad \xrightarrow[PBr_3]{Br_2} \quad CH_3CH_2CH_2-\underset{\underset{CH_3}{|}}{\overset{\overset{Br}{|}}{C}}-\overset{\overset{O}{\|}}{C}-Br$$

10. Claisen Condensation

$$C_6H_5-CH_2-\overset{\overset{O}{\|}}{C}-OCH_2CH_3 \quad \xrightarrow[CH_3CH_2OH]{CH_3CH_2O^-} \quad C_6H_5-CH_2-\overset{\overset{O}{\|}}{C}-\underset{\underset{C_6H_5}{|}}{\overset{\overset{H}{|}}{C}}-\overset{\overset{O}{\|}}{C}-OCH_2CH_3$$

$$CH_3O-\overset{\overset{O}{\|}}{C}-CH_2CH_2CH_2CH_2-\overset{\overset{O}{\|}}{C}-OCH_3 \quad \xrightarrow[2.\ H_3O^+]{1.\ CH_3O^-} \quad \text{(2-carbomethoxycyclopentanone)}\ CO_2CH_3$$

$$\text{(furyl)}CH_2-\overset{\overset{O}{\|}}{C}-OCH_3 + CH_3CH_2-\overset{\overset{O}{\|}}{C}-OCH_3 \quad \xrightarrow[2.\ H_3O^+]{1.\ CH_3O^-} \quad \text{(furyl)}CH_2-\overset{\overset{O}{\|}}{C}-\underset{\underset{CH_3}{|}}{\overset{\overset{H}{|}}{C}}H-\overset{\overset{O}{\|}}{C}-OCH_3$$

11. Aldol-Type Condensations of Acid Derivatives

$$\text{(3-Cl-C}_6H_4)\overset{\overset{O}{\|}}{C}-H + CH_2(CO_2CH_2CH_3)_2 \quad \xrightarrow{piperidine} \quad \text{(3-Cl-C}_6H_4)\overset{\overset{H}{|}}{C}=\underset{\underset{CO_2CH_2CH_3}{|}}{\overset{\overset{CO_2CH_2CH_3}{|}}{C}}$$

(furan-2-carbaldehyde) $+ \; Br-CH_2-\overset{\overset{\displaystyle O}{\|}}{C}-OCH_2CH_3 \;\xrightarrow[\text{benzene}]{\text{Zn}}\;$ (furan ring)$-\overset{*}{\underset{HO \;\; H}{C}}-CH_2-\overset{\overset{\displaystyle O}{\|}}{C}-OCH_2CH_3$

12. Acetoacetate Ester Synthesis

$CH_3-\overset{\overset{\displaystyle O}{\|}}{C}-\overset{\overset{\displaystyle H}{|}}{\underset{\underset{\displaystyle H}{|}}{C}}-\overset{\overset{\displaystyle O}{\|}}{C}-OCH_2CH_3 \;\xrightarrow[\text{2. }CH_3CH_2I]{\text{1. }CH_3CH_2O^-}\; CH_3-\overset{\overset{\displaystyle O}{\|}}{C}-\overset{\overset{\displaystyle H}{|}}{\underset{\underset{\displaystyle CH_2CH_3}{|}}{C}}-\overset{\overset{\displaystyle O}{\|}}{C}-OCH_2CH_3$

$CH_3-\overset{\overset{\displaystyle O}{\|}}{C}-\overset{\overset{\displaystyle H}{|}}{\underset{\underset{\displaystyle CH_2CH_3}{|}}{C}}-\overset{\overset{\displaystyle O}{\|}}{C}-OCH_2CH_3 \;\xrightarrow[\text{heat}]{H_3O^+}\; CH_3-\overset{\overset{\displaystyle O}{\|}}{C}-\overset{\overset{\displaystyle H}{|}}{\underset{\underset{\displaystyle CH_2CH_3}{|}}{C}}-H$

13. Malonate Ester Synthesis

$CH_3CH_2O-\overset{\overset{\displaystyle O}{\|}}{C}-\overset{\overset{\displaystyle H}{|}}{\underset{\underset{\displaystyle H}{|}}{C}}-\overset{\overset{\displaystyle O}{\|}}{C}-OCH_2CH_3 \;\xrightarrow[\text{2. }CH_3CH_2I]{\text{1. }CH_3CH_2O^-}\; CH_3CH_2O-\overset{\overset{\displaystyle O}{\|}}{C}-\overset{\overset{\displaystyle H}{|}}{\underset{\underset{\displaystyle CH_2CH_3}{|}}{C}}-\overset{\overset{\displaystyle O}{\|}}{C}-OCH_2CH_3$

$CH_3CH_2O-\overset{\overset{\displaystyle O}{\|}}{C}-\overset{\overset{\displaystyle H}{|}}{\underset{\underset{\displaystyle CH_2CH_3}{|}}{C}}-\overset{\overset{\displaystyle O}{\|}}{C}-OCH_2CH_3 \;\xrightarrow[\text{heat}]{H_3O^+}\; H-\overset{\overset{\displaystyle H}{|}}{\underset{\underset{\displaystyle CH_2CH_3}{|}}{C}}-\overset{\overset{\displaystyle O}{\|}}{C}-OH$

14. Michael Condensation of Acid Derivatives

$CH_3O-\overset{\overset{\displaystyle O}{\|}}{C}-\overset{\overset{\displaystyle H}{|}}{C}=CH_2 \;+\;$ dimethyl malonate $\;\xrightarrow[CH_3OH]{OH^-}\; CH_3O-\overset{\overset{\displaystyle O}{\|}}{C}-CH_2-CH_2-CH(CO_2CH_3)_2$

End of Chapter Exercises

Acidity of α Hydrogen Atoms

22.1 The pK_a of 2,4-pentanedione is 9. Calculate the equilibrium constant for the acid–base reaction of 2,4-pentanedione with sodium ethoxide. The pK_a of ethanol is 15.9.

Answer: The pK_a of 2,4-pentanedione is 1×10^{-9}, so it is a stronger acid than ethanol, whose K_a is approximately 1.3×10^{-16}. Thus, the reaction of 2,4-pentanedione with sodium ethoxide has $K_a = 7.7 \times 10^6$.

22.2 The pK_a of acetonitrile, CH_3CN, is 25. Calculate the equilibrium constant for the acid–base reaction of acetonitrile with LDA. The pK_a of isopropylamide is 40.

Answer: The K_a values of acetonitrile and diisopropylamine are 1×10^{-25} and 1×10^{-49}, respectively. Thus, acetonitrile is a stronger acid than diisopropylamine, and the reaction of acetonitrile with lithium diisopropylamide has $K_a = 1 \times 10^{15}$.

22.3 The pK_a of acetophenone is 16. Calculate the equilibrium constant for the acid–base reaction of acetophenone with LDA.

Answer: The K_a values of acetophenone and diisopropylamine are 1×10^{-16} and 1×10^{-49}, respectively. Thus, acetophenone is a stronger acid than diisopropylamine, and the reaction of acetophenone with lithium diisopropylamide has $K_a = 1 \times 10^{24}$.

22.4 The pK_a of nitromethane is 10.2. Calculate the equilibrium constant for the acid–base reaction of nitromethane with sodium ethoxide. The pK_a of ethanol is 15.9.

Answer: The K_a of nitromethane is 6.3×10^{-11}, so it is a stronger acid than ethanol, whose K_a is approximately 1.3×10^{-15}. Thus, the reaction of nitromethane with sodium ethoxide has $K_{eq} = 4.8 \times 10^5$.

22.5 The pK_a values of acetone and 3-pentanone, as measured in DMSO, are 26.5 and 27.1, respectively. Explain this order of values.

Answer: Acetone is the stronger acid, so its conjugate base is more stable. The conjugate base of acetone has a contributing resonance form with a negative charge on a primary carbon atom. For 3-pentanone, the charge of the carbanion is on a secondary carbon atom. Because primary carbanions are more stable than secondary carbanions, acetone gives a more stable conjugate base than 3-pentanone.

22.6 The pK_a values of acetone and 1-phenyl-2-propanone, as measured in DMSO, are 26.5 and 19.8, respectively. Explain this order of values.

Answer: 1-Phenyl-2-propanone is the stronger acid, so its conjugate base is more stable. The conjugate base of 1-phenyl-2-propanone has a contributing resonance form with a negative charge on a secondary carbon atom that is also benzylic. Thus, the charge is delocalized, and this conjugate base is resonance stabilized. For the conjugate base of acetone, the charge is localized on a primary carbon atom, so it is less stable.

Stability of Enols

22.7 Which ketone has the larger percent enol at equilibrium, cyclohexanone or cyclobutanone?

Answer: Both enols have the double bond within their respective rings in the enol form, and both have the same degree of substitution. However, the double bond of the enol of cyclobutanone increases the strain energy of the small ring, so this enol is less stable than the enol of cyclohexanone. Therefore, cyclohexanone has a larger percent of enol at equilibrium.

22.8 Which ketone has the larger percent enol at equilibrium, 1,3-cyclohexanedione or 1,4-cyclohexanedione?

Answer: Both enols have the double bond within a six-membered ring in the enol form. However, the double bond of the enol of 1,3-cyclohexanedione is conjugated with the second carbonyl group and is resonance stabilized. The enol of 1,4-cyclohexanedione is not resonance stabilized. Therefore, 1,3-cyclohexanedione has a larger percent of enol at equilibrium.

22.9 (a) Write the structures of the isomeric enols of 2,2-dimethyl-3-pentanone, and (b) rank them in order of their stability.

Answer: There is only one α-hydrogen atom, and it is located at C-4. However, two enols can form that are geometric isomers. The most stable enol has the large *tert*-butyl *trans* to the methyl group.

more stable, *trans* enol

22.10 (a) Write the structures of the isomeric enols of 2-methylcyclopentanone and (b) rank them in order of relative stability.

Answer: There are two α-carbon atom: C-2 and C-5. The double bond at C-2 is more substituted, and more stable.

more substituted,
more stable enol.

22.11 Which ketone has the larger percent enol at equilibrium, 1,2-diphenylethanone or 1,3-dipheny1-3-propanone?

Answer: The enol of 1,2-diphenylethanone has extended conjugation between the two phenyl rings through the carbon–carbon double bond. The enol of 1,3-dipheny1-3-propanone has only one ring only conjugated to the carbon–carbon double bond and is therefore less stable.

22.12 Write the structure for the enol tautomer of the following molecule. What structural features contribute to its stability?

Answer: The keto form has its carbonyl group conjugated with a benzene ring. However, the enol tautomer has a double bond that is part of the aromatic ring system of phenanthrene. The resonance stabilization of the fused ring system favors formation of the enol, which is a phenol.

Enolates

22.13 Write the resonance form with a negative charge on the oxygen atom for the enolates derived from each of the following compounds.

(a) 3,3-dimethyl-2-butanone   (b) acetophenone   (c) 2,2-dimethylcyclohexanone

Answer: In each case, there is only one α-carbon atom with enolizable hydrogen atoms, so only one enolate ion forms.

22.14 Write the resonance form with a negative charge on the oxygen atom for all possible enolates derived from each of the following compounds. Which enolate is the most stable in each case?

(a) 2-pentanone   (b) 1-phenyl-2-propanone   (c) 1,3-cyclohexanedione

Answer: (a) There are two α-carbon atoms with enolizable hydrogen atoms. C-1 gives a single enolate. C-3 gives a pair of enolates that are geometric isomers. The more substituted double bond of the enolates at C-3 is more stable than the enolate with a double bond at C-1. Of the two enolates involving at C-3, the isomer with the alkyl groups *trans* to one another is more stable.

enols of 2-pentanone in order of decreasing stability

Answer: (b) There are two α-carbon atoms with enolizable hydrogen atoms. C-3 gives a single enolate. C-1 gives a pair of enolates that are geometric isomers. The more substituted double bond of the enolates at C-1 is more stable than the enolate with a double bond at C-3 because the double bond is conjugated with the benzene ring. Of the two enolates involving the C-3 atom, the isomer with the alkyl and aryl groups *trans* to one another is more stable.

enols of acetophenone in order of decreasing stability

Answer: (c) There are two enolates. The enolate at C-6 has a localized double bond. The enolate at C-2 is conjugated with the second carbonyl group and is resonance stabilized.

enols of 1,3-cyclohexanedione in order of decreasing stability

352

22.15 Write the contributing resonance forms of the conjugate base of (a) acetonitrile (CH_3CN) and (b) nitromethane (CH_3NO_2).

Answers: (a)

$$:\bar{C}H_2\!-\!C\!\equiv\!N: \quad\longleftrightarrow\quad CH_2\!=\!C\!=\!\ddot{N}:^-$$

(b)

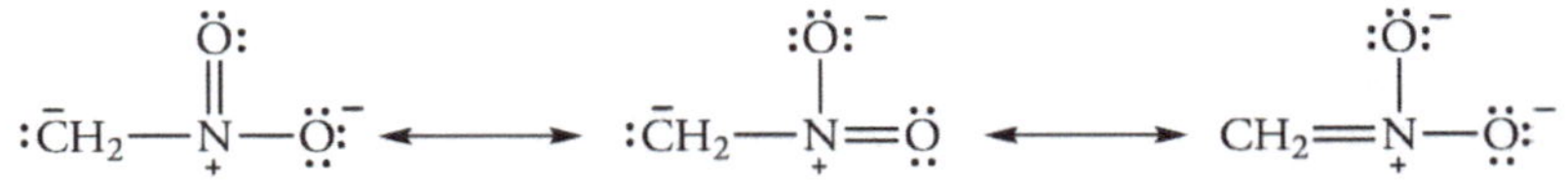

$$:\bar{C}H_2\!-\!\overset{+}{N}\!-\!\ddot{O}:^- \quad\longleftrightarrow\quad :\bar{C}H_2\!-\!\overset{+}{N}\!=\!\ddot{O} \quad\longleftrightarrow\quad CH_2\!=\!\overset{+}{N}\!-\!\ddot{O}:^-$$

22.16 Write the contributing resonance forms for all possible conjugate bases of 3,6,6-trimethyl-2-cyclohexenone.

Answers:

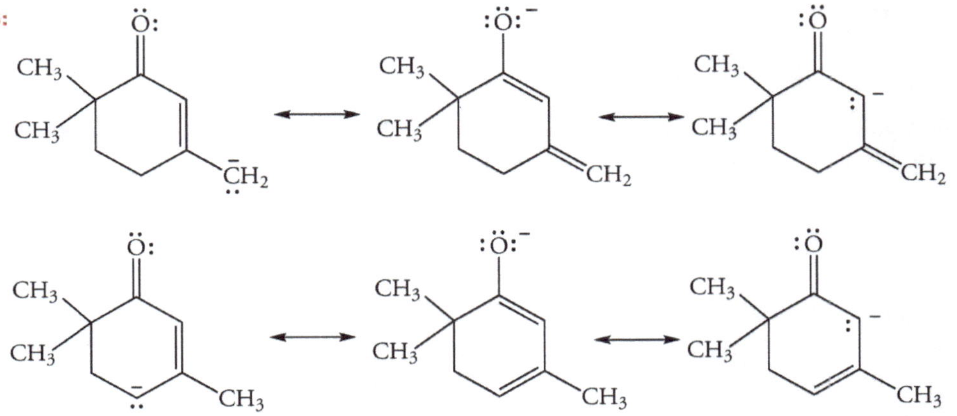

22.17 The following ketone gives a mixture of two enolates in approximately equal amounts (53:47). (a) Write the structures of the enolates and (b) explain why they are of comparable stability.

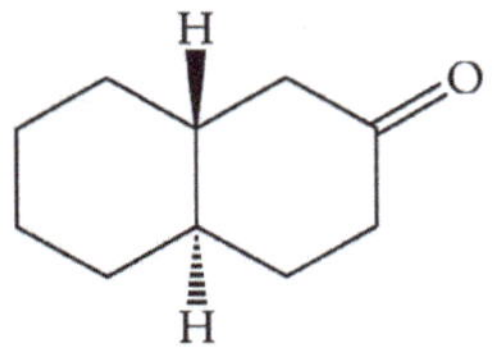

Answer: (a)

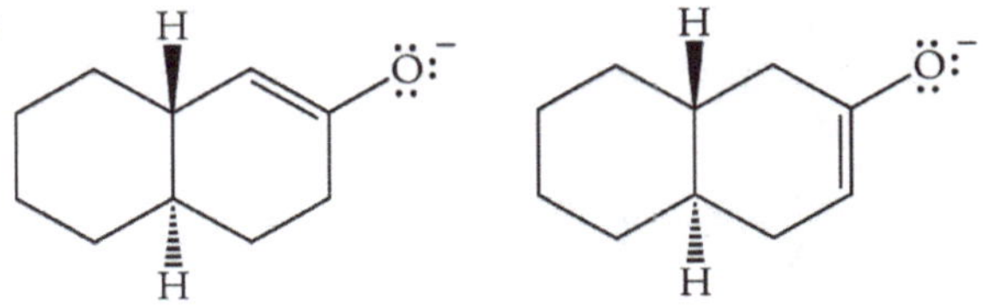

Answer: (b) The double bond of each enolate has the same degree of substitution.

22.18 2-Methylcyclopentanone gives a mixture of two enolates in a 94:6 ratio. (a) Write their structures and (b) assign their relative stabilities.

Answer: (a)

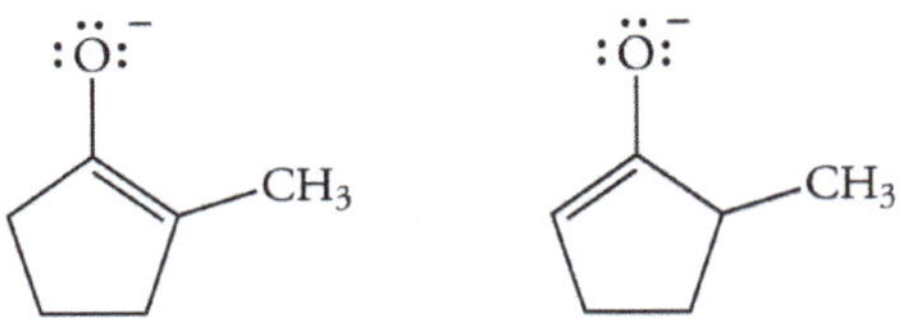

Answer: (b) There are two α-carbon atoms with enolizable hydrogen atoms. They are located at C-2 and C-5. The double bond of the enolate to C-2 has a higher degree of substitution and is more stable. It constitutes 94% of the mixture of enolates.

22.19   3-Pentanone gives a mixture of two enolates in a 84:16 ratio. (a) Write their structures and (b) assign their relative stabilities.

Answer: (a)

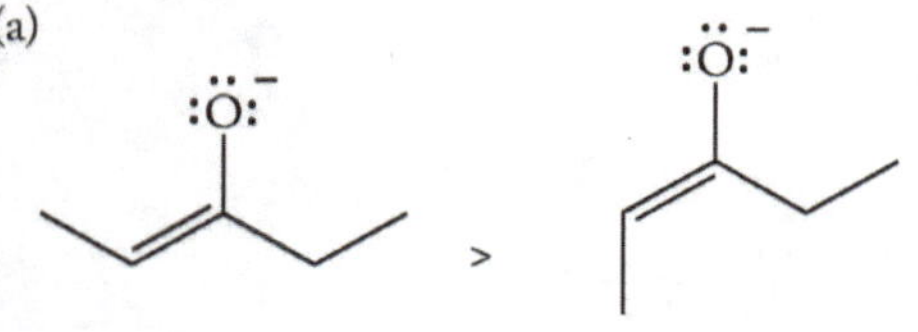

enols of 3-pentanone in order of decreasing stability

Answer: (b) There are two α-carbon atoms with enolizable hydrogen atoms, but they are equivalent. The C-2 atom gives a pair of enolates that are geometric isomers. Of the two enolates, the isomer with *trans* alkyl groups is more stable. It constitutes 84% of the enolate mixture.

22.20   2,2-Dimethyl-3-pentanone gives a mixture of two enolates. Based on the data in Exercise 22.19, predict how the ratio of the amounts of the two enolates would differ from the ratio for 1-pentanone.

Answer: (a)

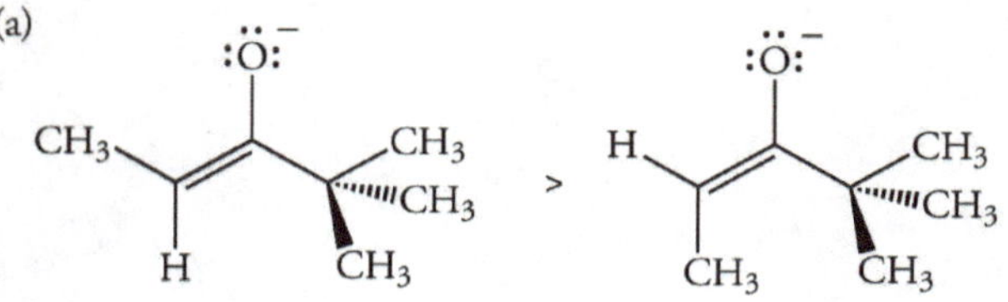

enols of 2,2-dimethyl-3-pentanone in order of decreasing stability

Answer: (b) There is only one α-carbon atom with enolizable hydrogen atoms. The C-3 atom gives a pair of enolates that are geometric isomers. Of the two enolates, the isomer with *trans* alkyl groups is more stable. However, in this compound the alkyl groups are a *tert*-butyl and a methyl group, so the steric hindrance is larger than for the ethyl and methyl groups of the enolate of 3-pentanone. Thus, the ratio of the two isomers is much larger, and the *trans* isomer constitutes more than 84% of the enolate mixture.

22.21   (a) Write the mechanism for the following isomerization reaction, which occurs using sodium ethoxide in ethanol. (b) Predict which isomer is more stable.

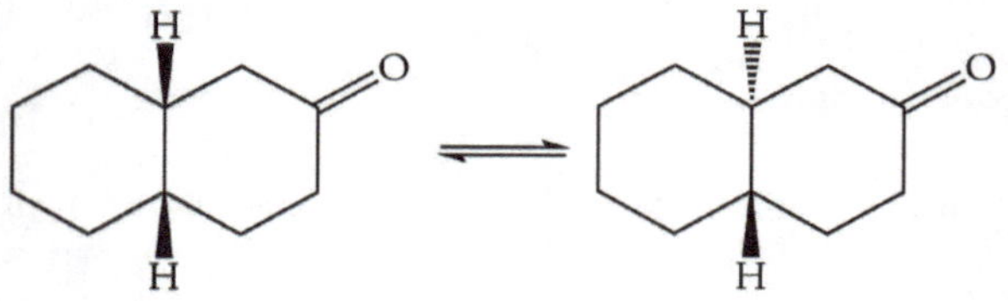

Answer: (a)

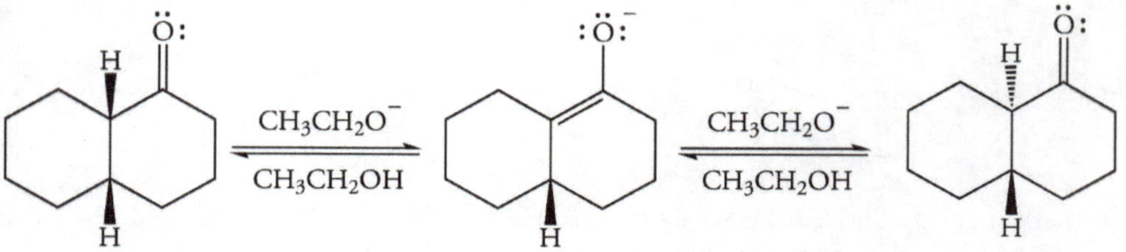

Answer: (b) Formation of an enolate occurs by abstraction of a proton from the bridgehead position by ethoxide ion. Protonation in a reverse step by ethanol can occur to give either the original ketone or its isomer. The isomer is a ketone derived from *trans*-decalin, which has its rings fused by diequatorial bonds, and is more stable than *cis*-decalin.

22.22 Write the mechanism for the following isomerization reaction, which occurs using sodium ethoxide in ethanol. Predict which isomer is more stable.

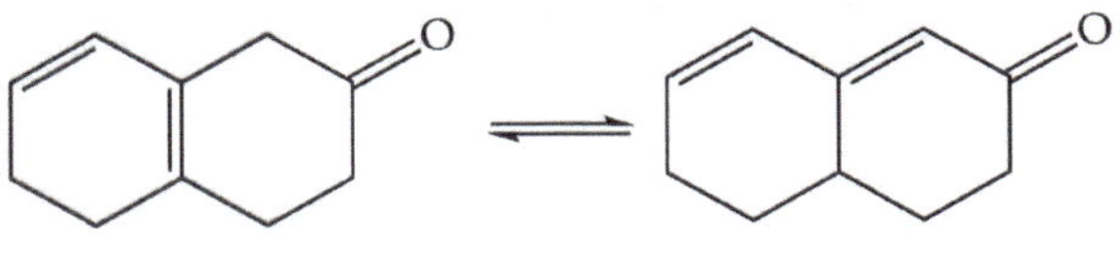

Answer: (a)

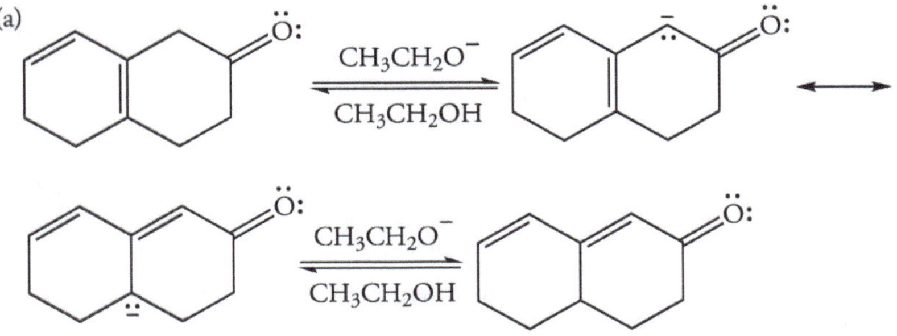

Answer: (b) The enolate is resonance stabilized, with charge distributed between a secondary and a tertiary carbon atom. Protonation can occur to give the original ketone or the isomer. This isomeric ketone is resonance stabilized by the extended conjugation of two double bonds with the carbonyl group, so it is more stable than the original ketone, whose carbonyl group is not conjugated with a double bond.

22.23 Write a mechanism for the base-catalyzed isomenzation of 3-cyclohexenone to 2-cyclohexenone.

Answer: (a)

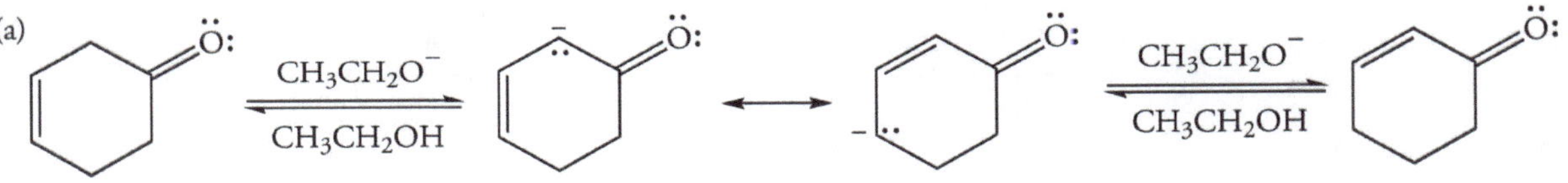

Answer: (b) The enolate formed is resonance stabilized, with charge distributed between two secondary carbon atoms. Protonation can occur to give the original ketone or the isomer. This isomeric ketone is not resonance stabilized by conjugation of the ketone with the double bond, so it is less stable than the original conjugated ketone. Nevertheless, at equilibrium there is a low concentration of the less stable 2-methyl-3-cyclopentenone.

22.24 Write a mechanism for the base-catalyzed isomerization of 5-methyl-2-cyclopentenone to 2-methyl-2-cyclopentenone. (Hint. A third isomeric unsaturated ketone is a required intermediate.)

Answer: (a) The enolate is resonance stabilized, with charge distributed between two secondary carbon atoms. Protonation can occur to give the original ketone or the isomer. This isomeric ketone is not resonance stabilized by conjugation of the ketone with the double bond, so it is less stable than the original conjugated ketone. Nevertheless, at equilibrium there is a low concentration of the less stable 2-methyl-3-cyclopentenone.

Answer: (a)

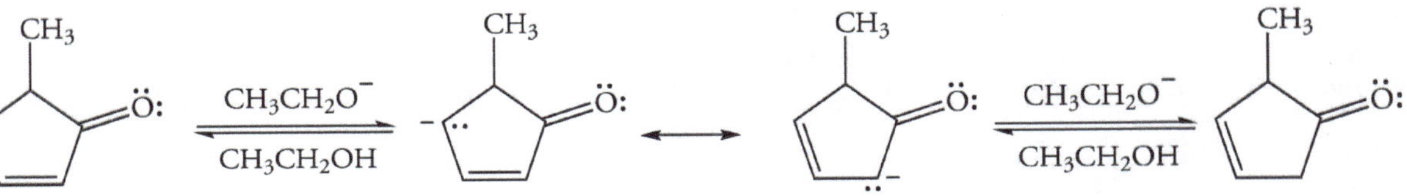

Answer: (b) A proton can be abstracted from either C-2- or C-5 of the nonconjugated product, 2-methyl-3-cyclopentenone. Loss of a proton from C-5 generates the resonance stabilized enolate of the above reaction, and leads to formation of the original 5-methyl-2-cyclopentenone. However, abstraction of a proton from C-2 gives a different resonance stabilized enolate with negative charge distributed between C-2 and C-4.

Answer: (b)

Answer: (c) Protonation can occur to give the original 2-methyl-3-cyclopentenone or an isomeric conjugated ketone. 2-Methyl-2-cyclopentenone is the most stable of the three isomers in equilibrium in this series of reactions because it is both conjugated and the more highly substituted ketone. Thus, 5-methyl-2-cyclopentenone is isomerized by the base into 5-methyl-3-cyclopentenone, which in turn is isomerized into 2-methyl-2-cyclopentenone. Although the number of the methyl group changes, it does not rearrange. The numbering is controlled by the location of the double bond relative to the ketone.

22.25   Write a mechanism that explains why a solution of (R)-2-methyl-1-pheny1-1-pentanone in ethanol containing sodium ethoxide gradually loses optical activity, but a solution of (R)-3-methyl-1-phenyl-1-pentanone does not.

Answer: (a) The chiral center of (R)-3-methyl-1-phenyl-1-pentanone is at the β-carbon atom relative to the ketone, so its hydrogen atom is not sufficiently acidic to be abstracted by ethoxide ion. Thus, its stereochemistry is not affected in the basic solution. The chiral center of the isomeric (R)-2-methyl-1-phenyl-1-pentanone is at the α carbon atom relative to the ketone, so its hydrogen atom is abstracted to form an enolate. Reprotonation can give either the (R) or (S) isomer.

Answer: (b)

22.26   Predict the change in the optical activity of each of the following in a solution of sodium ethoxide in ethanol.
   (a) (R)-2-methylcyclohexanone   (b) (R)-3-methylcyclohexanone   (c) (R)-2-ethyl-2-methylycyclohexanone

(a)

(R)-2-methylcyclohexanone

(b)

(R)-3-methylcyclohexanone

(c)

(R)-2-ethyl-2-methylcyclohexanone

Answer: (a) The chiral center of (R)-2-methylcyclohexanone is at the α carbon atom relative to the ketone, and it is abstracted to form an enolate. Reprotonation can give either the (R) or (S) isomer, so the solution gradually loses its optical activity.

Answer: (b) The chiral center of (R)-3-methylcyclohexanone is at the β carbon atom relative to the ketone, and its hydrogen atom is not sufficiently acidic to be abstracted by ethoxide ion. Thus, its stereochemistry is not affected in the basic solution.

Answer: (c) There is no hydrogen atom at the chiral center of (R)-2-ethyl-2-methylcyclohexanone, so an enolate can form only at C-6, which does not affect the chirality of the molecule.

Deuterium Exchange

22.27   (a) Explain why 7-bicyclo[2 2 1]heptanone does not undergo an exchange reaction using sodium hydroxide in D₂O, but 2-bicyclo[2.2.1]heptanone readily reacts. (b) Which hydrogen atoms are exchanged'?

7-bicyclo[2.2.1]heptanone   2-bicyclo[2.2.1]heptanone

Answer: (a) The α carbon atom of 7-bicyclo[2 2 1]heptanone is at a bridgehead position. The carbanion if formed could not form the carbon–carbon double bond of the enolate because the orbital of the carbanion and the orbitals of the carbonyl group are perpendicular to one another.

Answer: (b) However, the proton at C-3 of 2-bicyclo[2.2.1]heptanone is acidic because an enolate ion can form and is resonance stabilized. The proton at the bridgehead position is not acidic in this compound either. Deuterium exchange occurs at C-3.

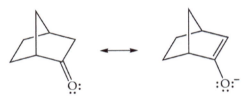

22.28   Explain why 3,3-dimethyl-2-bicyclo[2.2.1]heptanone does not undergo an exchange reaction using sodium hydroxide in D₂O.

3,3-dimethyl-2-bicyclo[2.2.1]heptanone

Answer: The α positions in this molecule are C-3 and C-1. There is no proton at C-3 of 3,3-bicyclo[2.2.1]heptanone. The proton at C-1 is at a bridgehead position, and is not acidic because a resonance-stabilized enolate would require a geometrically impossible double bond at the bridgehead position.

22.29 Explain how the following isomeric ketones could be distinguished using the base-catalyzed exchange reaction with deuterium.

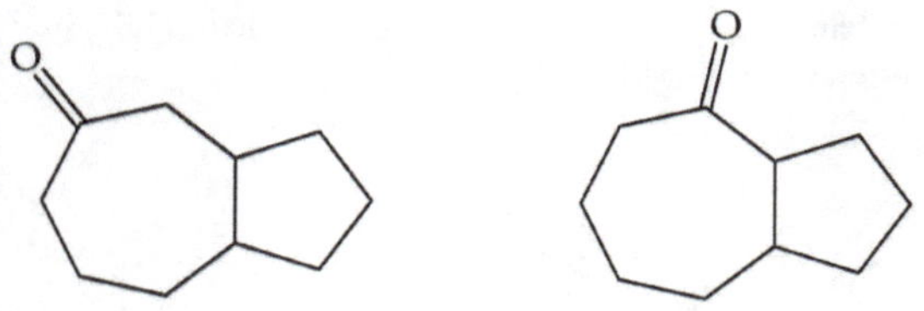

I

II

Answer: Compound I has two methylene carbon atoms at the α positions, so four hydrogen atoms can be exchanged by four deuterium atoms. Compound II has two C—H bonds at an α methylene carbon atom and one α C—H bond at a tertiary center. Thus, only three deuterium atoms can be incorporated into compound II.

22.30 Explain how 2-pentanone and 3-pentanone could be distinguished using the base-catalyzed exchange reaction with deuterium.

Answer: 2-Pentanone has a methyl carbon atom and a methylene carbon atom at the α positions, so five hydrogen atoms can be exchanged by five deuterium atoms. 3-Pentanone has two methylene carbon atoms at the α positions, so only four hydrogen atoms can be exchanged by deuterium atoms.

22.31 3-Methyl-2,4-pentanedione rapidly exchanges one hydrogen using sodium hydroxide and D_2O. After a long time, a total of seven hydrogen atoms are eventually exchanged. Explain these observations.

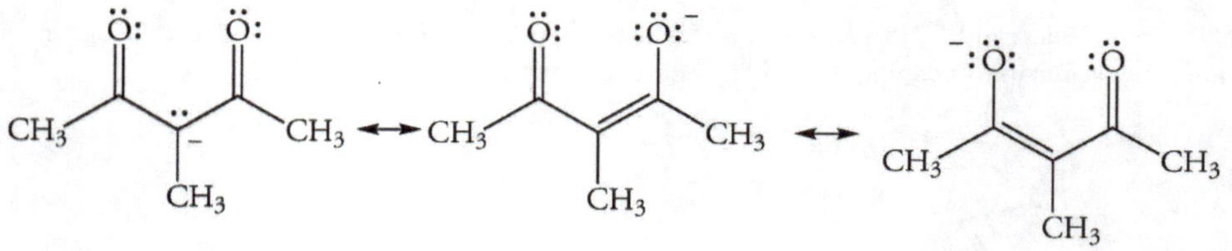

Answer: (a) The most acidic proton of 3-methyl-2,4-pentanedione is at C-3 because the enolate formed is resonance stabilized by conjugation with the second carbonyl group. Thus, this proton is exchanged rapidly.

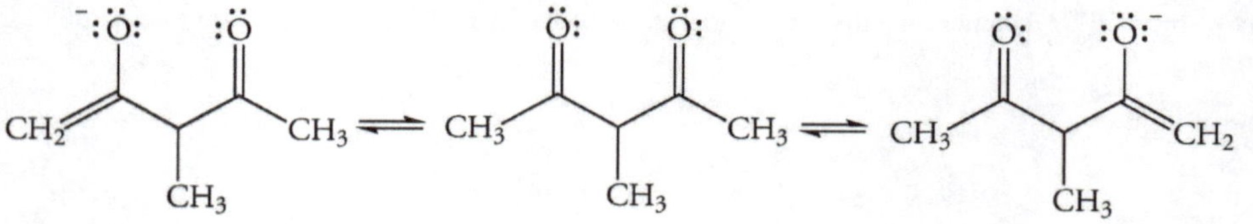

Answer: (b) The protons at C-1 and C-5 can also be exchanged, but they are not as acidic as the proton at C-3 because the enolate is not resonance stabilized. Eventually, all six hydrogen atoms of the two equivalent methyl groups are exchanged by deuterium atoms.

22.32 After a long time, 3-methyl-2-cyclohexenone exchanges a total of eight hydrogen atoms. (a) Identify the hydrogen atoms exchanged and (b) write a step showing the transfer of deuterium to an enolate that gives exchange at each of the required sites.

Answer: The two hydrogen atoms are C-6 are exchanged in the first enolate shown. The hydrogen atom at C-2 and those of the methyl group are exchanged in the second enolate. The two hydrogen atoms at C-4 are exchanged in the third enolate.

α-Halogenation Reactions

22.33 Reaction of 3-methyl-2,4-pentanedione with bromine under acidic conditions rapidly yields a monobromo derivative. (a) Write the structure of the product and (b) explain how it forms.

Answer: The most stable enol is produced by protonating one of the two equivalent carbonyl oxygen atoms and forming a double bond between C-2 and C-3. Bromination of the enol thus occurs at C-3.

22.34 Reaction of 3-methyl-2-butanone with bromine under acidic conditions yields a mixture of two monobromo derivatives in a 95:5 ratio. (a) Write the structure of the products and (b) explain why the high ratio of isomers occurs.

Answer: There are two possible enols. The one with a double bond between C-2 and C-3 is more stable because it is more highly substituted. This enol gives the major product. The enol with a double bond between C-1 and C-2 accounts for the minor product.

22.35 Which of the following compounds will give a positive iodoform test when treated with iodine in a basic solution?

Answer: The formation of CHI_3 in an iodoform reaction requires a methyl group bonded to a carbonyl carbon atom. Only (c) and (d) have such structures.

22.36 Write the structure of a compound with molecular formula $C_8H_{14}O_2$ that gives adipic acid when reacted with excess bromine in a basic solution.

Answer: Adipic acid is hexanedioic acid, a six-carbon dicarboxylic acid. The product has two fewer carbon than the reactant, and a dicarboxylic acid. This means that the reactant was a diketone with methyl groups bonded to each carbonyl carbon atom before the bromination reaction.

adipic acid (hexanedioic acid)

22.37   (a) Explain why the indicated hydrogen atom at the bridgehead carbon of the following compound is not replaced by bromine in basic solution. (b) What competing reactions may occur?

Answer: (a) An enolate cannot form because a double bond cannot be located at the bridgehead carbon atom. The orbital containing the electron pair of a possible bridgehead carbanion is approximately perpendicular to the plane of the p bond of the carbonyl groups, which is not suitable for π bond formation. Either of the two methylene groups that are α to the two carbonyl groups could be brominated.

22.38   Predict the structure of the dibromo derivative obtained from the following ketone in basic solution.

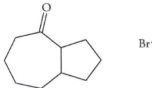

Answer: Bromination occurs at the site of the most acidic hydrogen atom, which in this case is the methylene group. It is secondary; the bridgehead carbon is tertiary and therefore cannot be brominated twice. The second bromination occurs at a faster rate than the first and occurs at the site of the first bromination.

22.39   Bromination of 4-*tert*-butylcyclohexanone under acidic conditions yields a mixture of two isomeric monobromo derivatives in approximately equal amounts. (a) Write the structures of the products and (b) explain why the ratio of the two compounds is approximately one.

Answer: The methylene groups at C-2 and C-6 are equivalent. The enol formed under acid conditions has a double bond that may be attacked by the electrophilic bromine from either side of the ring; *cis*- and *trans*-2-bromo-4-*tert*-butylcyclohexanone compounds are produced.

22.40   Write the structures of the four isomeric monobromo products that could result from bromination of the following ketone in acidic solution.

Answers:

Reactions at the α Carbon Atom

22.41 Write the structure of the product obtained by the reaction of 2,2-dimethyl-3-pentanone with sodium hydride followed by addition of 1-iodobutane.

Answer: Alkylation occurs at C-4, whose hydrogen atom is abstracted by base to form an enolate ion. The alkyl iodide is primary and undergoes substitution without significant competing elimination reaction.

Answer:

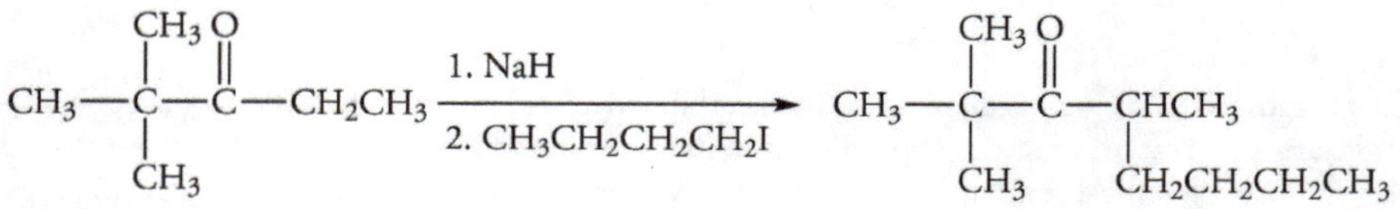

2,2-dimethyl-3-pentanone

22.42 Explain why reaction of cyclohexanone with LDA followed by the addition of 2-bromopropane gives only the original ketone upon aqueous workup.

Answer: The alkyl bromide is secondary, so it undergoes an elimination reaction in the presence of the strong conjugate base formed abstraction of the proton at C-2 of cyclohexanone. Protonation of the conjugate base gives the original ketone.

22.43 The enolate derived from reaction of LDA with 4-*tert*-butylcyclohexanone reacts with ethyl iodide to give a mixture of two monoalkylated products in approximately equal amounts. (a) Write the structures of the products. (b) Explain why the ratio of the two compounds is approximately one.

Answer:

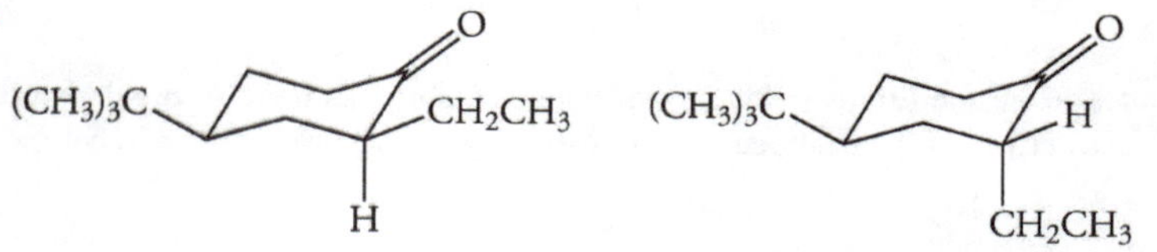

Answer: The methylene groups at C-2 and C-6 are equivalent. The enolate formed under basic conditions has electron density at C-2, which acts as a nucleophile that displaces iodide ion from iodoethane. The trigonal pyramidal carbanion can invert, so displacement can occur with the newly formed bond on either side of the ring. *Cis*- and *trans* isomers therefore form.

22.44 Write the structures of the four isomeric monobromo products that could result from bromination of the following ketone in acidic solution.

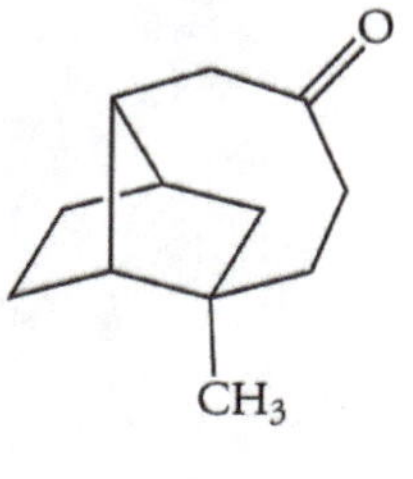

Answer:

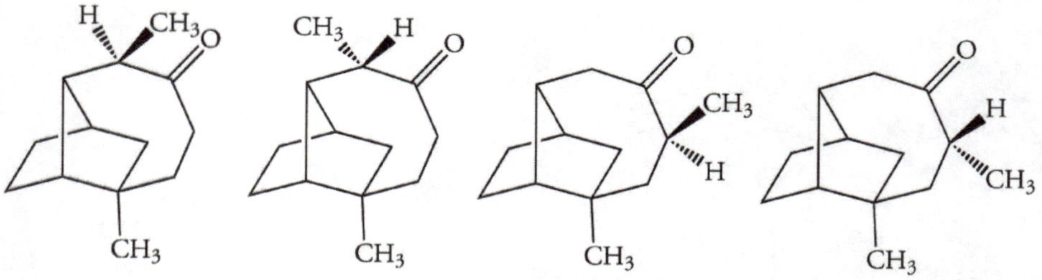

Answer: There are two possible enolates because two nonequivalent methylene groups are located α to the carbonyl group. Each can react with methyl iodide, and the methyl group may bond to either side of the ring.

22.45   Reaction of 6-bromo-3,3-dimethyl-2-hexanone with LDA gives a product with the molecular formula $C_8H_{14}O$. Write its structure.

Answer: The enolate has some negative charge at C-1, which acts as a nucleophile to displace bromide ion at C-6 in an intramolecular reaction.

Answer:

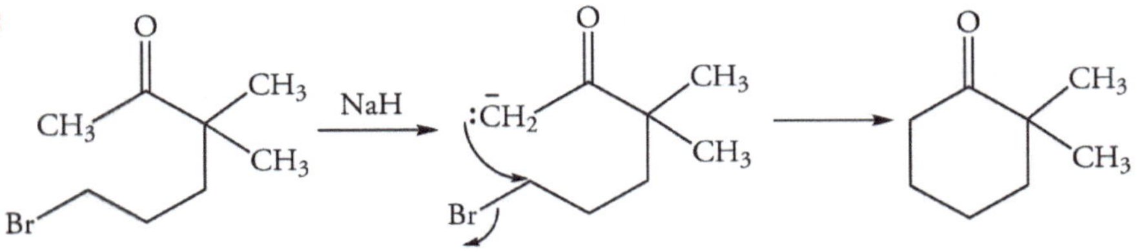

22.46   Reaction of the following ketone with a sterically hindered strong base gives a product with the molecular formula $C_{10}H_{14}O$. Write its structure.

Answer:

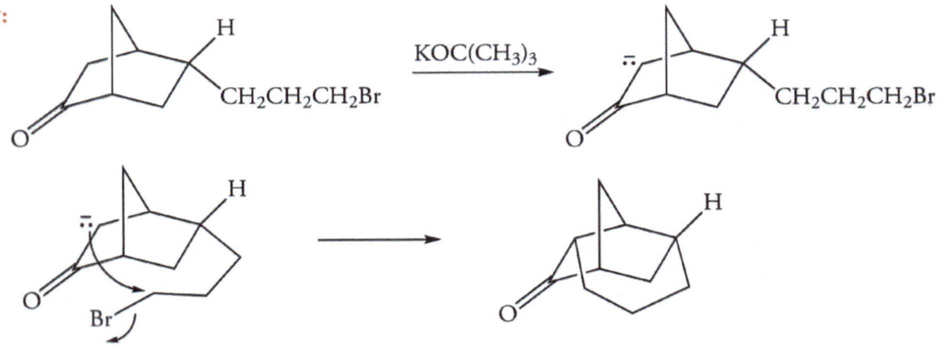

Answer: The enolate has some negative charge, which acts as a nucleophile to displace bromide ion in an intramolecular reaction.

22.47   Trimethylchlorosilane, $(CH_3)_3SiCl$, reacts with enolates exclusively at the oxygen atom to give trimethylsilyl enol ethers. When heated with triethylamine and trimethylchlorosilane, the silyl ethers I and II derived from 2-methylcyclohexanone occur in a 1:3 ratio. Which is more stable? Why is it more stable?.

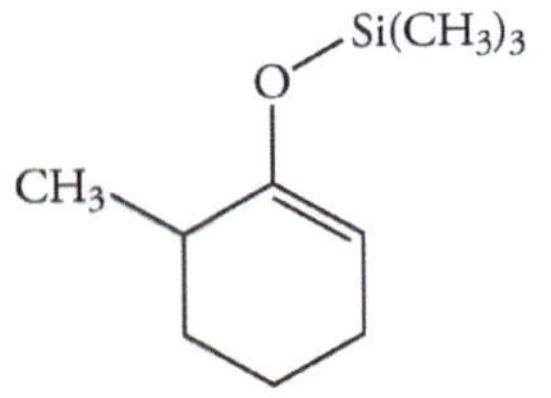

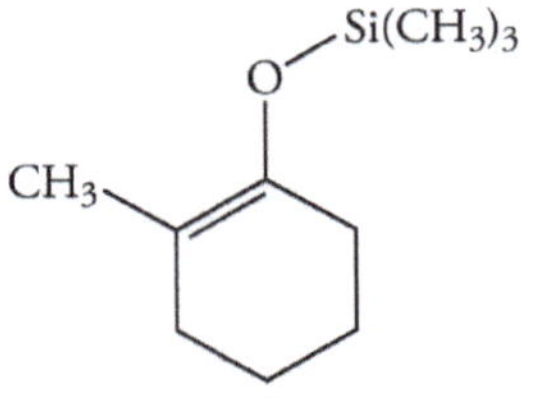

Answer: Compound II is more stable because the double bond is more highly substituted. Compound II is the major product because the enolates formed are in equilibrium with the weak base. The major product thus results from reaction of the major enolate in solution.

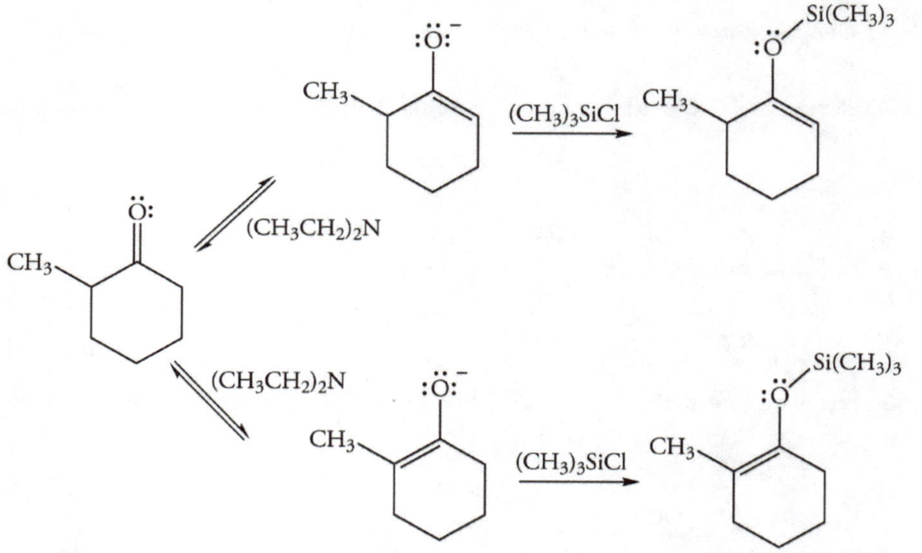

22.48 Using the data in Exercise 22.47, predict the structure of the maim product of the reaction of 2-pentanone with triethylamine and trimethylchlorosilane.

Answer:

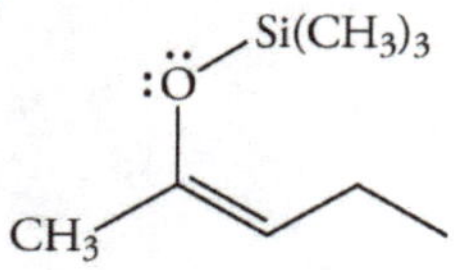

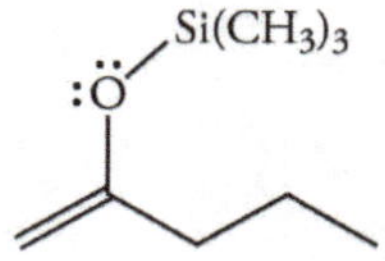

Answer: There are two isomeric enolates. The enolate derived from abstraction of a proton at C-3 is more highly substituted than the enolate derived from abstraction of a proton at C-1. The major product is the isomer with the double bond at C-2.

22.49 The reaction of 2-methylcyclohexanone with LDA in 1,2-dimethoxyethane at 0 °C yields a solution that, when subsequently reacted with trimethylchlorosilane and triethylamine, yields I and II in a 99:1 ratio. Explain why the indicated ratio occurs.

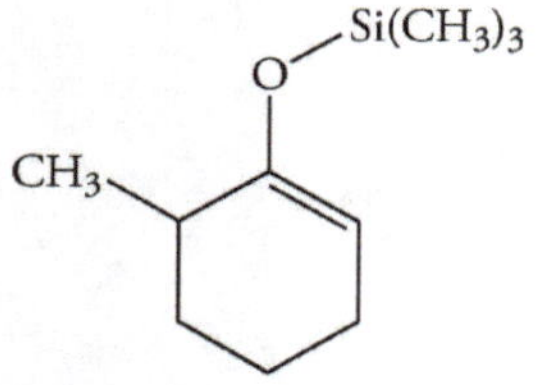

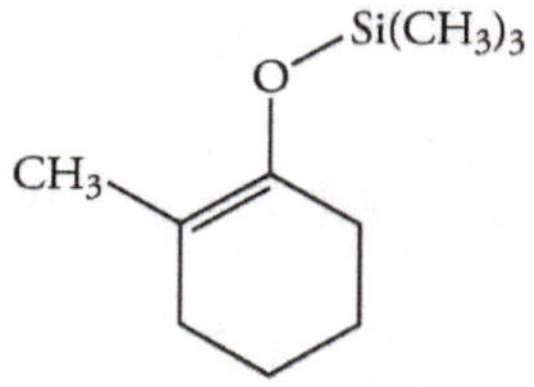

Answer: The strong base removes a proton from the more acidic secondary carbon atom at a faster rate than from the tertiary carbon atom. Because the base is very strong, neither enolate reverts to the carbonyl compound, and the two are not in equilibrium. The product is the result of kinetic control.

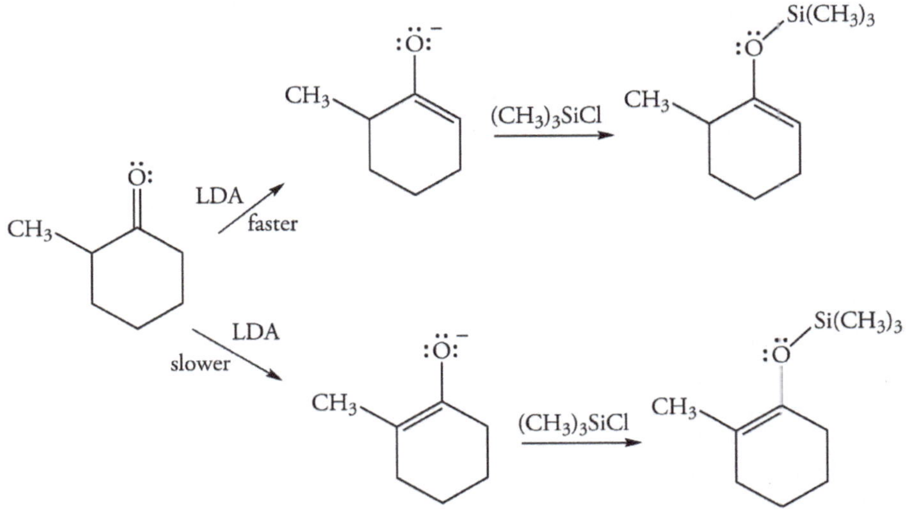

22.50 Based on the data in Exercise 22.49, predict the structure of the major product of the reaction of 2-pentanone with LDA in 1,2-dimethoxyethane at 0 °C followed by trimethylchlorosilane and triethylamine.

Answer:

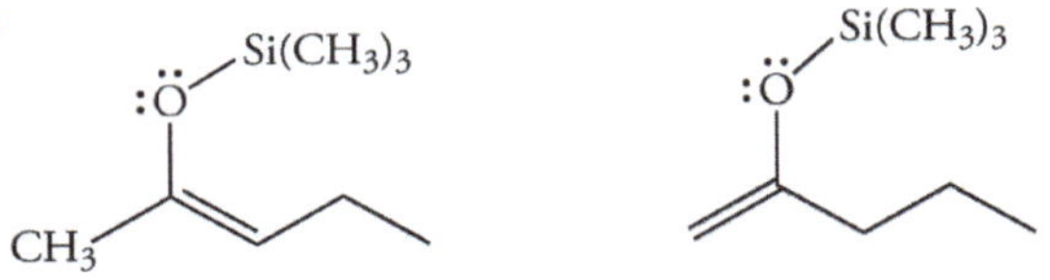

Answer: The strong base removes a proton from the more acidic primary carbon atom at a faster rate than from the secondary carbon atom. Because the base is very strong, neither enolate reverts to the carbonyl compound and the two are not in equilibrium. The major product, which has a double bond at C-1, is the result of kinetic control.

22.51 The reaction of 2-methylcyclohexanone with LDA in 1,2-dimethoxyethane at 0 °C yields a solution that reacts with benzyl bromide to give 2-benzyl-6-methylcyclohexanone and 2-benzyl-2-methylcyclohexanone in a 12:1 ratio. Explain why the indicated ratio occurs.

Answer: The strong base removes a proton from the more acidic secondary carbon atom at a faster rate than from the tertiary carbon atom. Because the base is very strong, neither enolate reverts to the carbonyl compound and the two are not in equilibrium. The major alkylation product is derived from the substitution at C-6 in the original compound. The product is named 2-benzyl-6-methylcyclohexanone because the benzyl group take alphabetic preference over the methyl group.

Answer:

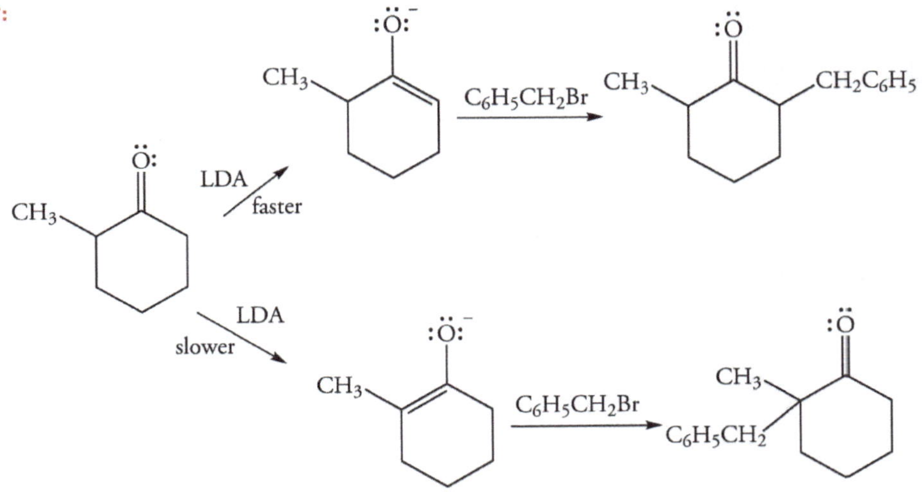

22.52   What experimental conditions would favor formation of 2-benzyl-2-methylcyclohexanone by alkylation of 2-methylcyclohexanone with benzyl bromide?

Answer: Reaction of 2-methylcyclohexanone with benzyl bromide using triethylamine as a base generates two isomeric enolates. They exist in equilibrium with the ketone in the weak base. The major enolate is the more highly substituted compound with a double bond at C-2. This enolate reacts with benzyl bromide gives 2-benzyl-2- methylcyclohexanone as the major product.

22.53   Explain why the reaction of the following ketone with LDA in THE at −60 °C yields the indicated bicyclic ketone.

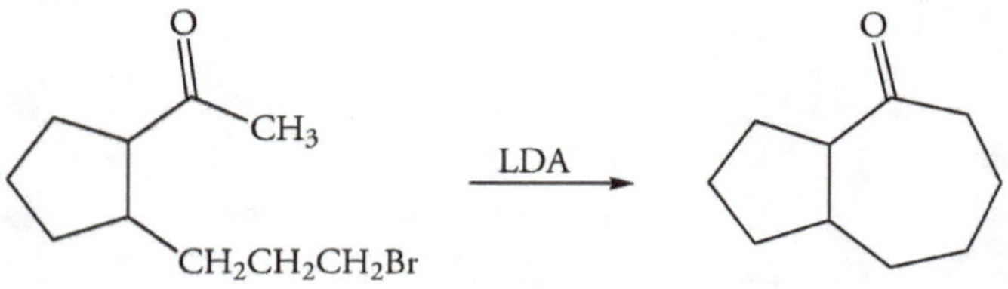

Answer: The strong base removes a proton from the more acidic primary carbon atom at a faster rate than from the tertiary carbon atom. Intramolecular attack of the primary carbanion at the primary bromoalkane center forms a seven-membered ring. Although formation of seven-membered rings is generally not favored, this process occurs because four of the bonds are restricted by the five-membered ring, and there is a larger probability of ring closure.

22.54   The ketone shown in Exercise 22.53 reacts with potassium *tert*-butoxide in *tert*-butyl alcohol to give a constitutional isomer of the bicyclic ketone shown above. (a) Write its structure and (b) explain its origin.

Answer: Potassium *tert*-butoxide in *tert*-butyl alcohol is a weaker base than LAD. It gives two enolates that are in equilibrium with the ketone, so two products can result. The enolate with a double bond to the atom of the five-membered ring can undergo an intramolecular reaction to produce a second five-membered ring.

Answer:

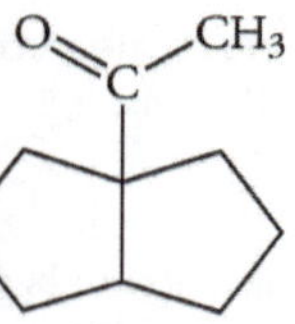

Aldol Condensations

22.55   Draw the structure of the product of the self-condensation of each of the following aldehydes in the presence of a catalytic amount of sodium hydroxide.
   (a) 2-methylpropanal   (b) phenylethanal   (c) octanal

Answers:

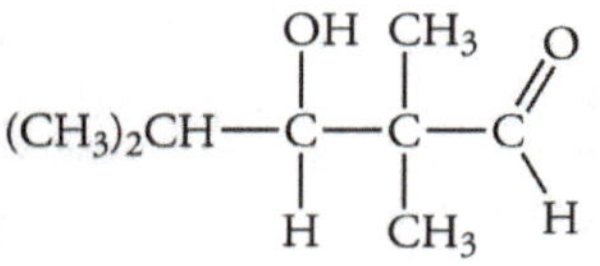

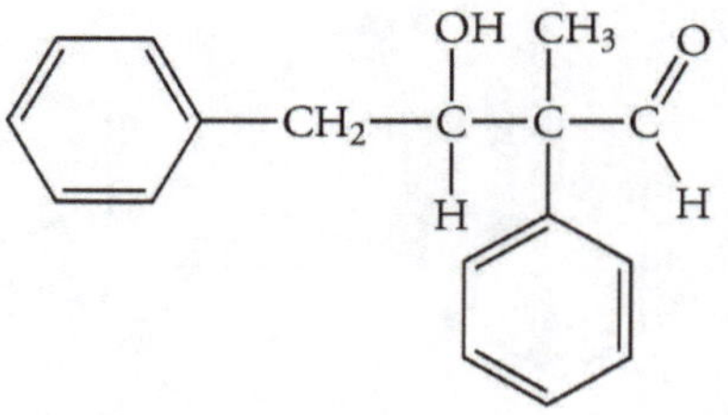

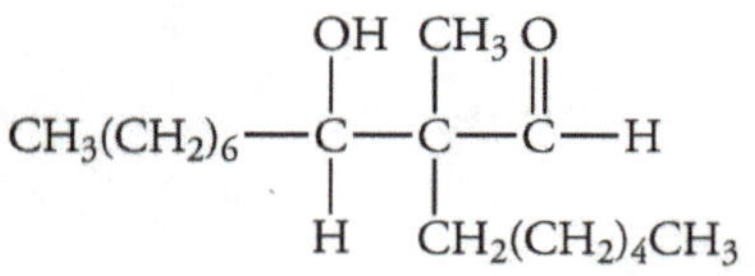

22.56   What reactants are required to give the following compounds by a mixed aldol reaction?

(a)

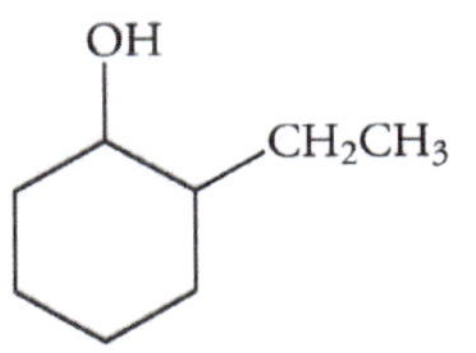

(b)

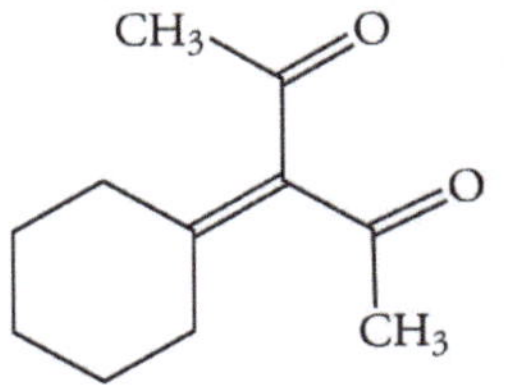

(c)

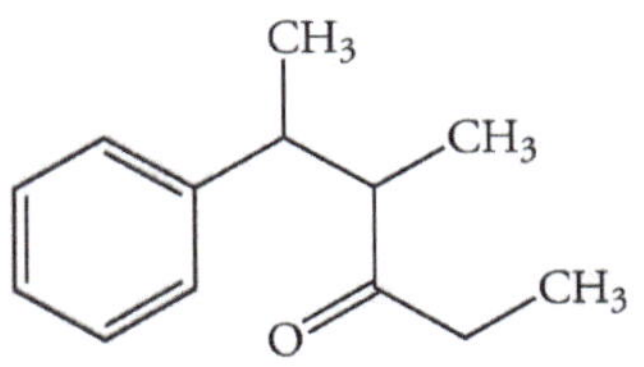

(d)

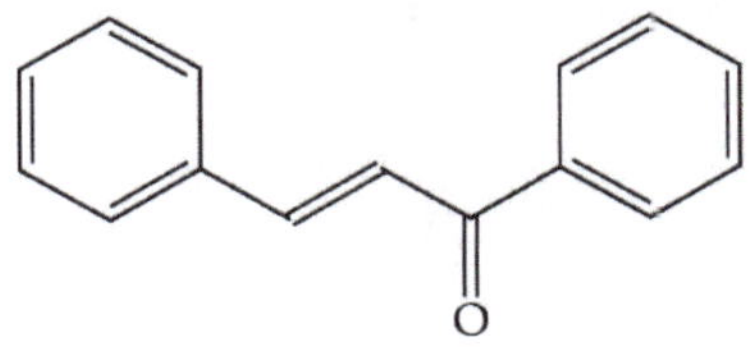

Answers: (a) cyclohexanone and ethanal   (b) cyclohexanone and 2,4-pentanedione
(c) acetophenone and 3-pentanone (d) benzaldehyde and acetophenone

22.57   Pseudoionone, a component of some perfumes, has the molecular formula $C_{13}H_{20}O$. It can be prepared by a mixed aldol reaction of citral and acetone. Write the structure of pseudoionone.

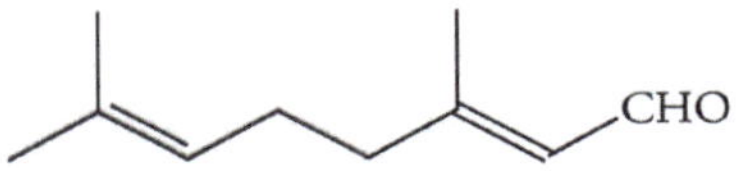

citral

Answer:

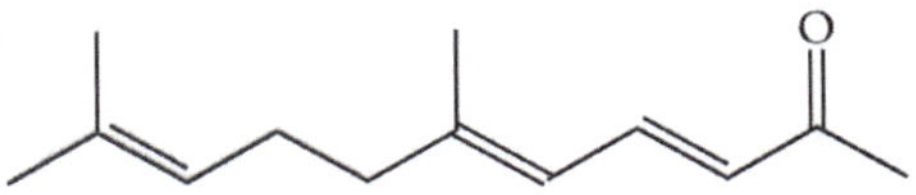

pseudoionone

22.58   A mixed aldol reaction between citral and 2-butanone yields two products. Write their structures.

Answers: Either C-1 or C-3 of 2-butanone can form an enolate. Subsequent reaction of each enolate at the carbonyl carbon of citral gives an unsaturated product.

Answer:

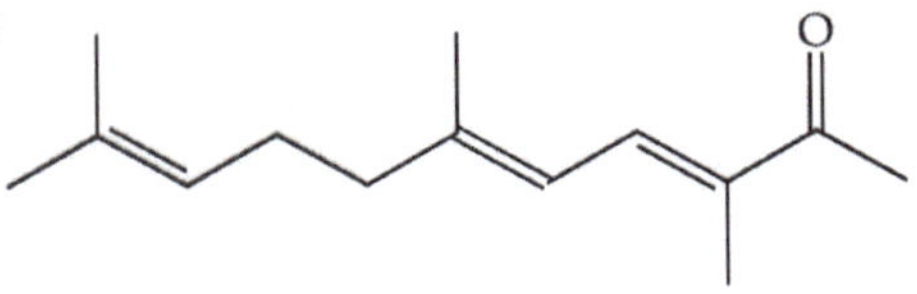

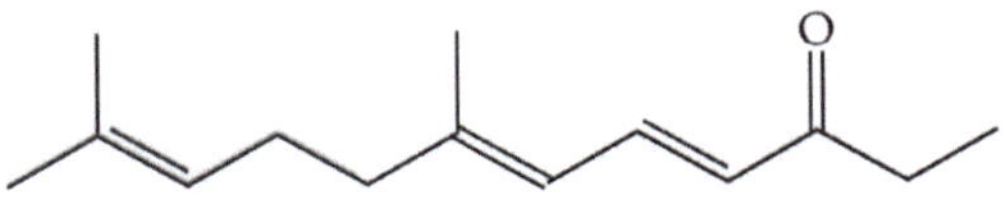

22.59 2,2-Dimethyl-l,3-propanediol can be synthesized by reduction of a mixed aldol product using sodium borohydride. (a) What is the aldol? (b) What two carbonyl compounds are required to produce it?

Answer: (a) The product is 3-hydroxy-2,2-dimethylpropanal. (b) It is the product of a mixed aldol condensation of 2-methylpropanal and methanal (formaldehyde).

3-hydroxy-2,2-dimethylpropanal

22.60 Suggest a synthesis of the following compound starting from acetophenone.

Answer: A mixed aldol condensation of acetophenone and methanal (formaldehyde) gives a product that can undergo a mixed aldol condensation two more times.

22.61 The favored products of the intramolecular aldol condensation of 2,5-hexanedione and 2,6-heptanedione are given in Section 22.9. (a) Write an alternative isomeric structure for each product. (b) Explain why it is not formed.

Answer: Formation of an enolate by abstraction of a proton at C-3 of 2,5-hexanedione followed by ring closure would give a cyclopropane ring. This process doesn't occur due to ring strain. Also, subsequent dehydration is disfavored because a double bond in the ring would further increase the ring strain. Formation of an enolate by abstraction of a proton at C-3 of 2,6-hexanedione followed by ring closure would give a cyclobutane. This process doesn't occur for the same reasons.

22.62 The intramolecular aldol condensation of 2,6-octanedione could yield two possible six-membered unsaturated products. (a) Write their structures. (b) Predict which isomer would be the major product.

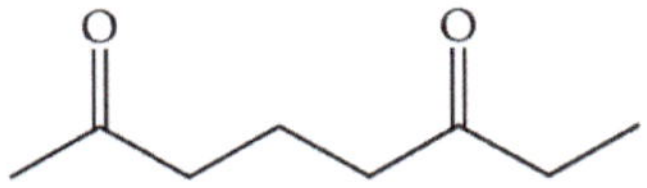

2,6-octanedione

Answer: (a) Enolates can form as the result of proton abstraction from any of four nonequivalent α carbon atoms. However, six-membered rings result only from the enolates derived from abstraction of protons at C-1 and C-7. All of the enolates are in equilibrium with the diketone. Thus, the products formed are the result of favorable rates of ring closure and their individual stability. Reaction of the enolate derived from C-1 with the C-6 carbonyl group gives the first of the two products listed below. Reaction of the enolate derived from C-7 with the C-2 carbonyl group gives the second product.

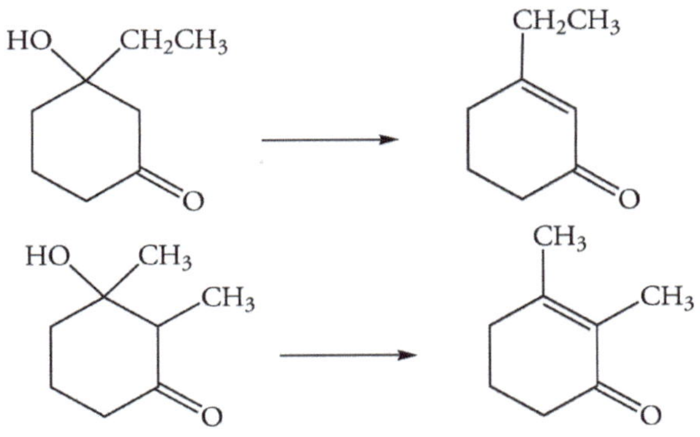

Answer: (b) Formation of the first product requires nucleophilic attack at the more hindered carbonyl carbon atom. Also, the dehydration product has the less substituted double bond. Thus, the second product is the major product.

22.63 (a) What diketone will yield the following as a product of an intramolecular aldol condensation? (b) What isomeric bicyclic compound could also form, but in smaller amount?

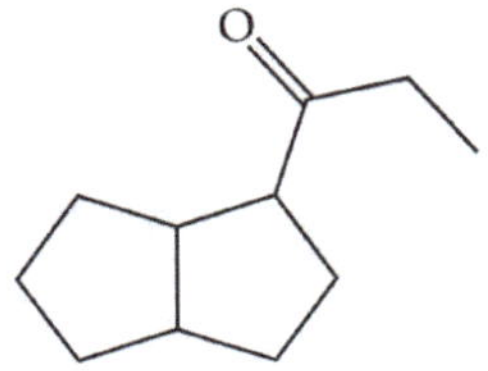

Answers:

(a)

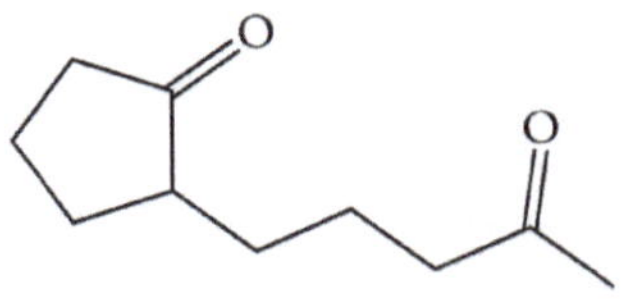

(b)

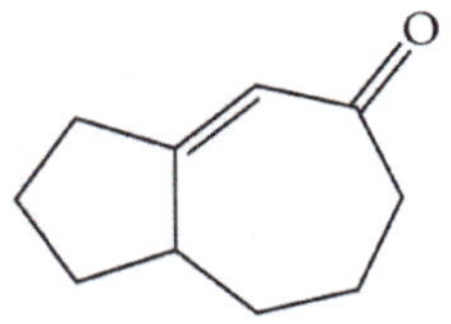

22.64 What reactant could yield the following product from an intramolecular aldol condensation?

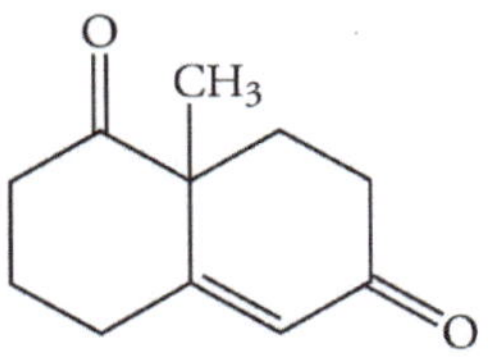

Answer:

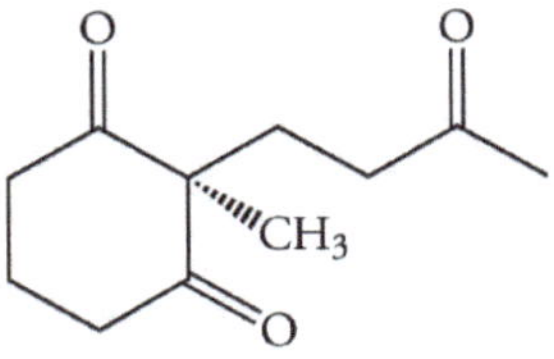

Conjugate Addition Reactions

22.65 Amines react with α,β-unsaturated ketones to give a conjugate addition products. Write the structure of the product for each of the following combinations of reactants.
(a) 2-cyclohexenone and $CH_3CH_2NH_2$
(b) 3-butenone and $(CH_3)_2NH$
(c) 4-methyl-3-penten-2-one and CH_3NH_2

Answers:

(a), (b), (c)

22.66 The conjugate addition of HCN to α,β-unsaturated ketones can be done using diethylaluminum cyanide, $(C_2H_5)_2Al$—CN, followed by acid workup. Write the structure of the addition product for each of the following reactants.

(a), (b), (c)

Answers:

(a), (b), (c)

22.67 (a) Write the structure of the addition product of 2-cyclohexenone with ethylmagnesium bromide after hydrolysis. (b) Do the same for the addition product of 2-cyclohexenone with lithium diethylcuprate.

Answer: (a) Addition of an ethyl Grignard reagent occurs 1,2 at the carbonyl group to give a tertiary alcohol. (b) Addition of lithium diethylcuprate gives a conjugate addition product.

Answers:

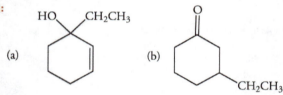

(a), (b)

22.68 What combination of an α,β-unsaturated ketone and a Gilman reagent is required to synthesize each of the following compounds?
(a) 3-phenylcycloheptanone
(b) 2-hexanone
(c) 3-vinylcyclohexanone

Answers: (a) 2-cycloheptenone and lithium diphenylcuprate
(b) 3-buten-2-one and lithium diethylcuprate
(c) 2-cyclohexenone and lithium divinylcuprate

22.69 Write the reaction sequence required to synthesize the following structures using a Michael addition reaction as one of the steps.

(a)

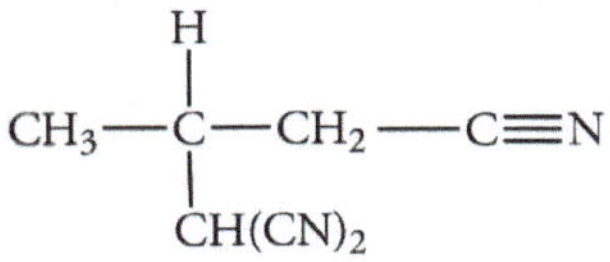

$$CH_3-\overset{\overset{\displaystyle H}{|}}{\underset{\underset{\displaystyle CH(CN)_2}{|}}{C}}-CH_2-C\equiv N$$

(b)

$$CH_3-\overset{\overset{\displaystyle O}{||}}{C}-\overset{\overset{\displaystyle H}{|}}{\underset{\underset{\displaystyle CH_3}{|}}{C}}-CH_2CH_2-\overset{\overset{\displaystyle O}{||}}{C}-CH_3$$

(c)

$$CH_3O-\overset{\overset{\displaystyle O}{||}}{C}-CH_2CH_2-\overset{\overset{\displaystyle CN}{|}}{\underset{\underset{\displaystyle CO_2CH_3}{|}}{C}}-CH_2-\overset{\overset{\displaystyle O}{||}}{C}-OCH_3$$

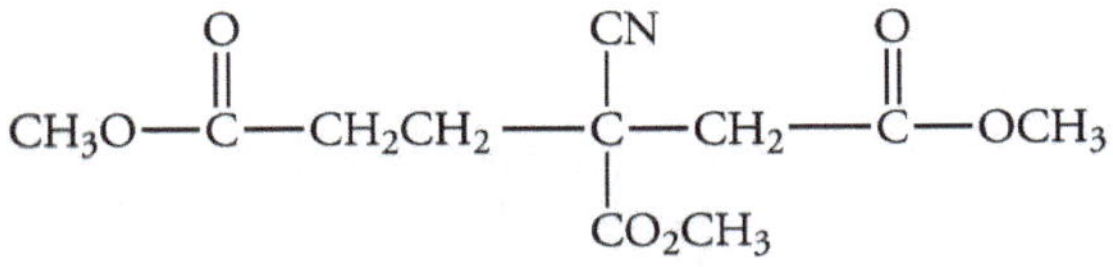

Answers:

(a)

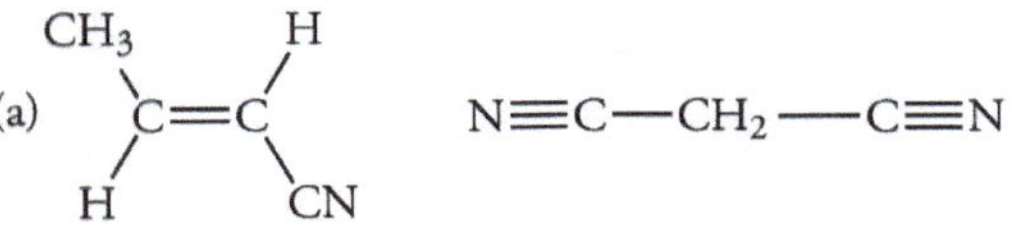

$$\overset{\displaystyle CH_3}{\underset{\displaystyle H}{}}C=C\overset{\displaystyle H}{\underset{\displaystyle CN}{}} \qquad N\equiv C-CH_2-C\equiv N$$

(b)

$$CH_3-\overset{\overset{\displaystyle O}{||}}{C}-\overset{\overset{\displaystyle H}{|}}{\underset{\underset{\displaystyle CH_3}{|}}{C}}-CO_2CH_2CH_3 \qquad CH_2=CH-\overset{\overset{\displaystyle O}{||}}{C}-CH_3$$

(c)

$$CH_3O-\overset{\overset{\displaystyle O}{||}}{C}-CH=CH_2 \qquad H-\overset{\overset{\displaystyle CN}{|}}{\underset{\underset{\displaystyle CO_2CH_3}{|}}{C}}-CH_2-\overset{\overset{\displaystyle O}{||}}{C}-OCH_3$$

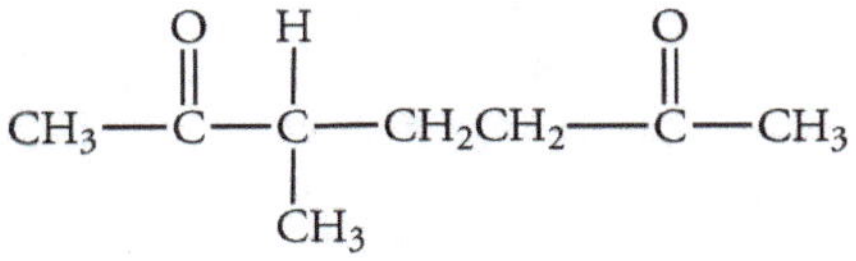

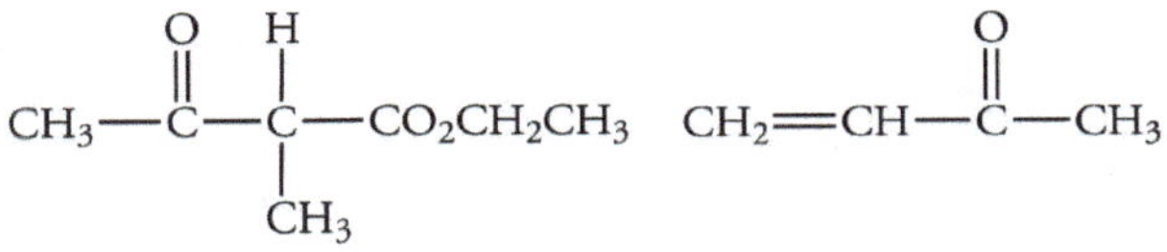

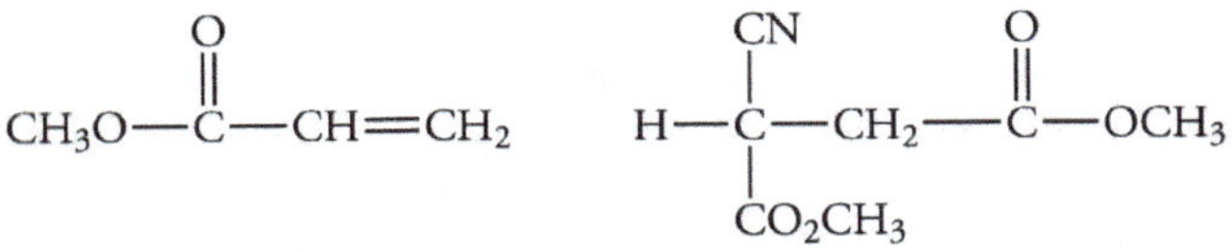

23 AMINES AND AMIDES

KEYS TO THE CHAPTER

23.1 Organic Nitrogen Compounds

Because nitrogen has five valence shell electrons, it can form three covalent bonds in neutral compounds, leaving one nonbonding electron pair. These bonds can be three single bonds as in amines and amides, one double bond and a single bond in imines, or a triple bond in nitriles.

Nitrogen is found in many biologically important compounds that have a wide range of physiological properties. However, once a nitrogen-containing functional group is identified, its chemical reactions can often be predicted since the functional groups in these compounds, whose structures are often complex, have the characteristic reactivities of much simpler compounds.

23.2 Bonding and Structure of Amines

Amines are pyramidal at the nitrogen atom, with approximately tetrahedral bond angles to all bonded atoms. The nitrogen atom in amines is sp³ hybridized. However, the configuration of an amine is not static. Amines undergo nitrogen inversion to give mixtures of mirror images. The process occurs via a planar transition state. The energy barrier to inversion is low, and nitrogen inversion is rapid so that amines with chiral nitrogen atoms cannot be isolated.

23.3 Classification and Nomenclature of Amines

Amines are classified according to the number of alkyl or aryl groups bonded to the nitrogen atom. Primary, secondary, and tertiary amines have 1, 2, and 3 groups bonded, respectively. Amides are classified the same way, with the acyl group counting as one of the carbon groups bonded to the nitrogen atom. Abbreviations for the classes of amines and amides are 1°, 2°, and 3°.

The common names of simple amines are based on the identity of the alkyl or aryl groups bonded to the nitrogen atom. The names of the alkyl groups are written in alphabetical sequence as one word, followed by the word amine.

The common name of a complex amine is based first on identifying the longest continuous chain containing an attached nitrogen atom. The chain is numbered to assign the lowest number to the carbon atom bonded to the nitrogen atom. The nitrogen atom may be contained in an amino group ($-NH_2$), an *N*-alkylamino group ($-NHR$), or an *N,N*-dialkylamino group ($-NR_2$). Aryl groups may be present in place of alkyl groups, and the same procedure is followed. In naming alkylamino or arylamino groups, the prefix *N*-indicates that the alkyl or aryl group is attached to the nitrogen atom, and not to the parent chain.

The IUPAC name of an amine is also based on the longest continuous chain containing an attached nitrogen atom. The -e ending of the parent alkane is changed to -amine. The chain is numbered to give the lowest number to the carbon atom bearing the nitrogen atom. Alkyl groups attached to nitrogen are designated with *N*-, but they are named along with other substituents on the parent chain.

Heterocyclic aromatic amines have rings that are numbered using a selected nitrogen atom as the number one atom. The Chemical Abstract System (CAS) of heterocyclic ring nomenclature has been accepted by the IUPAC. We use CAS names in this text.

23.4 Physical Properties of Amines

Amines may be gases, liquids, or solids depending on their molecular weight and structure. The boiling points of primary and secondary amines are higher than those for alkanes of similar molecular weight because these amines form intermolecular hydrogen bonds. Tertiary amines have lower boiling points than isomeric primary and secondary amines because tertiary amines do not have an N—H bond to form intermolecular hydrogen bonds. The lower molecular weight amines are soluble in water as a result of hydrogen bonding to water molecules.

23.5 Basicity of Amines

The basicity of amines of different classes do not follow a simple pattern because the number of groups bonded to nitrogen affects the electron density at the nitrogen atom. And, the stability of the conjugate acid in the solvent has a major affect on basicity. Thus, the basicity of amines can be explained only for amines with similar structures at the nitrogen atoms.

The basicity of an amine is increased by electron-donating groups and decreased by electron-withdrawing groups. Aryl amines are less basic than alkyl-substituted amines because some electron density provided by the nitrogen atom is distributed throughout the aromatic ring. Basicity is expressed using K_b values measured from the reaction of the amine with water. An alternate indicator of basicity is pK_b, which is $-\log K_b$. A strong base has a large K_b and a small pK_b. The basicity of amines is also expressed by the acidity of their conjugate acids. A strong base has a weak conjugate acid, as given by a small value of K_a and a large pK_a.

The basicity of heterocyclic amines depends on the location of the electron pair of the nitrogen atom, its hybridization, and whether or not resonance stabilization is possible. In pyrrole, the electron pair is part of the aromatic system. As a result, pyrrole is a very weak base. Pyridine is a weaker base than saturated amines of similar structure because its electron pair is in an sp²-hybridized orbital, and the electron pair is more tightly held by the atom. Protonation of a similar nitrogen atom in pyrimidine is more favorable because the charge is delocalized to the second nitrogen atom.

23.6 Solubility of Ammonium Salts

Formation of the conjugate acid of an amine gives a quaternary ammonium ion, an ionic substance that is more soluble in water than the original amine. Amines may be separated from other organic compounds that are not basic by adding acid to dissolve the amine. After physical separation and subsequent neutralization with base, the free amine separates from water.

23.7 Synthesis of Amines by Substitution Reactions

Ammonia and amines are nucleophiles, and they displace halide ion from haloalkanes to give more highly substituted amines. However, multiple substitution reactions are possible since each product successively acts as a nucleophile. Therefore, the reaction is not synthetically useful in the synthesis of a single product.

The Gabriel synthesis is used to convert a haloalkane into an amine in which the amino group replaces the halogen. However, only primary amines can be prepared because of competing elimination reactions in one of the steps. The steps required are:
1. Convert phthalimide into its conjugate base
2. Add a primary haloalkane to form an alkylated phthalimide
3. Release the amine using either strong base or hydrazine

23.8 Synthesis of Amines by Reduction

Any functional group containing nitrogen in a higher oxidation state can be reduced to give an amine, which contains nitrogen in its lowest oxidation state. These include azides, imines, nitriles, amides, and nitro compounds.

Azides are prepared by the S_N2 displacement of a halide from a haloalkane by the azide ion. Reduction of the azide by either hydrogen and platinum as catalyst, or lithium aluminum hydride in ether gives a primary amine.

The double bond of an imine can be reduced by hydrogen and Raney nickel, or by a metal hydride. Sodium borohydride reduces imines, so lithium aluminum hydride is seldom used. The **reductive amination** reaction forms an imine, by reaction of a primary amine with either an aldehyde or ketone, which is reduced immediately in the reaction by hydrogen in the presence of a nickel catalyst. Secondary amines also react by formation of an iminium ion followed by reduction.

Nitriles are reduced to primary amines using lithium aluminum hydride. The nitrile can be made by displacement of a halide ion from a haloalkane by cyanide ion. This method allows the formation of primary amines having one additional carbon atom.

Reduction of amides using lithium aluminum hydride is the most versatile way of producing amines. Amides are easily prepared by reaction of an acyl chloride and an amine.

However, the reaction is most versatile because primary, secondary, and tertiary amines can be synthesized using primary, secondary, and tertiary amides, respectively.

Reduction of nitroaromatic compounds is used to produce anilines. Tin and HCl is the usual reducing agent.

23.9 Hofmann Rearrangement

The Hofmann rearrangement occurs when a primary amide reacts with a basic solution of a halogen such as chlorine or bromine. In this process, the carboxyl carbon atom is lost as carbonate ion and a primary amine results. The rearrangement occurs when an alkyl group is transferred from the carboxyl carbon atom to the nitrogen atom in one of the several intermediates involved in the reaction mechanism. The rearrangement occurs with retention of configuration of the alkyl group.

23.10 Conjugate Addition Reactions

The reactions of amines are distinctly different than the reactions of alcohols. Amines are substantially stronger bases than alcohols. Amines are sufficiently basic to exist to some degree as the conjugate acid in water. Alcohols require strong acids to form the conjugate acid. Amines are much less acidic than alcohols—the pK_a values of amines and alcohols are 35 and 16, respectively.

Within a period of the periodic table, the nucleophilicity decreases from left to right for the elements in compounds of similar structure. Ammonia is a distinctly better nucleophile than water because nitrogen is less electronegative (EN 3.0) than oxygen (EN 3.5). Likewise, amines are better nucleophiles than alcohols. Usually, it is necessary to convert an alcohol to its alkoxide ion to make it sufficiently nucleophilic to displace a leaving group such as a halide ion from a haloalkane. The neutral amine is sufficiently nucleophilic for this type of displacement reaction.

Substitution reactions to replace oxygen as a leaving group occur if the oxygen is protonated to provide for water as a leaving group. The NH_2^- ion is a much stronger base than the hydroxide ion and is a much poorer leaving group. Even protonation of the amine to give an ammonium ion doesn't allow for the loss of ammonia as a leaving group. Elimination reactions of amines are similarly less likely than elimination reactions of alcohols.

Nitrogen moieties are more basic than the oxygen analogs and are poorer leaving groups.

Amines can be oxidized, but the resulting imines, which correspond to the carbonyl groups obtained by the oxidation of alcohols, are much more sensitive to further reaction.

23.11 Enamines

Enamines and enols are structural analogs. Enamines are formed by the reaction of secondary amines with carbonyl compounds. Common secondary amines used to form enamines include pyrrolidine, piperidine, and morpholine.

Enamines react as nucleophiles, resulting in alkylation at the position equivalent to the α carbon atom of the original carbonyl compound. The reaction is restricted to displacement of halide ion from primary alkyl halides. The product is an alkylated imine. Upon hydrolysis, it gives an α alkylated carbonyl compound.

One advantage of using enamines to alkylate carbonyl compounds is that strong base is not required. The second advantage is the formation of a singly alkylated product. In contrast, multiple alkylation of ketones occurs because proton exchange generates the enolate of the alkylated product.

23.12 Sulfonamides

Sulfonyl chlorides and acyl halides react with amines in much the same way. The resulting sulfonamides are more acidic than amides because the sulfonyl group is more electron withdrawing than an acyl group.

The **Hinsberg test** can be used to classify amines. Tertiary amines do not react with benzenesulfonyl chloride in the presence of base. There are no signs of a reaction. Secondary amines give a water insoluble sulfonamide because there are no acidic N—H bonds. Primary amines give a soluble conjugate base of the sulfonamide. Addition of acid results in protonation of the conjugate base, and the insoluble sulfonamide forms.

23.13 Quaternary Ammonium Salts

The reaction of an amine with a haloalkane does not stop after one step and eventually gives a quaternary ammonium ion. Reaction of an amine with methyl iodide to give a quaternary ammonium ion is termed **exhaustive methylation**.

Quaternary ammonium salts containing one long carbon chain are invert soaps because their polar end is positively charged in contrast to a negative charge on soaps. These compounds are effective against bacteria and are used in hospitals.

Quaternary ammonium hydroxide salts formed by exhaustive methylation followed by exchange of the halide ion by hydroxide undergo an elimination reaction called the **Hofmann elimination**. The elimination occurs by an *anti* periplanar transition state and gives the least substituted alkene. In cases where both E and Z isomers are possible, the E isomer predominates.

23.14 Spectroscopy of Amines

Infrared spectroscopy is usually not used to confirm the presence of the C—N bond because the stretching vibrations occur in a region complicated by other absorptions. The N—H stretching vibration of amines is easily seen as a broad absorption, similar to that found for the O—H vibration of alcohols, on the "left" of the spectrum in the 3200–3375 cm^{-1} region. Primary amines give two absorptions; secondary amines give one absorption.

In NMR spectra, the chemical shift of hydrogen atoms bonded to the carbon atom bearing the nitrogen atom of amines occurs in the 2–3 δ region. The N—H group has a variable chemical shift due to rapid exchange among various hydrogen bonding species whose identities are concentration dependent.

The α carbon atom of an amine has a ^{13}C chemical shift that reflects the smaller deshielding effect of the nitrogen atom relative to the more electronegative oxygen atom. The carbon absorptions are in the 30–50 δ region.

Summary of Reactions

1. Synthesis of Amines by Substitution Reactions of Haloalkane

2. Gabriel Synthesis

3. Synthesis of Amines by Reductive Methods

4. Hofmann Rearrangement

5. Alkylation of Enamines

6. Hofmann Elimination

End of Chapter Exercises

Bonding and Structure

23.1 Which compound has the greater N—H bond length, pyrrole or pyrrolidine?

Answer: The N—H bond of pyrrolidine is longer because the nitrogen atom is sp^3 hybridized. The nitrogen atom of pyrrole is sp^2 hybridized. Bond length decreases with increasing percent s character.

23.2 Which compound has the larger activation energy for the nitrogen inversion, *tert*-butyldimethylamine or trimethylamine?

Answer: The hybridization of the nitrogen atom changes from sp^3 in the ground state to sp^2 in the transition state for pyramidal inversion. Thus, groups bonded to the nitrogen atom are less sterically crowded in the transition state where the bond angles between groups is 120°. *tert*-Butyldimethylamine is more sterically hindered in the ground state than trimethylamine. Since it loses this steric strain in the transition state, *tert*-butyldimethylamine inverts at a faster rate.

Classification of Amines

23.3 Classify each of the following amines.

Answers:
(a) 2°, (b) 3°

(c) 2°, (d) 1°

(a) CH_3—N(H)—CH_2CH_3 (b) CH_3CH_2—N(CH_2CH_3)—CH_2CH_3

(c) CH_3—N(H)—C(H)(CH_3)—CH_3 (d) CH_3—C(H)(CH_3)—CH_2—NH_2

23.4 Classify each of the following amines.

Answers:
(a) 2°, (b) 2°

(c) 3°, (d) 2°

(a) piperidine N—CH_3 (b) diphenyl-N(H)

(c) cyclohexyl-N(CH_3)—CH_3 (d) phenyl-N(H)—CH_2CH_3

23.5 Classify the nitrogen-containing functional group in each of the following structures.
(a) methadone, a heroin substitute used in treating addicts

Answer:
3° amine

CH_3—CH_2—C(O)—C(phenyl)(phenyl)—CH_2—C(H)(CH_3)—N(CH_3)—CH_3

(a) The nitrogen atom located on the right in the structure has two methyl groups and a complex alkyl group bonded to it. The amine is tertiary.

379

(b) coniine, the hemlock poison that was used to execute Socrates

Answer:
2° amine

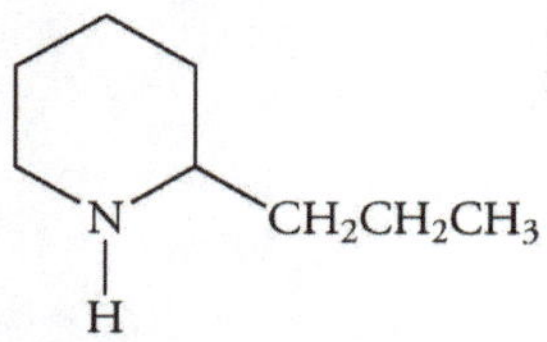

(b) The nitrogen atom located in the ring of the structure has two carbon atoms as part of the ring bonded to it. The amine is secondary.

coniine

(c) pantothenic acid, vitamin B$_5$

Answer:
2° amide

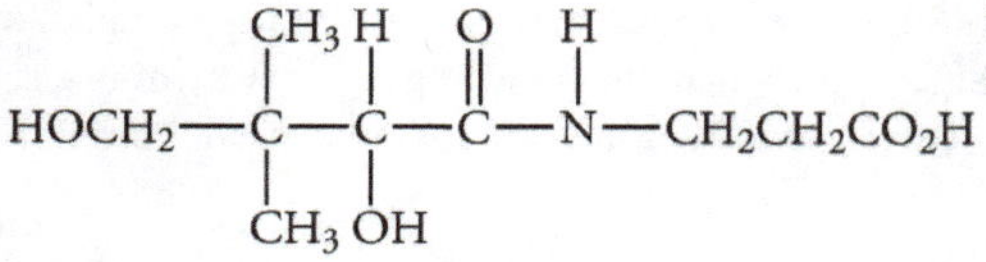

(c) The nitrogen atom is bonded an alkyl group and to a carbonyl group, so the functional group is a 2° amide.

pantothenic acid, vitamin B$_5$

23.6   Classify the nitrogen-containing functional group in each of the following structures.
(a) phencyclidine, a hallucinogen

Answer:
3° amine

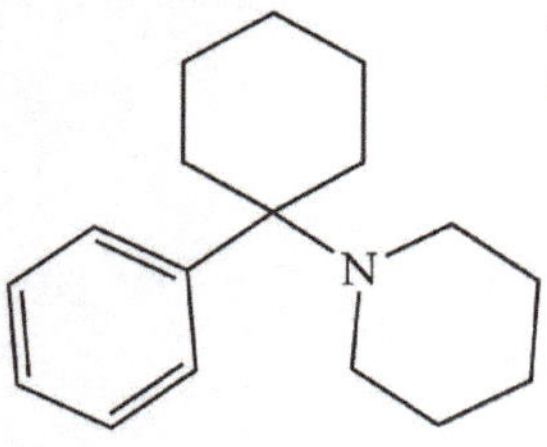

(a) The nitrogen atom located in the ring on the right in the structure has two methylene groups that are part of the ring and a quaternary carbon atom bonded to it. The amine is tertiary.

phencyclidine

(b) encainide, an antiarrhythmic drug

Answer:
3° amine
2° amide

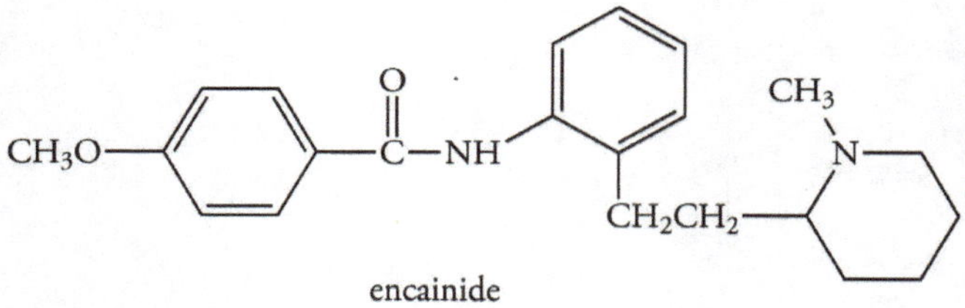

(b) The nitrogen atom bonded to the aromatic ring is also bonded to a carbonyl group, so the functional group is a 2° amide. The nitrogen atom located in the ring on the right in the structure has two methylene groups that are part of the ring and a methyl group bonded to it. The amine is tertiary.

encainide

(c) practolol, an antihypertensive drug

Answer:
2° amine
2° amide

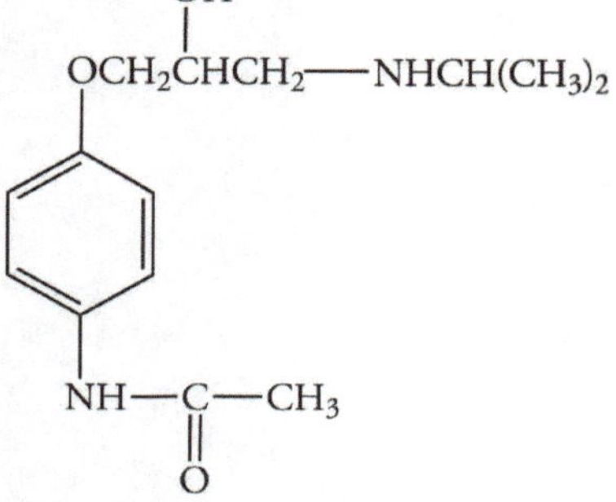

practolol

(c) The nitrogen atom located on the right of the structure has two alkyl groups and a hydrogen atom bonded to it. The amine is secondary. The nitrogen atom bonded to the aromatic ring is also bonded to a carbonyl group, so the functional group is a 2° amide.

Nomenclature

23.7 Give the IUPAC name for each of the following compounds.

(a) $CH_3CH_2\overset{\underset{\textstyle NH_2}{|}}{C}HCH_2CH_2CH_3$ (b) $CH_3CH_2CH_2CH_2\overset{\underset{\textstyle N}{|}}{\underset{\textstyle |}{\overset{\textstyle CH_3}{}}}\!-\!CH_3$

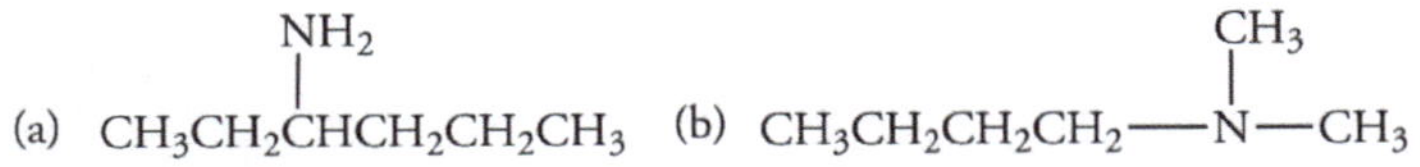

Answers:
(a) 3-hexanamine (b) *N,N*-dimethy1-1-butanamine

23.8 Give the IUPAC name for each of the following compounds.

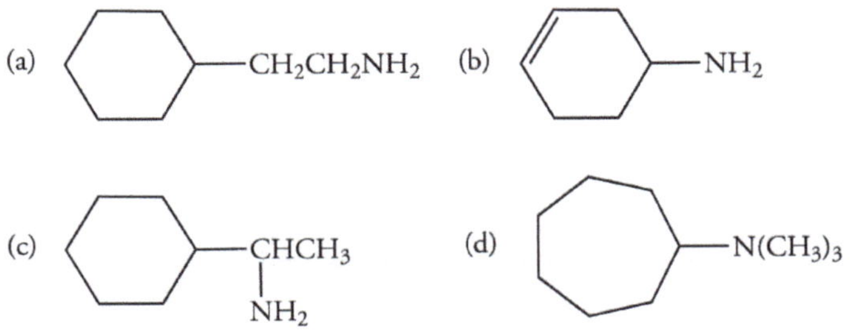

(a) —CH$_2$CH$_2$NH$_2$ (b) —NH$_2$

(c) —CHCH$_3$ with NH$_2$ (d) —N(CH$_3$)$_3$

Answers:
(a) 2-cyclohexy1-1-ethanamine (b) 3-cyclohexenamine
(c) 1-cyclohexy1-1-ethanamine (d) *N,N*-dimethylcycloheptanamine

23.9 An antidepressant drug is named *trans*-2-phenylcyclopropylamine. Draw its structure.

Answer:

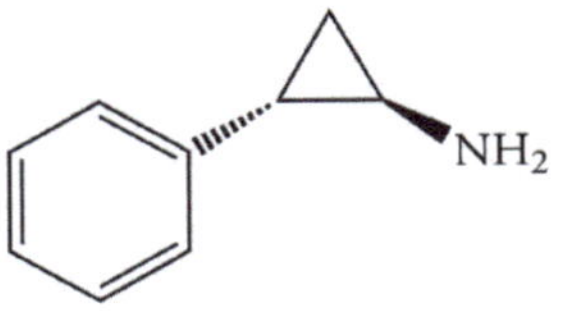

23.10 Tranexamic acid is a drug that aids blood clotting. Its IUPAC name is *trans*-4-(aminomethyl)cyclohexanecarboxylic acid. Draw its structure.

Answer:

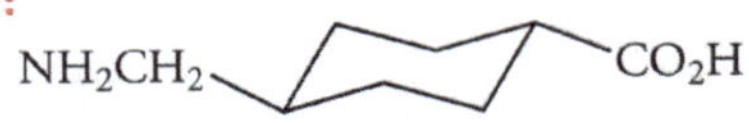

23.11 Draw the structure of each of the following compounds.
(a) 2-ethylpyrrole (b) 3-bromopyridine (c) 2,5-dimethylpyrimidine (d) 2,6,8-trimethylpurine

Answers:

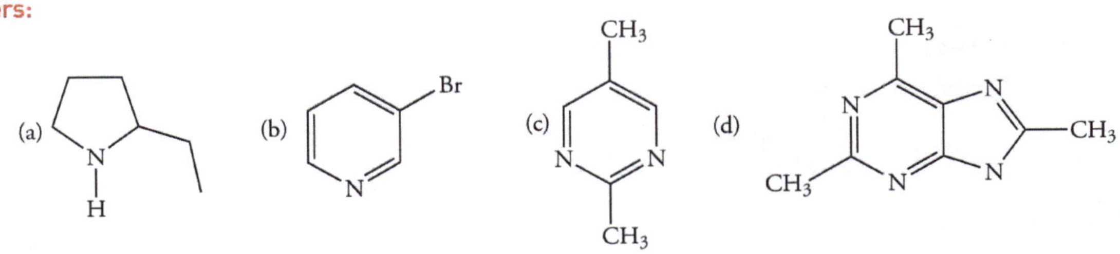

23.12 Name each of the following compounds.

(a)

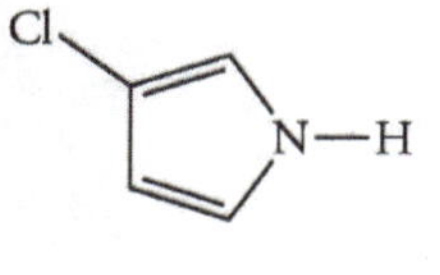

(b)

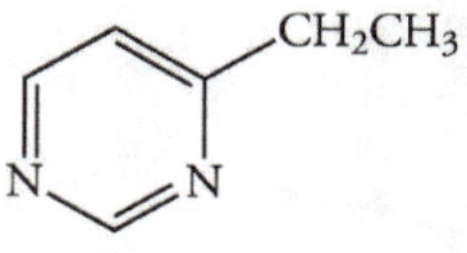

(c)

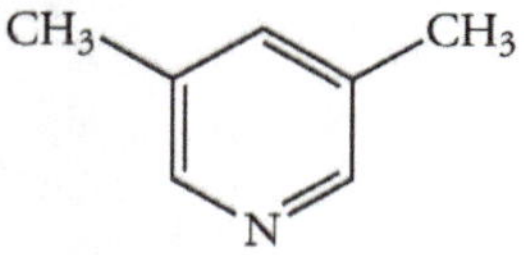

(d)

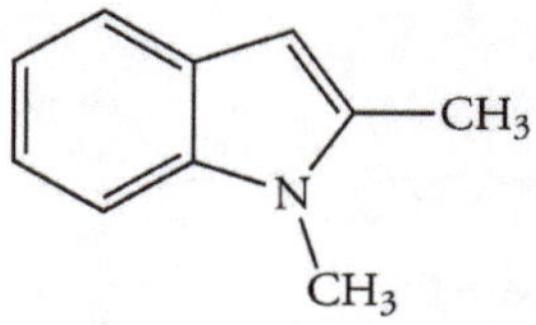

Answers:

(a) 3-chloropyrrole

(b) 4-ethylpyrimidine

(c) 3,5-dimethylpyridine

(d) 1,2-dimethylindole

Molecular Formulas of Amines

23.13 (a) What is the general molecular formula for a saturated amine? (b) What is the general molecular formula for a saturated cyclic amine?

Answer: A saturated amine has one more hydrogen atom than an alkane with the same number of carbon atoms. The general formula is C_nH_{2n+3}. A cyclic amine has two fewer hydrogen atoms, so the general formula is C_nH_{2n+1}.

23.14 How many isomers are possible for each of the following molecular formulas?
(a) C_2H_7N　　(b) C_3H_9N　　(c) C_3H_7N

Answers:

(a) two; they are ethylamine and dimethylamine
(b) four; they are propylamine, isopropylamine, ethylmethylamine, and trimethylamine
(c) eight; however, three are enamines

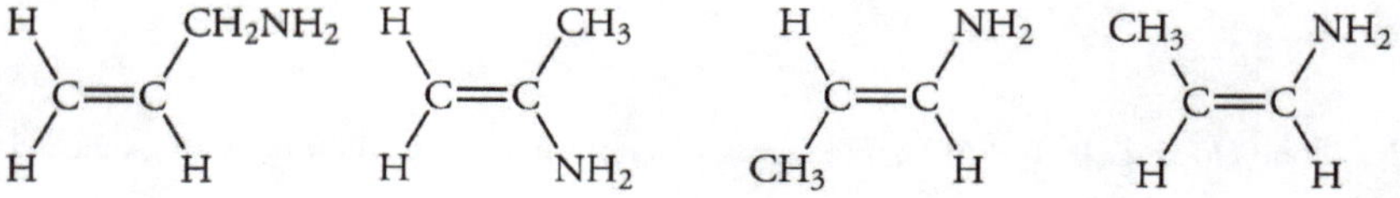

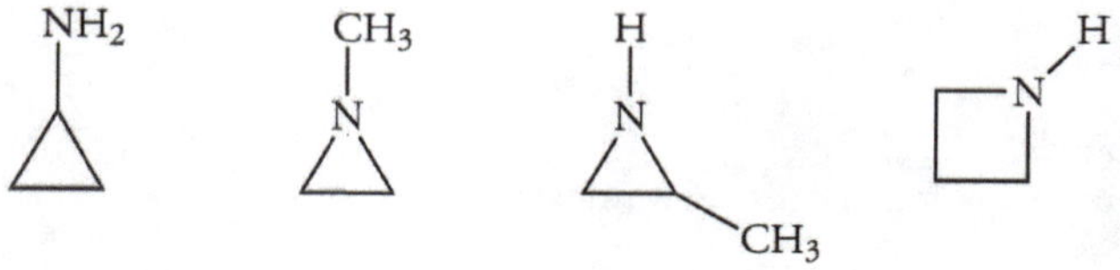

23.15 Draw the isomers of the primary amines with molecular formula $C_4H_{11}N$.

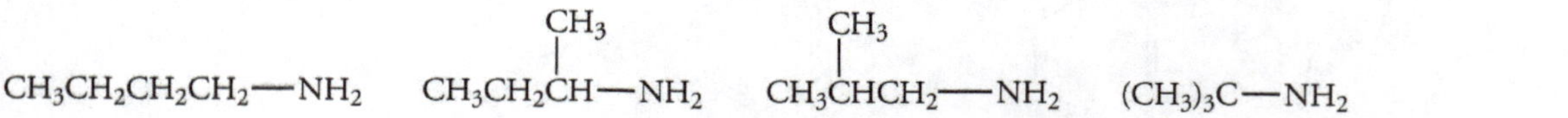

$$CH_3CH_2CH_2CH_2{-}NH_2 \qquad CH_3CH_2\overset{\overset{\displaystyle CH_3}{|}}{C}H{-}NH_2 \qquad CH_3\overset{\overset{\displaystyle CH_3}{|}}{C}HCH_2{-}NH_2 \qquad (CH_3)_3C{-}NH_2$$

Answer: Four alkyl groups can be part of the isomeric primary amines. The alkyl groups are butyl, *sec*-butyl, isobutyl, and *tert*-butyl.

23.16   Explain why oxazole is a weaker base than thiazole.

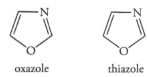

oxazole   thiazole

Answer: The site of protonation is at the sp² orbital of nitrogen in oxazole because nitrogen is more basic than oxygen. The oxygen atom of oxazole decreases the electron density at the nitrogen atom by an inductive effect. Thus, the oxazole is a weaker base than thiazole. Sulfur is less electronegative than oxygen and inductively is a weaker electron withdrawing atom.

23.17   Explain why pyrazole is a weaker base than imidazole.

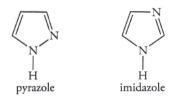

pyrazole   imidazole

Answer: The sp² orbital of nitrogen is the site of protonation in both pyrazole and imidazole. The nitrogen atom bonded to hydrogen is not basic because its lone pair of electrons is part of the aromatic ring system. However, that nitrogen atom withdraws electrons inductively from the atom where protonation occurs. The inductive effect is larger in pyrazole because the two atoms are closer. Inductive effects decrease with distance separating an electron withdrawing group and a reaction center.

Synthesis of Amines

23.18   Write the steps required for the synthesis of each of the following compounds starting from 1-pentanol.
(a) 1-butanamine   (b) 1-pentanamine

Answers:

(a) $CH_3CH_2CH_2CH_2CH_2OH$ $\xrightarrow[H_2SO_4]{CrO_3}$ $CH_3CH_2CH_2CH_2CO_2H$ $\xrightarrow[\text{pyridine}]{SOCl_2}$

$CH_3CH_2CH_2CH_2\overset{\displaystyle O}{\overset{\|}{C}}-Cl$ $\xrightarrow{NH_2}$ $CH_3CH_2CH_2CH_2\overset{\displaystyle O}{\overset{\|}{C}}-NH_2$

$\downarrow$ Br_2 / NAOH

$CH_3CH_2CH_2CH_2NH_2$

(b) $CH_3CH_2CH_2CH_2CH_2OH$ $\xrightarrow{PBr_3}$ $CH_3CH_2CH_2CH_2CH_2Br$ $\longrightarrow$

N—H $\xrightarrow{NaOH}$ N:⁻

$\downarrow$ $CH_3CH_2CH_2CH_2CH_2Br$

$CH_3CH_2CH_2CH_2CH_2NH_2$ $\xleftarrow{NaOH / H_2O}$ N—$CH_2CH_2CH_2CH_2CH_3$

23.19 Write the steps required for the synthesis of each of the following compounds starting from 3-methyl-l-butanol.
(a) 2-methyl-l-propanamine (b) 3-methyl-1-butanamine (c) 4-methyl-1-pentanamine

Answers:

(a) $CH_3CHCH_2CH_2OH$ (with CH_3 branch) $\xrightarrow[H_2SO_4]{CrO_3}$ $CH_3CHCH_2CO_2H$ (with CH_3 branch) $\xrightarrow[\text{pyridine}]{SOCl_2}$

$CH_3CHCH_2\overset{O}{\overset{\|}{C}}-Cl$ (with CH_3 branch) $\xrightarrow{NH_3}$ $CH_3CHCH_2\overset{O}{\overset{\|}{C}}-NH_2$ (with CH_3 branch) $\xrightarrow[NaOH]{Br_2}$ $CH_3CHCH_2NH_2$ (with CH_3 branch)

(b) $CH_3CHCH_2CH_2OH$ (with CH_3 branch) $\xrightarrow{PBr_3}$ $CH_3CHCH_2CH_2Br$ (with CH_3 branch)

phthalimide $\xrightarrow{NaOH}$ phthalimide anion $N:^-$

$\xrightarrow{CH_3CHCH_2CH_2Br}$

N-substituted phthalimide $N-CH_2CH_2CHCH_3$ (with CH_3 branch)

$CH_3CHCH_2CH_2NH_2$ (with CH_3 branch) $\xleftarrow{NaOH / H_2O}$

(c) $CH_3CHCH_2CH_2OH$ (with CH_3 branch) $\xrightarrow{PBr_3}$ $CH_3CHCH_2CH_2Br$ (with CH_3 branch) $\xrightarrow[DMF]{NaCN}$

$CH_3CHCH_2CH_2C\equiv N$ (with CH_3 branch) $\xrightarrow[\text{2. } H_3O^+]{\text{1. LiAlH}_4}$ $CH_3CHCH_2CH_2CH_2NH_2$ (with CH_3 branch)